Advances in Intelligent Systems and Computing

Volume 668

Series editor

Janusz Kacprzyk, Polish Academy of Sciences, Warsaw, Poland
e-mail: kacprzyk@ibspan.waw.pl

The series "Advances in Intelligent Systems and Computing" contains publications on theory, applications, and design methods of Intelligent Systems and Intelligent Computing. Virtually all disciplines such as engineering, natural sciences, computer and information science, ICT, economics, business, e-commerce, environment, healthcare, life science are covered. The list of topics spans all the areas of modern intelligent systems and computing.

The publications within "Advances in Intelligent Systems and Computing" are primarily textbooks and proceedings of important conferences, symposia and congresses. They cover significant recent developments in the field, both of a foundational and applicable character. An important characteristic feature of the series is the short publication time and world-wide distribution. This permits a rapid and broad dissemination of research results.

More information about this series at http://www.springer.com/series/11156

Subhransu Sekhar Dash
Paruchuri Chandra Babu Naidu
Ramazan Bayindir · Swagatam Das
Editors

Artificial Intelligence and Evolutionary Computations in Engineering Systems

Proceedings of ICAIECES 2017

 Springer

Editors
Subhransu Sekhar Dash
Electrical and Electronics Engineering
SRM Engineering College
Kattankulathur, Tamil Nadu
India

Paruchuri Chandra Babu Naidu
Department of Electrical and Electronics
 Engineering
Madanapalle Institute of Technology and
 Science
Madanapalle, Andhra Pradesh
India

Ramazan Bayindir
Department of Electrical and Electronics
 Engineering
Gazi University
Ankara
Turkey

Swagatam Das
Indian Statistical Institute
Kolkata
India

ISSN 2194-5357   ISSN 2194-5365 (electronic)
Advances in Intelligent Systems and Computing
ISBN 978-981-10-7867-5   ISBN 978-981-10-7868-2   (eBook)
https://doi.org/10.1007/978-981-10-7868-2

Library of Congress Control Number: 2017962973

Printed on acid-free paper

This Springer imprint is published by Springer Nature
The registered company is Springer Nature Singapore Pte Ltd.
The registered company address is: 152 Beach Road, #21-01/04 Gateway East, Singapore 189721, Singapore

Preface

This AISC volume contains the papers presented at the International Conference on Artificial Intelligence and Evolutionary Computations in Engineering Systems (ICAIECES 2017) held during April 27–29, 2017, at Madanapalle Institute of Technology and Science, Madanapalle, Andhra Pradesh, India. ICAIECES 2017 is an international conference aiming at bringing together the researchers from academia and industry to report and review the latest progresses in the cutting-edge research on various research areas of electronic circuits, power systems, renewable energy applications, image processing, computer vision and pattern recognition, machine learning, data mining and computational life sciences, management of data including big data and analytics, distributed and mobile systems including grid and cloud infrastructure, information security and privacy, VLSI, antenna, intelligent manufacturing, signal processing, intelligent computing, soft computing, Web security, privacy and optimization, communications, smart wireless and sensor networks, networking and information security, mobile computing and applications, industrial automation and MES, cloud computing, green IT, and finally to create awareness about these domains to a wider audience of practitioners.

ICAIECES 2017 received 250 paper submissions from India and five foreign countries. All papers were peer-reviewed by experts, and comments have been sent to the authors of accepted papers. Finally, 69 papers were accepted for oral presentation in the conference. This corresponds to an acceptance rate of 31% and is intended to maintain the high standards of the conference proceedings. The papers included in this AISC volume cover a wide range of topics in intelligent computing and algorithms and their real-time applications in problems from diverse domains of science and engineering.

The conference was inaugurated by Dr. M. Y. S. Prasad, Indian Scientist and the former Director of the Satish Dhawan Space Centre, Sriharikota. The conference featured distinguished keynote speakers as follows: Dr. Ramazan Bayindir, Gazi University; Dr. Roman Senkerik, Tomas Bata University in Zlin; Dr. Bhekisipho Twala, University of Johannesburg; Dr. Sanjeevi Kumar Padmanaban, University of Johannesburg; Dr. Ranjan Kumar Behera, IITP; Dr. Subhransu Sekhar Dash, SRM University; and Dr. Swagatam Das, ISI, Kolkata, India.

We take this opportunity to thank the authors of the submitted papers for their hard work, adherence to the deadlines, and patience with the review process. The quality of a refereed volume depends mainly on the expertise and dedication of the reviewers. We are indebted to the Technical Committee members, who produced excellent reviews in limited time. We are highly indebted to Honorable Dr. N. Vijaya Bhaskar Choudary, Ph.D., Secretary and Correspondent; and Honorable Sri. N. Krishna Kumar, Chairman, MITS, India, for supporting our cause and encouraging us to organize the conference. In particular, we would like to express our heartfelt thanks for providing us with the necessary financial support and infrastructural assistance to hold the conference. Our sincere thanks to Dr. C. Yuvaraj, Principal, Mrs. M. Prathibha, Sr. Administrative Officer (G.A.D), Madanapalle Institute of Technology and Science (MITS), Andhra Pradesh, India, for their continuous support and guidance. We specially thank Dr. R. Rajkamal, Associate Professor, Program Co-Chair, Madanapalle Institute of Technology and Science (MITS), Andhra Pradesh, India, for his excellent support and arrangements. We thank International Advisory Committee members for providing valuable guidance and inspiration to overcome various difficulties in the process of organizing this conference. We would also like to thank the participants of this conference. The members of faculty and students of Madanapalle Institute of Technology and Science (MITS), Andhra Pradesh, India, deserve special thanks. Finally, we thank all the volunteers who made great efforts in meeting the deadlines and arranging every detail to make sure that the conference could run smoothly. We hope the readers will find the papers inspiring and enjoyable.

Kattankulathur, India   Subhransu Sekhar Dash
Madanapalle, India   Paruchuri Chandra Babu Naidu
Ankara, Turkey   Ramazan Bayindir
Kolkata, India   Swagatam Das
April 2017

Contents

About the Editors

Dr. Subhransu Sekhar Dash is presently Professor in the Department of Electrical and Electronics Engineering, SRM Engineering College, SRM University, Chennai, India. He received his Ph.D. degree from College of Engineering, Guindy, Anna University, Chennai, India. He has more than 20 years of research and teaching experience. His research areas are power electronics and drives, modeling of FACTS controller, power quality, power system stability, and smart grid. He is a Visiting Professor at Francois Rabelais University, POLYTECH, France. He is the Chief Editor of International Journal of Advanced Electrical and Computer Engineering.

Dr. Paruchuri Chandra Babu Naidu is presently Senior Assistant Professor in the Department of Electrical and Electronics Engineering, Madanapalle Institute of Technology and Science, Madanapalle, Andhra Pradesh, India. He received his Ph.D. degree from SRM University, Chennai, India. He has more than five years of research and teaching experience. His research interests include power quality, power electronics, FACTS devices, and solar system.

Dr. Ramazan Bayindir worked as a Research Assistant and Assistance Professor in Electrical Education Department at Gazi University. He became an Associate Professor at Electrical and Electronics Engineering Department at Gazi University. Currently, he is working as a Professor in the same department. His areas of research include renewable energy sources, microgrids, distributed generation, smart grid applications, and industrial automation with programmable logic controller (PLC). He served as a program chair in the annual conference of the International Conference on Renewable Energy Research and Applications (ICRERA 2012–2016), 16th International Power Electronics and Motion Control Conference and Exposition (PEMC 2014), and International Conference on Power Engineering, Energy and Electrical Drives, (POWERENG 2013).

Dr. Swagatam Das received the B.E. (Tel.E.), M.E. (Tel.E—control engineering specialization), and Ph.D. degrees, all from Jadavpur University, India, in 2003, 2005, and 2009, respectively. He is currently serving as Assistant Professor at the Electronics and Communication Sciences Unit of the Indian Statistical Institute, Kolkata, India. His research interests include evolutionary computing, pattern recognition, multi-agent systems, and wireless communication. He has published one research monograph, one edited volume, and more than 200 research articles in peer-reviewed journals and international conferences.

Testing the Functionality of Firewall in Software-Defined Networking

Adebayo Oluwaseun Adedayo and Bhekisipho Twala

Abstract Software-defined Networking (SDN) is a new network architecture that separates the control plane from the data plane in a computer network environment. SDN uses OpenFlow protocol in the control plane to achieve a more flexible operation, monitoring and networking management system. Although SDN offers various advantages over a traditional network, one of the challenges facing the use of this technology is the limited amount of knowledge on implementing various aspects of a network in a SDN, as well as the increasing number of platforms that may be used in the implementation. This paper addresses some of this concern by describing the implementation and functionality of firewalls in SDN environment. The paper uses a software firewall application based on OpenFlow protocol, built on top of Ryu controller. It shows some of the firewall functionalities in SDN without the need for hardware. Using the Mininet network emulator on virtual machine, the experiments conducted in the research describe how the purpose of a firewall set-up can be at different layers of the OSI model. The paper also discusses the efficiency of a SDN firewall by describing the latency and throughput of the emulated networks.

Keywords Software-defined Networking · OpenFlow · Mininet
Ryu · Firewall

A. O. Adedayo (✉) · B. Twala
Faculty of Engineering and Built Environment, Institute for Intelligent
Systems, University of Johannesburg, PO Box 524, Auckland Park,
Johannesburg 2006, South Africa
e-mail: bayoadedayo1@gmail.com

B. Twala
e-mail: btwala@uj.ac.za

© Springer Nature Singapore Pte Ltd. 2018
S. S. Dash et al. (eds.), *Artificial Intelligence and Evolutionary Computations
in Engineering Systems*, Advances in Intelligent Systems and Computing 668,
https://doi.org/10.1007/978-981-10-7868-2_1

1 Introduction

The exponential increase in the number of connected devices on the Internet [1] and the continuous use of virtual machines in many data centres have led to complexities in the management of today's networks. One aspect of this complexity involves the fact that managing and configuring network devices such as routers and switches in many traditional networks have become more cumbersome and expensive due to the rapid growth of Internet usage and the amount of effort required in managing the devices.

In many cases, networking devices run proprietary software that is vendor specific. Network administrators must learn each software to configure multiple devices in line with the configuration guide provided by the vendors—a process that is often time-consuming, expensive and sometimes complex. Although the traditional network has been effective in handling networking requirements for the past decades, various authors have described the challenges that traditional network will face in accommodating next-generation networking trends and expansions [2]. The application of firewall, QoS, load balancing and other network management system is becoming more expensive due to increase in the Operational Expense (OpEx), and the Capital Expenditure (CapEx) required to run the network smoothly [1–3].

Software-defined Networking (SDN) came into existence out of a need to provide ways of overcoming proprietary software that is limited to particular vendor devices, to meet the current networking demands, plan for future expansion and create a network environment that can be easily configured and managed from a logically centralised location. Another advantage of SDN is that the switches (OpenFlow switches) are cheaper than proprietary ones [2, 4].

Software-Defined Networking was established in 2006 as a result of research efforts by academic researchers desiring to run experiments on their campus network without disrupting the network [5–8]. SDN is a new network technology that decouples the control plane from the data plane to have a logically centralised controller and network view [4, 9]. This separation enables the network intelligence inside the control plane to be easily managed and monitored [4, 5]. SDN aims to solve the problems affecting traditional network architecture such as complexity, inability to scale and vendor dependency by bringing a new technology into networking that is easier to configure, flexible, scalable, programmable, vendor independent and less expensive [4]. Although SDN is designed to be used as a networking technology, it also provides academic researchers and industry experts a way of experimenting with new ideas and creating solutions to persistent problems in a network without affecting the operations of the network [8, 10].

As with traditional networks, SDN allows the implementation of a firewall in monitoring and controlling access to the network. This is mostly achieved through the OpenFlow protocol on which SDN is based. The OpenFlow protocol which is standardised by the Open Network Foundation (ONF) is the first protocol employed in SDN for communication between networking devices [4]. This paper describes how the implementation of a firewall in SDN can be achieved using approaches (controller and technology) that have not been previously considered by other researchers.

The rest of the paper is organised as follows: Sect. 2 presents overview of firewall and firewall classification. Section 3 discusses OpenFlow protocols and gives a brief explanation of how the protocol is used in setting up a SDN firewall. In Sect. 4, we talked about the methodology used in this paper. Section 5 represents the experimental set-up, while Sect. 6 presents the results and discussion of different scenarios in the experiment. Section 7 talks about the latency and the throughput in the network. Finally, Sect. 8 concludes the paper and also gives details about future work.

2 Firewall

Firewall is a network security system used to monitor and control access to a network environment [11]. A firewall may be either a dedicated hardware or software, and it is primarily used in an organisation to secure and prevent unauthorised access to the network based on predetermined security rules [11]. It controls traffic flowing in and out of a network by granting or denying permission to packets flowing in the network. The rules or policies set up for a firewall may include parameters such as IP address, MAC address, port or protocol. Permission within a network or between two networks is only given to a packet that matches information in the access control list (ACL). The firewall serves as a security barrier by protecting the network resources from unwanted access or illegal visitors. A firewall can be classified into three categories depending on the technology used, these are as follows: packet filtering, circuit-level gateway and application-level gateway.

A. Packet Filtering

Packet filtering technology firewall can also be called network layer firewall. This type of firewall works at the network layer by filtering IP addresses. IP packet header entering or leaving the network is checked to validate the IP source address, IP destination, TCP or UDP source and destination port number and ICMP messages. Depending on the filtering rules, the decision is made using the ACL to forward or discard the packet. The advantages of packet filtering-based firewall are its simplicity, speed, transparency and minimal effect on the network performance. Although it lacks auditing and verification and poses difficulty in managing the filtering rules, this is the most commonly used firewall category. For this reason, the experiments conducted in the paper are based on this firewall category [12].

B. Circuit-Level Gateway

In this category, firewall is implemented at the layer 5 session layer. The security policies are applied when the TCP or UDP connection is established. It is fast and has low maintenance but has limited logging and does not provide further analysis after connection has been established [12].

C. **Application-Level Gateway**

This type of firewall classification is implemented at the layer 7 application layer. It is also referred to as a proxy firewall. A proxy server is installed on the firewall host where specific TCP/IP role is analysed with the corresponding application. The proxy server serves as a middlebox, and packets or data are first sent to the proxy server for analysis before being forwarded to the destination. It has strong user authentication, flow monitoring, filtering, recording and reporting facilities and also shields the internal network from the external network. The disadvantages, however, are high maintenance and low performance [12].

Javid et al. [13] describe a layer 2 firewall by implementing a firewall based on MAC address in layer 2 using POX controller based on Python programming language. The firewall allows (or restricts) traffic based on the source MAC address and destination MAC address which is not efficient because it only blocks layer 2 traffic. Suh et al. [14] also use POX controller in their work and check a few packet header attributes. However, their result did not go into details and only show the ICMP protocol.

This paper contributes to the knowledge of firewalls in SDN by describing the use of the Ryu controller also based on Python programming language [15]. The experiments include the ability to add, modify and delete rules to be utilised in the firewall and use one OpenvSwitch and eight hosts in conjunction with the Ryu controller [15]. In addition, we also discuss the implementation of a firewall at layer 2, layer 3 and layer 4 of the OSI model. In achieving the objective of this paper, some of the match fields utilised in our experiments and the network layer at which they are applicable are listed in Table 1.

3   OpenFlow Protocol

As mentioned earlier, implementing a firewall in a SDN is mostly achieved through the OpenFlow protocol. In any SDN, the protocol must be enabled on both the controller software and the associated network devices. The controller in SDN uses OpenFlow to manipulate the forwarding table of the networking devices (also referred to as datapath) by remotely inserting, modifying or deleting flow entries on the flow table residing in each device. The idea of flows is used to identify network

Table 1 Match fields

Match fields	Layer
TCP, UDP, ICMP	Transport layer 4
IPv4, ARP	Network layer 3
MAC address	Data link layer 2

traffic based on predefined match rules which can be reactive or proactive. The flow entries consist of fields that are matched with incoming packets and the associated actions with each match. When a switch receives a packet, the packet header is analysed against the match fields in the flow entries which is installed on the flow table. Following a found match, the associated action is performed. If no match is found, depending on the configuration, the packet may be dropped or forwarded to the controller for further processing. Otherwise, it is forwarded to the destination.

Flows can be either reactive (dynamic) or proactive (static). Reactive flows are flow entries that are added by the controller to the flow tables in the networking devices dynamically after the arrival of a packet and the execution of corresponding actions. On the other hand, proactive flows are added to the flow table by a network administrator prior to the arrival of a packet with predefined actions to be taken when such packet arrives. Some of the fields that may be employed in a flow entry include the Ingress port, Ethernet source address, Ethernet destination address, Ethernet type, VLAN_Id, VLAN priority, IP protocol, the IP protocol's source address, IP protocol's destination address, TCP source port, TCP destination port, UDP source port and the UDP destination port. These fields cover layer 2, layer 3 and layer 4 traffic of the OSI model and as such allow a SDN firewall to operate in these layers. For any flow entry, the actions that may be carried out include [16]:

- ALL/FLOOD—flood the packet on all switch's port other than the incoming port
- CONTROLLER—send the packet to the controller for decision
- PORT—send out the packet on specific port
- DROP—drop the packet without any further action.

The controller uses the information in the flow table to communicate with the switches (OpenFlow switches) on the network via a secure channel that uses SSL or TLS. These OpenFlow switches are capable of acting as either a switch or a router depending on the configuration. Since the inception of SDN and OpenFlow, various controllers have been developed with variations in the programming language used and the market focused. Some of the available open-source controllers include NOX, POX, Trema, Beacon, Floodlight, Ryu and OpenDaylight. The most common OpenFlow software switch is the OpenvSwitch and as such is used in experiments described in this paper.

4 Methodology

As mentioned earlier, OpenFlow controller can communicate with the switches or router in the network either reactively or proactively. In the first case, the switch forwards packets received to the controller for decision making, and the controller forwards the decision back in the form of flow entry generated on the flow table on the switch. The action from the decision may be to flood the packet on all other ports except the incoming port, send the packet to the controller or to send the packet to a specific port or drop the packet. In the latter case, the flow entries are pre-installed on the devices by the controller before the arrival of packets. Actions regarding each flow entry are taken on the arrival of the packets on the network devices.

The firewall application described in this paper is built upon the Ryu controller and operates in a proactive mode for hosts in the network to communicate. Although the reactive mode is considered more efficient because the flow entries are automatically generated, it has some setbacks. One of these setbacks includes the fact that it involves some overhead in the amount of time required by the switch in acquiring the flow entries to populate the flow table and subsequently use it. Also, in a situation where the connection between the switch and the controller is lost, operating in a reactive mode implies that switches will not be able to forward the packet. This situation could also occur when the flow entry times out and there is a need for a new flow entry.

By using the proactive mode, we overcome these challenges making communication faster (once the flow entries have been installed) since there is no overhead incurred from the generation of the flow entries. Another advantage of the proactive mode is that in the event of a loss of connection, services are not disrupted and the switch will still forward traffic as normal. Since the original entries are known to the network administrator, the proactive mode enables the network administrator to monitor flow entries that are installed by the controller.

The flow entries defined contain the firewall rules in the form of ACL. The experiments explore the implementation of firewalls at layer 2, layer 3 and layer 4. Some of the functionalities of SDN firewalls that we investigate in this paper include the following:

- Allowing ICMP traffic while blocking HTTP
- Blocking certain hosts from accessing a Web server
- Blocking and (or) allowing hosts to communicate based on source or destination IP address
- Measurement of the latency and throughput in the SDN.

5 Experimental Set-Up

The experiments discussed in the following section were carried on a virtual machine (VM) running Ubuntu 14.04 LTS and installed on Oracle VirtualBox with an Intel Core i5 quad core and 16 GB of RAM host. Applications including Mininet, Ryu controller [15], Iperf, Wireshark and Postman were installed on the Ubuntu VM.

Mininet network emulator allows the creation of a realistic virtual network and runs a collection of end hosts, switches and links on a single Linux kernel. Ryu controller, which is written in Python programming language and maintained by

NTT [15]. Iperf is a command line tool used for measuring the maximum achievable bandwidth on the network. Wireshark is a network packet analyser used to capture and examine details of communication on the network. Lastly, we use Postman to get a coordinated response from the switch instead of getting the results through cURL.

The network topology consists of 1 controller, 1 OpenFlow switch (OpenvSwitch) and eight hosts and is created using Mininet. Running the command via Mininet:

```
sudo mn -topo single,9 -mac -switch ovsk -controller remote -x
```
creates the network. The controller is then named c0, the switch s1, and the hosts h1, h2, h3, h4, h5, h6, h7 and h8. The topology is shown in Fig. 1. Switch s1 is set to use OpenFlow specification 1.3 using the command:

```
ovs-vsctl set Bridge s1 protocols = OpenFlow13
```

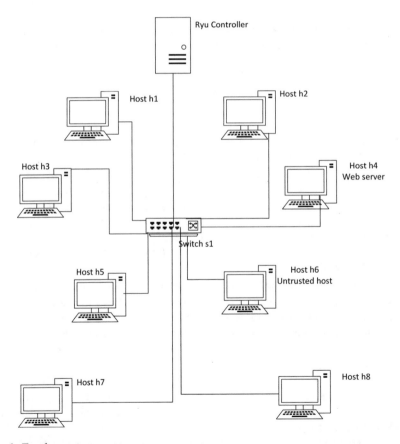

Fig. 1 Topology set-up

To start the firewall application (on c0), we issued the command:
`ryu-manager ryu.app.rest_firewall ryu.app.ofctl_rest`.

The ofctl_rest application started with the firewall allows easy management and viewing of the network and device configurations (in JSON format) using a Web browser or Postman. In this experiment, we use Postman to view the rules and states of a device for easy readability.

The configured firewall is enabled on the switch using:

`curl -X PUT` http://localhost:8080/firewall/module/enable/0000000000000001
. The state of the switch is obtained through Postman, using: http://localhost:8080/firewall/module/status.

6 Result and Discussion

To verify the functionality of the SDN firewall, we investigated the operation of the firewall in different scenarios. The scenarios tested are described below.

6.1 Scenario 1

The first aspect of the experiments investigates the operation of the firewall at layer 2 using the MAC address of hosts. The rules set-up first allows the hosts to ping each other. This is achieved by specifying a flow entry to allow communication from host h1 to h2 and vice versa by specifying their MAC addresses in the *sourceMAC* and *destinationMAC* using the following command format.

`curl -X POST -d '{"dl_src": "sourceMAC", "dl_dst": "destinationMAC", "nw_proto": "ICMP", "actions": "ALLOW", "priority": "priority number"}'` http://localhost:8080/firewall/rules/0000000000000001.

This rule specifies that communication from the specified source MAC address to the specified destination MAC address should be allowed, with a given priority value when the ICMP protocol is used. The rules were then updated to disallow communication between the two hosts. This is also achieved by specifying a new flow entry to disallow communication from h1 to h2 and vice versa using the command formats:

`curl -X POST -d '{"dl_src": "sourceMAC", "dl_dst": "destinationMAC", "nw_proto": "ICMP", "actions": "DENY", "priority": "priority number"}'` http://localhost:8080/firewall/rules/0000000000000001.

To ensure that the new rules take precedence over the previous rule, a higher priority value is given to the new flow entries to disallow communication between the two hosts. This is confirmed by pinging both hosts, which shows that they were unreachable from each other.

6.2 Scenario 2

Firewall rules used in this experiment is based on layer 3 IP address and layer 4 protocols. Firewall rules were added to allowing pinging using the ICMP messages between host h1 and host h2. The network source address and the destination address are used while permission is given to allow ICMP network protocol communication between the two hosts. The format for the command used is as follows:

```
curl  -X  POST  -d  '{"nw_src": "sourceIP", "nw_dst":
"destinationIP", "nw_proto": "ICMP", "actions": "ALLOW",
"priority": "priority number"}'  http://localhost:8080/firewall/rules/
0000000000000001.
```

Two commands are entered with source and destination IP addresses and then interchanged the source with destination address on the second command. Priority number is also increased to overwrite the existing rule used in the earlier experiment. Figure 2 shows the ARP and the ICMP requests and response as the packets flow through the network after executing ten ping requests. Even though the flow entry has already been proactively added to the flow table earlier before, a broadcast is still sent using the address resolution protocol (ARP) followed by response from the concerned host. The use of proactive flow entry makes the ARP response to be faster than reactive flow entry as seen in the pinging results. It is also seen that new ARP message is sent after the fifth reply message, but after analysing the packet,

	Time	Source	Destination	Protocol
1	0.000000000	00:00:00_00:00:01	Broadcast	ARP
2	0.000494833	00:00:00_00:00:02	00:00:00_00:00:01	ARP
3	0.000506753	192.168.10.1	192.168.10.2	ICMP
4	0.000629380	192.168.10.2	192.168.10.1	ICMP
5	1.001558091	192.168.10.1	192.168.10.2	ICMP
6	1.001615253	192.168.10.2	192.168.10.1	ICMP
7	2.002348178	192.168.10.1	192.168.10.2	ICMP
8	2.002417778	192.168.10.2	192.168.10.1	ICMP
9	3.004268300	192.168.10.1	192.168.10.2	ICMP
10	3.004337491	192.168.10.2	192.168.10.1	ICMP
11	4.006105577	192.168.10.1	192.168.10.2	ICMP
12	4.006163328	192.168.10.2	192.168.10.1	ICMP
13	5.004813899	00:00:00_00:00:02	00:00:00_00:00:01	ARP
14	5.004899790	00:00:00_00:00:01	00:00:00_00:00:02	ARP
15	5.005094624	192.168.10.1	192.168.10.2	ICMP
16	5.005125755	192.168.10.2	192.168.10.1	ICMP
17	6.004559760	192.168.10.1	192.168.10.2	ICMP
18	6.004626420	192.168.10.2	192.168.10.1	ICMP
19	7.006338251	192.168.10.1	192.168.10.2	ICMP
20	7.006393609	192.168.10.2	192.168.10.1	ICMP
21	8.008214278	192.168.10.1	192.168.10.2	ICMP
22	8.008268163	192.168.10.2	192.168.10.1	ICMP
23	9.010180895	192.168.10.1	192.168.10.2	ICMP
24	9.010248355	192.168.10.2	192.168.10.1	ICMP

Fig. 2 ARP and ICMP messages

we discover that this ARP message does not consume bandwidth. This message serves as a keep alive message sent by host h2 to host h1 to confirm whether host h1 is still communicating.

6.3 Scenario 3

In this experiment, host h4 was made a Web server using command 'python -m SimpleHTTPServer 80 &'. All hosts in the network are granted permission to access the Web server except for host h6. Layer 3 IP address and layer 4 protocols were used with a higher priority number in the firewall rules.

The first command below shows that access was granted to all hosts on the network after which the second command is then issued to deny host h6 from accessing the Web server. This is done by specifying the source IP address, destination IP address, port number, network protocol and associated action to be taken.

(a) `curl -X POST -d '{"nw_src"`: `"192.168.10.0/24"`, `"nw_dst"`: `"192.168.10.0/24"`, `"nw_proto"`: `"ICMP"`, `"actions"`: `"ALLOW"`, `"priority"`: `"205"}'` http://localhost:8080/firewall/rules/000000000000 0001

(b) `curl -X POST -d '{"nw_src"`: `"192.168.10.6/32"`, `"nw_dst"`: `"192.168.10.4/32"`, `"nw_proto"`: `"TCP"`, `"tp_dst"`: `"80"`, `"actions"`: `"DENY"`, `"priority"`: `"206"}'` http://localhost:8080/firewall/rules/0000000000000001.

Hosts h5 and h6 were picked to access the Web server. Host h5 successfully connect and retrieve information from the Web server residing on host h4. Host h6 on the other hand could not secure a connection to the Web server. This can be seen in Fig. 3 as the host is only displaying connecting for a long time and retrying again after an unsuccessful connection. It can also be seen that firewall is functioning very well in SDN, and this is confirmed in the controller terminal. The controller terminal shows that the host and packets are blocked.

The second command grant pinging access is to host h5 and host h6. This enables both hosts to be able to ping the Web server. Pinging is not possible between the two hosts to the Web server before, and this is because there is no rule to allow the pinging. Pinging between host h5 and host h6 is then blocked while that of host h6 remains, this is done by either deleting the "rule_id" or adding another rule with a higher priority number. This still allows host h5 to access the Web server but not being able to ping the Web server, while host h6 can still ping but does not have access to any information on the Web server.

From all the experiments in the three scenarios, it is seen that firewall in SDN can function more like firewall used in traditional network. It can be used to secure the network by blocking access to the Web server from specific IP address or group of IP addresses. It can also be used to secure intranet where permission is only

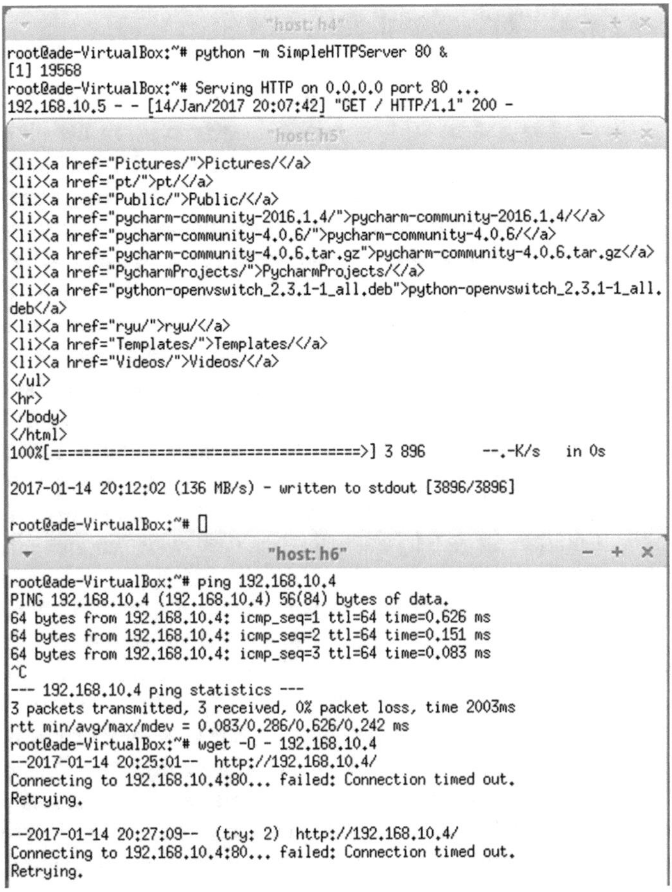

```
                              "host: h4"                      -  +  x
root@ade-VirtualBox:~# python -m SimpleHTTPServer 80 &
[1] 19568
root@ade-VirtualBox:~# Serving HTTP on 0.0.0.0 port 80 ...
192.168.10.5 - - [14/Jan/2017 20:07:42] "GET / HTTP/1.1" 200 -
                              "host: h5"                      -  +  x
<li><a href="Pictures/">Pictures/</a>
<li><a href="pt/">pt/</a>
<li><a href="Public/">Public/</a>
<li><a href="pycharm-community-2016.1.4/">pycharm-community-2016.1.4/</a>
<li><a href="pycharm-community-4.0.6/">pycharm-community-4.0.6/</a>
<li><a href="pycharm-community-4.0.6.tar.gz">pycharm-community-4.0.6.tar.gz</a>
<li><a href="PycharmProjects/">PycharmProjects/</a>
<li><a href="python-openvswitch_2.3.1-1_all.deb">python-openvswitch_2.3.1-1_all.
deb</a>
<li><a href="ryu/">ryu/</a>
<li><a href="Templates/">Templates/</a>
<li><a href="Videos/">Videos/</a>
</ul>
<hr>
</body>
</html>
100%[====================================>] 3 896      --.-K/s   in 0s

2017-01-14 20:12:02 (136 MB/s) - written to stdout [3896/3896]

root@ade-VirtualBox:~# []
                              "host: h6"                      -  +  x
root@ade-VirtualBox:~# ping 192.168.10.4
PING 192.168.10.4 (192.168.10.4) 56(84) bytes of data.
64 bytes from 192.168.10.4: icmp_seq=1 ttl=64 time=0.626 ms
64 bytes from 192.168.10.4: icmp_seq=2 ttl=64 time=0.151 ms
64 bytes from 192.168.10.4: icmp_seq=3 ttl=64 time=0.083 ms
^C
--- 192.168.10.4 ping statistics ---
3 packets transmitted, 3 received, 0% packet loss, time 2003ms
rtt min/avg/max/mdev = 0.083/0.286/0.626/0.242 ms
root@ade-VirtualBox:~# wget -O - 192.168.10.4
--2017-01-14 20:25:01--  http://192.168.10.4/
Connecting to 192.168.10.4:80... failed: Connection timed out.
Retrying.

--2017-01-14 20:27:09--  (try: 2)  http://192.168.10.4/
Connecting to 192.168.10.4:80... failed: Connection timed out.
Retrying.
```

Fig. 3 Firewall confirmation

granted to specific host in the network. Even though SDN is still in early stage, firewall can still be implemented and can function effectively in an SDN environment. The use of firewall application in SDN can save some amount on Capital Expenditure (CapEx) incurred on the purchasing of firewall equipment.

All hosts have access to the Web server; Fig. 3 shows that host h5 still has access to the Web server, while host h6 did not have access but can ping the server. Another rule was then added to block the pinging from host h6 to the Web server.

Pinging access from host h6 to the Web server was later denied by deleting the rule_id from the firewall rule using the following command:

curl -X DELETE -d '{"rule_id": "5"}' http://localhost:8080/firewall/ rules/0000000000000001 . Deleting the rule_id then gives host h6 pinging access back, but it did not have access to the Web server.

7   Latency and Throughput

Latency in networking can be seen as the time taken for a packet to travel from a source to a destination and back. It is mostly measured using the ping tool. We measured latency without the introduction of firewall and with the firewall running on the network. The result achieved is then plotted on a graph and shown in Fig. 4. From the result, it is seen that firewall introduces minimal overhead compared to a network without firewall. From the analysis of the network using Wireshark, it is seen that a unicast ARP message is sent from host h2 to host h1 asking for the owner of 192.168.10.1 and host h1 also replies back. This unicast message does not consume any bandwidth and therefore does not create additional overhead on the network but has a lower latency at the sixth sent packet. Even though both latencies have their flow entries installed proactively, the firewall still has a little more latency than the network without firewall. The latency without firewall has an average of 0.138 ms, while the average latency with firewall is 0.155 ms. This little difference can be ignored in a network environment. The latency results show that firewall implementation in SDN is functioning very well without affecting the quality of the network.

Throughput is measurement of data transfer rate from one location to the other over a given amount of time. Network throughput is measurement of maximum achievable bandwidth on a network. This can be measured as maximum achievable data transfer rate from one host to the other over a specified period of time.

In other to test for the throughput, one host is made a server while others a client. We run the test for different number of times. The throughput with firewall and without firewall is plotted on the graph as shown in Fig. 5. The result shows that throughput without firewall is slightly higher than with firewall. The resulted average of throughput without firewall is 65.1 Mbps out of the 100 Mbps, while the one with firewall is 59.9 Mbps which is still acceptable.

Fig. 4 Network latency

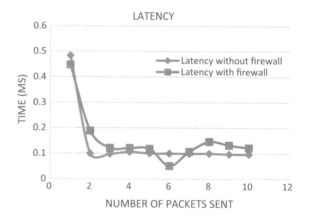

Fig. 5 Network throughput

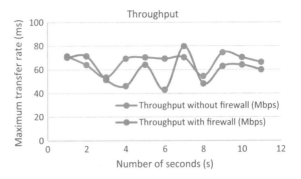

8 Conclusion

In testing the functionality of firewall in SDN environment, this paper shows that the objectives were achieved. This includes the use of firewall in securing the network by blocking untrusted host or allowing trusted ones using ICMP, TCP and HTTP protocols. We were able to measure the latency and the throughput in the network with firewall in place and without firewall. The firewall should be put in place to block any attack within the network or from outside the network. We were able to perform different experiments using Mininet, OpenvSwitch and Ryu controller to set up the network topology to achieve the aim of this paper. The firewall used in this paper functions by focusing on the packet header of flows in the network rather than the data. The firewall operates from layer 2 to layer 4 on the OSI model for matching of packet header against the firewall rules. Packet filtering technology firewall was used, and rules were added proactively to achieve the goal of this paper. The results acquired show that sources and destination not listed in the policy are blocked while those listed and matched with the packet header are granted permission based on the associated actions attached. Some of our key findings include the ability to use SDN firewall to secure the network by granting permission or denying access to network services. Considering latency and throughput, we conclude that SDN firewall implementation only has minimal overhead compared to SDN network without firewall. The use of firewall in an SDN environment will reduce the capital expenditure incurred in purchasing firewall hardware in organisations. The limitation of this experiment is that it is only tested using one controller, the use of multiple controllers may increase overhead on the network, and there are also security concerns regarding the connection of multiple controllers to switches on the network. This security concern can be recommended for future work to check security issue that may affect secured connection between Ryu controllers and the switches. Another future work is to test the functions of the firewall using application layer 7 on proxy server by using proxy technology.

References

1. European Commission, *Consultation on Future Network Technologies research and Innovation in HORIZON2020* (European Commission, Brussels, 2012)
2. P. Goransson, C. Black, *Software Defined Networks: A Comprehensive Approach* (Morgan Kaufman, Waltham, 2014)
3. Arthur D Little, Bell Labs, *Reshaping the future with NFV and SDN* (Arthur D Little, 2015)
4. ONF, Software defined networking: the new norm for networks. 13 April 2012. [Online]. Available: https://www.opennetworking.org/images/stories/downloads/sdn-resources/white-papers/wp-sdn-newnorm.pdf. [Accessed 4 Jan 2017]
5. S. Azodolmolky, *Software Defined Networking with OpenFlow* (Packt Publishing Ltd, Birmingham, 2013)
6. Stanford University, Clean slate design for the internet. May 2006. [Online]. Available: http://www.cleanslate.stanford.edu/research_project_ethane.php. [Accessed 9 Jan 2017]
7. M. Casado, M. J. Freedman, J. Pettit, J. Luo, N. McKeown, S. Shenker, Ethane: Taking Control of the Enterprise, in *SIGCOMM'07* (Kyoto, 2007)
8. N. McKeown, T. Anderson, H. Balakrishnan, G. Parulkar, L. Peterson, OpenFlow: enabling innovation in campus networks. SIGCOMM Comput. Commun. Rev. **38**(2), 69–74 (2008)
9. N. Feamster, J. Rexford, E. Zegura, The road to SDN—an intellectual history of programmable networks. SIGCOMM Comput. Commun. Rev. **44**(2), 87–98 (2014)
10. G. Dasmalc, UnderstandingSDNTechV1.pdf. December 2014. [Online]. Available: https://www.sdxcentral.com/wpcontent/uploads/2014/12/UnderstandingSDNTechV1.pdf. [Accessed 11 Jan 2017]
11. S. Morzhov, I. Alekseev, M. Nikitinskiy, Firewall Application for Floodlight SDN controller, in *2016 International Siberian Conference on Control and Communications* (Moscow, 2016)
12. Karunsubramanian.com, Firewall classification and architecture, Karunsubramanian.com, [Online]. Available: http://www.karunsubramanian.com/security/firewall-classifications-and-architectures/. [Accessed 16 Jan 2017]
13. T. Javid, T. Riaz, A. Rasheed, A layer2 firewall for software defined network, in *Conference on Information Assurance and Cyber Security (CIACS)* (Rawalpindi, 2014)
14. M. Suh, S. H. Park, B. Lee, S. Yang, Building firewall over the software-defined network controller, in *16th International Conference on Advanced Communication Technology* (Phoenix Park, 2014)
15. Ryu, Ryu SDN Framework. [Online]. Available: https://osrg.github.io/ryu/. [Accessed 16 Dec 2016]
16. Open Networking Foundation, openflow-switch-v1.3.4.pdf, 27 March 2014. [Online]. Available: https://www.opennetworking.org/images/stories/downloads/sdn-resources/onf-specifications/openflow/openflow-switch-v1.3.4.pdf. [Accessed 10 Jan 2017]

An Extension of 2D Laplacian of Gaussian (LoG)-Based Spot Detection Method to 3D

Matsilele Mabaso, Daniel Withey and Bhekisipho Twala

Abstract Investigations into the 3D interior of cells and other organisms have taken a significant role in biological and medical research, thanks to the advancement in fluorescence microscopy imaging which made it possible to image and visualize 3D intracellular particles within a cell and other structures. Studying the relationship between these intracellular particles, and their functions, is a long-term research effort which could reveal and help in understanding certain aspects of the cell. Automated tools for microscopy image analysis play a crucial role in advancing the understanding of various aspects of the cell and can help in extracting the interactions between molecules and address various issues in biological research. The initial step in microscopy image analysis involves the detection of these intracellular particles. This step is involves computing the particle's position and other properties which can assist in further analysis. In this work, we present an automated 3D Laplacian of Gaussian (LoG) method for the detection of spots. The proposed method is an extension of the 2D method proposed by Raj et al. (Nat Methods 5:877–879, 2008, 1). Experiments were conducted using realistic synthetic images with a varying signal-to-noise ratio, and F_{score} was used to evaluate the performance of the method on each of the synthetic images used. The performance of the LoG method was then compared to isotropic undecimated wavelet transform (IUWT). The experimental results clearly show that the proposed method yields similar results in no-background synthetic images and higher F_{score} values for synthetic images with a real background, confirming its viability as a spot detection method for 3D images.

M. Mabaso (✉) · D. Withey
MDS (MIAS), Council for Scientific and Industrial Research, Pretoria, South Africa
e-mail: MMabaso@csir.co.za

D. Withey
e-mail: dwithey@csir.co.za

M. Mabaso · B. Twala
Department of Electrical and Mining Engineering, University of South Africa,
Pretoria, South Africa
e-mail: Twalab@unisa.ac.za

© Springer Nature Singapore Pte Ltd. 2018
S. S. Dash et al. (eds.), *Artificial Intelligence and Evolutionary Computations
in Engineering Systems*, Advances in Intelligent Systems and Computing 668,
https://doi.org/10.1007/978-981-10-7868-2_2

Keywords Microscopy images · 3D realistic synthetic images
Spot detection · 3D Laplacian of Gaussian · Isotropic undecimated wavelet
transform

1   Introduction

Recent advances in 3D fluorescence microscopy and fluorescent labeling technique
have made it possible to study the interaction of intracellular particles to assist in
answering open questions in molecular and cell biology [2].

Data produced by fluorescence microscopy is extensive and requires automated
tools to extract meaningful information encoded in the images. The analysis of
images obtained via fluorescence microscopy involves the detection and tracking of
various intracellular particles called spots. These spots are bright particles with
intensity higher than that of its immediate background as shown in Fig. 1, and they
are formed when the imaged light is emitted by fluorescent molecules, e.g.,
fluorophores that are attached to the objects of interest.

The main aim of the detection and tracking is to acquire meaningful information
from the vast amount of images. The initial step in the bioimage analysis is the spot
detection. This step extracts important features of spots in a given image.
Fluorescence microscopy generates a vast amount of image data which needs
reliable automated methods for spot detection. As a result, a well-performing
detection method will enable accurate feature extraction that could lead to more
precise further analysis [3]. Currently, the analysis of these images is done manually
which is both time-consuming and laborious [4]. To overcome these, a number of
automated methods have been proposed for the detection of spots. Despite such
developments in automated detection methods, the detection of spots is still a

Fig. 1 A sample of real
fluorescence image with
bright particles obtained using
confocal microscopy

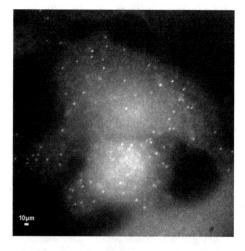

challenging task due to the complex nature of biological sequences. The other factor which decreases the performance of these methods is a high degree of background noise. This condition causes many automated detection methods to fail or generate many spurious features.

To minimize the number of spurious detections, vigorous testing of these automated methods is critical to establish their limits. Testing these automated methods is mainly done using synthetic images that mimic real fluorescence images and help practitioners to choose a suitable method and developers to understand shortcomings of their approaches. Most of these methods are tested on unrealistic synthetic datasets, and as a result, they give a poorer performance on real cases. With increased need for automated methods, the use of realistic synthetic images is critical. Synthetic images are of importance because they contain the ground-truth data. Until recently, there were no benchmark datasets with ground truth to help in evaluating various automated spot detection methods. Developers in image analysis were required to annotate their real images manually to obtain ground-truth information to evaluate their methods. Due to the complexity of biological images, this procedure is laborious and prone to errors. The development of automated image analysis methods has drawn much attention to developers in microscopy image analysis as a solution to overcome the limitations of the manual analysis.

The framework for spot detection usually consists of three steps, namely noise filtering, enhancement, and binarization [4]. The existing methods include these steps in one form or another with the signal enhancement being the distinguishing part. The first step involves reducing the background noise resulting in a denoised image, I_f. Methods for noise reduction range from simple filtering approaches such as median filter. Then, the denoised image is further enhanced in the second step to highlight spots and suppress the background and other unnecessary objects. The output, grayscale classification map represents the likelihood of the spots in an image. The last step involves applying the thresholding on the classification map to separate spots from background structures. In this step, objects which are above a given threshold are marked with 1 (spot), and everything below is marked with a zero (background). There are many different automated approaches developed for the quantification of spots position in microscopy images [5]. These methods range from simple thresholding techniques such as Otsu [6] to more sophisticated methods such as Wavelets [7]. The existing method can be classified into two categories, supervised and unsupervised detection methods [8]. The supervised methods are those who require training while the unsupervised do not require training. A study by Smal et al. [4] showed that at the very low SNR ($\approx$2), the supervised methods perform better than the unsupervised. However, with regard to the unsupervised methods, their differences were small. At higher SNR's, the difference in performance for all methods studied was negligible. The most popular are based on mathematical morphology, which detects spots based on their shape and intensity [9] and adaptive thresholding methods which analyze the spot height and size [10]. Other sophisticated methods include the multiscale approaches such as Fourier space and wavelets decomposition to identify spots [3, 7]. Most of these methods are restricted to 2D images with 3D methods increasing in popularity. Another approach proposed by Raj et al. based Laplacian of Gaussian and threshold selection, gained popularity recently [11]. The approach is based on applying the LoG filter to reduce noise and enhance background structures; thus, after filtering a multiple thresholding is applied, and a number of spots for each threshold are computed, and this relationship is plotted. The presence of a plateau region in the plot indicates a region in which the number of detected spots does not vary significantly. The plateau region is a correct threshold to choose, as the spots detected at this threshold correspond. The manual threshold selection is sometimes time-consuming but offers a better chance to see the detected spots and adjust the threshold accordingly. This step is also critical as it boosts the performance of the method compared to other detection methods which lack this step. The limitation of the method is that it was designed for 2D images, and with recent advances in 3D microscopy imaging, the analysis of 3D images is critical.

In this work, we present a 3D method for the detection of spots in microscopy images. The method is inspired by the 2D Laplacian of Gaussian (LoG) developed by Raj et al. [1] for the extraction of mRNA spots. The approach extends the version of Raj to 3D to enable the extraction of spots in (X, Y, Z) space. Compared to the original method, the extension method also outputs the spots' coordinates and also shows the detected spots for each threshold. We have used 3D synthetic images for the experiments, where the images were formed by overlaying ground-truth spot positions onto real image backgrounds [12]. The MATLAB implementation of the 3D LoG method is available on request.

This paper is divided into four parts. In Sect. 2, we describe the methodology of this study, and Sect. 3 evaluates our algorithm and reports some validation results. Finally, in Sect. 4, conclusions for the study are provided.

2 Methodology

2.1 Original 2D Laplacian of Gaussian (LoG)

The LoG method as proposed by Raj et al. [1] is based on applying a 2D LoG filter to the original image, $I(x, y)$, in order to suppress noise and enhance particles. The 2D log filter is defined as

$$\text{LoG}(x, y) = \frac{1}{\pi \sigma^4} \left(\frac{x^2 + y^2}{2\sigma^2} - 1 \right) e^{-\frac{x^2 + y^2}{2\sigma^2}} \tag{1}$$

where (x, y) are the coordinates and σ is the standard deviation of the filter. The noise reduction and spot enhancement are based on convolving the image $I(x, y)$ with a filter $\text{LoG}(x, y)$ as shown in Eq. (2)

$$I_f(x, y) = \text{LoG}(x, y) * I(x, y). \tag{2}$$

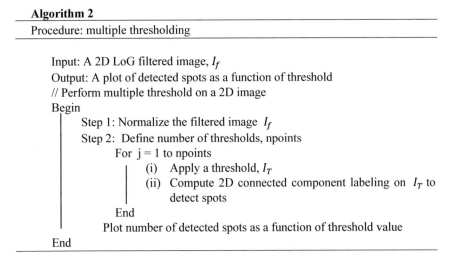

Algorithm 1

Function: 2D_LoGFilter(2D image)

Input: A 2D gray scale image, $I(x, y)$ of size XY
Output: A filtered image, I_f.
// Filter the input image to reduce noise and enhance particles
Begin
    Step 1: Create a LoG kernel, H using equation (1).
    Step 2: Convolve image the input image using equation (2).
    Return, I_f
End

The second step involves the multiple thresholding of the filtered image, I_f and threshold selection.

Algorithm 2

Procedure: multiple thresholding

Input: A 2D LoG filtered image, I_f
Output: A plot of detected spots as a function of threshold
// Perform multiple threshold on a 2D image
Begin
    Step 1: Normalize the filtered image I_f
    Step 2: Define number of thresholds, npoints
        For $j = 1$ to npoints
            (i)   Apply a threshold, I_T
            (ii) Compute 2D connected component labeling on I_T to detect spots
        End
    Plot number of detected spots as a function of threshold value
End

Upon threshold selection, the method outputs the number of spots detected at that threshold.

2.2 3D Laplacian of Gaussian (LoG)

The Laplacian of Gaussian method was proposed by Raj et al. [1] for the detection of spots in microscopy images. This method gives a strong response in connected regions with high intensity compared to the background which makes it suitable for detecting spots in microscopy images. The proposed method is an extension of the previous 2D Laplacian of Gaussian method proposed by Raj et al. [1] to 3D. The extension involves the 3D LoG filter as described in Eq. (3) and consists of number of steps described below.

Noise filtering. The first step is the noise suppression process, which involves the reduction of noise in an input image using a 3D LoG filter.

$$\text{LoG}(x, y, z) = A\left(\frac{x^2}{\sigma_x^4} - \frac{1}{\sigma_x^2} + \frac{y^2}{\sigma_y^4} - \frac{1}{\sigma_y^2} + \frac{z^2}{\sigma_z^4} - \frac{1}{\sigma_z^2}\right)e^{-\frac{1}{2}\left(\frac{x^2}{\sigma_x^2} + \frac{y^2}{\sigma_y^2} + \frac{z^2}{\sigma_z^2}\right)} \tag{3}$$

where A is given as;

$$A = 1/\left((2\pi)^{\frac{3}{2}}(\sigma_x\sigma_y\sigma_z)\right) \tag{4}$$

The parameters, σ_x, σ_y and σ_z, are associated with the size of the LoG filter, and position estimation of each spot is obtained by computing the center of mass of each connected component.

Algorithm 3

Function: 3D_LoGFilter(3D image)

Input: A 3D gray scale image, $I(x, y, z)$ of size XYZ
Output: A filtered image, I_f.
// Filter the input image to reduce noise and enhance particles
Begin
    Step 1: Create a LoG kernel, H using equation (3).
    Step 2: Transform H to 3D, H_3.
    Step 3: Convolve image, I with a 3D LoG kernel H_3.
        $I_f = I * H_3$
    Return, I_f
End

After the noise filtering and particle enhancement, the enhanced image is thresholded multiple times as described in Algorithm 4.

Algorithm 4

Procedure: multiple thresholding

Input: A 3D LoG filtered image, I_f
Output: A plot of detected spots as a function of threshold
// Perform multiple threshold on a 3D image
Begin

Step 1: Normalize the filtered image I_f using $I_N = {I_f}\big/{\max(I_f)}$.

Step 2: Define number of thresholds, npoints
  For j = 1 to npoints
    (iii) Apply a threshold, $I_T = I_N > {j}\big/{npoints}$.
    (iv) Compute 3D connected component labeling on I_T to detect spots (use 26 connectivity)
    (v) Compute the centroid of each connected component
  End
  Plot number of detected spots as a function of threshold value
End

The third step involves the manual threshold selection. The selection is performed by an operator and is based on inspecting a plateau on the threshold versus detected plot as shown in Fig. 2c. The perfect threshold value is in the plateau region, a region in which the detected spots is less sensitive to the threshold value. After the threshold selection at the plateau region the method outputs bounding circles for each detected spot, as shown in Fig. 2d.

2.3 3D Synthetic Image Datasets

Synthetic images play a crucial role when testing the performance of any automated detection method. In order to test the performance of proposed method, we created two types of synthetic image scenarios. The first scenario consisted of images of no background structures (named NOBGND), and the second scenario consisted of synthetic images with real fluorescence background structures (named BGND), as shown in Fig. 3. These synthetic images were created using a 3D extension [12] to the framework proposed by Mabaso et al. [13]. All the scenarios were corrupted by Gaussian noise, with the mean of zero and varying standard deviation {2, 2.86, 4, 5, 10, 20}. The following signal-to-noise ratio (SNR) levels were explored {10, 7, 5, 4, 2, 1} where the spot intensity was 20 gray levels. The signal-to-noise ratio is defined as of spot intensity, $S_{\max}$, divided by the noise standard deviation, σ_{noise},

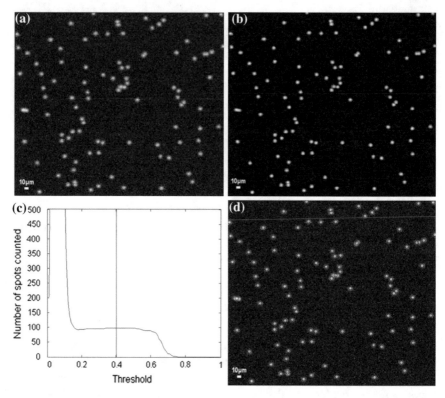

Fig. 2 Identification of spots using the proposed 3D Laplacian of Gaussian method. **a** Original 3D input synthetic image with spots (all slices merged), **b** LoG-filtered image to reduce noise and enhance spots, **c** the number of detected spots found upon thresholding the LoG-filtered image. The presence of plateau indicates a region where the number of spots in constant and vertical redline shows the selected threshold. **d** Showing the results of the selected threshold

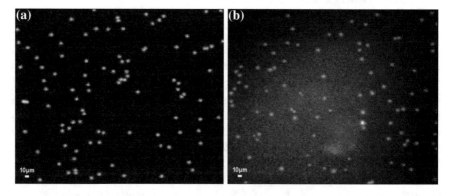

Fig. 3 Examples of 3D synthetic images of SNR = 10 (merged z slices). **a** NOBGND, **b** BGND

$$SNR = \frac{S_{max}}{\sigma_{noise}} \tag{5}$$

The number of slices varied between {12 and 20} for each synthetic image. Each synthetic image slice was of size 256 by 256 pixels. The number of spots in each synthetic image was fixed at {100}. The spot positions were randomized using Icy plugin [14] to mimic the kinds of properties in real microscopy images. MATLAB was used to add spots, and the OMERO.matlab-5.2.6 toolbox [15] was used to read and save images.

2.4 Performance Measures

In order to test the performance of the four detection methods, we computed several measures: true positives (TP), false positives (FP), and false negatives (FN). True positives are detected spots that correspond to the ground-truth spots. If the detected spot does not correspond to the ground truth, it is considered as a false positive. A missed ground-truth spot is considered as a false negative. In order to compare the performance of different methods on the generated synthetic images, we computed the centroid of every detected connected component in a binary map, resulting in a set of locations $\{d\}$. Then, following the criteria suggested by [4], a particle ω of a given ground truth is correctly detected if and only if: its nearest neighbor d in the sets of detected centroids is closer than 200 nm away, and ω is also a nearest neighbor d in the ground-truth set of locations. Two performance measures are considered in this study, recall and precision [4, 13].

Recall measures the ratio of correctly detected spots over all ground-truth spots, and precision measures the ratio of correctly detected spots among all detected spots:

$$Recall = \frac{N_{TP}}{N_{TP} + N_{FN}}, \quad Precision = \frac{N_{TP}}{N_{TP} + N_{FP}} \tag{6}$$

where N_{TP} is the number of true positives, N_{FN} is the number of false negatives, and N_{FP} is the number of false positives. Then, the F_{score} measure is computed as a weighted average of the two measures, precision and recall:

$$F_{score} = 2 \times \frac{precision \times recall}{(precision + recall)} \tag{7}$$

A good detection method should have the value of F_{score} approaching one.

3   Experimental Results

We evaluate the performance of the proposed 3D detection method using realistic
synthetic images. The first scenario consisted of images with no background
structures, NOBGND. This will help with the evaluation of the performance of the
algorithm as a function of image noise (SNR). The second to seventh scenario
consisted of image sequences with a real background (BGND). The second
experimental scenario was used to evaluate the performance of the method as a
function of real background and noise. All the synthetic images used were of 3D.
The proposed method was compared with the isotropic undecimated wavelet
transform (IUWT) [7] method.

Figure 4 presents the F_{score} versus SNR plot for the LoG and IUWT methods
applied to different kinds of synthetic images. The plot shows higher F_{score} values
for NOBGND synthetic images from SNR (4–10) compared to BGND images. This
indicates that the performance of the method decreases when real backgrounds are
introduced to the synthetic images. As SNR decreases <4, the F_{score} values drop
significantly for all synthetic images. The reduction in performance could be caused
by the increase in background noise which then reduces the quality of synthetic

Fig. 4 Curves of F_{score}
versus SNR for the LoG and
IUWT detection methods
applied to two different
synthetic images,
a NOBGND and **b** BGND
synthetic images

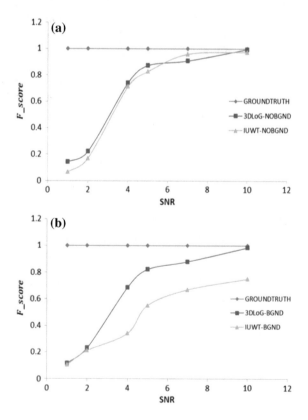

images in a way that the intensities of a spot and noise are nearly equal. The performance of the two methods on NOBGND is relatively similar for all SNR levels, but BGND shows different behavior.

The LoG method on Fig. 4b gives higher F_{score} values for SNR > 2; however, for SNR < 2, both methods perform similarly. The higher F_{score} values for LoG method are as a result of the manual threshold selection step. This step gives the user an opportunity to interact with the method before final results by looking at the plateau on the threshold versus number of spots plot.

4   Conclusions

In this paper, we have presented an extension of Laplacian of Gaussian-based method to 3D and tested its performance using 3D realistic synthetic images generated using the framework proposed in [12]. These synthetic images with varying noise and background enabled us to investigate the performance of the method on a given degradation (noise or background). The success rate of the LoG filter was expressed in terms of F_{score} and was then compared to 3D IUWT to investigate its performance against a benchmark method. The experimental results clearly indicate that the proposed method yields similar results in no-background synthetic images and higher F_{score} values for synthetic images with real background.

Bibliography

1. A. Raj, P. van den Bogaard, S.A. Rifkin, A. van Oudenaarfen, S. Tyagi, Imaging individual mRNA molecules using multiple singly labeled probes. Nat. Methods 5(10), 877–879 (2008)
2. C. Yao, J. Zhang, G. Wu, H. Zhang, Motion analysis of live objects by super-resolution fluorescence microscopy. Comput. Math. Methods Med. 2012, 1–12 (2012)
3. A. Genovesio, T. Liendl, V. Emiliana, W.J. Parak, M. Coppey-Moisan, J.-C. Olivo-Marin, Multiple particle tracking in 3d+t microscopy: method and application to the tracking of endocytosed quantum dots. IEEE Trans. Image Process. 15(5), 1062–1070 (2006)
4. I. Smal, M. Loog, W. Niessen, E. Meijering, Quantitative comparison of spot detection methods in fluorescence microscopy. IEEE Trans. Med. Imag. 29(2), 282–301 (2010)
5. A. Basset, J. Boulanger, J. Salamero, P. Bouthemy, C. Kervrann, Adaptive spot detection within optimal scale selection in fluorescence microscopy images. IEEE Trans. Image Process. 24(11), 4512–4527 (2015)
6. N. Otsu, A threshold selection method from gray-level histograms. IEEE Trans. Syst. Man Cybern. 9(1), 62–66 (1979)
7. J.-C. Olivo-Marin, Extraction of spots in biological images using multiscale products. Pattern Recogn. 35(9), 1989–1996 (2002)
8. K. Štěpka, P. Matula, P. Matula, S. Wörz, K. Rohr, M. Kozubek, Performance and sensitivity evaluation of 3D spot detection methods in confocal microscopy. Cytom. Part A 87(8), 759–772 (2015)

9. M. Gue, C. Messaoudi, J.S. Sun, T. Boudier, Smart 3d-fish: automation of distance analysis in nuclei of interphase cells by image processing. Cytom. Part A **67A**(1), 18–26 (2005)
10. H. Netten, I.T. Young, L.J. van Vliet, H.J. Tanke, Fish and chips: automated of fluorescent dot counting in interphase cell cuclei. Cytometry **28**(1), 1–10 (1997)
11. M.N. Cabili, M.C. Dunagin, P.D. McClanahan, A. Biaesch, O. Padovan-Merhar, A. Rgev, J. L. Rinn, A. Raj, Localization and abundance analysis of human lncRNAs at single-cell and single-molecule resolution. Genome Biol. **16**(20), 1–16 (2015)
12. M. Mabaso, D. Withey, B. Twala, Generation of 3D realistic synthetic image datasets for spot detection evaluation, in *Joint International Conference on Artificial Intelligence and Evolutionary Computations in Engineering Systems (ICAIECES-2017) & Power, Circuit and Information Technologies (ICPCIT-2017)* (Madanapalle, 2017)
13. M. Mabaso, D. Withey, B. Twala, A framework for creating realistic synthetic fluorescence microscopy image sequences, in *Proceedings of the 9th International Joint Conference on Biomedical Engineering System and Technologies* (Rome, Italy, 2016)
14. N. Chenouard, Particle tracking benchmark generator. Institut Pasteur, 2015. [Online]. Available: http://icy.bioimageanalysis.org/plugin/Particle_tracking_benchmark_generator. [Accessed 1 Nov 2016]
15. The Open Microscopy Environment, 2016. [Online]. Available: http://www.openmicroscopy. org/site/support/omero5.2/developers/Matlab.html. [Accessed 15 Nov 2016]

Distributive MPPT Approach Using ANFIS and Perturb&Observe Techniques Under Uniform and Partial Shading Conditions

Adedayo M. Farayola, Ali N. Hasan, Ahmed Ali
and Bhekisipho Twala

Abstract PV systems work under different weather conditions such as uniform and partial shading weather conditions. This causes inconsistent power in PV systems. This paper presents a reconfigurable interconnections approach that uses and compares between two powerful maximum power point tracking (MPPT) techniques of artificial neuro-fuzzy inference system [ANFIS front-end distributive MPPT (DMPPT)] technique and Perturb&Observe distributive MPPT technique. This approach is introduced in order to decrease the partial shading and mismatch effect caused by varying light falling on the PV arrays, which will lead to extract more power from the PV modules. The PV systems were configured as series-connected PV string that uses Perturb&Observe MPPT technique and as a PV series-connected system that uses ANFIS-MPPT technique. The proposed PV systems were tested under uniform and partial shading weather conditions. The results show that MPPT could be tracked accurately with the ANFIS-DMPPT for both cases of uniform irradiance and partial shaded irradiance conditions.

Keywords Perturb&Observe · ANFIS · DC to DC converter · Distributive
MPPT · Mismatch · PV system

A. M. Farayola (✉) · A. N. Hasan · A. Ali · B. Twala
Department of Electrical Engineering Technology, University of Johannesburg,
PO Box 524 Auckland Park, Johannesburg 2006, South Africa
e-mail: lordfaraday@yahoo.com

A. N. Hasan
e-mail: alin@uj.ac.za

A. Ali
e-mail: axmedee@live.com

B. Twala
e-mail: btwala@uj.ac.za

© Springer Nature Singapore Pte Ltd. 2018   27
S. S. Dash et al. (eds.), *Artificial Intelligence and Evolutionary Computations
in Engineering Systems*, Advances in Intelligent Systems and Computing 668,
https://doi.org/10.1007/978-981-10-7868-2_3

1 Introduction

Photovoltaic (PV) solar system is a non-pollutant, renewable energy that can be used to produce electricity when light falls on PV cells [1]. The PV cells produce higher voltage when connected in series and higher current when it is connected in parallel under uniform irradiance [2]. When power is demanded at large quantity, for example, stand-alone PV system for solar water pumps and geezers, several PV modules are interconnected to form an array. PV array is greatly disturbed when the light falling on it is not uniform. This is referred to as partial shading, a case where the overall PV power is determined by the PV module with the lowest insolation and the power generated by the module(s) with higher irradiance is wasted as a form of heat [3]. For a series–parallel PV configuration referred to as strings, another diode called blocking diode can be connected to the terminals of each string for protection against current imbalance between the strings [4].

Maximum power point tracking is a technique that is used to extract maximum and stable power from a PV system. The technique aims to set the input impedance (Z_i) to be equal to the load impedance (Z_o) [5].

Partial shading is defined as "a condition where the quantity of sun is less than partial sun, but greater than shade." This occurs when the light falling on PV cells or modules has different insolation levels while uniform irradiance is referred to as "full sun," a condition when the light intensity falling on PV cells or modules is fixed [6]. Several of the used MPPT techniques such as Perturb&Observe, incremental conductance, and techniques based on artificial intelligence such as ANFIS work fine when operated under a uniform irradiance and temperature. However, only few of these MPPT algorithms can track the true global MPP under partial shading and mismatch conditions as most MPPT techniques resolute on locating the local maxima power, thereby leading to a loss in the extracted power [7].

Distributed maximum power point tracking (DMPPT) approach has been recently used to overcome the effect of partial shading and cell mismatch in a PV system [8]. DMPPT can be implemented using several configurations such as PV voltage equalization, series-connected DMPPT, parallel-connected DMPPT, differential or minimal power processing, etc. DMPPT technique can also be presented as a distributed power conditioning where MPP tracking is achieved at the unit level by means of module integrated DC-DC converters (MIC), positioned at each PV module's front end as it is concluded that the front-end DMPPT can harvest more power under deviating irradiance compared to the ordinary MPPT [9].

Recent research has introduced the use of Perturb&Observe distributive MPPT technique and particle swarm optimization DMPPT technique for partial shading case. In this paper, Artificial Neuro-Fuzzy Inference system (ANFIS) DMPPT was used to track the global MPP, and the results were compared with the results obtained using Perturb&Observe DMPPT approach. Also, series-connected PV modules using the Perturb&Observe MPPT and the ANFIS-MPPT were implemented under non-shaded (uniform) and shaded environmental conditions.

This work was introduced in order to explore the feasibility of using ANFIS techniques for distributive MPPT system.

The structure of this paper is organized as follows: Sect. 2 will present the used MPPT techniques. In Sect. 3, a report of the experiments setup and procedure is provided. Section 4 will present the results, and Sect. 5 will include the conclusions.

2 MPPT Techniques

2.1 Perturb&Observe (P&O) Technique

P&O is a hill-climbing MPPT algorithm. It is used because it is cheap and easy to implement. P&O only requires two sensors (power/current sensor and voltage sensor). The system is made by perturbing a small rise in voltage (ΔV) of PV and observing the variation in power (ΔP). If ΔP is positive, the operating point (perturbation) will shift near maximum power point, and if ΔP is negative, the operating point is moved away from maximum point [10]. The issue with this technique is the oscillatory power around MPP (drift) which can be minimized by decreasing the step size of perturbation [11, 12].

2.2 Adaptive Neuro-Fuzzy Inference System (ANFIS)

ANFIS is a combination of two offline techniques (fuzzy logic and neural network MPPT). ANFIS is designed using basic procedures such as data gathering, training, testing, validating, fuzzification, inference, rules, and defuzzification [13]. ANFIS can be used for optimization and its training data can be obtained using real-time system or using simulation approach by developing a dynamic photovoltaic system [14].

2.3 Series-Connected Distributed Maximum Power Point Tracking (DMPPT)

Series-connected DMPPT is a common MPPT method that is used to tackle partial shading and mismatch effect as it is cheaper and more efficient compared to the parallel-connected DMPPT. Series-connected DMPPT can be implemented by connecting MPPT and DC-DC converter separately to each of the PV panels. The DC-DC converter's outputs are then connected in series to increase the voltage and current and have more power extracted from the PV system [15].

3 Simulation Model

In order to validate the performance of the four machine-learning techniques (ANFIS distributive MPPT, Perturb&Observe distributive MPPT, ANFIS-MPPT, and standard Perturb&Observe MPPT) under uniform and partial shading conditions, four experimental tests (case study) were conducted to determine the accuracy of the MPPT techniques. The PSIM professional tool was used to generate our offline data (instances), where 129 instances were acquired and used for training, testing, and validation of the ANFIS used in this work. Table 1 illustrates the specification of the Soltech 1STH-215-P PV panel and the modified CUK DC-DC converter used in this experiment. The simulation time $t = 0.2$ s at a discrete sampled period, $T_s = 1e-6$ s, was used, and Eq. (1) was used to obtain the efficiency of the PV strings at these stated environmental conditions, where $P_{pv(max)}$ is the output PV power and $P_{pv(mpp)}$ is the rated output power at STC $= 213.15 * 3 = 639.45$ W (three Soltech 1STH-215-P PV modules power).

$$P_{pv(efficiency)} = \int^t P_{pv(max)}/P_{pv(mpp)} dt \qquad (1)$$

Figure 1 is the block diagram of the ANFIS distributive MPPT (ANFIS-DMPPT) technique using three similar Soltech 1STH-215-P panels that were connected independently. ANFIS technique is used as the MPPT controller. ANFIS is connected to each of the three series-connected MCUK DC-DC converters having a duty cycle or pulse signal D_1, D_2, and, D_3, respectively. In designing each of the three ANFIS-MPPT algorithms, the data (129 instances) were divided to 70% for training, 15% for testing, and the remaining 15% for validation. The ANFIS model

Table 1 PV specification and modified CUK specification

Solar panel specification		MCUK DC-DC converter specification	
PV model	1STH-215-P	L_1	4 mH
Maximum voltage (V_{mp})	29.0 V	L_2	4 mH
Maximum current (I_{mp})	7.35 A	C_1	100 μF
Maximum power (P_{mp})	213.15 W	C_2	100 μF
Number of cells in series (N_s)	60	C_0	270 μF
Short-circuit current (I_{sc})	7.84 A	R_0	20 Ω
Open-circuit voltage (V_{oc})	36.3 V		
Temperature coefficient of I_{sc}	−0.36099%/°C		
Temperature coefficient of V_{oc}	0.102%/°C		
Diode ideality factor (A)	0.98117		
Series resistance (R_s)	0.39383 Ω		
Shunt resistance	313.3991 Ω		

has two inputs, the irradiance (G) and the temperature (T), and one predicted output (I_{ref}). The output response (I_{ref}) was compared with the PV current (I_{pv}) and the measured error was tuned by a discrete PI controller and outputs the duty cycle (D). The duty cycle signal was conveyed using pulse width modulation scheme as pulse at frequency ($F = 50$ kHz). The pulse signal was used to switch ON the MOSFET gate of the non-isolated modified CUK DC-DC converter. The ANFIS is optimized using a hybrid optimization [combination of the least-square estimation (LSE) (forward pass) and the back-propagation gradient descent methods (backward pass)]. The duty cycle method used is the current reference control method.

For the Perturb&Observe distributive MPPT (PO-DMPPT) case, the same Fig. 1 blocks were used to implement the model, but the MPPT controllers were replaced with the Perturb&Observe algorithms. The PO-DMPPT model used the direct duty cycle method where the inputs to each of the three Perturb&Observe algorithms are voltage, power, and output duty cycles (D_1, D_2, and D_3) to the MCUK DC-DC converters, respectively.

For the Perturb&Observe MPPT (PO-MPPT), Fig. 2 shows the block of the PO-MPPT where the three PV modules at different irradiances G_1, G_2, and G_3 were connected in series to form a string or an array. In this configuration, only one Perturb&Observe MPPT and a single MCUK are required. The inputs to the PO-MPPT are the PV voltage (V_{pv}) and the PV current (I_{pv}) and outputs the duty cycle D. The direct duty cycle method is also implemented here.

For the series-connected PV module using ANFIS-MPPT, Fig. 3 shows the complete system. The ANFIS-MPPT algorithm is achieved using the ANFIS-DMPPT method but with one MCUK DC-DC converter, and the PV panels in this model were connected in series. In this case, voltage control reference duty cycle is implemented.

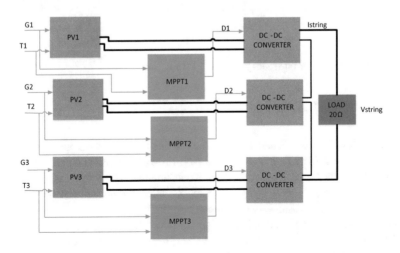

Fig. 1 Distributive MPPT technique for three PV modules

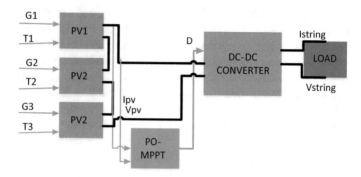

Fig. 2 Series-connected PV MPPT

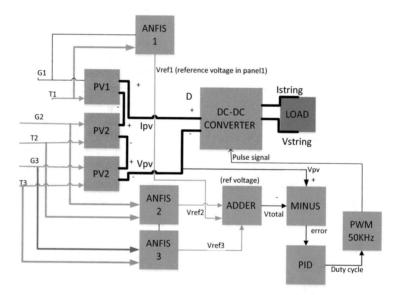

Fig. 3 ANFIS-MPPT for three series-connected PV modules

4   Results and Discussion

Figures 4, 5, 6, 7, 8, 9, 10, and 11 show the obtained input and output power graphs for the ANFIS-DMPPT, PO-DMPPT, ANFIS-MPPT, and standard PO-MPPT technique results using the four-varied environmental condition case study at a fixed temperature, $T_1 = T_2 = T_3 = T = 25$ °C throughout the experiments, and Table 2 shows the measured experimental results.

From Table 2 experimental results, for the case 1 (uniform irradiance), ANFIS-DMPPT happens to overperform the other used techniques and has a predicted efficiency of 99.98% at the PV end and a predicted efficiency of 95.86% at

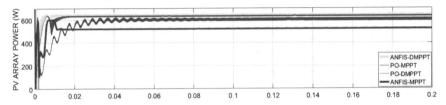

Fig. 4 Graph of power input for case 1 environmental conditions ($G_1 = G_2 = G_3 = 1000$ W/m^2)

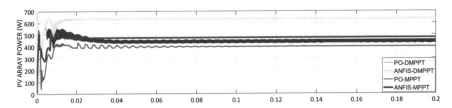

Fig. 5 Graph of power input for case 2 environmental conditions

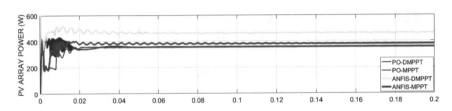

Fig. 6 Graph of power input for case 3 environmental conditions

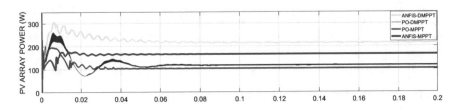

Fig. 7 Graph of power input for case 4 environmental conditions

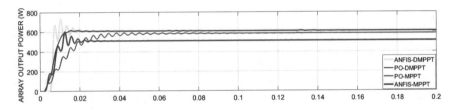

Fig. 8 Graph of power output for case 1 environmental conditions

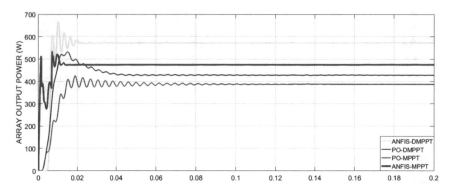

Fig. 9 Graph of power output for case 2 environmental conditions

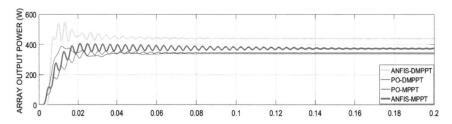

Fig. 10 Graph of power output for case 3 environmental conditions

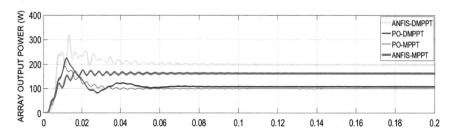

Fig. 11 Graph of power output for case 4 environmental conditions

the DC-DC converter load end while ANFIS-MPPT technique underperformed with a predicted efficiency of 81.09% at the PV end and a predicted efficiency of 79.76% at the DC-DC converter load. For the case 2 (partial shading at PV2), ANFIS-DMPPT overperformed with a predicted efficiency of 93.52% at the PV end and 89.45% at the load end while PO-MPPT underperformed with a predicted efficiency of 62.40% at the PV end and 60.52% efficiency at the DC-DC. For the case 3 (partial shading at PV1 and PV3), ANFIS-DMPPT overperformed with a predicted efficiency of 72.17% at the PV end and 68.65% at the load end while PO-MPPT floundered with an efficiency 55.99% at the PV end and a predicted

Table 2 Experimental result of PV modules

Weather conditions	Measurement	ANFIS-DMPPT	ANFIS-MPPT	PO-DMPPT	PO-MPPT
$G_1 = 1$ kW/m^2 $G_2 = 1$ kW/m^2 $G_3 = 1$ kW/m^2 $T_1 = T_2 = T_3 =$ 25 °C CASE 1 uniform insolation	Panel1 P_1 (W)	213.1 W	173.9 W	212.5 W	202.3 W
	Panel2 P_2 (W)	213.1 W	173.9 W	212.5 W	202.3 W
	Panel3 P_3 (W)	213.1 W	173.9 W	212.5 W	202.3 W
	PV power ($P_1 + P_2 + P_3$)	639.3 W	518.5 W	638.0 W	607.0 W
	Output current (A)	−6.534 A	−5.05 A	−5.519 A	−5.398 A
	Output voltage (V)	−110.7 V	−101.0 V	−110.4 V	−108.0 V
	Output power (P_o)	613.0 W	510 W	609 W	582.8 W
	MCUK losses (W)	26.3 W	8.50 W	29.0 W	25.8 W
	PV efficiency (%)	99.98%	81.09%	99.77%	94.93%
	Load efficiency (%)	95.86%	79.76%	95.24%	91.14%
$G_1 = 1$ kW/m^2 $G_2 = 0.8$ kW/m^2 $G_3 = 1$ kW/m^2 $T_1 = T_2 = T_3 =$ 25 °C CASE 2 partial shading	Panel1 P_1 (W)	213.1 W	160.7 W	203.0 W	197.0 W
	Panel2 P_2 (W)	171.8 W	153.5 W	51.1 W	4.60 W
	Panel3 P_3 (W)	213.1 W	160.7 W	203.0 W	197.0 W
	PV power ($P_1 + P_2 + P_3$)	598.0 W	474.7 W	456.0 W	399.0 W
	Output current (A)	−5.349 A	−4.821 A	−4.625 A	−4.397 A
	Output voltage (V)	−107.0 V	−96.42 V	−92.5 V	−87.94 V
	Output power (P_o)	572.0 W	464.9 W	428.0 W	387.0 W
	MCUK losses (W)	26.0 W	9.80 W	28.0 W	12.0 W
	PV efficiency (%)	93.52%	74.24%	71.31%	62.40%
	Load efficiency (%)	89.45%	72.70%	66.93%	60.52%
$G_1 = 0.8$ kW/m^2 $G_2 = 1$ kW/m^2 $G_3 = 0.6$ kW/m^2 $T_1 = T_2 = T_3 =$ 25 °C CASE 3 partial shading	Panel1 P_1 (W)	172.0 W	128.2 W	172.0 W	172.9 W
	Panel2 P_2 (W)	213.0 W	131.5 W	186.0 W	188.5 W
	Panel3 P_3 (W)	76.5 W	119.6 W	0.9 W	−3.8 W
	PV power ($P_1 + P_2 + P_3$)	461.5 W	379.3 W	359.0 W	358.0 W
	Output current (A)	−4.684 A	−4.326 A	−4.114 A	−4.151 A
	Output voltage (V)	−93.68 V	−86.51 V	−82.28 V	−83.02 V
	Output power (P_o)	439.0 W	374.2 W	339.0 W	345.0 W
	MCUK losses (W)	22.5 W	5.10 W	20.0 W	13.0 W
	PV efficiency (%)	72.17%	59.2%	56.14%	55.99%
	Load efficiency (%)	68.65%	58.52%	53.01%	53.95%
$G_1 = 0.4$ kW/m^2 $G_2 = 0.8$ kW/m $G_3 = 0.2$ kW/m^2 $T_1 = T_2 = T_3 =$ 25 °C CASE 4 partial shading	Panel1 P_1 (W)	212.4 W	13.0 W	89.1 W	109.1 W
	Panel2 P_2 (W)	2.9 W	78.9 W	7.0 W	−1.3 W
	Panel3 P_3 (W)	0.0 W	−1.3 W	0.0 W	−1.3 W
	PV power ($P_1 + P_2 + P_3$)	215.4 W	165.7 W	120.0 W	104.0 W
	Output current (A)	−3.143 A	−2.844 A	−2.321 A	−2.230 A
	Output voltage (V)	−62.87 V	−56.87 V	−46.41 V	−44.60 V
	Output power (P_o)	198.0 W	161.7 W	108.0 W	99.5 W
	MCUK losses (W)	17.4 W	4.0 W	12.0 W	5.5 W
	PV efficiency (%)	33.69%	25.91%	18.77%	16.26%
	Load efficiency (%)	30.96%	25.29%	16.89%	15.56%

efficiency of 53.95% at the DC-DC load. For the case 4 (partial shading at PV1 and PV3), ANFIS-DMPPT overperformed with a predicted efficiency 33.69% at the PV end and 30.96% efficiency at the DC-DC load end while PO-MPPT underperformed with an efficiency of 16.26% at the PV end and 15.26% efficiency at the DC-DC converter load end. Also, for the case 4 environmental condition, at extremely low irradiance ($G3 = 200$ W/m^2), PV panel 3 (PV3) acts as a short circuit ($P_3 = 0$ W) using the distributive MPPT techniques (ANFIS-DMPPT and PO-DMPPT), whereas a negative power was obtained, $P_3 = -1.26$ W (power dissipation), with the conventional series-connected PV modules using PO-MPPT technique and $P_3 = -1.32$ W using ANFIS-MPPT technique.

5 Conclusion

This paper presents a machine-learning scheme that is suitable for maximum power point tracking under varied environmental conditions by using reconfigurable ANFIS distributive MPPT technique. Obtained experimental results showed that the ANFIS-DMPPT method gives an overall the best performance under uniform irradiance and partial shading environmental conditions when compared to the other used techniques. ANFIS-DMPPT method achieved the highest accuracy for all four different weather conditions and could locate the global maximum power point (GMPP) successfully during uniform and partial shading cases, thereby allowing more power to be extracted from the PV modules. Also, it can be noticed that the PO-DMPPT and the PO-MPPT techniques applied using the conventional hill climbing (Perturb&Observe) technique are suitable for uniform irradiance conditions as the performance declined under partial shading environmental conditions, whereas the ANFIS-DMPPT and ANFIS-MPPT are well-matched for partial shading and mismatch conditions.

References

1. A.M. Farayola, A.N. Hasan, A. Ali, Comparison of modified incremental conductance and Fuzzy logic MPPT algorithm using modified CUK converter, in *8th IEEE International Renewable Energy Congress (IREC) 2017*, Amman, Jordan (2017)
2. S.S. Mohammed, D. Devaraj, T.P.I. Ahamed, Modeling, simulation and analysis of photovoltaic modules under partially shaded conditions, Indian J. Sci. Technol. **9**(16) (2016)
3. L.E. Johnathan, *Methods for Increasing Energy Harvest with PV Module Integrated Power Converters* (University of Illinois, Urbana, 2013)
4. L.V. Vincenzo, D.L. Manna, E.R. Sanseverino, D.D. Vincenzo, P. Romano, P. di Buono, P. Maurizio, M. Rosario, C. Giaconia, Proof of concept of an irradiance estimation system for reconfigurable photovoltaic arrays, Energy 2015, **8**, 6641–6657; ISSN 1996-1073 (2016)
5. A.R. Reisi, H.M. Mohammad, J. Shahriar, Classification and comparison of maximum power point tracking techniques for photovoltaic system, a review. Renew. Sustain Energ. Rev. **19**, 433–443 (Elsevier, 2013)

6. A.S. Martha, *Stepping Stones to Perennial Garden Design* (University of Illinois, USA, 2013)
7. M. Seyedmahmoudian, B. Horan, R. Rahmani, A.M. Than, A. Stojcevski, Efficient photovoltaic system maximum power point tracking using a new technique, Energy **9**(3) (2016)
8. C. Cheng-Wei, C. Yaow-Ming, Analysis of the series-connected distributed maximum power point tracking PV system, in *IEEE Applied Power Electronics Conference and Exposition (APEC) 2015*, pp. 3083–3088 (2016)
9. G.R. Walker, P.C. Sernia, Cascaded DC–DC converter connection of photovoltaic modules, in *IEEE 33rd Annual Power Electronics Specialists Conference, PESC 02, 2002*, pp. 24–29 (2002)
10. A. Ahmed, A.N. Hasan, T. Marwala, Perturb and observe based on Fuzzy logic controller maximum power point tracking (MPPT), in *IEEE International Conference on Renewable Energy Research 2014*, Milwaukee, USA (2014)
11. R. Abdul-Karim, S.M. Muyeen, A. Al-Durra, Review of maximum power point tracking techniques for photovoltaic system. Glob. J. Control Eng. Technol. **2**, 8–18 (2016)
12. H.J. El-Khozondar, R.J. El-Khozondar, K. Matter, T. Suntio, A review study of photovoltaic array maximum power tracking algorithms. Open Renew. Wind, Water, Sol. (Springer, 2016)
13. M.Y. Worku, M.A. Abido, Grid connected PV system using ANFIS based MPPT controller in real time, in *International Conference on Renewable Energy and Power Quality (ICREPQ' 16), Renewable Energy and Power Quality Journal (RE & PQJ)* (2016)
14. A.M. Farayola, A.N. Hasan, A. Ali, Curve fitting polynomial technique compared to ANFIS technique for maximum power point tracking, in *8th IEEE International Renewable Energy Congress (IREC) 2017*, Amman, Jordan (2017)
15. S. Vijipriyadharshini, C.A. Vennil, Distributed MPPT system for non-uniform solar irradiation. Middle-east J. Scientific Res. **24**(3), 927–936 (2016)

Performance of MPPT in Photovoltaic Systems Using GA-ANN Optimization Scheme

Ahmed Ali, Bhekisipho Twala and Tshilidzi Marwala

Abstract Researchers all over the world are currently moving toward using solar energy resulting from large energy demand and sources of energy as well as the environmental problems, such as dynamic weather conditions. The control of maximum power point tracking (MPPT) meteorological conditions is an essential portion of improving solar power systems. In this paper, we introduce an elastic controller depend on artificial neural network for regulating the MPPT. This controller is employed to the buck–boost DC-to-DC converter using the MATLAB/Simulink software program. This paper proposes a design that maximizes the performance of GA-ANN scheme, and compared with ANN scheme, efficiency of PV module is shown as well as the saving power for both schemes. The results show that GA-ANN has performance about 45% over ANN scheme.

Keywords Neural networks · Genetic algorithm · Buck–Boost DC
Photovoltaic systems · MPP tracker

1 Introduction

Presently, photovoltaic (PV) power from solar is a common source of renewable energy because of its numerous advantages, such as the low functional cost, low servicing, as well environment-amiable properties. Although the solar modules are expensive, grid-connected PV power generation systems have been popularized in several countries' cause of their long-term benefits [1]. Moreover, decent fiscal

A. Ali (✉) · B. Twala · T. Marwala
Faculty of Engineering and Built Environment, University of Johannesburg,
PO Box 524, Auckland Park, Johannesburg 2006, South Africa
e-mail: axmedee@live.com

B. Twala
e-mail: btwala@uj.ac.za

T. Marwala
e-mail: T.Marwala@uj.ac.za

© Springer Nature Singapore Pte Ltd. 2018
S. S. Dash et al. (eds.), *Artificial Intelligence and Evolutionary Computations in Engineering Systems*, Advances in Intelligent Systems and Computing 668,
https://doi.org/10.1007/978-981-10-7868-2_4

schemes, such as the feed-in tariff [2] and the subsidized policy [3], have been instituted by different countries, producing in the fast industry growth. Large array PV modules must be optimized for full utilization. Hence, the maximum power point tracker (MPPT) is generally used with power converters (i.e., DC-to-DC converters or/and inverters). These approaches have the different complexities, speeds, and accuracies. Each approach can be classified depending on the types of control variables, and each utilizes: (1) the duty cycle, (2) the current, and (3) the voltage. Three main methods were used to maximize the extraction of power in any medium and system scale. These methods are the MPPT, sun tracking, and bath. The MPPT is common in small-scale frameworks because of economic reasons.

The contributions of this paper can be summarized as follows:

(i) The design of PV system that solves the problem of the unstable outputs of the traditional PV systems which resulted from the unstable sun irradiances.
(ii) The design of genetic algorithm (GA) optimization schemes together with the artificial neural network (ANN) for MPPT controller that utilizes a DC-to-DC boost converter under varying solar radiation conditions and temperatures.

2 Related Works

Over the last two decades, the PV technology has developed rapidly from small to large scale. This section presents a few previous studies that focused mainly on the PV technology, the MPPT, and using ANN in solving several MPPT problems. The small perturbation can reduce oscillations, but with speed-tracking expenditure. Another considerable P and O drawback is the probability of the algorithms losing their direction while following actual MPP during the fast fluctuation of isolation [4]. Numerous studies have employed the methods of artificial intelligence, like neural networks (NNs) and fuzzy logic controllers (FLC) [5] to address the aforementioned drawbacks [6]. These approaches are effective with the not denoting properties of the curves of $I-V$ and are requiring comprehensive computations. For instance, FLC works with fuzzifications, rule-based storages, inference mechanisms, and defuzzification functions. For NNs, the huge amount training data needed is a significant constraints' source. Moreover, the MPPT responds to these variations in isolation as well as temperature dynamics in actual time because the wind condition of PV systems varies continuously. Low-cost processors cannot be utilized in similar systems. The use of evolutionary algorithms (EA) is an alternative method. EA is perceived capable of handling the problem of the MPPT because of its capability of managing nonlinear objective functions [7, 8]. Particle swarm optimizations through the EA algorithms are highly powerful because of their clear structure, simple implementation, and fast computation capabilities [9].

3 Proposed Model

The proposed system has photovoltaic generator as a main component that could be a cell, an array, or a module. The PV generator consists of various units, protective modules, photovoltaic cells, supports, and connections. One of the essential components of the PV or the solar cells is the PN junction which is made in the thin semi-conductor layer. The solar panel is the first step in the system. The properties of this panel relay on solar irradiance as well as temperature. This is the corresponding point on the PV curves to the maximum power point (MPP). The proposed PV generator and MPPT block diagram are also presented in Fig. 1; the PV panel can charge the battery by utilizing the MPPT circuit. As shown in Fig. 1, MPP can be optimized using ANN and then the updated version of ANN is done using GA to enhance the efficiency of PV module.

3.1 Modeling of Buck–Boost DC-to-DC Converter

The DC-to-DC power converter can convert the supply voltage source to both lower and higher voltages to a load terminal. However, the voltage V may be smaller or greater than the input V, relaying on the value S duty cycle D. Moreover, Fig. 2 represents the circuit of the Buck–Boost inverting converter.

Mainly, the power switch aims at modeling the energy transfer to the load from the input by changing the D duty cycle. However, equation that follows shows how the input and output V relate in the proposed power converter.

$$\frac{V_0}{V_g} = -\frac{D}{1-D} \tag{1}$$

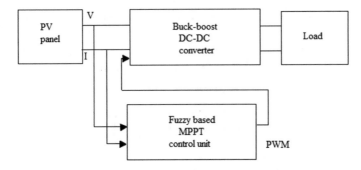

Fig. 1 Circuit diagram for the proposed system

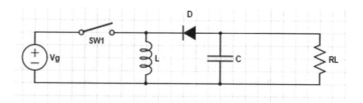

Fig. 2 Circuit of the Buck–Boost inverting converter

Whereas, the output voltage V_0 as well as the output current I_o are both negative. Nevertheless, in this work simulation, the results are all with positive signs in the plot diagram without any effect on the sign of the output power.

3.2 Artificial Neural Network Maximum Power Point Tracking Controller (ANN-MPPT)

ANN system combines the artificial systems' qualities with the employ of various neural networks (NN). This ANN simplifies transforming the NN's computational power, linking configurations, as well as low-level learning into high-level learning and artificial systems that simplify integrating the particularistic knowledge as well as analyzing artificial systems to NNs.

Figure 3 shows the major ANN system diagram. ANN system has four layers: rule layer, input and output layer as well as the membership layer. Input layer is utilized to transfer the input that is determined via the user. The layer of the membership has various nodes that are employed to map the input to a possible allocation path. Each node can be responsible for only one input variable. The rule layer has nodes in which its output is achieved from the suggested two layers. The output layer gives the final results from all layers. Weights' links connect these layers that are preset through the user or edited by employing various mathematical methods.

$$E = \frac{1}{N} \sum_{k=1}^{N} \sum_{j=1}^{M} h_{j,N(x^{(k)})} \left\| W_j - X^{(k)} \right\|^2 \tag{2}$$

$$h_{i,j(t)} = \exp\left(-\frac{\left\| r_j - r_i \right\|^2}{2\sigma_t^2} \right) \tag{3}$$

$$\sigma(t) = \sigma_0 \exp(-t/T) \tag{4}$$

In the presented equation above, the used unites for mapping are represented by M, the neurone owning the closest exhortation that relates to the data sample is

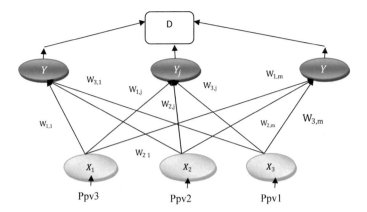

Fig. 3 ANN system model

represented by $N(X^{(k)})$, h on the other hand is Gaussian neighborhood function, while i and J represents input sample and the map unit, respectively. $\|r_j - r_i\|^2$ represents the distance separating the map unit and input sample, σ_t represents radius of neighborhood taken at t time, t stands for iteration number used in learning the network, and the number of maximum iteration at training process which is considered as training length are represented by T at this point, calculation of the distance separating the vector of mapping and map neurons done. The winner at the competition phase is the neural with the minimum distance regarding to X_k input sample as shown in Eq. 5.

$$N(X_k) = \arg_{1 < j \leq m} \min \|W_j - X_k\|^2 \tag{5}$$

The argument within the above equation is used to make a decision and select the neuron that has the minimum distance value with X_k input sample. This decision is performed after calculating all neurons corresponding distances with (X_k), and then returning the neuron with the minimum distance value as the winner of the competition stage. After initiating the learning process, σ_t characterizes with large value. The radius is reduced with the increase in algorithm time. And with the end of competition phase, uploading of the winner's weight vector is done. In the update process, the winner vector is included, and they are located on $R\ (N(X_K))$ neighborhood radius. Now,

$$W_{j(t+1)} = \begin{cases} W_j(t) + \alpha(t)h_{j,N(x_k)}(t)_k \left(x(t) - W_j(t)\right); & \text{when} W_j \in R^{N(X_i)} \\ W_j(t) & ; \quad \text{otherwise} \end{cases} \tag{6}$$

$h_{j,N(x_k)}$ represents the neighborhood function that taken at t time, and $\alpha(t)$ represents the linear factor of learning which can be calculated as given below in Eq. 7

$$\alpha(t) = \alpha_0(1 - t/T) \qquad (7)$$

The use of ANN was in determination of the real-time reference voltage, which is dependent upon T and G. The ANN training dataset used was acquired by use of experimental measures, and the establishment of relations between inputs (T and G) and that of the output (VMPP) done. In this model, there were two input layers which included the cell temperature (T) and the solar irradiance passed through five hidden layers to produce the output layer of the voltage at VMPP. The hidden layer was obtained based on the method of trial and error.

The above process can be illustrated by computing the mean square error (MSE) over the network as shown in Fig. 4.

3.3   Genetic Algorithm Maximum Power Point Tracking Controller (GA-MPPT)

To acquire smaller effective input dataset, the use of GA is necessary as it optimizes the input dataset in ANN. Each individual population in the optimization of GA input dataset stands for feature subsets that require being solved for problem optimization. N-bit binary vectors represent individuals in the GA [10]. Mutation is used to achieve some mathematical form of GA in order to solve the optimization problem to detect any unstable dataset or noise for irradiance and temperature of PV module. For that, two main constraints should be applied to all time processing in PV module, as shown below.

$$V(n+1) - V(n) < \Delta V \qquad (8)$$

$$\frac{P(n+1) - P(n)}{P(n)} > \Delta P \qquad (9)$$

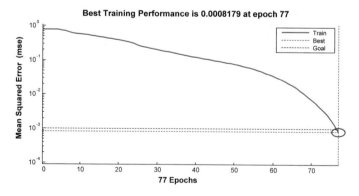

Fig. 4 MSE in ANN

where V and P are related to voltage and power of PV modules, while Δn and ΔP are related to changes in voltage and power over two time instants. Levenberg–Marquardt (LM) algorithm is used to train the neural network implemented. In the optimization of the input data, the GA parameters employed were a single point of crossover, 0.03 mutation operators Pm, the population size was 15 and 350 maximization number of generations as well as 0.85 crossover operator, Pm with selection type being Roulette. The optimization algorithm of input dataset implementation was done in MATLAB environment. From parameters used, the best object value of the generation GA was done as shown from above factual figures and the new dataset was obtained containing 197 samples. Using optimaziation of GA and ANN, there was obtained better results those where optimization of GA was not used. Treatment of ANN without using the GA optimization, values obtained for errors of cross-validation and training were 0.2802 and 0.2137, respectively. On the other hand, when GA optimization was used in ANN training, the new dataset for cross-validation error and training were 0.2381 and 0.1806, respectively. For the two training algorithm, the early termination method was used to end training. By this method, when it was noticed that the cross-validation errors were increasing, termination of training was done to prevent overturning of ANN.

4 Simulation and Numerical Results

In the proposed model, a 100 kW photovoltaic array is used. This array utilizes 330 sun-power modules with a power equals to 305.2 kW and includes 66 strings of 5 modules that are connected in series. These strings are connected in parallel. The total power equals to $66 * 5 * 305.2 = 100.7$ kW. Each module has four specifications: 96 cells that are connected in series, open circuit voltage that equals to 64.2 V, short circuit current that equals to 5.96 A, and maximum power voltage equals to 54.7 V and maximum power current equals to 5.58 A (Table 1).

Figure 5 shows $V–I$ curves under different temperature and sun irradiances values; as shown, a cutoff value occurred at 20 V; as expected while increasing T and IRR, $V–I$ curves are enhanced. Figure 6 shows the PV panel output power and voltage where its improvement is applied by increasing T and IRR inputs values.

Table 1 System parameter	PV module type	KC200GH-2P
	At 1000 W/m², 25 °C	(STC)
	Maximum power	200 W
	Maximum power voltage	26.3 V
	Maximum power current	7.61 A
	Open circuit voltage (I_{OC})	32.9 V
	Short circuit current (I_{SC})	8.21 A

Fig. 5 PV panel output
current and voltage

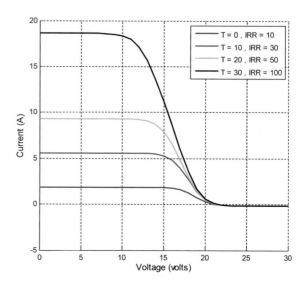

Fig. 6 PV panel output
power and voltage

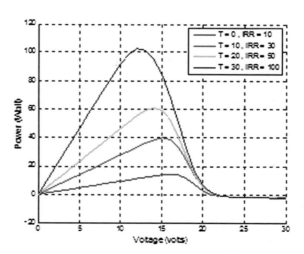

In order to compare between traditional ANN for optimization MPPT and updated scheme of GA-ANN, Fig. 7 illustrates the energy efficiency of both schemes. As shown, at the beginning of iteration size, MPPT-ANN provides performance than GA-ANN; this is due to that GA still searching on best population size with fitness function, after that, GA-ANN begins provide better performance than ANN; and finally, the enhancement of GA-ANN over ANN about 45%.

On the other hand, GA-ANN can provide high power saving of PV module, which can be converted to positive gain over ANN scheme. Figures 8 and 9 show

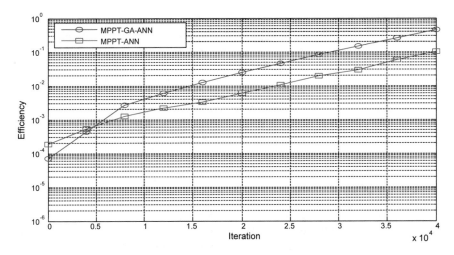

Fig. 7 PV panel efficiency per iteration

the power saving of two schemes at different situation in irradiance and temperature. As shown, GA-ANN provides better saving power compared to ANN scheme; for example, at 4000 iteration in Fig. 9, GA-ANN needs 1 W for operation while ANN needs about 1.4 W. Note that, at the beginning of the iterations, ANN provides better saving power, since GA needs more power to operate the main three parts of GA. Finally, by increasing T and G, the two schemes changed accordingly, since for high input values of G and T, more power should be used to operate MPPT in stable region.

Fig. 8 PV saving power per iteration ($T = 10$, $G = 30$)

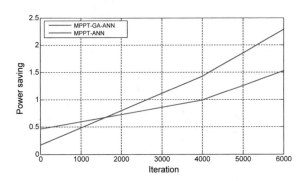

Fig. 9 PV saving power per iteration ($T = 30$, $G = 100$)

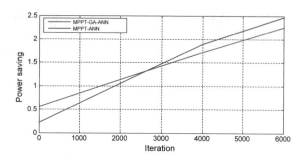

5 Conclusion

Solar PV is believed to be a common source of renewable energy because of its many advantages, such as lower operational cost, low maintenance, and environment-friendly properties. The results show that the designed GA-ANN-MPPT controller has a suitable output performance compared with the ANN-MPPT. Results show that, at first few steps in iterations, ANN provides more performance than GA-ANN, due to GA need more time to operate the three parts of GA scheme. Efficiency of GA-ANN was applied and compared with ANN scheme; also, saving power of GA-ANN and ANN were shown.

References

1. L. Bangyin, D. Shanxu, C. Tao, Photovoltaic DC-building-module based BIPV system-concept and design considerations. IEEE Trans. Power Electron. **26**(5), 1418–1429 (2011)
2. J. Young-Hyok, J. Doo-Yong, K. Jun-Gu, K. Jae-Hyung, L. Tae-Won, W. Chung-Yuen, A real maximum power point tracking method for mismatching compensation in PV array under partially shaded conditions. IEEE Trans. Power Electron. **26**(4), 1001–1009 (2011)
3. I.H. Rowlands, Envisaging feed-in tariffs for solar photovoltaic electricity: European lessons for Canada. Renew. Sustain. Energy Rev. **9**, 51–68 (2005)
4. A. Ali, A.N. Hasan, T. Marwala, Perturb and observe based on fuzzy logic controller maximum power point tracking (MPPT), in *IEEE International Conference on Renewable Energy Research 2014*, Milwaukee, USA (2014)
5. A. Safari, S. Mekhilef, Simulation and hardware implementation of incremental conductance MPPT with direct control method using cuk converter. IEEE Trans. Ind. Electron. **58**(4), 1154–1161 (2011)
6. A.K. Rai, N.D. Kaushika, B. Singh, N. Agarwal, Simulation model of ANN based maximum power point tracking controller for solar PV system. Solar Energy Mater. Solar Cells **95**, 773–778 (2011)
7. K. Ishaque, Z. Salam, An improved modeling method to determine the model parameters of photovoltaic (PV) modules using differential evolution (DE). Sol. Energy **85**, 2349–2359 (2011)

8. K. Ishaque, Z. Salam, H. Taheri, A. Shamsudin, A critical evaluation of EA computational methods for Photovoltaic cell parameter extraction based on two diode model. Sol. Energy **85**, 1768–1779 (2011)
9. A.M. Farayola, A.N. Hasan, A. Ali, Comparison of modified incremental conductance and fuzzy logic MPPT algorithm using modified CUK converter, in *8th IEEE International Renewable Energy Congress (IREC) 2017*, Amman, Jordan (2017)
10. A. Ali, I. Boulkaibet, B. Twala, T. Marwala, Hybrid optimization algorithm to the problem of distributed generation power losses, in *IEEE International Conference on Systems, Man and Cybernetics*. https://doi.org/10.1109/smc.2016.7844485

The Use of Multilayer Perceptron to Classify and Locate Power Transmission Line Faults

P. S. Pouabe Eboule, Ali N. Hasan and Bhekisipho Twala

Abstract This paper investigates the use of multilayer perceptron (MLP) technique for locating and detecting faults in a power transmission line. MLP was used twice in this paper to locate and to detect faults. The experiments were conducted on a 600-km-length, three-phase power transmission line data which include the required faults to detect and locate the fault. Matlab was used to perform the experiments. Results show that MLP achieved high prediction accuracy for fault type detection of 98% and a prediction accuracy of 78% for fault location.

Keywords Faults location · Faults classification · Faults detection
MLP · Power transmission line

1 Introduction

Transmission line protection is considered a significant matter in electrical power systems since 85–87% of power system faults are taking place in transmission lines [1]. In power transmission lines, the utilization of artificial intelligence techniques begins to be important due to its superior performance and simplicities [2–5]. Analysis of an electrical network has been proven complicated, especially when it is done on a very high-voltage power transmission line and with a fairly long distance [6, 7]. The use of artificial intelligence techniques such as neural network has given great support, in terms of the faults detection and faults classification as well as identifying the fault location [8, 9].

P. S. Pouabe Eboule · A. N. Hasan (✉) · B. Twala
Department of Electrical Engineering Technology, University of Johannesburg,
PO Box 524 Auckland Park, Johannesburg 2006, South Africa
e-mail: alin@uj.ac.za

P. S. Pouabe Eboule
e-mail: sercho2004@gmail.com

B. Twala
e-mail: btwala@uj.ac.za

© Springer Nature Singapore Pte Ltd. 2018
S. S. Dash et al. (eds.), *Artificial Intelligence and Evolutionary Computations
in Engineering Systems*, Advances in Intelligent Systems and Computing 668,
https://doi.org/10.1007/978-981-10-7868-2_5

Artificial neural networks (ANNs) have been used to detect faults on a short power transmission line such as 240-km transmission line [6, 10, 11]. In power transmission line, in a 240-km power transmission line, the line acts as a juxtaposition of multiple lines which have equal distance. Analysis of such a power grid may prove to be much more complex [12]. Usually simulations are conducted using range of resistance between 0.001 and 100 Ω [13].

MLP is one of the most popular ANN-used techniques due to its capacity to adjust input by reducing the error [7]. It is a feed-forward backpropagation neural network that uses nonlinear activation function [10, 13]. MLP structure includes the following: input layers, hidden layers, and output layers. Data are fed from the input layers and then processed by the hidden layers to be transmitted to the output layers. The sum of all different values is received, and then, the error is calculated. Thus, backpropagation is used to feedback the output to adjust the input [14].

Recently, several researches based on the usage of backpropagation neural network to classify and locate faults for transmission line have been conducted; however, these studies focused on analyzing relatively short-length lines with maximum of 10 identified faults [13, 15]. In addition, there was a lack of research about classifying triple line to ground fault. Also, there was a few research on relatively long transmission lines (more than 240 km); however, none experimented on a length of 600 km [13, 15].

This work was conducted in order to monitor a long transmission line system of 600 km length and predict a total of 11 identified fault types and locations. This could lead to reduce the time consumed for maintaining the faults and improve the protection process for the distribution system.

This paper is organized as follows. Section 2 introduces the experimental setup. Section 3 explains the experimental results, and Sect. 4 contains the conclusions.

2 Experimental Setup

The experiments were conducted on an electric generation system that supplies 750 kV and a 600-km-length transmission line with different electrical loads, and the reactance effect of the generator was excluded from this experiment. Figure 1 shows the transmission system with a fault on the transmission line at (D) distance from the power generator.

The total length of the line (600 km) was divided into five equal zones, each of 20% of the line total length. This was done in order to identify the fault type and the fault location more accurately. Types of faults that were considered for this specific experiment are 11 faults as follows:

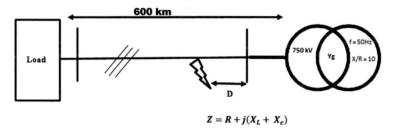

$$Z = R + j(X_L + X_c)$$

Fig. 1 Power transmission system with the fault on the line

- Single Line A to Ground (SLAG)
- Single Line B to Ground (SLBG)
- Single Line C to Ground (SLCG)
- Double Line A and B (DLABG)
- Double Line B and C (DLBC)
- Double Line A and C (DLAC)
- Double Line A, B and Ground (DLABG)
- Double Line B, C and Ground (DLBCG)
- Double Line A, C and Ground (DLACG)
- Triple Line A, B, C (TLABC)
- Triple Line A, B, C and Ground (TLABCG).

Values of short-circuit voltage and short-circuit current were used to identify the fault type and location. Then, faults were created using the Simulink fault breaker block.

In general, faults are symmetrical and asymmetrical. When the fault occurs on the line with a fault resistance greater than 10 Ω, the short-circuit voltage increases exponentially; therefore, the value of the short-circuit voltage (fault voltage) depends on fault resistance value. Figure 2 presents the SLAG fault that occurred at 600 km location.

After the faults data were generated and normalized for all 11 faults, MLP technique was used to classify and thus detect the fault type and the fault location. Processed, normalized values were generated by the simulation of the power transmission line system using MATLAB, for each post-fault obtained.

Values of 3 phases line voltages Va, Vb, Vc and the short-circuit currents Ia, Ib, Ic at the moment of fault. Occurrence has been normalized. A set of binary numbers (1, 0) have been assigned as output variables (Figs. 3 and 4).

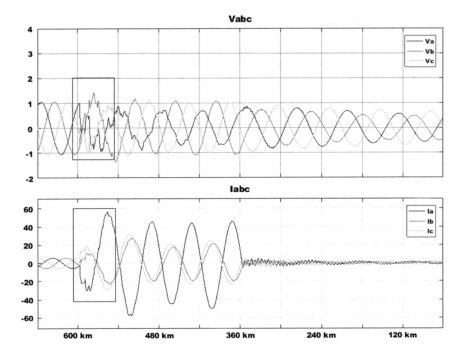

Fig. 2 SLAG with $R = 0.001\ \Omega$

3   Experimental Results

For fault type detection, the average achieved prediction accuracy for all fault types was approximately 98%. One example on how MLP algorithm detects faults is shown in Fig. 5. It can be seen that fault type SLAG is detected according to binary codes assigned to it in Table 1. It can be seen from Table 1 that predicted results using MLP technique for each fault type were close to the desired output for LA, LB, LC, and G with an average prediction error of 2%.

For faults location prediction, MPL achieved total prediction accuracy for the whole line length of 78%. Table 2 shows the desired and the predicted output in terms of faults location. Each location was given a set of binary numbers.

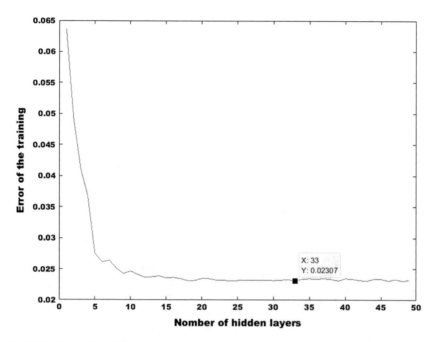

Fig. 3 Number of the hidden layers required for the MLP for fault detection

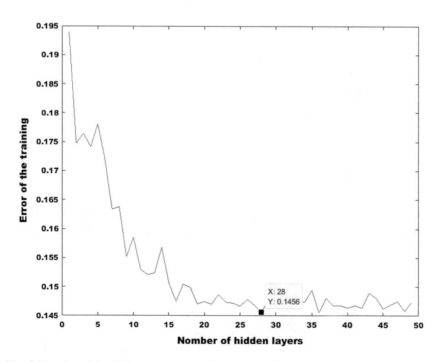

Fig. 4 Number of the hidden layers required for the MLP for fault location

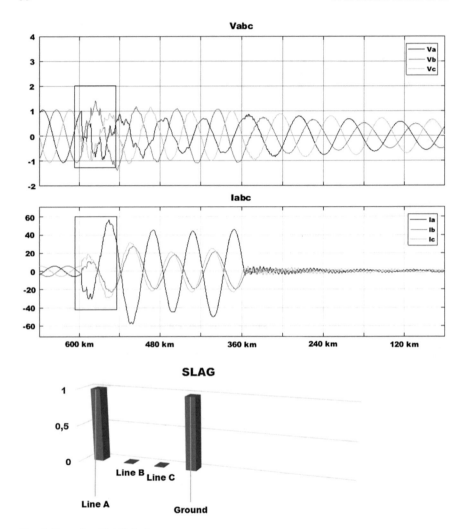

Fig. 5 Detecting SLAG fault type

In Fig. 6, it can be observed that the fault SLAG was located using the binary code (1 0 1 0). This code is assigned for the 600 km location as shown in Table 2.

Table 1 Predicted and desired outputs

Faults	Desired results				MLP predicted results			
	LA	LB	LC	G	L1	L2	L3	G
SLAG	1	0	0	1	1.00E+00	7.88E−05	2.40E−05	0.999999
SLBG	0	1	0	1	4.85E−06	1.00E+00	0.00020725	0.999948
SLCG	0	0	1	1	2.90E−05	3.75E−11	1.00E+00	0.999972
DLAB	1	1	0	0	1.00E+00	0.9999998	6.86E−05	0.499743
DLBC	0	1	1	0	1.58E−05	1.00E+00	0.99999319	0.499260
DLAC	1	0	1	0	1	2.65E−08	1.00E+00	0.500015
DLABG	1	1	0	1	1.00E+00	0.9999998	6.86E−05	0.499743
DLBCG	0	1	1	1	1.58E−05	1.00E+00	0.99999319	0.499260
DLACG	1	0	1	1	1	2.65E−08	0.99997631	0.5000159
TLABC	1	1	1	0	0.9999585	0.9999644	0.99987386	0.501496
TLABCG	1	1	1	1	0.9999585	1.00E+00	1.00E+00	0.501496

Table 2 Faults location obtained and desired

Fault distance (km)	Desired output				Predicted location output accuracy				E (%)
120	0	0	1	0	0.02353	0.07822	0.99997	3.04E−14	2.35
240	0	1	0	0	0.05370	0.95986	1.26E−05	0.997	1.49
360	0	1	1	0	9.15E−08	0.99949	0.99999	0.00027	10.09
480	1	0	0	0	0.97619	0.03493	0.04929	6.36E−07	3.43
600	1	0	1	0	1	4.23E−13	0.99980	1.27E−19	1.54

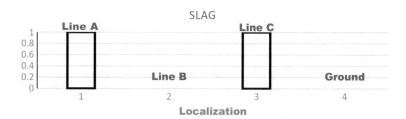

Fig. 6 SLAG fault type localization at 600 km

4 Conclusions

In this paper, the case of identifying and locating faults on a long power transmission line of 600 km length using backpropagation multilayer perceptron was investigated. For fault type detection, 11 faults were identified and predicted. The MLP technique achieved a high prediction accuracy of a total of 98% which makes MLP a reliable method for transmission lines faults detection and identifying. However, for the fault location prediction experiment, the total prediction accuracy reached only a 78%. Although using MLP has an advantage of being fast approach for fault locating, the achieved prediction accuracy was not as accurate as required; therefore, it may need to be improved, and this may be done by forming an ensemble of neural networks or by using other powerful artificial intelligence techniques. Then, for the future work, the reactance effect will be included to compare with the results obtained.

References

1. M. Singh, B.K. Panigrahi, R.P. Maheshwari, Transmission line fault detection and classification, in *IEEE International Conference on Emerging Trends in Electrical and Computer Technology (ICETECT)* (2011)
2. N. Zhang, M. Kezunovic, Coordinating fuzzy art neural networks to improve transmission line fault detection and classification, in *IEEE Power Engineering Society General Meeting* (2005), pp. 1–7
3. E. Koley, A. Jain, A.S. Thoke A. Jain, S. Ghosh, Detection and classification of faults on six phase transmission line using ANN, in *2011 2nd International Conference on Computer and Communication Technology (ICCCT)* (IEEE, 2011), pp. 100–103
4. L. Tekli, B. Filipović-Grčić, I. Pavičić, in *Artificial Neural Network Approach for Locating Faults in Power Transmission System* (IEEE, 2013), pp. 1425–1430
5. K. Lout, R.K. Aggarwal, A feedforward artificial neural network approach to fault classification and location on a 132 kV transmission line using current signals only, in *47th International Universities Power Engineering Conference* (UPEC) (2012), pp. 1–6
6. A.N. Hasan, B. Twala, Improving single classifiers prediction accuracy for underground water pump station in a gold mine using ensemble techniques, in *IEEE International Conference on Computer as a Tool (EUROCON)* (2015), pp. 1–7
7. A.N. Hasan, B. Twala, T. Marwala, Moving towards accurate monitoring and prediction of gold mine underground dam levels, in *International Joint Conference on Neural Networks (IJCNN)* (2014), pp. 2844–2849
8. R.C. Bansal, Optimization methods for electric power systems: an overview. Int. J. Emerg. Electr. Power Syst. 2(1) (2005)
9. S. Azhar, Application of artificial neural network in fault detection study of batch esterification process. Int. J. Eng. Technol IJENS **10**(03), 36–39 (2010)
10. M.M Ismail, M.M. Hassan, Distance relay protection for short and long transmission line, in *Proceedings of International Conference on Modelling, Identification & Control (ICMIC)* (2013), pp. 204–211
11. M. Gilany, E.S.T. El Din, M.M. Abdel Aziz, D.K. Ibrahim, An accurate scheme for fault location in combined overhead line with underground power cable, in *IEEE Power Engineering Society General Meeting* (2005), pp. 2521–2527
12. M.B. Hessine, H. Jouini, S. Chebbi, Fault detection and classification approaches in transmission lines using artificial neural networks, in *17th IEEE Mediterranean Electrotechnical Conference (MELECON)* (2014), pp. 515–519
13. M. Tayeb, Faults detection in power systems using artificial neural network. Am. J. Eng. Res. AJER **02**(06), 69–75 (2013)
14. P. Subbaray, B. Kannapiran, Artificial neural network approach for fault detection in pneumatic valve in cooler water spray system. Int. J. Comput. Appl. **9**(7), 44–52 (2010)
15. P. Ray, Fast and accurate fault location by extreme learning machine in a series compensated transmission line, in *IEEE Power and Energy Systems Conference: Towards Sustainable Energy* (2014), pp. 1–6

Generation of 3D Realistic Synthetic Image Datasets for Spot Detection Evaluation

Matsilele Mabaso, Daniel Withey and Bhekisipho Twala

Abstract Automated image analysis provides a powerful tool for detecting and tracking fluorophore spots in fluorescence microscopy images. The validation of automated spot detection methods requires ground-truth data. Here, a simple framework is proposed for generating 3D fluorescence microscopy images with real background and synthetic spots, forming realistic, synthetic images with ground-truth information. Similarity between synthetic and real images was evaluated using similarity criteria, such as visual comparison, central moments with Student's t test and intensity histograms. Student's t test shows that there is no statistical difference between central moment features of the real and synthetic images and the intensity histograms exhibit similar shapes, demonstrating high similarity between real images and the synthetic images. The performance of four detection methods using synthetic images (with real background and no background) created using the proposed framework was also compared. F_{score} values were higher on synthetic images with no background compared to those with a real background indicating that the presence of the background reduces the effectiveness of the spot detection methods.

Keywords Image analysis · Microscopy images · 3D realistic synthetic images Spot detection

M. Mabaso (✉) · D. Withey
Council for Scientific and Industrial Research, Pretoria, South Africa
e-mail: MMabaso@csir.co.za

D. Withey
e-mail: DWithey@csir.co.za

M. Mabaso · B. Twala
Department of Electrical and Mining Engineering,
University of South Africa, Pretoria, South Africa
e-mail: Twalab@unisa.ac.za

© Springer Nature Singapore Pte Ltd. 2018
S. S. Dash et al. (eds.), *Artificial Intelligence and Evolutionary Computations in Engineering Systems*, Advances in Intelligent Systems and Computing 668,
https://doi.org/10.1007/978-981-10-7868-2_6

1   Introduction

Advances in microscopy imaging have made it possible to acquire a large amount of image data leading to increased research in automated bio-image analysis to help in extracting information from large numbers of microscopy images, since manual analysis is not feasible [1]. In fluorescence microscopy imaging, the imaged light is emitted by fluorescent molecules, called fluorophores, that are attached to the object of interest. In this way, small particles can be observed as bright spots.

The typical automated image analysis pipeline includes the following steps: image pre-processing, detection of bright particles and quantifying of features from the detected particles [2]. Images acquired from a microscope are pre-processed to reduce the errors resulting from a measurement system, such as uneven illumination, noise and compression artefacts.

This is followed by extracting the objects of interest, e.g. spots, in the pre-processed image. Finally, different kinds of features are quantified from the extracted spots and stored for further analysis. In this study, a spot refers to a bright particle which has a local intensity maximum that is significantly different from the local neighbourhood, as shown in Fig. 1.

Obtaining ground-truth information often involves the manual annotation procedure where an expert manually labels the images to obtain the ground-truth data. The obtained ground truth is then compared to the results obtained by the automated image analysis method. However, the manual labeling of images is exhausting especially in 3D cases, may include uncertainty, and the ground-truth

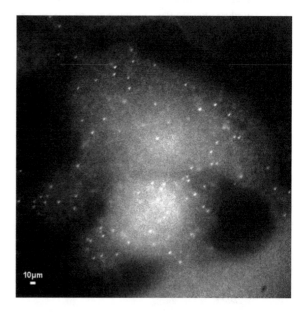

Fig. 1 A sample of real fluorescence image with bright particles obtained using a confocal microscopy

results are based on user opinion. As an example, one expert might label a moderate intensity as a spot, while another can label that as noise or background residues. These issues make the manual annotation unreliable.

1.1 Previous Work

A number of approaches have been developed to avoid the problem of manual annotation, and synthetic images were introduced to simulate real microscopy images; for example, [3, 4] have introduced the use of synthetic images to simulate real microscopy images. The use of synthetic images became popular because they contain the ground-truth data and give the opportunity to compare and validate the results of automated methods. A number of studies have adopted the use of synthetic images when developing and validating their automated tools, e.g. [5]. Considerable research is still ongoing on the development of synthetic fluorescence microscopy images. However, there exist few studies for the creation of synthetic images with bright spots. A popular synthetic image generator was developed by Lehmussola [6]. This tool can generate different kinds of synthetic images with round or elongated spots and varying backgrounds. However, there are a number of disadvantages. First, it restricted to only 2D images, and secondly, the background is also simulated. Another simulation tool named SIMCEP was proposed by [1]; the tool can create different types of synthetic images and can mimic cell structures and spot shapes. This tool is also limited to 2D images and cannot mimic the effect of non-homogeneity in background intensity. Another study [7] generated synthetic images using a mixture of Gaussians to form the background. Their study modelled some image properties; however, it lacked the properties of real background structures. Similar to [7, 8] proposed a framework which can model the movements of spots in microscopy images. However, [8] did not take into account the background in microscopy images. There exist few methods which can model the effects of image noise, spot motion and realistic background in synthetic microscopy images. The work by Rezatifighi [4] uses HDome transformation [9] to estimate the background in real microscopy image sequences. Although their study can model spot motion and noise, it still lacks important characteristics of real data.

In this work, we describe a novel framework for creating 3D realistic synthetic images of fluorescence microscopy. The presented technique is an extension of our previous 2D framework [10]. The framework is flexible and can be modified to generate different kinds of synthetic images. The approach presented in this study is based on the use of real microscopy images, unlike other frameworks that simulate the entire image. Instead of learning the background and trying to simulate it, our framework makes use of real 3D microscopy images and adds 3D synthetic spots onto the image. To simulate our spots, we place a Gaussian profile directly into the real image with parameters that mimic the properties of real images, resulting in partially synthetic images. The main contributions of the paper are as follows: (1) we propose a mathematical framework for generating 3D realistic synthetic images of fluorescence microscopy with accurate models to mimic real images; (2) a comparison of four 3D detection methods using synthetic images generated by the proposed framework. The main motivation is the need for synthetic images with realistic background because realistic background affects operation of the spot detector.

The rest of the paper is organized as follows: Sect. 2 explains the framework strategy, then Sects. 3 and 4 present the experimental set-up and results, and, finally, Sect. 5 concludes the study.

2 Simulation Framework

Simulations of microscopy images provide a valuable tool when developing and validating automated image analysis tools. Figure 2 illustrates the main steps of the proposed framework for the creation of synthetic images. A detailed description of each step is given below. To generate synthetic images that resemble the real fluorescence microscopy images, access to real images is important. The development of the framework presented in this study was based on the use of real fluorescence images in different z slices. An example of a real microscopy image with mRNA spots is shown in Fig. 1.

2.1 Background Modelling

Microscopic images suffer from a lot of factors such as errors and artefacts which originate from the imaging system. One common artefact is the non-uniformity in illumination which can degrade the acquired image by varying the background image intensity. Previous studies have made simple assumptions when developing synthetic images such as images with no background structures and constant signal-to-noise ratio. Other studies try to estimate the background by using either

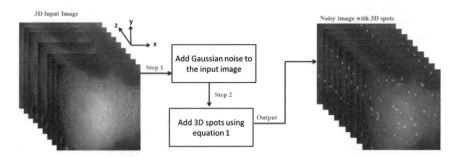

Fig. 2 A diagram showing the steps involved in our framework for the creation of synthetic image sequences

HDome [9] or Gaussian mixture model [7]. These assumptions ease the procedure involved in developing synthetic images as they do not reflect the complexity involved in microscopy images. The disadvantage of estimating the background is that it will still be different from the real background. In our framework, instead of modelling the effect of non-uniformity we make use of the real fluorescence microscopy 3D background image (without spots) and add the spots. The real images were obtained from our collaborator, the Synthetic Biology Research Group at the CSIR. The real images that are used as background contain no spots and were of channel CY3 with magnification of 100× of HeLa cells.

2.2 Spot Appearance

Fluorescence microscopy images contain a number of bright particles (spots) superimposed on an uneven background, as shown in Fig. 1. Modelling the spot appearance can be a challenging task. The most common approach to model these spots is to fit a Gaussian intensity profile [11]. In this work, we considered a 3D Gaussian function with five parameters, the positions, (x_0, y_0, z_0) standard deviation and peak intensity. The model for a single spot is given by:

$$G(x, y, z) = Ae^{-\left(\frac{(x-x_0)^2}{2\sigma_x^2} + \frac{(y-y_0)^2}{2\sigma_y^2} + \frac{(z-z_0)^2}{2\sigma_z^2}\right)} \tag{1}$$

The parameters, $\sigma_x, \sigma_y, \sigma_z$, describe the width of the spot, and, A, the peak intensity. In order to model an isotropic spot, the parameters, σ_x, σ_y and σ_z, were set to be equal. When the shape model is defined accurately, the model parameters are estimated from the real data, and the simulated images will be realistic. Modifying these parameters allows the generation of different types of synthetic images with varying characteristics.

2.3 Noise Generation

Noise in fluorescence microscopy images is caused by external factors during the image acquisition process and leads to random variations in pixel intensities. There exist many noise sources in microscopy imaging which affect the image quality. To simulate the kind of noise found in microscopy imaging, we used additive Gaussian noise with a mean of zero and varying standard deviation, $N \sim N(\mu = 0, \sigma_{noise})$. Gaussian models are commonly used models in microscopy imaging. The final realistic synthetic image is given as follows:

$$A(x, y, z) = G(x, y, z) + \epsilon_\epsilon + N_b \tag{2}$$

where the quantity ϵ_ϵ is assumed to be white noise with mean μ_ϵ and standard deviation σ_ϵ, and N_b describes the background contribution, in this case a real fluorescence image.

2.4 Signal-to-Noise Ratio

The quality of images can be expressed in terms of its signal-to-noise ratio (SNR). The SNR measures the amount of noise affecting the image and is widely used in image processing. The signal-to-noise ratio was defined as the ratio of spot intensity, $S_{\max}$, divided by the noise standard deviation, σ_{noise},

$$\text{SNR} = \frac{S_{\max}}{\sigma_{\text{noise}}} \tag{3}$$

3 Experimental Set-up

3.1 Synthetic Images

The framework presented in this study is capable of simulating different kinds of microscopy images. There exists no straightforward solution to qualitatively evaluate the realistic synthetic images with real fluorescence images. Therefore, to study the effect of real background on synthetic image sequences, we created two types of synthetic image scenarios. The first scenario consisted of images of no background structures (named NOBGND), and the second scenario consisted of synthetic images with real fluorescence background structures (named BGND). For the second scenario, six realistic synthetic images (named BGND1, BGND2, BGND3, BGND4, BGND5 and BGND6) were created using six different backgrounds, as shown in Fig. 3. All the scenarios were corrupted by Gaussian noise, with the mean of zero and varying standard deviation {2, 2.86, 4, 5, 10, 20}. The following signal-to-noise ratio (SNR) levels were explored {10, 7, 5, 4, 2, 1} where the spot intensity was 20 grey levels. The number of slices varied between {12 and 20} for each synthetic image. Each synthetic image created was of size 256 by 256 pixels.

The number of spots in each synthetic image was fixed at {100}. The spot positions were randomized using Icy plugin [12] to mimic the kinds of properties in real microscopy images. MATLAB was used to add spots, and the OMERO. matlab-5.2.6 toolbox was used to read and save images.

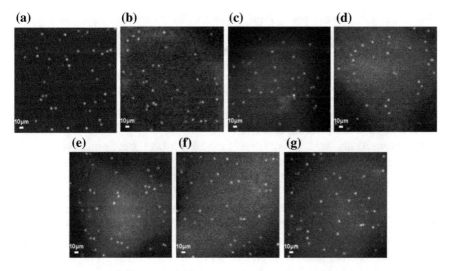

Fig. 3 Examples of synthetic images of SNR = 10 created using the proposed framework. **a** NOBGND, **b–g** BGND

3.2 Evaluation

The quality of the generated synthetic images using the proposed framework needs to be evaluated. However, there exist no universal procedures for assessing the quality of synthetic images. A standard approach is to visually compare between the synthetic and real images which require an expert biologist. This method seems attractive, but it requires a lot of time and resources. In this work, we focus on the use of central moments. These features are computed from real and synthetic images and then compared to check for similarity based on Student's t test.

Central Moments. Central moments in image analysis assist in the identification of particular characteristics in an image. The mth central moment is a statistical measure evaluated at a given discrete random variable.

$$m_n = \sum_{i=0}^{j-1} (x_i - \mu)^n p(x_i) \qquad (4)$$

where $\mu = \sum_{i=0}^{j-1} x_i p(x_i)$

The parameter, x_i, represents the intensity level of the image. Each moment (m_n) resembles a specific meaning of a given data. The first moment (m_0) is the mean, m_1 is the variance, m_2 is the skewness, and the fourth m_3 is the kurtosis which measures the sharpness of the distribution.

Student's t test. The Student's t test checks the separateness of two groups and gives an indication of whether two sets of measures are statistically different.

This test considers only two groups, however, for multiple groups; one needs to compare each pair of groups. In this study, we chose the α-value to be 0.05.

Null hypothesis: The two sets of measurements are similar.

We will reject the null hypothesis if the p-value less than or equal 0.05, otherwise we fail to reject.

Intensity Histogram. The intensity histogram is a graph showing the number of pixels with the same intensity within a given image. If two images are nearly the same, their histogram should be similar.

Detection Methods. In order to study the effect of real background on synthetic image, we compared results from four spot detection methods applied to our synthetic images. These methods were chosen based on their 3D implementation availability, and they were also being used in different comparison studies [13, 14]. The detection algorithms compared are isotropic undecimated wavelet transform (IUWT) [15], feature point detection (FPD) [16], HDome transformation (HD) [9] and Laplacian of Gaussian (LoG) [17]. In selecting the optimal parameters, each of the method was tested for performance on SNR = 5 synthetic images and these optimal parameters were then kept constant for all experiments.

Performance Measure. In order to test the performance of the four detection methods, we computed several measures: true positives (TP), false positives (FP) and false negatives (FN). True positives are detected spots that correspond to the ground-truth spots. If the detected spot does not correspond to the ground truth, it is considered as a false positive. A missed ground-truth spot is considered as a false negative. In order to compare the performance of different methods on the generated synthetic images, we computed the centroid of every detected connected component in a binary map, resulting in a set of locations $\{d\}$. Then, following the criteria suggested by [14], a particle ω of a given ground-truth is correctly detected if and only if: its nearest neighbour, d, in the sets of detected centroids is closer than 200 nm away, and ω is also a nearest neighbour d in the ground-truth set of locations. Two performance measures are considered in this study, recall and precision [10, 14].

Recall measures the ratio of correctly detected spots over all ground-truth spots, and precision measures the ratio of correctly detected spots among all detected spots:

$$\text{Recall} = \frac{N_{\text{TP}}}{N_{\text{TP}} + N_{\text{FN}}}, \quad \text{Precision} = \frac{N_{\text{TP}}}{N_{\text{TP}} + N_{\text{FP}}} \tag{5}$$

where N_{TP} is the number of true positives, N_{FN} is the number of false negatives, and N_{FP} is the number of false positives. Then, the F_{score} measure is computed as a weighted average of the two measures, precision and recall:

$$F_{\text{score}} = 2 \times \frac{\text{precision} \times \text{recall}}{(\text{precision} + \text{recall})} \tag{6}$$

A good detection method should have the value of F_{score} approaching one.

4 Experimental Results

To show that the synthetic image represents real images, we computed central moments (m_0 to m_3) over each image to get a feature vector. These features are then compared to real image features based on Student's t test. The Student's t test gives an indication of how different the two measurements are. Table 1 presents the averaged central moments for both synthetic (BGND) and real images (RI). To perform a Student's t test, we grouped all synthetic images (BGND and NOBGND) with real images (RI) to form seven pairs (BGND1 and RI, BGND2 and RI, etc.). The Student's t test at 5% level of significance shows that there is no significant different between the features (for both real and synthetic images). As a result, this indicates that the synthetic images resemble the real images.

The histogram analysis in Fig. 4 shows there exist some similarities between synthetic and real images, and these indicate that the synthetic images produced by our framework resemble the real images.

We also evaluate the performance of four detection methods using synthetic images consisting of seven experimental scenarios. The first scenario consisted of images with no background structures, NOBGND. This will help with the evaluation of the performance of the algorithms as a function of image noise (SNR). The second to seventh scenario consisted of image sequences with a real background (BGND1, BGND2, BGND3, BGND4, BGND5 and BGND6). The second to seventh experimental scenarios were used to evaluate the performance of the methods as a function of real background and image noise. All synthetic images used in the experiment were of 3D. For each method, the performance measure, F_{score}, was computed.

Figure 5a–d presents the F_{score} versus SNR plots for the four detections methods (HD, LoG, IUWT and FPD) respectively applied to two synthetic image scenarios (NOBGND and BGND). The only difference between the two scenarios is the presence of a real background on BGND synthetic images, and all other properties remained the same. The plots show higher F_{score} values for NOBGND synthetic

Table 1 Evaluation of central moments over real and synthetic images

	Central moments (m_n)			
	m_0	m_1	m_2	m_3
Real image (RI)	89.88 ± 1.23	8.59 ± 0.13	8.51 ± 0.32	18.65 ± 0.89
BGND1	90.78 ± 8.09	8.79 ± 0.27	9.02 ± 0.42	18.58 ± 0.75
BGND2	92.48 ± 4.17	8.73 ± 0.76	8.99 ± 1.41	19.16 ± 2.11
BGND3	89.17 ± 5.04	8.82 ± 0.16	8.76 ± 0.74	18.00 ± 1.09
BGND4	92.78 ± 3.83	8.80 ± 0.55	8.86 ± 2.16	18.53 ± 0.91
BGND5	90.32 ± 7.89	8.50 ± 0.82	8.53 ± 1.61	18.67 ± 4.58
BGND6	89.33 ± 7.12	8.72 ± 0.56	8.97 ± 0.69	18.58 ± 1.13
NOBGND	85.36 ± 10.25	8.54 ± 0.45	8.42 ± 0.64	17.78 ± 1.84

Each measure represents the average of the whole dataset including standard deviation

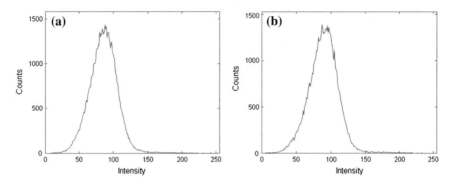

Fig. 4 Intensity histogram for **a** realistic synthetic image, and **b** real image

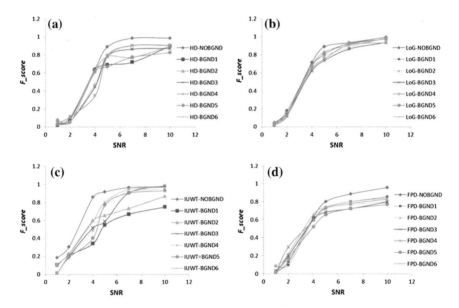

Fig. 5 Curves of F_{score} versus SNR of the four detection methods applied to two synthetic image scenarios. **a** HD, **b** LoG, **c** IUWT and **d** FPD

images from *SNR* (5–10) compared to BGND images. This shows that the performance of the algorithms decreases when the real background is introduced. However, as the SNR decreases, <4, the F_{score} values drop for all synthetic images. The decrease in performance of the algorithms could be caused by the fact that when the background noise increases, then the quality of a given image decreases in such a way that the intensities of a spot and noise are nearly equal.

In all experiments, the change in performance of LoG method is relatively small. This is caused by the manual selection of threshold in the algorithm. The methods,

HD, IUWT and FPD, show a change in performance on the two synthetic scenarios. This indicates that the presence of real background increased the complexity of the synthetic images.

5 Conclusions

In this article, we presented a framework for the simulation of 3D fluorescence microscopy images and also study the effect of real background on synthetic images. The framework is an extension of our previous work that improves the modelling of real microscopy image sequences by including realistic spots, realistic noise and realistic motion with real image background. The framework provides the simulation package capable of simulating diverse images of fluorescence microscopy with different background and quality. The synthetic image sequences created using this framework offer a better way to evaluate different automated image analysis tools since the ground truth is available. Our evaluation results showed that the generated synthetic images using the proposed framework resemble real images. The results also indicate that the performance of four detection methods is reduced when tested with synthetic images exhibiting realistic background, compared to the sequences which had no background. This showed that the real background has an effect on spot detection algorithm performance. The performance of the detection methods is reduced in the presence of background structures.

Acknowledgements Financial support was received from Council for Scientific and Industrial Research (CSIR) and the Electrical and Electronic Engineering Department at the University of Johannesburg. We would also like to thank the Synthetic Biology research group at the CSIR for providing us with real microscopy images.

References

1. A. Lehmussola, P. Ruusuvuori, J. Selinummi, H. Huttunen, O. Yli-Harja, Computational framework for simulating fluorescence microscopy images with cell population. IEEE Trans. Med. Imag. **26**(7), 1010–1016 (2007)
2. R.S. Wilson, L. Yang, A. Dun, A.M. Smyth, R.R. Duncan, C. Rickman, W. Lu, Automated single particle detection and tracking for large microscopy datasets. R. Soc. Open Sci. **3**(5), 1–13 (2016)
3. P. Ruusuvuori, A. Lehmussola, J. Selinummi, T. Rajala, H. Huttunen, O. Yli-Harja, Benchmark set of synthetic images for validating cell image analysis algorithms, in *Proceedings of the European Signal Processing Conference* (Lausanne, Switzerland, 2008)
4. S.H. Rezatifighi, W.T. Pitkeathly, S. Goud, R. Hartley, K. Mele, W.E. Hughes, J.G. Burchfield, A framework for generating realistic synthetic sequences of total internal reflection fluorescence microscopy images, in *Proceedings of the IEEE International Symposium on Biomed Imaging* (San Franscisco, USA, 2013)
5. N. Chenouard, S. Ihor, F. de Chaumont et al., Objective comparison of particle tracking methods. Nat. Methods **11**(3), 281–290 (2014)

6. I. Smal, Online, Sept 2009. Available: http://smal.ws/wp/software/synthetic-data-generator/. Accessed 8 Nov 2016
7. A. Genovesio, T. Liendl, V. Emiliana, W.J. Parak, M. Coppey-Moisan, J.-C. Olivo-Marin, Multiple particle tracking in 3d + t microscopy: method and application to the tracking of endocytosed quantum dots. IEEE Trans. Image Process. **15**(5), 1062–1070 (2006)
8. J.W. Yoon, A. Bruckbauer, W.J. Fitzgerald, D. Klenerman, Bayesian inference for improved single molecule fluorescence tracking. Biophys. J. **94**, 4932–4947 (2008)
9. L. Vincent, Morphological grayscale reconstruction in image analysis: applications and efficient algorithms. IEEE Trans. Image Process. **2**, 176–201 (1993)
10. M. Mabaso, D. Withey, B. Twala, A framework for creating realistic synthetic fluorescence microscopy image sequences, in *Proceedings of the 9th International Joint Conference on Biomedical Engineering System and Technologies* (Rome, Italy, 2016)
11. B. Zhang, J. Zerubia, J.-C. Olivo-Marin, Gaussian approximations of fluorescence microscope PSF models. Appl. Opt. **46**(10), 1–34 (2007)
12. N. Chenouard, Particle tracking benchmark generator, in *Institut Pasteur* (2015). Online. Available: http://icy.bioimageanalysis.org/plugin/Particle_tracking_benchmark_generator. Accessed 1 Nov 2016
13. P. Ruusuvuori, T. Äijö, S. Chowdhury, C. Garmaendia-Torres, J. Selinummi, M. Birbaumer, A.M. Dudley, L. Pelkmans, O. Yli-Harja, Evaluation of methods for detection of fluorescence labeled subcellular objects in microscope images. BMC Bioinform. **11**, 1–17 (2010)
14. I. Smal, M. Loog, W. Niessen, E. Meijering, Quantitative comparison of spot detection methods in fluorescence microscopy. IEEE Trans. Med. Imag. **29**(2), 282–301 (2010)
15. J.-C. Olivo-Marin, Extraction of spots in biological images using multiscale products. Pattern Recogn. **35**(9), 1989–1996 (2002)
16. I.F. Sbalzarini, P. Koumoutsakos, Feature point tracking and trajectory analysis for video imaging in cell biology. J. Struct. Biol. **151**(2), 182–195 (2005)
17. A. Raj, P. van den Bogaard, S.A. Rifkin, A. van Oudenaarfen, S. Tyagi, Imaging individual mRNA molecules using multiple singly labeled probes. Nat. Methods **5**(10), 877–879 (2008)

Modified Newton's Method in the Leapfrog Method for Mobile Robot Path Planning

Belinda Matebese, Daniel Withey and Mapundi K. Banda

Abstract The problem of determining an optimal trajectory for an autonomous mobile robot in an environment with obstacles is considered. The Leapfrog approach is used to solve the ensuing system of equations derived from the first-order optimality conditions of the Pontryagin's Minimum Principle. A comparison is made between a case in which the classical Newton Method and the Modified Newton Method are used in the shooting method for solving the two-point boundary value problem in the inner loop of the Leapfrog algorithm. It can be observed that with this modification there is an improvement in the convergence rate of the Leapfrog algorithm in general.

Keywords Trajectory planning · Obstacle avoidance · Leapfrog algorithm
Pontryagin's Minimum Principle · Modified Newton algorithm

1 Introduction

Advances in optimal control provide the necessary tools to determine optimal trajectories for problems such as those arising from applications in robotics. An amount of research is done on optimal control with application to mobile robot path planning [1–3]. Most of these problems are nonlinear and can today be solved using numerical methods. The general numerical approaches for solving optimal control

B. Matebese (✉) · D. Withey
Mobile Intelligent Autonomous Systems, Council for Scientific and
Industrial Research, Brummeria, Pretoria, South Africa
e-mail: bmatebese@csir.co.za

D. Withey
e-mail: dwithey@csir.co.za

M. K. Banda
Department of Mathematics and Applied Mathematics,
University of Pretoria, Pretoria, South Africa
e-mail: mapundi.banda@up.ac.za

© Springer Nature Singapore Pte Ltd. 2018
S. S. Dash et al. (eds.), *Artificial Intelligence and Evolutionary Computations
in Engineering Systems*, Advances in Intelligent Systems and Computing 668,
https://doi.org/10.1007/978-981-10-7868-2_7

problems can be classed into direct and indirect methods [4]. In the direct method, the differential equations and their integrals are discretized. The drawback of direct methods is that they suffer from lower accuracy. Indirect methods on the other hand have outstanding precision and have a possibility of verifying necessary conditions. The necessary conditions of optimality are formulated through Pontryagin's Minimum Principle (PMP) resulting to a system of boundary value problem. The disadvantage of using the indirect methods is that a good approximation of the initial guess for the costates is required.

The Leapfrog method proposed in [5] can be viewed as an indirect method for solving TPBVP. In the method, a feasible path given from the starting and the final states is subdivided into segments where local optimal paths are computed. Application of the Leapfrog algorithm to path planning of a two-wheeled mobile robot was introduced in [6] to determine optimal paths for kinematic model. Later on, the work was extended to finding optimal trajectories for a mobile robot while avoiding obstacles [7, 8]. The algorithm gave promising results for path planning in both obstructed and unobstructed workspace.

With the Leapfrog method, one does not need a good approximation for initial guesses for the costates along a trajectory. As part of the algorithm, an affine approximation helps in choosing good initial guesses for the costates. In [9], it was shown that the approximation approach provides good guesses for at least some of the costates. The focus of their work was finding a way to choose initial guesses for the costates. Numerical and experimental solutions were done to validate the proposed approach.

Even though initial guess of the costates is not crucial in the Leapfrog method, the guesses are needed for the success of local shooting method. In addition, the algorithm is convergent to a critical trajectory provided that these local shooting methods produce optimal trajectories. As mentioned in [5], improvements to the simple shooting method may be necessary for more difficult optimal control problems. This may reduce the computation time that the Leapfrog takes to execute, especially in the first and last iterations. Moreover, this may improve the stability of the Leapfrog numerical solutions.

In this paper, the Modified Newton Method (MNM) implemented in [10] is incorporated in simple shooting [5] in an attempt to improve the convergence of the Leapfrog method for solving mobile robot path planning problem. A general discussion on the convergence of the MNM is given in [11]. For simulation purposes, the work done in [6–8] is revisited to evaluate the MNM with simple shooting. A comparison is made between a case in which the classical Newton Method used in our previous work and the Modified Newton Method are used in the shooting method. The computational time along the path and cost for each example are given in Table 1.

The paper organization is as follows: in Sect. 2, the Leapfrog algorithm is described. Section 3 presents the simulations with Modified Newton's Method, and Sect. 4 provides discussion and conclusions, respectively.

2 Optimal Control

A general optimal control system can be modelled by

$$\min_{u \in U} \int_{t_0}^{t_f} L(x(t), u(t)) \, dt \tag{1}$$

subject to state equation

$$\dot{x} = f(x(t), u(t)), \tag{2}$$

$$x(t_0) = x_0, \quad x(t_f) = x_f.$$

In the equations above, x is the state variable, u is a control input, t_0 and x_0 are initial time and state.

Following the Pontryagin's Minimum Principle (PMP), necessary conditions of optimality are formulated as

$$\dot{x} = \partial H / \partial \lambda \tag{3}$$

$$\dot{\lambda} = -\partial H / \partial x \tag{4}$$

$$0 = \partial H / \partial u \tag{5}$$

with the Hamiltonian function

$$H(t, \lambda, x, u) = L + \lambda^{\mathrm{T}} f(t, x, u) \tag{6}$$

where λ is for the costates variable. Solving for $u(t)$ in Eq. (5) and substituting it to (3) and (4) reduces to a TPBVP which is solved using indirect numerical method.

2.1 Leapfrog Method

Given a feasible path μ_z^k, from the initial state x_0 and the final state x_f, the Leapfrog method starts by dividing the path into q partitions with

$$z_0^{(k)}, z_1^{(k)}, \ldots, z_{q-1}^{(k)}, z_q^{(k)}$$

as partition points. On each iteration, k a sub-problem

$$\min_{u \in U} \int_{t_{i-1}}^{t_{i+1}} f(x(t), u(t)) dt$$

$$\dot{x}(t) = f(x(t), u(t)) \qquad (7)$$

$$x(t_{i-1}) = z_{i-1}, \quad x(t_{i+1}) = z_{i+1}$$

is solved by updating the initial feasible path towards the optimal path. This is achieved by midpoint mapping scheme, where a point z_i^{k+1} is selected as the point which is reached roughly half the time along the optimal trajectory between z_{i-1}^k and z_{i+1}^k. This continues for all the partitions until the maximum number of iterations is attained. It is noted that for every update on z_i^{k+1} and t_i^{k+1}, the states and costates are also updated. On each iteration of the Leapfrog algorithm, the cost decreases. Hence, the number of partition points decreased as the algorithm is executing. If the partitions are reduced accordingly, the convergence towards optimal trajectory proceeds efficiently. However, when $q = 2$, a simple shooting method is used to determine the optimal solution between the starting and final states z_0 and z_f. This causes a slow convergence for the algorithm on the final iteration.

2.2 Modified Newton's Method

The Newton Method (NM) is a standard root-finding method which uses the first few terms of Taylor series of a function $f(x)$. With the assumption that $f(x)$ is continuous and is a real-valued, the method finds numerical solution of $f(x) = 0$.

An iterative NM is given by

$$x_{j+1} = x_j - \frac{f(x_j)}{f'(x_j)} \qquad (8)$$

Given the initial guess x_j, one can find the next approximation of x_{j+1}. In [11], it is shown that the NM converges, provided the initial guess is close enough to the estimated point. It is stated that the Newton Method may however diverge. Hence, the Modified Newton Method was introduced to solve large class of function f in which global convergence can be proven. The Modified Newton Method (MNM) introduces an extra term, λ, and Eq. (8) becomes

$$x_{j+1} = x_j - \lambda_j s_j \qquad (9)$$

where $s_j = d_j \equiv \frac{f(x_j)}{f'(x_j)}$ is a search direction.

The steps for the MNM [11] are as follows:

1. Choose a starting point x_0
2. For each $j = 0, 1, \ldots$, define x_{j+1} from x_j as follows

   (a) Set

$$d_j = \mathrm{DF}(x_j)^{-1} F(x_j),$$

$$\gamma_j = \frac{1}{\mathrm{cond}\left(\mathrm{DF}(x_j)\right)},$$

   and let $h_j(\tau) = h(x_j - \tau\, d_j)$, where $h(s) = F(s)^\mathrm{T} F(s)$. Determine the smallest integer $m \geq 0$ satisfying

$$h_j(2^{-m}) \leq h_j(0) - 2^{-m} \frac{\gamma_j}{4} \|d_j\| \|Dh(x_j)\|.$$

   (b) Determine λ_j, so that $h(x_{j+1}) = \min_{0 \leq \kappa \leq m} h_j(2^{-\kappa})$ and let $x_{j+1} = x_j - \lambda_j d_j$.

A theoretical analysis for Modified Newton's Method can be found in [11]. The method was used in [12, 13], and the implementation can be found in [14].

3 Simulation Results

In this section, a comparison is made between the case in which the classical Newton Method (NM) and Modified Newton's Method (MNM) are used in Leapfrog algorithm to find optimal solutions of mobile robot path planning. The simulations presented utilize the set of examples adapted from our previous work. The arrows in Figs. 1, 2, 3 and 4 indicate the orientations of the robot at key positions.

From the simulations shown above, the computational time along the path and cost for each example are given in Table 1.

4 Conclusion and Future Work

The numerical simulations showed that the Leapfrog method with MNM is capable of finding optimal paths. It was also noted that the final cost was the same for both implementation of the classical NM and MNM. On the last iteration of the Leapfrog algorithm, the convergence is normally slower due to the simple shooting method, but after implementing the MNM, a faster convergence was noted. In the majority of the test cases in this study, the MNM took less time to compute the optimal path, as compared to the time required when simple shooting with NM was used. This

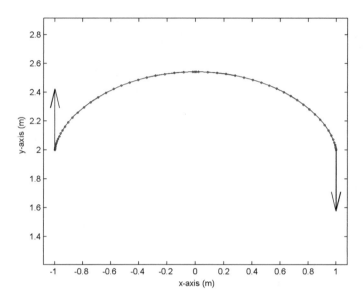

Fig. 1 Optimal path produced by Leapfrog where the robot is considered to be moving from the initial state $\begin{bmatrix} -1 & 2 & \pi/2 \end{bmatrix}$ to the final state $\begin{bmatrix} 1 & 2 & -\pi/2 \end{bmatrix}$

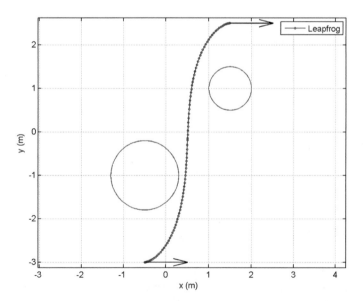

Fig. 2 Optimal path returned by Leapfrog in the presence of two obstacles. The initial and final states of the robot are $\begin{bmatrix} 0 & 0 & 0 \end{bmatrix}$ and $\begin{bmatrix} 3.5 & 3.5 & 0 \end{bmatrix}$, respectively

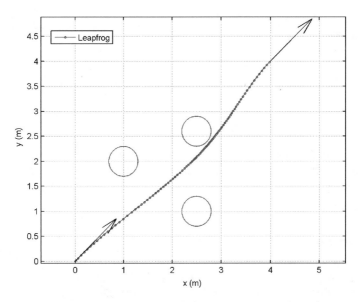

Fig. 3 Optimal path obtained from Leapfrog in the presence of three obstacles with the robot initial state $\begin{bmatrix} 0 & 0 & \pi/4 \end{bmatrix}$ and final state $\begin{bmatrix} 1 & 1 & \pi/4 \end{bmatrix}$

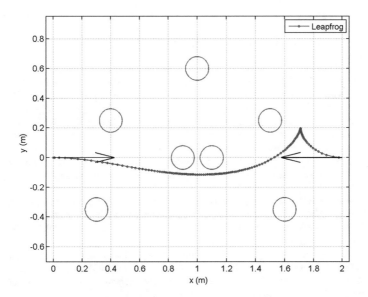

Fig. 4 Optimal path generated by Leapfrog in the presence of seven obstacles. The robot is considered to move from the initial state $\begin{bmatrix} 0 & 0 & 0 \end{bmatrix}$ to the final state $\begin{bmatrix} 2 & 0 & 0 \end{bmatrix}$

Table 1 Simulation results for simple shooting and Modified Newton's Method

Case	No. of obstacles	Cost	Computational time	
			NM (s)	MNM (s)
1	0	8.64	34.07	32.61
2	2	5.98	78.91	77.80
3	3	3.15	95.66	86.49
4	7	2.79	113.36	106.75

indicates that an advantage is gained when using the MNM with the simple shooting. For future work, a more sophisticated simple shooting method with the MNM incorporated will be considered to improve the Leapfrog algorithm.

Bibliography

1. H. Sussmann, G. Tang, *Shortest paths for the Reeds-Shepp car: a worked out example of the use of geometric techniques in nonlinear optimal control* (Piscataway, NJ, 1991)
2. J.-D. Boissonnat, A. Cérézo, J. Leblond, Shortest paths of bounded curvature in the plane. J. Intell. Rob. Syst. **11**, 5–20 (1994)
3. F.E. Bertolazzi, M.D. Lio, Symbolic-numeric indirect method for solving optimal control problems for large multibody systems. Multibody Sys. Dyn. **13**, 233–252 (2005)
4. A. Rao, A survey of numerical methods for optimal control, in *AAS/AIAA Astrodynamics Specialist Conference* (2009)
5. C. Kaya, J. Noakes, Leapfrog algorithm for optimal control. SIAM J. Numer. Anal. **46**, 2795–2817 (2008)
6. B.T. Matebese, D.J. Withey, M.K. Banda, Application of the Leapfrog method to robot path planning, in *IEEE International Conference on Information and Automation* (2014)
7. B. Matebese, D. Withey, K.B. Mapundi, Path planning with the Leapfrog method in the presence of obstacles, in *IEEE International Conference on Robotics and Biomimetics* (2016)
8. B. Matebese, D. Withey, M.K. Banda, Initialization of the leapfrog algorithm for mobile robot path planning, in *Pattern Recognition Association of South Africa and Robotics and Mechatronics International Conference* (2016)
9. A. Vamsikrishna, A.D. Mahindrakar, S. Tiwari, Numerical and experimental implementation of Leapfrog algorithm for optimal control of a mobile robot, in *Indian Control Conference* (2017)
10. R. Holsapple, R. Venkataraman, D. Doman, A modified simple shooting method for solving two-point boundary-value problems, in *IEEE Aerospace Conference* (2003)
11. J. Stoer, R. Burlirsch, *Introduction to Numerical Analysis*, 2nd edn. (Springer, New York, 1993), pp. 272–286
12. R. Holsapple, R. Venkataraman, D. Doman, A new, fast numerical method for solving two-point boundary value problems. J. Guid. Control Dyn. **26**(3), 505–508 (2004)
13. A. Trent, R. Venkataraman, D. Damon, Trajectory generation using a modified simple shooting method, in *IEEE Aerospace Conference* (2004)
14. A. Trent, Trajectory generation using a modified simple shooting method, in *Thesis, Master of Science, Texas Tech University* (2004)

Impact of Poor Data Quality in Remotely Sensed Data

Thembinkosi Nkonyana and Bhekisipho Twala

Abstract Poor data quality impacts negatively on various classification approaches. Classification is part of an important aspect of research, and often there are missing values found in data, which poses a problem in critical decision-making in terms of accuracy. Often selecting the best approach to handle missing data can be a difficult task, as there are several conditions to consider. Remote sensing as a discipline is quite susceptible to missing data. The objective of this study is to evaluate the robustness and accuracy of four classifiers when dealing with the incomplete remote sensing data problem. Two remote sensing data sets are utilised for this task with a four-way repeated-measures design employed to analyse the results. Simulation results suggest k-nearest neighbour as a superior approach to handling missing data, especially when regression imputation is used. Most classifiers achieve lower accuracy when listwise deletion is used. Nonetheless, RF is much less robust to missing data compared to other classifiers such as ANN and SVM.

Keywords Missing data · Remote sensing · Classifier · Imputation
Machine learning

1 Introduction

Remote sensing (RS) has been widely used in many research areas and applied to real-world applications. RS plays a major role as a tool to achieve sound decision-making especially in image processing applications. In order to achieve

T. Nkonyana (✉) · B. Twala
Department of Electrical and Electronics Engineering Science, University of Johannesburg,
P.O. Box 524, Auckland Park, Johannesburg 2006, South Africa
e-mail: tnnkonyana@uj.ac.za

B. Twala
e-mail: btwala@uj.ac.za

© Springer Nature Singapore Pte Ltd. 2018
S. S. Dash et al. (eds.), *Artificial Intelligence and Evolutionary Computations
in Engineering Systems*, Advances in Intelligent Systems and Computing 668,
https://doi.org/10.1007/978-981-10-7868-2_8

such reliable information, real-world data sets are used to come up with solution to related or specific problem areas and domains. Classification (viz a viz prediction) plays a huge role in interpreting and coming up with accurate results. In most cases when dealing with real-world data, poor data quality issues have become a problem for classification tasks, hence, in part, having huge impact on classification accuracy. It is for this reason that approaches to handling missing values in remote sensing became an essential part of this research.

RS applications can be used in various fields such as biodiversity, agriculture, geography, earth observation, energy and many more. Despite the lack of remote sensing data, the problem we have is that tasks that often require classification are hugely impacted on performance of classifiers due to missing values in data, thus raising a concern in analysis of data [8]. RS as a domain treats the issue of missing data presence as a crucial problem in multi-temporal and multi-sensor remote sensing image analysis which could be as a result of transmission problems or faulty sensors in the case of satellite images [1, 14]. Other contributing factors could be due to atmospheric conditions, sensor restrictions from satellite, data entry process, heavy clouds and failure of instruments in the process of data recording. In other words, incomplete data occur when important features and values are found missing.

There are various techniques for handling missing data, although each technique would have its own advantages and disadvantages. However, before any application of a specific missing data technique to address the incomplete data problem, it is important to understand the mechanism driving the missingness as each MDT is based on its own assumptions. There are various classes of missing data (also known as missing data mechanisms). These classes describe the underlying cause of missing data and are distinguished as follows: missing at random (MAR), missing completely at random (MCAR) and not missing at random (NMAR). In most scenarios, MAR is preferred by most methods [10]. This occurs when the value of the feature variable is missing, but maybe conditional on some other feature variable observed in the data set which is not in the feature variable of interest.

For example, the fact that the spectral attribute has missing values could have been as a result of the spatial attribute having missing values as well. MCAR is observed when missing value is independent of the feature variable being considered or the other feature variables within the data set (there could be no specific reason why the spectral attribute has missing values). NMAR is found where patterns of missing data are not random, and it is impossible to predict the missing data using the rest of the variables in the data set (the missing values on the spectral attribute could have been as a result of project managers withholding specific information related to spectral). In this particular scenario, ignoring missing values can result in biased decision because the missing values may contain important information which is crucial to the final result.

While some remote sensing researchers may have altogether ignored (and not reported) the incomplete data problem, several researchers have examined various techniques to solve the problem. Specific results from previous studies include applications in both the spatial and spectral domains [4, 5], while other applications involve the fusion of data with different resolutions (e.g. panchromatic and multi-spectral satellite images).

The major contribution of this paper is to evaluate the robustness and accuracy of four classifiers in terms of dealing with the incomplete data problem. Our goal is to identify the best classifier in terms of dealing with missing data (accuracy) and further identify the missing data method that particular classifier is more robust too. Four missing data techniques are also utilised for this task. Our study assumes the MAR approach in all our experiments of which the mechanism was artificially simulated from an original complete data set. The empirical evaluation is based on five performance measures: accuracy (AC), mean absolute error (MAE), mean-squared error (MSE), precision and receiver on curve (ROC).

This paper is organised as follows: Sect. 2 presents a brief description of missing data techniques used in all our experiments; this is followed by a brief discussion of four machine learning techniques (classifiers). Section 4 presents simulations in terms of experimental set-up and then results, while Sect. 5 summarises the key findings and direction for future research.

2 Missing Data Techniques

2.1 Listwise Deletion

Listwise deletion (LD) removes any instances that have one or more missing values and performs an analysis based on the remaining data [3, 10, 13]. One of the advantages of LD is its user-friendliness. In fact, it is the default strategy for most software packages. Its major weakness lies in its reduction in the sample size due to the elimination of all instances with missing values. This approach is based on the MCAR assumption which most of the time is rare to support.

2.2 Mean–Mode Imputation

This technique uses a process of replacing values of all missing numerical attributes with the mean of the observed values, while for missing categorical attributes, the mode (or most frequent value) is used [3, 10]. The advantage of this method is that it is easy to use and can use complete-case analysis methods, but it has been shown to reduce variability and it exhibits biased parameter estimates, such as means and correlations (unless the data are MCAR).

2.3   Nearest Neighbour Imputation

This type of imputation method searches for the nearest neighbour cases to estimate and replace missing data [2]. One major strength of *k*-nearest neighbour (*k*-NN) is its robustness to noisy training data (especially if we use inverse square of weighted distance as the "distance"). *k*-NN is also effective when the training set is large. [2]. However, one needs to determine the value of parameter *k* (number of nearest neighbours). Also, the computation cost of *k*-NN is quite high because we need to compute distance of each query instance to all training samples.

2.4   Regression Imputation

This is a type of imputation technique actually replaces the missing values with predicted score from a regression equation [10]. What this means is that it predicts the values of missing values from a case with complete values. The advantage of this method is that it uses information which is obtained from observed data. However, it has the limitation of overestimating model fitting and correlation estimates and weakening the variance.

3   Machine Learning Techniques

3.1   Random Forest

A random forest (RF) is a meta-estimator that fits a number of decision tree (DT) classifiers on various subsamples of the data set and uses averaging to improve predictive accuracy. DTs are non-parametric methods which portray tree-like graph or model of decisions and their possible outcomes [2, 14, 15]. One strength of DTs is their simplicity and easy to use; they are also robust to nonlinear and noisy relations among input features and class labels; and they are computationally efficient.

3.2   k-*nearest Neighbour*

k-nearest neighbour (*k*-NN) is a technique that uses the value of *k* to find the nearest neighbours. In terms of computing new instances, the Euclidean distance is run between instances and each training instances stored new instances are assigns value of *k* [14]. This method is one of the simplest methods that can be applied in classification. *k*-NN is a non-parametric method which makes use of searching for the closest number of its neighbours by using a distance function using the value of *k*.

3.3 Artificial Neural Networks

This type of method is widely used and popular, as it mimics human brain [14]. Artificial neural network (ANN) is a non-parametric classifier. The multilayer perceptron is commonly used in neural network and remote sensing applications, as it applied three or more layers of neurons that consist of nonlinear function. They are very easy to train depending on the size of the data.

3.4 Support Vector Machines

Having been introduced by Vapnik [12], support vector machine (SVM) is a technique that has a set of linear decision hyperplanes, meaning it requires that the training set must be linear separable and uses a hyperplane to separate the training patterns of two classes [11, 14]. SVM classifiers have been commonly used and applied in various field of study including remote sensing.

4 Study Methodology

4.1 Experimental Set-up

The objective of our experiments is to evaluate the robustness and accuracy of four classifiers (RF, *k*-NN, ANN, SVM) against four missing data methods (LD, MMI, NNI, RI) and five performance measures. Two data sets (obtained from the machine learning repository UCI [9]) are used in all our 1600 experiments (i.e., 4 classifiers × 4 missing data methods × 5 performance measures × 2 data sets × 10 replicates). The first data set is a multi-temporal remote sensing data, used for a forested area in Japan [7]. The objective of the data set is to map different forest types using spectral data. The data set is made up of 326 instances and 27 attributes. The second data set is a high-resolution remote sensing data set using QuickBird [6]. The Wilt data set is made up of six attributes and 4889 instances. The goal of this data set is intended for classification, so as to detect diseased trees in QuickBird imagery.

All experiments were run using the WEKA library [9]. To evaluate the performance of each of the classifiers described in previous sections, we performed a set of ten cross-validations by using each classifier in classifying each of the data sets. To this end, each data was split into three parts: 70% training; 20% testing; and 10% validating. In this way, the classifiers were evaluated with the use of independent data for each step. In order to have control over the missing data, 20% of missing values are artificially generated in all the testing data sets, based on the MAR assumption. This is because missing at random data has been shown to have more impact when they occur in the testing set than in either the training or both training and test sets. The simulation design is a three-way repeated-measures design, in which each main effect is tested against its interaction with data sets. The main effect factors are as follows: four classifiers, four missing data techniques and five performance measures. The only random effect factor is the two data sets (as the data set levels were randomly sampled).

4.2 Experimental Results

The results on the robustness and accuracy of four classifiers against four missing data methods are summarised in Table 1. The overall results are based on both data sets; other words, the results are averaged over the two data sets (for the sake of space and generalisation purposes). From these experiments, which suggest important differences, the following results are observed as follows:

- Firstly, all the main effects were found to be significant at the 5% level of significance ($F = 57.34$, df = 4 for classifiers; $F = 191.30$, df = 3 for missing data methods; $F = 184.37$, df = 3 for performance measures; p-value < 0.05 for each main effect). Only the three-way interaction effects between classifiers were found to be significant at the 5% level ($F = 23.4$, df = 63 for missing data methods and performance measures).
- The results show all classifiers as being more robust to missing data when regression imputation is used. This is the case for all the performance (with a few exceptions for RMSE). All the classifiers are less robust to missing data when listwise deletion is used. Overall, k-NN achieves bigger accuracy rates when dealing with MAR data. The second best method is ANN, while RF exhibits the worst performance.

Table 1 Remote sensing problem

	RF	k-NN	ANN	SVM
Accuracy				
Listwise deletion	90.919	86.704	89.469	91.010
Mean–mode imputation	79.619	60.498	71.482	69.978
Nearest neighbour imputation	91.206	89.610	87.222	91.515
Regression imputation	**95.212**	**92.464**	**94.776**	**93.371**
MAE				
Listwise deletion	0.089	0.086	0.070	0.140
Mean–mode imputation	0.251	0.293	0.293	0.248
Nearest neighbour imputation	0.080	0.067	0.067	0.078
Regression imputation	**0.063**	**0.057**	**0.057**	**0.069**
RMSE				
Listwise deletion	0.183	0.213	0.203	**0.226**
Mean–mode imputation	0.346	0.476	0.395	0.413
Nearest neighbour imputation	0.174	**0.193**	0.219	**0.226**
Regression imputation	**0.148**	0.195	**0.145**	0.246
Precision				
Listwise deletion	0.914	0.869	0.889	0.909
Mean–mode imputation	0.762	0.612	0.632	0.620
Nearest neighbour imputation	0.913	0.896	0.867	0.913
Regression imputation	**0.952**	**0.929**	**0.949**	**0.922**
ROC				
Listwise deletion	0.882	0.731	0.851	0.707
Mean–mode imputation	0.716	0.632	0.668	0.648
Nearest neighbour imputation	0.931	0.783	0.827	0.710
Regression imputation	**0.974**	**0.821**	**0.979**	**0.727**

Bold values highlight the highest performing classifiers after imputation on the results

5 Conclusion

The objective of the study was to evaluate the accuracy and robustness of four machine learning techniques in terms of dealing with incomplete remotely sensed data. Selecting methods for dealing with missing data are difficult, since the same procedure can give higher predictive accuracy rates in certain circumstances and not in others. It has been found that k-NN is the best performance method, especially when regression imputation is used. All classifiers performed badly when listwise deletion was used. However, k-NN has competitors like ANN and SVM. The poor performance of RF was not expected as it has the average reduction strategy. Finally, so far we have restricted our experiments to only single imputation methods and only two data sets. It would be interesting to carry out a comparative study using multiple imputation methods and on more than two data sets. We leave the above issues for future work.

Acknowledgements This work was funded by the Institute for Intelligent Systems at the University of Johannesburg, South Africa. The authors would like to thank Ahmed Ali and anonymous reviewers for their useful comments and to the UCI for making the data sets available.

References

1. S. Aksoy, K. Koperski, C. Tusk, G. Marchisio, Land cover classifcation with multi-sensor fusion of partly missing data. Photogram. Eng. Rem. Sens. **75**(5), 557–593 (2009)
2. L. Baretta, A. Santaniello, Nearest neighbor imputation algorithms: a critical evaluation. BMC Med. Inf. Decis. Making **16** (2016). https://doi.org/10.1186/s12911-016-0318-z
3. G. Chechik, G. Heitz, G. Elidan, Max-margin classification of data with absent features. J. Mach. Learn. Res. **9**, 1–21 (2000)
4. Y. Ding, J.S. Simonoff, An investigatio of missing data methods for classification trees. J. Mach. Learn. Res. **11**, 131–170 (2010)
5. K.W. Hsu, Weight-adjusted bagging of classification algorithms sensitive to missing values. Int. J. Inf. Educ. Technol. **3**(5) (2013)
6. B. Johnson, R. Tateishi, N. Hoan, A hybrid pansharpening approach and multiscale object-based image analysis for mapping deseased pine and oak trees. Int. J. Rem. Sens. **34**(20), 6969–6982 (2013)
7. B. Johnson, R. Tateishi, Z. Xie, Using geographically-weighted variables for image classification. Rem. Sens. Lett. **3**(6), 491–499 (2012)
8. W.M. Khedr, A.M. Elshewey, Pattern classification for incomplete data using PPCS and KNN. J. Emerg. Trends Comput. Inf. Sci. **4**(8) (2013). ISSN 2079-8407
9. M. Lichman, *UCI Machine Learning Repository.* http://archive.ics.uci.edu/ml. (University of California, School of Information and Computer Science, Irvine, CA, 2013)
10. R.J.A. Little, D.B. Rubin, *Statistical Analysis with Missing Data* (Wiley, New York, 1987)
11. G. Mountrakis, J. Im, C. Ogole, Support vector machines in remote sensing: a review. ISPRS J. Photogram. Rem. Sens. **66**, 247–259 (2011)
12. T. Nkonyana, T. Twala, An empirical evaluation of machine learning algorithms for image classification, in *Advances in Swarm Intelligence. ICSI 2016*, ed. by Y. Tan, Y. Shi, L. Li. Lecture Notes in Computer Science, vol 9713 (Springer, Cham, 2016)
13. N.R. Pimplikar, A. Kumar, A.M. Gupta, Study of missing value imputation methods—acomparative approach. Int. J. Adv. Res. Comput. Sci. Sofw. Eng. **4**(3) (2014). ISSN 2277 128X
14. B. Twala, T. Nkonyana, Extracting supervised learning classifiers from possibly incomplete remotely sensed data, in *Computational Intelligence and 11th Brazil Congress on Computation Intelligence (BRICS-CCI & CBIC), 2013 Brics Congress.* https://doi.org/10.1109/brics-cci-cnic.2013.85
15. L. Xuerong, X. Qianguo, K. Lingyan, Remote sensing image method based on evidence theory and decision tree. Proc. SPIE **7857**, 7857Y-1 (2010)

Image Enrichment Using Single Discrete Wavelet Transform Multi-resolution and Frequency Partition

Chinnem Rama Mohan and Siddavaram Kiran

Abstract The role of image fusion is very prominent in current trend in different fields. The main aim of multi-sensor is to generate better description of scene further which is useful for human and machine perception. The main problem in image fusion is that how to determine the best procedure to combine multiple images. There are so many classical techniques available to perform the above. The main factors of image fusion are entropy and root mean square error. The description of scene is estimated with the above factors. The proposed work concentrates on enhancement of above factors with DWT maximum rule combining with DCT-FP fusion. The DWT maximum rule is considered for single image in this work.

Keywords SDWT · Maximum rule · Frequency partition · Multiple images
Sensors · Perception

1 Introduction

Fusion is a method of retrieving relevant data from multiple images into a solitary image. The purpose of fusion techniques is to improve the description of scene more qualitatively. The general formal solution in image fusion is multi-sensor data fusion. In remote sensor applications, observational constraints are necessary. Generally, sensing images do not have that much of clarity to obtain observational constraints, and enhancing clarity is one of the possible solutions of image fusion.

C. R. Mohan (✉)
Department of CSE, VTU, Belgaum, Karnataka, India
e-mail: ramamohanchinnem@gmail.com

S. Kiran
Department of CSE, Y.S.R. Engineering College of YVU, Proddatur,
Andhra Pradesh, India
e-mail: rkirans125@gmail.com

© Springer Nature Singapore Pte Ltd. 2018
S. S. Dash et al. (eds.), *Artificial Intelligence and Evolutionary Computations in Engineering Systems*, Advances in Intelligent Systems and Computing 668,
https://doi.org/10.1007/978-981-10-7868-2_9

In pixel-level fusion [1], it is important to combine the set of pixels which are necessary for future level representation [1]. In decision-level fusion [1] from the merging information with higher level of abstraction, final fusion image is yielded.

1.1   Image Fusion

To fuse the image, it is required to consider two images and both the images must be similar. They must have relevant information, but the process of image fusion [2] always generates high-resolution images when compared with source images. The high-resolution images are very important in medical field for diagnosis of diseases because the resultant image gathers information from different sources. To adopt quality of image, it is essential to merge multi-sensor and multi-temporal images.

1.2   Standard Image Fusion Methods

To obtain quality of information, image fusion is one of the methods in digital image processing. This method can be divided into two groups. They are as follows:

1. Spatial approach
2. Transform approach

In spatial approach, operations are performed directly on original pixels. The main disadvantage of spatial technique is that classification of spectral distortion. In the case of transform domain image, first it is morphed into another form, and then different actions are performed on morphed values. In transform domain, different kinds of transform approaches are available such as DCT, DWT, FFT, DFT, and its inverse. Further for better fusion, SDWTMR method is applied. Especially for remote images, it is essential to enhance the resolution. For this, multi-resolution method is one of the ways. Most of the results prove that transform domain is better than the spatial domain.

2   DCT-Frequency Partition

The energy coefficients such as AC and DC are generated with DCT [3]. Frequency partitioning [3] is one of the methods included within DCT where frequency level ranges from 0.0 to 1.0. Computationally, a fixed-level approach has been performed and other transformation techniques, but DCT is having an approach of level-oriented image acquisition to enhance the intensity of an image.

2.1 Procedure for DCT-FP

1. Fix the frequency value ranging from 0 to 1 with 0.1 interval.
2. Generate multi-focus image with source images.
3. Find the DCT-FP for the multi-focused images.

   a. Find image frequency for multi-resolution images with respect to rows and columns.
   b. Find out the average of image frequency.
   c. Multi-resolution DCT frequency is calculated as a product of rows, columns, and frequency value multiplication. If data is real, it is rounded further to find out multi-resolute DCT to gain more information with given frequency.
   d. Retaining the multi-resoluted image with respect to exact rows and columns of source image.
   e. Finally generates multi-fused image.

4. After the generation of multi-fused image with the above steps, find out the parameter metrics.

3 Proposed Methodology

Due to the limitations in transformation, wavelets are introduced in image fusion. Wavelets focus on depth of optical lens relevant to object. To obtain relevant information from the source image, it is essential to focus on all the possible views for human visual perception. The field of signal processing is the base for wavelet transformation but these are applicable for images. Wavelets [4–7] provide to find out good localization properties in the image. Wavelets are represented with 1-D and 2-D forms. To find out 2-D decomposition levels of image, wavelets are useful. Generally, wavelets generate frequency components LL, LH, HL, and HH which have been depicted in Fig. 1.

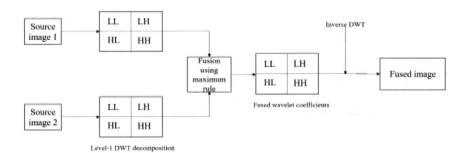

Fig. 1 Wavelet fusion method

Further each frequency component is again divided into the above-said frequency levels. There exist simple average, select minimum, and select maximum rules to find out decomposition levels of image with respect to wavelets. In maximum rule, it focuses on highest intensity pixels region which results in highly focused image output. If in a particular region each image pixel is compared with another, further, only the highest pixel value is assigned to corresponding lowest pixel value. The wavelet transform is applied on single image. It can correspond to by the subsequent Eq. (1). Wavelet coefficients are fused by means of the fusion maximum rule Eq. (2).

Inverse single discrete wavelet transform maximum rule (ISDWTMR) is functional on the fused wavelet coefficients to achieve the fused image $I_F(M_1, N_2)$ specified by subsequent Eq. (3).

$$I(M_1, N_2) = W(I_1(M_1, N_2)) \tag{1}$$

$$\begin{aligned} a3(x,y) = \max(a1(x,y)) \quad & b3(x,y) = \max(b1(x,y)) \\ c3(x,y) = \max(c1(x,y)) \quad & d3(x,y) = \max(d1(x,y)) \end{aligned} \tag{2}$$

$$I_F(M_1, N_2) = W^{-1}(\varphi(W(I_1(M_1, N_2)))) \tag{3}$$

A schematic illustration of the wavelet fusion method of one input source image $I_1(M_1, N_2)$ is shown in Fig. 2.

The proposed work combined with DCT-FP along with SDWTMR to get the better fusion for source images. Figure 3 depicts a flow graph of the proposed method.

3.1 Algorithm: DCT-FP with SDWTMR

Input: Source Images
Output: Fused Image
Steps:

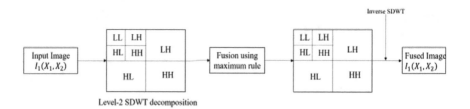

Fig. 2 Single wavelet fusion method

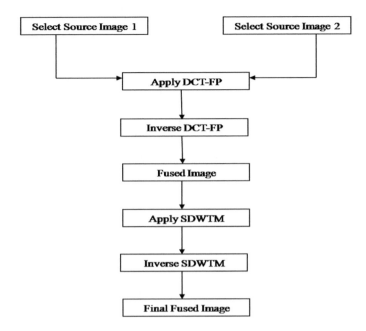

Fig. 3 Proposed algorithm

1. Select input images
2. Apply DCT-frequency partition on input images
3. Find out inverse DCT-FP to obtain the fused image
4. The result of DCT-FP is applied in SDWT with maximum rule
5. Find out inverse SDWTMR
6. Required fused image is generated as output.

In addition to maximum rule to extract the required description scene with required intensity values, DCT-FP is applied for source images which are an added flavor in the proposed work. These impacts on the results measure entropy, RMSE, PSNR, and others.

4 Performance Evaluation

Fusion quality can be appraised optically. Human discernment determines fusion quality. Human object evaluators provide grade for the corresponding image (fused) and average the grade. This sort of appraisal has some disadvantages such as the grade is not accurate and the degree of accuracy can be decided by the human observer's ability. To eliminate the demerits, quantitative measures should be used for accuracy and empirical evaluation of the fused objects [8].

4.1 With Allusion Image

If the reference of image is available, the performance of image synthesis algorithms can be assessed by using the below proportions.

4.1.1 Root Mean Square Error (RMSE)

Computed as the root mean square error of the resultant pixels in the allusion image
and synthesized image. A lesser RMSE value means better image synthesis quality.
It is computed as follows:

$$\text{RMSE} = \text{sqrt}(\text{sum}((I_r(i,j) - \text{If}(i,j)).^2)/(M*N)) \qquad (4)$$

where

M	the number of rows
N	the number of columns
(i, j)	pixel index
I_r	reference image
I_f	synthesized image
$I(i, j)$	gray value at pixel (i, j).

4.1.2 Peak Signal to Noise Ratio (PSNR)

Its value will be far above the ground when the synthesized and allusion images are
related. A higher PSNR value means better image synthesis quality. It is computed
as follows:

$$\text{PSNR} = 10 * \log 10(L^2/\text{RMSE}) \qquad (5)$$

where L is the number of gray levels in the image.

4.1.3 Mean Absolute Error (MAE)

Computed as the mean absolute error of the resultant pixels in allusion and synthesized images. A lower MAE value means better image synthesis quality. It is
computed as follows:

$$\text{MAE} = \text{sum}(\text{sqrt}((I_r(i,j) - \text{If}(i,j)).^2))/(M*N) \qquad (6)$$

4.1.4 Percentage Fit Error (PFE)

Percentage fit error is calculating the norm of the difference between the corresponding pixels of the allusion and synthesized image to norm of the allusion image. If the calculated value is zero, then both the allusion and the synthesized images are same and numbers will be increased when the combined image is not same to the allusion image. A lower PFE value means better image synthesis quality. It is computed as follows:

$$PFE = 100 * norm(I_r(i,j) - If(i,j))/norm(I_r(i,j)) \qquad (7)$$

where norm is the operator to calculate the biggest singular value.

4.1.5 Signal to Noise Ratio (SNR)

A superior SNR value means enhanced image synthesis quality. It is computed as follows:

$$SNR = 10 * \log 10(ST/NTF) \qquad (8)$$

where

ST   mean (mean $((double(I_r)).\char94 2))$
NTF   mean (mean $((double(I_r - I_f)).\char94 2))$

4.1.6 Correlation (CORR)

This depicts the correlation between the allusion and synthesized images. The ideal value is one when the allusion and synthesized are exactly alike and it will be less than one when the dissimilarity increases. A higher CORR value means better image synthesis quality. It is computed as follows:

$$CORR = 2 * RTF/(RT + RF) \qquad (9)$$

where

RTF   sum (sum($I_r. * I_f$))
RT   sum (sum($I_r. * I_r$))
RF   sum (sum($I_f. * I_f$))

4.2 Without Allusion Image

If allusion image is not available, then recital of image synthesis methods can be evaluated by means of below measurements.

4.2.1 Entropy (EN)

It is useful to measure the information content of a synthesized image. The large entropy value indicated by synthesized image has high information contents. It is computed as follows:

$$\text{Entropy}(I_f) \tag{10}$$

5 Results and Discussion

Image fusion is performed by means of DCT-FP and DCT-FP with SDWTM, their recital is calculated in terms of entropy (EN), RMSE, PSNR, PFE, SNR, MAE, and correlation (CORR) performance metrics and the results of DCT-FP and DCT-FP with SDWTMR are shown in Table 1 and Figs. 4, 5, 6, 7, 8, 9, 10, and 11. In all aspects, SDWT maximum rule with FP is showing better results when compared with DCT-FP.

6 Comparison of Image Enrichment Results Using DCT-FP and DCT-FP with SDWTM

Figures 5, 6, 7, 8, 9, 10, and 11 show the assessment of enrichment results of sample multi-focus images of IDS1, IDS2, IDS3, IDS4, IDS5, IDS6, IDS7, IDS8, IDS9, and IDS10 using DCT-FP and DCT-FP with SDWTMR.

Table 1 Performance evaluation metrics to evaluate image fusion methods

Input images	Reference image entropy	Methods	Entropy	RMSE	PSNR	SNR	PFE	MAE	CORR
Image data set 1	6.7933	DCT-FP	5.8877	19.6312	35.2353	19.8959	10.1205	17.1187	0.9953
		DCT-FP + SDWTM	7.3282	10.3963	37.996	25.4173	5.3596	7.6344	0.9986
Image data set 2	4.3348	DCT-FP	4.0926	15.1098	36.3722	23.5979	6.6085	12.9455	0.9979
		DCT-FP + SDWTM	5.8537	8.2438	39.0035	28.8606	3.6056	6.2648	0.9993
Image data set 3	7.7196	DCT-FP	7.6422	16.5076	35.988	18.3356	12.1121	15.4318	0.9934
		DCT-FP + SDWTM	7.7773	6.4672	40.0576	26.475	4.7452	4.4298	0.9989
Image data set 4	7.8033	DCT-FP	7.5695	16.3784	36.0221	19.2405	10.9138	15.538	0.9946
		DCT-FP + SDWTM	7.829	5.5871	40.6929	28.5822	3.723	3.916	0.9993
Image data set 5	7.8527	DCT-FP	7.3517	43.2442	31.8055	9.8703	32.0984	42.8016	0.9611
		DCT-FP + SDWTM	7.8912	6.0154	40.3721	27.0036	4.465	4.4351	0.999
Image data set 6	7.7228	DCT-FP	7.6071	33.2771	32.9433	11.0132	28.141	32.5792	0.9687
		DCT-FP + SDWTM	7.7362	6.808	39.8346	24.7958	5.7572	5.1422	0.9984
Image data set 7	7.7115	DCT-FP	7.6454	13.5817	36.8352	21.2274	8.6822	11.852	0.9965
		DCT-FP + SDWTM	7.6694	7.7574	39.2677	26.0923	4.9589	5.7026	0.9988
Image data set 8	7.8097	DCT-FP	7.6034	22.2391	34.6936	16.9983	14.1281	19.7357	0.9911
		DCT-FP + SDWTM	7.8194	10.8039	37.829	23.269	6.8635	8.4551	0.9976
Image data set 9	6.0167	DCT-FP	4.4592	23.9767	34.3669	19.3441	10.7843	21.2012	0.9947
		DCT-FP + SDWTM	5.9979	10.697	37.8722	26.3546	4.8114	7.0316	0.9988
Image data set 10	7.0403	DCT-FP	7.1139	26.8893	33.869	12.0375	25.0108	26.5612	0.9751
		DCT-FP + SDWTM	7.1032	4.227	41.9045	28.1084	3.9317	3.1702	0.9992

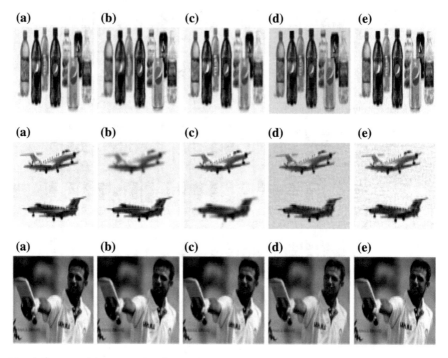

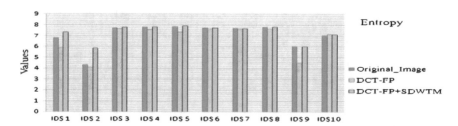

Fig. 4 Image enrichment results of multi-focus image data sets using DCT-FP and DCT-FP with SDWTMR **a** reference image, **b** input image 1, **c** input image 2, **d** enrichment using DCT-FP, **e** enrichment using DCT-FP with SDWTMR

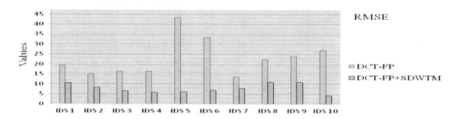

Fig. 5 Comparison of DCT-FP and DCT-FP with SDWTMR versus entropy

Fig. 6 Comparison of DCT-FP and DCT-FP with SDWTMR versus RMSE

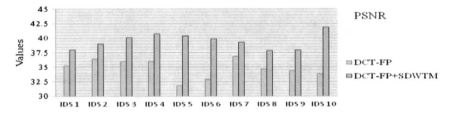

Fig. 7 Comparison of DCT-FP and DCT-FP with SDWTMR versus PSNR

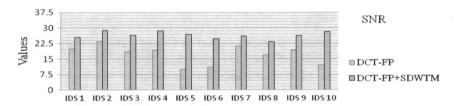

Fig. 8 Comparison of DCT-FP and DCT-FP with SDWTMR versus SNR

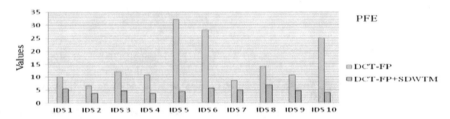

Fig. 9 Comparison of DCT-FP and DCT-FP with SDWTMR versus PFE

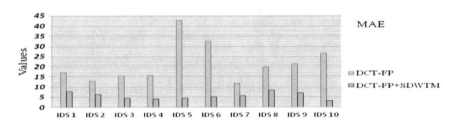

Fig. 10 Comparison of DCT-FP and DCT-FP with SDWTMR versus MAE

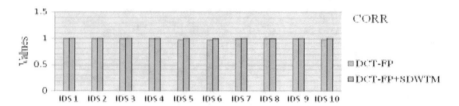

Fig. 11 Comparison of DCT-FP and DCT-FP with SDWTMR versus CORR

7   Conclusion

Fusion algorithm selection is always dependent on type of problem. The main goal of fusion is to achieve high spatial resolution but there exists a problem of blurring in spatial domain. The wavelet maximum rule is a kind of technique which is achieving high quality of content combing with DCT-FP method. The quality of image is represented with the evaluation of different metrics. All the metrics are shown in better content most of the times. In further investigation, different kinds of wavelets with texture features need to be applied for image fusion.

References

1. A. Kanagasabapathy, A. Vasuki, Image fusion based on wavelet transform. Int. J. Biomed. Sig. Process **2**(1), 15–19 (2011)
2. Z. Wang, D. Ziou, C. Armenakis, D. Li, Q. Li, A comparative analysis of image fusion methods. IEEE Trans. Geosci. Remote Sens. **43**(6) (2005)
3. V.P.S. Naidu, Novel image fusion techniques using DCT. Int. J. Comput. Sci. Bus. Inform. ISSN **5**(1), 1694–2108 (2013)
4. Y. Yong, Multiresolution image fusion based on wavelet transform by using a novel technique for selection coefficients. J. Multimedia **6**(1) (2011)
5. V.P.S Naidu, J.R Raol, Pixel Level image fusion using wavelet and principal component analysis. Defence Sci. J. **58**(3), 338–352 (2008)
6. H.H. Wang, A new multiwavelet-based approach to image fusion. J. Math. Imaging Vis. **21**, 177–192 (2004)
7. M. Heng, C. Jai, S. Liu, Multisource image fusion based on wavelet transfrom. Int. J. Inf. Technol. **11**(7) (2005)
8. P. Jagalingam, A.V. Hegde, A Review of quality metrics for fused image. Elsevier Trans. Aquat. Procedia **4**, 133–142 (2015)

An Experiential Metrics-Based Machine Learning Approach for Anomaly Based Real Time Prevention (ARTP) of App-DDoS Attacks on Web

K. Munivara Prasad, A. Rama Mohan Reddy and K. Venu Gopal Rao

Abstract The Internet is often victimized to the distributed denial of service (DDoS) attack, in which purposefully occupies the bandwidth and computing resources in order to deny that services to potential users. The attack situation is to flood the packets hugely to the target system. If the attack is from a single source, then the attack is called as denial of service (DoS) and if attack is from divergent servers, then it is called as DDoS. Over a decade, several researchers succeeded to deliver few significant DDoS detection and prevention strategies by considering the detection and prevention of DDoS attack as research objective. In present level of Internet usage, "how fast and early detection of DDoS attack" is done in streaming network transactions which is still a significant research objective. Unfortunately, the current benchmarking DDoS attack detection strategies are failed to justify the objective called "fast and early detection of DDoS attack." In order to this, we devised an anomaly based real time prevention (ARTP) of application-layer DDoS attacks (App-DDoS attacks) on Web that is in the aim of achieving fast and early detection. The ARTP is a machine learning approach that is used to achieve the fast and early detection of the App-DDoS by multitude request flood. The experiments were carried out on benchmarking LLDoS dataset, and the results delivered are boosting the significance of the proposed model to achieve the objective of the paper.

Keywords Denial of service (DoS) attacks · Distributed DoS (DDoS) attacks
Application-layer DDoS (App-DDoS) · LLDoS dataset · Detection of App-DDoS

K. Munivara Prasad (✉) · A. Rama Mohan Reddy · K. Venu Gopal Rao
JNTUH, Hyderabad, Telangana, India
e-mail: prasadkmv27@gmail.com

A. Rama Mohan Reddy
e-mail: ramamohansvu@gmail.com

K. Venu Gopal Rao
e-mail: kvgrao1234@gmail.com

© Springer Nature Singapore Pte Ltd. 2018
S. S. Dash et al. (eds.), *Artificial Intelligence and Evolutionary Computations in Engineering Systems*, Advances in Intelligent Systems and Computing 668,
https://doi.org/10.1007/978-981-10-7868-2_10

1 Introduction

In the modern society, the Internet has become increasingly important in the day-to-day life. It is bringing drastic change in the communication, business mode, and even everyday life of people in the society. The major threats to the Internet are denial of service (DoS attack) and Distributed denial of service (DDoS attack). A DoS is an cyber-attack which interrupts the operations of a victim system, typically a Web server, from providing services to legitimate users without physically violating the network or server. On the other hand, distributed denial of service (DDoS) attacks are a cyber-attack launched from multiple devices distributed across the Internet to produce a huge volume of traffic loads which exhaust the resources on the target systems and subsequently degrades the performance of the network which leads to deny service to legitimate users.

DDoS attacks [1] are classified into two major categories: network-layer and application-layer DDoS attacks. In network-layer DDoS attacks, attacker uses technique such as IP spoofing and sends a huge volume of fake packets toward the victim system. In application-layer DDoS attacks (App-DDoS attack), attackers may flood the target victim system through a genuine requests. In App-DDoS attack model, a TCP connection has to be established by the zombie machine with the victim system by providing a genuine IP address else the connection will not be established. One of the well-known examples of App-DDoS attack is HTTP flooding.

The strategy of App-DDoS attack is to make the host server of the online application, purposely occupied by multiple sessions of requests and multiple cooperative sources, such that it blocks different users to realize access to that host server. The recent acquainted victims of DDoS attack are explored, and techniques for successful attack mitigating are explored in [2]. The HTTP flood is meant to abuse the HTTP requests and generates the flood of request packets to occupy the resources of the target server [3, 4]. During this flood, the target servers are going to be unable to differentiate the payload from the jammed HTTP requests.

2 Related Work

In the recent literature, there are few DDoS defensing ways associated with HTTP flood [5], that are largely relied on application-layer data. DDoS shield [6] is one, that defense the HTTP flood by analyzing the session time of arrival intervals and request time of arrival intervals. Another important model, HTTP Flood Defensing by Page access behavior [7, 8], analyzes the data size, browsing time, and their correlation. These two models are inadequate to manage the packet sending rates because the offender involves botnet [9]. During this context, the role of the user is overloaded by a CAPTCHA-based probabilistic authentication technique. This typically reflects as denial of service if users aggravated, due to the task of finding graphic puzzles.

The contribution from [10] specifies the hidden semi-Markov model (HsMM) [11] which notifies the normal user access behavior by using the pattern of request order. Further, the state of incoming users is assessed based on the thresholds obtained from HSMM. This model ends with frequent false alarms, since the request pattern can change with high frequencies due to the divergent browsing contexts of legitimate users; the pattern of requests is not a considerable metric. As proof to the present argument, the incident wherever users will click the external net links, pass URLs on to a relevant request, or use a mix of browsers could cause false alarms.

Assessing the transaction similarity with normal and flood data is the main aim of the proposed technique. Unlike existing benchmarking approaches, the proposed model the features are extracted from the observed request stream in an absolute time interval rather from the user sessions. The rest of this paper is organized as follows. In Sect. 3, the proposed ARTP model is explored followed by experimental study and performance analysis of ARTP in Sect. 4 and concluding the article in Sect. 5.

3   An Experiential Metrics-Based Machine Learning Approach for App-DDoS Attacks on Web

In this section, we developed a machine learning based approach known as an anomaly based real time prevention (ARTP) of App-DDoS attacks on Web, which is used to attain the "fast and early detection" of the App-DDoS by mass request flood.

The projected model ARTP is focused on defining set of individual metrics, namely "request chain length, ratio of packet count, ratio of packet types, ratio of request intervals, and request chain context." The key reason of the proposal is unlike many of the benchmarking models, which considered sessions or requests as input to find the anomalies. The LLDoS dataset was used to carry experiments, and the performance is analyzed through the statistical analysis of the metrics, namely scalability and robustness of the defined model. The projected model is extremely substantial in App-DDoS attack detection to accept by the present state of Web applications with full requests that is impressively overstated to petabytes that associated to the past Web request load in gigabytes.

3.1 Discovering Time Frame Length

Consider CS be the pool of user sessions $CS = \{s_1, s_2, \ldots s_C\}$ and every session consists of a regular transactions $\{t \exists t \in s_i \wedge s_i \in CS\}$ characterized as N (normal) or D (DDoS attack). The collected total transactions CS is separated into CS_N (normal) and CS_D (DDoS attack), respectively. Further, the heuristic metrics explored (in Sect. 3.2) will be assessed on these datasets.

Consider the dataset CS (which is the collection of CS_N and CS_D), separation the sessions CS_N (normal) and CS_D (DDoS attack) and cluster the sessions by using K-Means algorithm [12] individually for CS_N and CS_D. Apply K-Means algorithm [12] to discover the number of clusters in CS_N and CS_D, then calculate time frames approximately, similar for both CS_N and CS_D individually.

Consider $C = \{c_1, c_2, \ldots, c_K\}$ be the group of clusters with K size for the collected normal transactions CS_N. Then for each cluster C_i $\{c_i \exists c_i \in C \wedge i = 1, 2, 3, \ldots K\}$, the time frame is measured as the difference between the end time of maximum session and begin time of least session.

For the cluster C_i $\{c_i \exists c_i \in C \wedge i = 1, 2, 3, \ldots K\}$ consider $SB_N(c_i) = \{sb_1, sb_2, \ldots sb_{|c_i|}\}$ the bunch of session begin times of cluster c_i in ascending order and sb_1 contains maximum session begin time and $sb_{|c_i|}$ contains least begin time. Consider $SE_N(c_i) = \{se_1, se_2, \ldots se_{|c_i|}\}$ be the set of session end times of cluster c_i in descending order; se_1 contains maximum session end time, and $se_{|c_i|}$ contains least end time. Then the time frame $(tf(c_i))$ of the cluster c_i is as follows:

$$tf(c_i) = \sqrt{(se_1 - sb_1)^2}$$

Then from all the clusters, the average time frame length is calculated as follows:

$$\langle tf(C) \rangle = \frac{\sum_{i=1}^{K} tf(c_i)}{K} \tag{1}$$

The time *frame absolute deviation* (*tfAD*) of all the clusters is defined as

$$tfAD = \frac{\sqrt{\sum_{i=1}^{K} (\langle tf(C) \rangle - tf(c_i))^2}}{K} \tag{2}$$

Finally, the time frame (tf) is calculated as the sum of the average time frames length $(\langle tf(C) \rangle)$ and time frame absolute deviation (*tfAD*).

$$tf = \langle tf(C) \rangle + tfAD \tag{3}$$

3.2 Experiential Metrics and Their Heuristics

3.2.1 Request Chain Length (RCL)

DDoS attack is based on sending numerous requests to the Web server. However, Web sites can also be attacked by sending huge HTTP requests, and same can happen when legitimate users send request that are much larger than expected size. These cause memory issues on the server, and this threat is mitigated by defining the length of each requests in a given time frame for N (normal) as CS_N and D (DDoS attack) as CS_D. In (normal set) CS_N, the sequence of transactions are partitioned into time frames as $TS(CS_N) = \{ts_1, ts_2, \ldots ts_{|TS(CS_N)|}\}$ where ts_i contains the transactions perceived in the ith time frame and the length of each time frame is tf (see Sect. 3.1).

- The average chain length of transactions detected in all time frames of CS_N is calculated as follows:

$$\langle TS(CS_N) \rangle = \sum_{i=1}^{|TS(CS_N)|} \{|ts_i| \exists ts_i \in TS(CS_N)\} \tag{4}$$

- The request chain length absolute deviation ($clAD$) for all time frame groups of CS_N is calculate as follows:

$$clAD = \frac{\sqrt{\sum_{i=1}^{|TS(CS_N)|} (\langle TS(CS_N) \rangle - (|ts_i| \exists ts_i \in TS(CS_N)))^2}}{|TS(CS_N)|} \tag{5}$$

- The request chain length (RCL) is calculated as the sum of the average chain of transactions $\langle TS(CS_N) \rangle$ and RCL absolute deviation ($clAD$) observed in all time frames of CS_N

$$rcl(CS_N) = \langle TS(CS_N) \rangle + clAD \tag{6}$$

3.2.2 Ratio of Packet Count

The DDoS attacks generate extremely huge volumes of traffic flows in order to block the target. Packet count is one of the important properties used for every flow and used to detect spoofed attacks, some non-spoofed DDoS attacks and, most essentially, it act as an indicator of similarity between unrelated flows. For the collected transactions considered for training, the number of packets observed for every time frame is used to find the ratio of packet count under the sequence of requests in N (normal) CS_N and D (DDoS attack) CS_D. The ratio of packets for each time frame ts_i $\{ts_i \exists ts_i \in TS(CS_N) \wedge i = 1, 2, \ldots |TS(CS_N)|\}$ is calculated as

$$rp(ts_i) = \sum_{i=1}^{|TS(CS_N)|} \frac{|P(ts_i)|}{\sum_{k=1}^{|TS(CS_N)|} |P(ts_k)|} \tag{7}$$

The time frame level packets support absolute deviation (*tflpsAD*) for all time frames in CS_N is calculated as

$$tflpsAD = \frac{\sqrt{\sum_{j=1}^{|TS(CS_N)|} \left(1 - rp\left(ts_j\right)\right)^2}}{|TS(CS_N)|} \tag{8}$$

Finally, the ratio of packet count of CS_N is defined as the sum of average of all time frames ratio of packets and time frame level packets support absolute deviation.

$$rpc(TS(CS_N)) = \frac{\sum_{i=1}^{|TS(CS_N)|} rp(ts_i)}{|TS(CS_N)|} + tflpsAD \tag{9}$$

3.2.3 Ratio of Intervals Between Requests

From the group of transactions collected for training, the intervened time between continues requests of the same session in sequence are measured as access time. For all pairs of requests of the same session in sequence of each time frame is measured to assess the ratio of intervals between requests as normal and DDoS attack.

For each time frame local mean $lm(ts_i)$ of the intervals between requests describes the ratio of intervals between requests for ts_i $\{ts_i \exists ts_i \in TS(CS_N)\}$. For each time frame ts_k $\{ts_k \exists ts_k \in TS(CS_N) \wedge k = 1, 2, 3, \ldots |TS(CS_N)|\}$ local mean is calculated as the average of set of intervals observed in the time frame tf_k.

$$lm(ts_k) = \frac{\sum_{j=1}^{|I(ts_k)|} i_j}{|I(ts_k)|} \tag{10}$$

The global mean $gm(CS_N)$ of the intervals of time frames defines the ratio of intervals between requests in CS_N. For the dataset CS_N, global mean is calculated as the average set of intervals observed in all the time frames of CS_N.

$$gm(CS_N) = \frac{\sum_{j=1}^{|I(CS_N)|} i_j}{|I(CS_N)|} \tag{11}$$

The interval absolute deviation (iAD) of all time frames in CS_N is calculated as

$$iAD = \frac{\sqrt{\sum_{j=1}^{|TS(CS_N)|} \left(gm(CS_N) - lm(ts_j) \right)^2}}{|TS(CS_N)|} \qquad (12)$$

Finally, the ratio of interval for the training set CS_N is defined as the sum of average of all intervals observed in CS_N and interval absolute deviation (iAD).

$$ri(CS_N) = \frac{\sum_{i=1}^{|TS(CS_N)|} lm(ts_i)}{|TS(CS_N)|} + iAD \qquad (13)$$

3.2.4 Ratio of Packet Types in Fixed Time Frame

The ratio of packet types of a given time frame under the sequence of requests is defined as the packet ratio threshold for different packet types such as HTTP, FTP, FTP-DATA, SMTP, POP3, DNS and SSL under N (normal) as CS_N and D (DDoS attack) as CS_D transactions.

The local support of divergent packets of ts_k $\{ts_k \exists ts_k \in TS(CS_N) \wedge k = 1, 2, 3, \ldots |TS(CS_N)|\}$ of CS_N defines the threshold for each packet type in sequence of all the requests observed in ts_k.

Let PT be the packet types, $PT = \{pt_1, pt_2, \ldots pt_{|PT|}\}$ and $P(ts_i) = \{p_1, p_2, \ldots, p_{|P(ts_i)|}\}$ are the packets observed in the time frame ts_i of CS_N. For each time frame ts_i $\{ts_i \exists ts_i \in TS(CS_N)\}$ and for each packet type pt_j $\{pt_j \exists pt_j \in PT \wedge j = 1, 2, \ldots, |PT|\}$ of tf_i, the number of occurrences of each dissimilar pt_j is calculated as

$$c_{ts_i}(pt_j) = \sum_{k=1}^{|P(ts_i)|} \left\{ 1 \exists t(p_k) \equiv pt_j \right\} \qquad (14)$$

Finally, the local support for packet types of tf_i is calculated as

$$ls_{ts_i}(pt_j) = \frac{c_{ts_i}(pt_j)}{|P(ts_i)|} \qquad (15)$$

The global support of divergent packets in the given training set CS_N defines the threshold values for each packet type observed in the sequence of all time frames in CS_N. For each packet type pt_j $\{pt_j \exists pt_j \in PT \wedge j = 1, 2, \ldots, |PT|\}$ observed in sequence of all time frames in CS_N, the global support is calculated as

$$gs(pt_j) = \frac{c(pt_j)}{\sum_{k=1}^{|TS(CS_N)|} |P(ts_k)|} \tag{16}$$

Packet type support absolute deviation (*ptsAD*) between local support and global support for each packet type pt_j $\{pt_j \exists pt_j \in PT \wedge j = 1, 2, \ldots, |PT|\}$ is observed in sequence of all time frames in CS_N which is calculated as follows:

$$ptsAD(pt_j) = \frac{\sqrt{\sum_{i=1}^{|TS(CS_N)|} \left(gs(pt_j) - ls_{ts_i}(pt_j)\right)^2}}{|TS(CS_N)|} \tag{17}$$

Finally, ratio of pt_j, $\left(rpt(pt_j)\right)$ is measured as the sum of average threshold of packet type and packet type support absolute deviation (*ptsAD*) observed in sequence of all time frames of training set CS_N.

$$rpt(pt_j) = \frac{\sum_{i=1}^{|TS(CS_N)|} ls_{ts_i}(pt_j)}{|TS(CS_N)|} + ptsAD(pt_j) \tag{18}$$

3.2.5 Request Chain Context or Order of Requests

From group of collected transactions for training, we isolate the order of requests of time frame that are normal and DDoS attack. In training transaction set CS_N, formulate request pair set $rps_N = \{p_1, p_2, \ldots, p_{|rps_N|}\}$ as pair p_i is the neighboring two requests in sequence of CS_N. Local support $ls_{ts_j}(p_i)$ of p_i $(p_i \exists p_i \in rps_N \wedge i = 1, 2, \ldots, |rps_N|)$ is the occurrences of pair p_i in ts_j. Global support $gs(p_i)$ of each pair p_i $(p_i \exists p_i \in rps_N \wedge i = 1, 2, \ldots, |rps_N|)$ is indicating the number of occurrences of a pair in all chain of requests observed in the training dataset CS_N.

Request pair support absolute deviation (*rpsAD*) of the local supports of pair p_i $(p_i \exists p_i \in rps_N \wedge i = 1, 2, \ldots, |rps_N|)$ from global support is defined as

$$rpsAD = \sqrt{\frac{\sum_{j=1}^{|TS(CS_N)|} \left(gs(p_i) - ls_{ts_j}(p_i)\right)^2}{|TS(CS_N)|}} \tag{19}$$

The request chain context $rcc(p_i)$ of $(p_i \exists p_i \in rps_N \wedge i = 1, 2, \ldots, |rps_N|)$ is measured as the sum of mean of each pair support value, and request pair support absolute deviation (*rpsAD*) of all pairs is observed in sequence of all time frames of training set CS_N.

$$rcc(p_i) = \frac{\sum_{j=1}^{|TS(CS_N)|} ls_{ts_j}(p_i)}{|TS(CS_N)|} + rpsAD \tag{20}$$

4 Experimental Results and Performance Analysis

The experiments are carried out for assessing detection accuracy, scalability, robustness, and process complexity of the proposed model ARTP.

4.1 Experimental Results

The LLDOS 2.0.2 [13, 14] is used for App-DDoS attack requests for normal and attack scenario. The total number of transactions considered for experiments was 216,150 which includes N (normal) and D (DDoS attack). The total transactions are partitioned for training and testing into 60 and 40%, respectively.

Each metric is calculated individually on the dataset CS which includes N (normal) as CS_N and D (DDoS attack) as CS_D, and its detection accuracy is assessed. The total number of transactions in CS_N is 123,750 in which 60% of transactions, i.e., 74,250 are considered for the training process and 40% of transactions, i.e., 49,500 for the testing process. Training and testing datasets of CS_N are converted into sessions with each session size of 60 s. Then, K-Means algorithm is applied on the training and testing sessions separately to prepare clusters. The same process is repeated for the attack dataset CS_D to convert into sessions and prepare the clusters for training and testing sets. Further, the stream of sessions is processed as time frames. The time frame length for the CS_N and CS_D of training and testing is evaluated individually (see Sect. 3.1) (Table 1).

a. **Request chain length (RCL)**

For the transactions, CS_N and CS_D of training, request chain length is measured as possible length of requests and is defined as the average length of requests frequently observed from the clients in a time frame which is labeled as an attack or normal. Details are shown in Table 2.

b. **Ratio of packet count**

For transactions in training, the ratio of packet count is identified as the number of packets received for each time frame under the chain of requests as CS_N and CS_D. In the training and testing process of CS_N and CS_D sets, we observed the count of different packet types such as HTTP, FTP, FTP-DATA, SMTP, POP3, DNS and SSL and measured the ratio of packet in sequence of all time frames in CS_N and CS_D. The values are given in Table 2.

c. **Ratio of intervals between requests**

For transactions labeled as CS_N and CS_D, the access time is considered to be the elapsed time between the sequences of requests in the same session. The access time in all pairs of requests in same session of each time frame will be considered to assess the ratio of intervals between requests. From the given training sets of CS_N and CS_D,

Table 1 Training and testing dataset details and the time frame length for normal and DDoS attack

The total number of transactions (CS)				216,150
		Training (60%)	Testing (40%)	Total
N (Normal) CS_N	Transactions	74,250	49,500	123,750
	Sessions	2772	1848	4620
	Clusters	277	184	461
	Time frame length (tf)	606	615	1221
	Number of time frames	232	156	388
D (DDoS Attack) CS_D	Transactions	55,440	36,960	92,400
	Sessions	2448	1632	4080
	Clusters	243	163	406
	Time frame length (tf)	734	768	1502
	Number of time frames	254	176	430

Table 2 Values for metrics request chain length (RCL) and ratio of intervals between requests, ratio of packet count and packet count of different types of packets observed in CS_N and CS_D

Packet types	N (normal) CS_N packet count		D (DDoS Attack) CS_D packet count	
	Train	Test	Train	Test
HTTP	33,117	20,470	41,088	29,106
FTP	10,146	9798	7440	6210
FTP-DATA	5957	3427	2208	882
SMTP	17,813	12,397	3936	1746
POP3	7068	4462		
DNS	285	207		
SSL	143	138		
SSH	57	23		
Ratio of packet count	0.667	0.607	0.434	0.407
Request chain length (RCL)	21.254	20.735	29.408	28.10
Ratio of intervals between requests	0.251	0.240	0.123	0.115

we observed different time intervals 11 and 14, respectively and the ratio of intervals between requests for testing and training of CS_N and CS_D is given in Table 2.

d. *Ratio of packets types*

The packets of type observed in a time frame for the chain of requests as CS_N and CS_D of training and testing process are measured to assess the ratios used to define the threshold for each packet type like HTTP, FTP, FTP-DATA, SMTP, POP3, DNS, and SSL. From the training sets of CS_N and CS_D, ratio of packet types thresholds for different packet types that are most commonly used packet types in application layer are given in Table 3.

4.2 Performance Analysis

The statistical metrics such as precision, sensitivity (true positive rate), specificity (true negative rate), accuracy, and F-measure [15] are used to assess to detection accuracy of ARTP. The statistics of ARTP process are discovered in Table 5. The precision specifies predicted the requests as normal which are actually normal or predicted the attack as attack which are actually attack. Sensitivity detected from the experiments directs true positives that are predicted as positive. Specificity indicates the true negatives predicted as negatives. Accuracy is the rate of classifying the requests accurate. The value perceived for sensitivity is greater than the value observed for specificity, which specifies that the chance of an attack scenario recognized as normal is much smaller than the probability of a normal situation recognized as an attack. The rate of classifying an attack as normal requests is 0.0145 (1-sensitivity), and the rate of classifying a normal requests as an attack is 0.0859 (1-specificity), which is much superior to the earlier. Hence, it is evident to estimate that the proposed ARTP is further delicate toward attack detection and the statistical performance metrics and observed values are given in Table 4.

Jyothsna and Prasad proposed FAIS [15] and FCAAIS [16] for detecting DDoS attacks. The experiments are conducted on the same dataset and results are indicating that these models are also scalable and robust towards forecasting the DDoS attacks scope of a network transaction (observed detection accuracy is approx. 91%), but the major obstacle observed these models are that compared to the proposed model is process complexity, which influence the statistical metrics defined for measuring the performance. As per these results, the accuracy of our proposed model ARTP is improved the performance that compared to FAIS, FCAAIS and also retains the maximum prediction accuracy which is shown in Table 5 and Fig. 1.

Table 3 Values for the metric, ratio of packet types for various types of packets

Packet types	Ratio of packets types (see Sect. 3.2.4)					
	N (normal) CS_N		Test	D (DDoS Attack) CS_D		Testing
	Training			Training		
	Lower limit	Upper limit		Lower limit	Upper limit	
HTTP	0.4285	0.45	0.37	0.69	0.74	0.76
FTP	0.1249	0.14	0.14	0.11	0.14	0.11
FTP-DATA	0.0726	0.08	0.06	0.02	0.04	0.03
SMTP	0.2267	0.25	0.22	0.052	0.08	0.06
POP3	0.0883	0.10	0.09			
DNS	0.0026	0.003	0.02			
SSL	0.0008	0.002	0.02	0.03	0.05	0.47

Table 4 Performance metrics for ARTP and observed values

Total number of records considered for training and testing					216,150
The total number of records used in training					129,690
Total number of records tested					86,460
tp (True positive)	The transactions identified as normal, which are actually normal				36,531
fp (False Positive)	The number of transactions identified as intruded, which are actually normal				4244
tn (True Negative)	The number of transactions identified as intruded, which are actually intruded				45,144
fn (False Negative)	The number of transactions identified as normal, which are actually intruded				541
Precision	$\frac{tp}{tp+fp}$	0.897	Accuracy	$\frac{(tp+tn)}{(tp+tn+fp+fn)}$	0.944
Recall/sensitivity	$\frac{tp}{tp+fn}$	0.985	F-Measure	$2 \times \frac{(precision*recall)}{(precision+recall)}$	0.938
Specificity	$\frac{tn}{fp+tn}$	0.914			0.914

Table 5 Comparison of proposed approach ARTP with FAIS and FCAAIS

	ARTP	FAIS	FCAAIS
Precision	0.895	0.889	0.869
Sensitivity/recall	0.985	0.935	0.942
Specificity	0.914	0.485	0.894
Accuracy	0.944	0.851	0.917
F-Measure	0.938	0.911	0.855

Fig. 1 Comparison of ARTP, FAIS, and FCAAIS

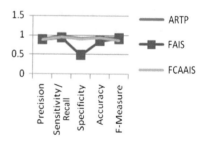

5 Conclusion

This paper has illustrates an experiential metrics-based machine learning approach for detecting and preventing anomaly based real time App-DDoS attacks for Web applications to defend App-DDoS attacks in an efficient manner. The contribution of this paper is disseminated into three levels. First contribution classifies the feature metrics to the request stream to analyze the intension is attack or not. Further, the experimental threshold values calculated for the defined metrics in ARTP were used to discover the behavior of stream is a flood or not. The performance of the ARTP which is tested is on third contribution using benchmarking LLDoS dataset. The metrics developed in ARTP enhanced the detection accuracy with nominal process complexity and most speed. The investigation of the results concludes that the metrics devised from training records dataset to determine the state of request stream are promising and significant. Further, in an absolute time interval the request stream is estimated as App-DDoS attack or not based upon the threshold values calculated on the training dataset. ARTP is analyzed to be strong with minimal process complexity and maximal speed. Hence, the model developed in this paper has expressively retains the maximal prediction accuracy and minimized the computational overhead.

References

1. K. Munivara Prasad, A.R.R. Reddy, K.V.G. Rao, DoS and DDoS attacks: defense, detection and traceback mechanisms—a survey. Global J. Comput. Sci. Technol. **14**(7-E) (2014)
2. S.M. Lee, in *Distributed Denial of Service: taxonomies of Attacks, Tools, and Countermeasures*, Proceedings of the international workshop on security in parallel and distributed systems (San Francisco, 2004), pp. 543–550
3. S. Byers, A.D. Rubin, D. Kormann, Defending against an internet based attack on physical world. ACM Trans. Internet Technol. 239–254 (2004)
4. J.M. Estevez-Tapiador, P. García-Teodoro, J. Díaz-Verdejo, in *Detection of Web-Based Attacks Through Markovian Protocol Parsing*, 10th IEEE symposium on computers and communications (2005), pp. 457–462
5. V. Jyothsna, V.V.R. Prasad, A review of anomaly based intrusion detection systems. Intern. J. Comput. Appl. 26–35 (2013)
6. T. Yatagai, T. Isohara, I. Sasase, in *Detection of HTTP-GET Flood Attack Based on Analysis of Page Access Behaviour*, Proceedings IEEE Pacific RIM conference on communications, computers, and signal processing (2007), pp. 232–235
7. S.S. Sindhu, Decision tree based light weight intrusion detection using a wrapper approach. Expert Syst. Appl. 129–141 (2012)
8. A. Shevtekar, N. Ansari, Is it congestion or a DDoS attack? IEEE Commun. Letters 546–548 (2009)
9. S. Kandula, D. Katabi, M. Jacob, A. Berger, in *Botz-4-Sale: surviving Organized DDoS Attacks That Mimic Flash Crowds*, Proceedings of the 2nd conference on symposium on networked systems design & implementation (2005), pp. 287–300
10. C. Katar, Combining multiple techniques for intrusion detection. Intern. J. Comput. Sci. Netw. Secur. 208–218 (2006)
11. Y. Xie, S.Z. Yu, A large-scale hidden semi-markov model for anomaly detection on user browsing behaviors. IEEE/ACM Trans. Netw. 54–65 (2009)
12. J.A. Hartigan, Algorithm AS 136: "A k-means clustering algorithm". J. Roy. Stat. Soc.: Ser C (Appl. Stat.) 100–108 (1979)

13. M.I. MIT, in *Darpa Intrusion Detection Evaluation*. Retrieved from Lincoln Laboratory: https://www.ll.mit.edu/ideval/data/1998data.html
14. D.M. Powers, in *Evaluation: from Precision, Recall and F-measure to ROC, Informedness, Markedness and Correlation*, 23rd international conference on machine learning (Pitsburg, 2006)
15. V. Jyothsna, V.V. Rama Prasad, Anomaly based network intrusion detection through assessing feature association impact scale (FAIS). Intern. J. Inform. Comput. Secur. (IJICS) (*in forthcoming article). Inderscience (2016)
16. V. Jyothsna, V.V. Rama Prasad, *FCAAIS: anomaly based network intrusion detection through feature correlation analysis and association impact scale* (ICT Express, The Korean Institute of Communications Information Sciences, Elsevier, 2016)

Lifetime Enhancement of a Node Using I—Leach Protocol in WSN

D. Venkatesh and A. Subramanyam

Abstract The developments in the place of wireless networks and their integration in unique sectors effectively caused the emergence of a new own family of networks known as wireless sensor networks. The various demanding situations the maximum studied in WSN are the energy intake and the life of the nodes. An excellent node placement deploying is one of the issues which may be exploited to reap the maximum most effective layout for saving the strength. This paper proposes improved leach (I—leach) protocol which increases network lifetime, and it will improve the quality of QoS. The experimental results show that the proposed technique is giving better results in terms of energy consumption than existing techniques.

Keywords Wireless sensor · Lifetime · Clusters · k-means · Power

1 Introduction

Latest advances have allowed the improvement of increasingly more effective sensors that may now manipulate complex data like multimedia. This evolution makes the Wi-fi sensor one of the main additives of the net of things. It will become possible for network users to study and engage continuously with a wide variety of objects of their environment. Wireless sensor networks (WSNs) are ad hoc Wi-fi, huge scale networks deployed to measure the parameters of the surroundings and offer a feedback to at least one or more locations. The drift of facts from sensors will

D. Venkatesh (✉)
Department of Computer Science and Engineering, Rayalaseema University,
518 007, Kurnool, Andhra Pradesh, India
e-mail: dvvenkatesh@yahoo.co.in

A. Subramanyam
Department of CSE, Annamacharya Institute of Technology and Sciences,
Rajampet, Kadapa, Andhra Pradesh, India
e-mail: smarige@gmail.com

© Springer Nature Singapore Pte Ltd. 2018
S. S. Dash et al. (eds.), *Artificial Intelligence and Evolutionary Computations in Engineering Systems*, Advances in Intelligent Systems and Computing 668,
https://doi.org/10.1007/978-981-10-7868-2_11

generate a tremendous boom in bandwidth requirements, particularly if those sensors accumulate statistics in actual time, providing users a fixed of autonomy and direct interaction with the surroundings. However, sensors show off a massive wide variety of constraints, especially due to hardware boundaries. They are characterized by low capability in phrases of reminiscence, processing and battery. Wireless sensors are typically small and feature four simple factor devices: a sensing unit, a processing unit, a communiqué unit and a battery. The processing unit is often related to a small garage unit wherein all statistics captured by the sensing unit are amassed. The processing unit analysis approaches and transmits records to the communiqué unit. The communiqué unit typically has a radio transceiver in price of emitting and receiving information produced through the processing unit and allowing sensors to speak with different community compo-nents. Each sensor has a battery to power its additives; however, because of its tiny size, the energy will become a crucial resource for the reason that battery lifestyles directly influence that of sensors. Those characteristics lead to the development of protocols primarily enabling self-organization and decreasing energy consumption. With the emergence of critical packages for the WSNs, new wishes rise up. Several procedures are provided to compute the top-rated direction in multi-hop routing protocols. Power is the most important resource for wireless sensors, mainly in environments in which changing or recharging a sensor's batteries is not possible. Therefore, energy efficient routing protocol is the principle goal for wireless sensor networks (WSNs). However, in lots of current applications of WSNs, which include forest hearth detection, records need to be transmitted from assets to a sink within a restricted quantity of time for the facts be beneficial. For this reason, an exchange-off exists among minimizing electricity consumption and minimizing end-to-cease put off. Positive proposals are based totally on the shortest path in terms of distance to the bottom station (BS). Others are primarily based on the energy level in the nodes along the path by means of favouring the nodes with maximal quantities of strength. Others nonetheless choose a finest direction with the aid of favouring a few nodes whose presence in the path allows eating less energy. We are inquisitive about constraints imposed by routing and management of strength in the WSNs to extend their lifetime. As shown in Fig. 1, the sensor node works.

2   Related Work

In the wireless sensor networks, the energy present at each node is limited. As soon as the power of specific sensor node drops down the nodes are not rechargeable and run at lots powered batteries, the lifetime of the sensor node also decreases. This sort of constraints routinely reduces the overall performance of sensor nodes, and the conversation postpone may additionally occur with the aid of this. It is able to be prevented with the aid of many strategies consisting of k-method facts relay clustering algorithms [1], distributed extraction algorithm and Novel affiliation rule

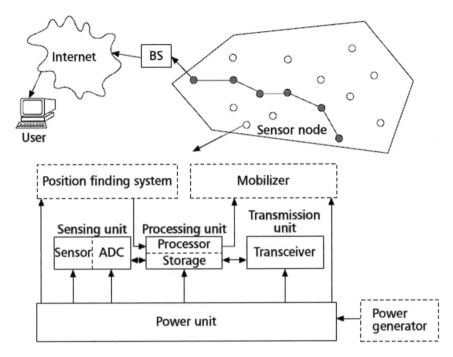

Fig. 1 Mechanism of a sensor node [9]

mining [2], records aggregation strategies [3], etc. Large location insurance is likewise one among numerous troubles of Wi-fi sensor community. Sensor network is a huge ranged network in which large range of nodes distributed autonomously. Due to the large disbursed location, there can also occur many troubles like hyperlink failure, more bandwidth usage, malicious attacks and so on. These types of issues are addressed by way of the usage of many strategies such as prediction strategies [4]. The problem of occasion detection has been tackled from distinctive perspectives. The primary idea of event detection can be best described by a threshold value. An alarm is generated while sensor readings decreases and reaches to a predefined threshold values. Those kind of problems are addressed by the usage of many techniques inclusive of guide vector system and category [5], reputation based totally voting, selection tree class method and device getting to know [6]. The sensor nodes that have less energy are treated as outlier sensor nodes and will be discarded. Class strategies [7] are an assignment of assigning new item into a class of predefined item classes. That is found out the usage of the set of training records and classifies new statistics into one of the learned class. Type is properly idea-out as a prevalence of supervised learning, i.e. getting to know wherein a training set of nicely identified annotations is offered. By using the supervised learning strategy it is easy to identify and reduce the nodes with less energy. There is a particular fact named "energy hole" in sensor networks [8] which leads to reduce the efficiency of energy value.

Little physical size and low power utilization: At any point in mechanical advancement, size and power compel the preparing, stockpiling and interconnect capacity of the essential gadget. It is understood that energy efficiency plays a vital role in WSN. In like manner, the product must make proficient utilization of processor and memory while empowering low power correspondence.

Simultaneousness serious operation: The essential method of operation for these gadgets is to stream data from place to put with a humble measure of preparing on the fly, as opposed to acknowledge a charge, stop, think and react. For instance, data might be at the same time caught from sensors, controlled and gushed onto a system. On the other hand, information might be gotten from different hubs and sent in multi-jump directing or crossing over circumstances. There is minimal interior stockpiling limit, so buffering a lot of information between the inbound and the outbound streams is ugly. In addition, each of the streams for the most part includes countless level occasions interleaved with more elevated amount preparing. A portion of the abnormal state handling will reach out finished various constant occasions.

Constrained physical parallelism and controller hierarchy: The quantity of free controllers, the abilities of the controllers and the complexity of the processor-memory-switch level interconnect are much lower than in traditional frameworks. Commonly, the sensor or actuator gives a primitive interface specifically to a solitary chip smaller scale controller. Conversely, regular frameworks convey the simultaneous preparing related to the accumulation of gadgets over various levels of controllers interconnected by an intricate transport structure. Space and power requirements and constrained physical configurability on chip are probably going to drive the need to help simultaneousness serious administration of moves through the installed microchip.

Differing qualities in design and usage: Networked sensor gadgets will have a tendency to be application particular, as opposed to universally useful, and convey just the accessible equipment bolster really required for the application. As there is an extensive variety of potential applications, the variety in physical gadgets is probably going to be vast. On a specific gadget, it is imperative to effectively gather only the product parts required to incorporate the application from the equipment segments. In this manner, these gadgets require an irregular level of programming measured quality that must likewise be extremely effective. A non-specific advancement condition is required which enables particular applications to be developed from a range of gadgets without heavyweight interfaces. In addition, it ought to be normal to move segments over the equipment/programming limit as innovation advances.

Powerful operation: These gadgets will be various, to a great extent unattended, and anticipated that would shape an application which will be operational a huge rate of the time. The utilization of conventional repetition systems to improve the unwavering quality of individual units is restricted by space and power. In spite of the fact that repetition crosswise over gadgets is more appealing than inside gadgets, the correspondence taken a toll for traverse is restrictive. Along these lines, improving the dependability of individual gadgets is basic. Moreover, we can expand the dependability of the application by enduring individual gadget disappointments. Keeping that in mind, the working framework running on a solitary hub ought to be hearty, as well as ought to encourage the advancement of dependable conveyed applications.

3 Proposed System

The proposed scheme is an improved leach protocol. Here, it is a clustering algorithm. In this technique, we are using improved k-means algorithm. Hence, cluster head and cluster formation modified. Here, a multi-hop routing strategy is used to forward data to the base station (BS). Clustering is one of the basic approaches to for designing energy efficient distributed WSNs. Clustering can be performed in centralized way by BS or in distributed way where each node decides separately about its position. Cluster configuration can be either static or dynamic according to whether the network is varied or homogenous. Since energy is the major anxiety, balanced energy consumption is important in energy conservation. The process of the future scheme is divided into rounds; each round consists of set-up phase and data transmission phase. The cluster formation, CHs choice and multi-hop paths established order are accomplished successively in set-up section. While in statistics transmission segment, the nodes feel and transmit facts to CHs and then CHs forward data to BS after aggregation. In LEACH, the cluster heads are and then participants joint their nearest CHs to construct clusters in set-up phase improvement in terms of electricity consumption, and it is vital in scaling the networks. Cluster formation is done by using the improved k-means algorithm as in Fig. 2.

CH selection primarily based on the remaining strength, percentage of CHs in the network and the variety of times the node has been CH. In the proposed algorithm, the CH choice equation of LEACH was prolonged with the aid of placing the residual power and the distances to BS and cluster centroid. In statistics transmission section each body all nodes send statistics to their CHs or BS. The CHs open their radios to acquire statistics from nodes. Everyday nodes are placed into sleep mode, and each one opens its radio in its personal time slot. Each node saves its residual strength when it is sending facts to CH or to the BS. While any cluster head dies, the BS informs the network nodes approximately the beginning of the cluster reformation step. Then, the information transmission segment is terminated and the complete system actions to the set-up phase. Because while some nodes die, the total number of nodes decreases and the distribution of the nodes modifications, consequently, the range of cluster heads should be adjusted to keep the adaptively of the set of rules. The device continues these rounds until each node's strength has been depleted.

4 Simulation

For simulation environment, 100 nodes are deployed randomly over an area of
100 × 100 m, and the BS is located in the sensing field. The control and data
message lengths are 200 and 6400 bits, respectively.

Figure 3 shows the energy over rounds. X axis represents no of rounds, and Y
axis represents energy. It is clearly observed that proposed technique uses less
energy than existing technique.

Figure 3 shows the comparison of energy and no of rounds. In Fig. 4, we try to
give the comparison of energy percentage with no of days.

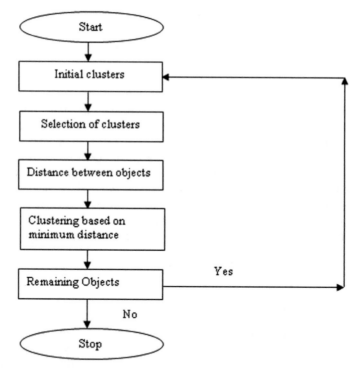

Fig. 2 Cluster formation

Fig. 3 Energy utilization

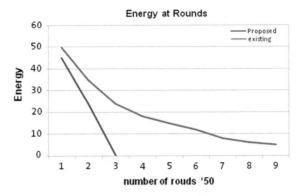

Fig. 4 Energy in system

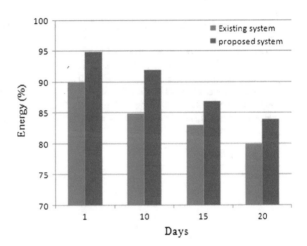

5 Conclusion

The number one intention of protocols designed for sensor networks has been electricity performance a good way to extend the community lifetime and make WSN applications economically feasible. With the arrival of new sophisticated programs of sensor networks, WSN customers want protocols which might be self prepared; self recovery; fault tolerant; latency green; have predictable overall performance; can adapt to topology adjustments and varying site visitors necessities. This paper proposes a new protocol which enhances existence time of node in wireless sensor network.

References

1. N.K. Jassi, S.S Wraich, An enhanced K-means clustering technique with Hopfields artificial neural network based on reactive clustering protocol, in *International Conference—Confluence the Next Generation Information Technology Summit*-2014 978-1-4799-4236-7/14/IEEE
2. A. Das, G. Das, Distributed extraction and a novel association rule mining mechanism for WSN: an empirical analysis. ICETACS-2013,978-1-4673-5250-5/13/2013-IEEE
3. C. Zhao, W. Zhang, Y. Yang, S. Yao, Treelet-based clustered compressive data aggregation for wireless sensor networks. IEEE Trans. Veh. Technol. 0018-9545© 2013
4. M.B. Thaker, U. Nagaraj, P.D. Ganjewar, Data reduction techniques in wireless sensor network: a servey. Int. J. Innovative Res Comput Commun. Eng. **2**(11) (2014)
5. Y. Singh, S. Saha, U. Chugh, C. Gupta, Distributed event detection in wireless sensor networks for forest fires, *15th International Conference on Computing Modelling and Simulation*-2013 978-0-7695-4994-1/13/IEEE
6. S. Rashid, U. Akram, S. Qaisar, S.A. Khan, E. Felemban, Wireless sensor network for distributed event detection based on machine learning, in *IEEE International Conference on Internet of Things, Green Computing and Communications and Cyber-Physical-Social Computing*-2014,978-1-4799-5967-9/14/IEEE
7. G. Sahni, S. Sharma, Study of various anomalies detection methodologies in wireless sensor network. Int. J. Adv. Res. Comput. Sci. Softw. Eng. **3**(5), May 2013, ISSN:2277128X
8. X. Liu, A novel transmission range adjustment strategy for energy hole avoiding in wireless sensor networks. J. Netw. Comput. Appl. **67**, 43–52 (2016)
9. I.F. Akyildiz et al., A survey on sensor networks. IEEE Commun. Mag. **40**(8), 102–114 (2002)

Self Regulating Power Saving System for Home Automation

**Nesarani Abraham, Ramalakshmi Ramar
and Jayalakshmi Ramachandran**

Abstract More energy is required by the home appliances like light, air conditioners, or fan (heating, cooling) that are physically kept under control, which leads to power consumption. By using smart automatic controlling system, we can save some amount of power. In this paper, we are preventing the consumption of energy in an adequate and cost-effective manner. We proposed a system for automatic control of appliances as well as power reduction module. When the user enters into the room, the light and fan will turn ON automatically, and while the user exits, it will turn OFF automatically using PIR sensor. Based on the presence of the user, the power reduction module switches the appliances in the span. It also controls the speed of fan based on room temperature with LM-35 sensor. By using ACS712 Current Sense Module, we control the current flow. And also we can control the appliances through mobile when the user is not at home. All the data can be stored in cloud for future reference.

Keywords Arduino Nano · GSM module · PIR sensor · LM-35
ACS712 · Solid-state relay

1 Introduction

Internet of Things (IoT) is internetworking of mobile, residence, and fixed device that are linked to the Internet, integrating computing abilities and using data analytics to extract expensive information. All kinds of net-connected machines with conversation to one another are known as Internet of Things.

N. Abraham (✉) · R. Ramar · J. Ramachandran
Department of CSE, Kalasalingam University, Krishnankoil, Tamil Nadu, India
e-mail: a.nesarani@klu.ac.in

R. Ramar
e-mail: rama@klu.ac.in

J. Ramachandran
e-mail: jayacse1@gmail.com

© Springer Nature Singapore Pte Ltd. 2018
S. S. Dash et al. (eds.), *Artificial Intelligence and Evolutionary Computations
in Engineering Systems*, Advances in Intelligent Systems and Computing 668,
https://doi.org/10.1007/978-981-10-7868-2_12

The IoT applications provides services such as security, energy saving, automation, and communication with integrated shared user interface.

There is increasing public awareness about the changing paradigm in energy supply, infrastructure, and consumption. The power saving is based on improved user awareness of energy consumption in future energy management. The smart meter can give information about the instant energy consumption to the user, allowing for identification and elimination of energy wastage by devices for energy consumption.

Smart home can be incorporated within advanced intelligent systems and internal network to provide the inhabitants with sophisticated monitoring and control. The intelligent network can be built through wired or wireless communication techniques between actuators and sensors. The Internet control defines the entire house being monitored by Internet services. It can be used in industries, home, and offices anywhere. In home automation, we can decide how a device should react, when it should react, and why it should react. It may involve control and automation appliances like illumination, heat, exposure to air, security locks for doors, gates, and other systems.

The objective of this paper is to save the power at places like homes, library where lighting is very important for the people. The user has to switch off all the lights, fans, and electrical equipment when they leave the home to save power, avoid short circuit, etc. But sometimes, we forget to switch off them, for that we have to come back home to switch off. This provides wastage of time, power, money, and it creates a lot of anxious feelings. To avoid such type of situation, the latest technology coming up worldwide is the smart home.

2   Related Works

2.1   Existing Home Automation Technologies

The IoT allows real-world devices to be linked with each other and communicate between people and objects like smartphone. Some commercial smartphones have invisible spectrum modules; we are able to use those modules to control smart house.

It allows users to control an electric tool by positioning the smartphone to it, the phone's screen can automatically blow up the control panel of that device that the user can manage instructions directly using wireless communication technology [1].

Smart home model is a term which combines several parts of information and communication technology, where it allows controlling of any devices through voice reorganization technology [2]. They used WLAN 802.11/b to control the home appliances through speech recognition software.

The home appliances can be controlled through voice and touch system using Bluetooth module [3]. The devices can be controlled by fingerprinting process, after the double verification to grant access to the home [4]. On the other hand, we can remotely control the appliances by sending commands through email; that time Internet should be active at home and user end [5]. User with login user id and password causes the time delay, and due to many spam messages, the emergency mail could not be viewed properly.

The devices can be controlled and monitored by using a Web portal and same can be used to control in an android smartphone for any emergency from remote places using Internet protocol [6].

2.2 Analysis of the Existing System

The existing system may have the

- Lack of Internet interoperability—Home appliance control will fail due to irregular Internet connectivity between end user and home.
- Complex and scalability—The existing system incorporates the network management with remote access and monitoring system that works in the short range.
- User Movement: The majority of the system depends on motion sensing of the user where the adjustment of user requirement is not possible.The system also fails if there is less user movement.
- Cost of Technology—One more difficulty is the price of wireless network.

2.3 Feature of Proposed Work

The proposed work is stand-alone, flexible, and low-cost PIR, GSM-based home automation. This method allows consumer to control and observe the connected device automatically when the user enters into the home and vice versa. Consumer may distantly control and monitor their home using GSM module. The most important aim of this paper is to avoid consumption of power.

3 Proposed Work

3.1 Smart Home Energy Management

In smart home, many appliances can participate in the efficient usage of electric energy through thermal loads, electric vehicle, light, fan, and air conditioning. We can control those appliances by using intelligent controlling system, and it can provide leverage for energy and cost saving.

(1) Lighting

Smart home works when the user enters and exits the room. This helps to preserve energy and avoid power wastage, the lighting can be switched ON and OFF automatically with PIR sensor.

(2) Air Conditioning (Fan)

The fan speed is automatically adjusted with temperature sensor based on the room temperature. It reduces the usage of electricity and hence reduces the money.

(3) Smart Appliances

The appliances can be controlled remotely using GSM technology, when we are not present at home. By using current sensor, we may know how much electricity is used in a particular time.

3.2 Hardware Components Used in the System

Arduino Nano

The Arduino Nano is a small, complete and open-source electronics platform based on ATmega328. It has same connectivity and specs of Arduino Nano board. It is programmed based on Arduino (IDE).

Features of Arduino Nano:

- 5 V Operating voltage
- 7–12 V Input voltage
- 14 Digital I/O and 6 Analog input pins
- 40 mA DC per I/O pin
- 32 KB of which 0.5 KB used boot loader (Fig. 1).

Passive Infrared Sensor:

A passive infrared sensor allows sensing motion, used to detect whether a person has stimulated in or out of the sensor's range. They are undersized, cheap,

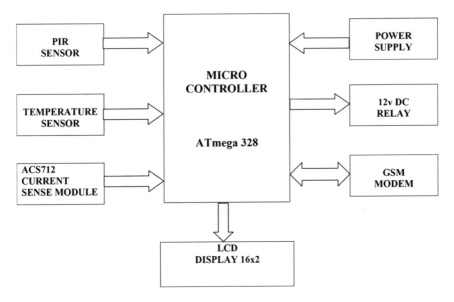

Fig. 1 Block diagram of the system [5]

low-power, and simple to use. PIRs are essentially made of pyroelectric sensor, which can detect levels of infrared radiation.

Features of PIR Sensor:

- Wide 5 m × 5 m, 60° detection pattern.
- Complete, fully functional motion detection.
- Operates from 2.7 to 3.6 V power supply.
- Simple 8-pin interface.

Temperature Sensor:

The temperature sensor measures the coolness or hotness of the room. Generally, this is used to convert heat into electricity to analyze the magnitude. The temperature sensor is made of semiconductor material used to change its resistance to electrical power by considering the temperature.

Features of Temperature Sensor:

- The range of temperature sensing can be −40 to 185 °F (−40 to 85 °C).
- It can be designed only for indoor usage.
- It is available in thermocouple.
- It can be mounted either horizontally or vertically.

ACS712 Sensor Module:

The ACS712 provides inexpensive and accurate solutions for DC or AC sensing in commercial and communication systems. It allows easy implementation by the

Fig. 2 ACS712 sensor module

customer. The ACS712 measures current in two directions, it means we can sample fast enough and long enough to find the peak in one direction and peak in another direction (Fig. 2) [6].

Features of ACS712 Sensor Module:

- The bandwidth is 80 kHz.
- The signal supply operation is 5 V.
- Low voice and analog signal.
- External stable output offset GSM Modem.

GSM Modem:

GSM (global system for mobile communication) is an open digital mobile telephony technology used for mobile data transmitting services. It supports voice call and data transfer speed up to 9.6 kbps, with the transmission of SMS (short message service).

Features of GSM Modem:

- Low power consumption.
- Single supply voltage 3.2–4.5 V.
- Supported SIM card.

Relay

Relay is an electromagnetic device, used to separate two circuits electrically and attach them magnetically. It is used to control the switch OFF or ON electrical circuit at high AC voltage by means of a low DC control voltage. Relay is a very useful tool for allowing one circuit to control another one, whereas they are totally disconnected. They are used to interface a low voltage electronic circuit with a high voltage electrical circuit.

SSR (Solid-State Relay)

A solid-state relay (SSR) is an electronic relay; it allows switching device ON or OFF for small outside electrical energy and is useful across its manage terminals. It

Fig. 3 Solid-state relay

is easy to use and permit to switching a load circuit controlled by low-power. It provides high quantity of reliability, extended life, and magnetic field. SSR has fast switching speed compared with electromechanical relays. The relay can be designed to control moreover DC or AC load (Fig. 3) [7].

Features of Solid-State Relay

- 230 V AC can be controlled with 8-bit parallel communication.
- 230 V AC can be controlled with TTL-based serial communication.

3.3 Design and Implementation Smart System

The working diagram of smart system is shown in Fig. 4. Here, the control of home appliances is based on PIR sensor. At first, the complete system is in disable state. The Arduino Nano has power input; when the PIR sensor detects the person, it will send input to microcontroller and it will give the output through relay and the light and fan will be switched ON.

This system may also control the speed of fan through solid-state relay, that is PIC 8-bit voltage controller. The normal room temperature is taken about 20–25 °C with an average of 23 °C (i.e. 73.4 °F). If temperature goes above 24 °C, the fan speed has to increase, and if it goes below 23 °C, the fan speed should decrease. The microcontroller also displays the power unit through LCD.

In case the user is not at home but needs to control the appliance, GSM modem is used for transmission of data to the mobile automatically to control the appliances. Every time when the appliances run, the GSM sends a message to the user. The user commands to the GSM modem, thereby we can control the home appliances [8, 9].

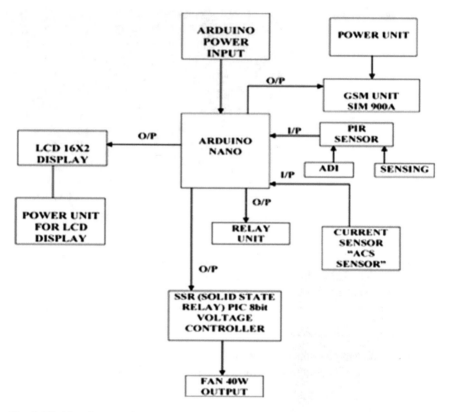

Fig. 4 Working diagram of smart system

We can also measure the current for power consumption using current sensor ACS712. The ACS712 mechanism which gives a cost-effective and exact way of sensing DC along with AC is based on Hall effect.

3.4 Energy Calculation for Smart System

The experiment is conducted for 72 h, and energy consumption is calculated for every 1 h. In our experiment, we used 60 W bulb and 100 W fan and measured with power of 230 V (Table 1) [10].

- (Wattage * hour used per day) % 1000 = daily kilowatt hour (kWh)
- 1 kWh = 1 unit.

Table 1 Power usage measurement (h)

Power usage reading for 72 h						
Appliances	Without smart system			With smart system		
	24 h	48 h	72 h	24 h	48 h	72 h
Fan	1.36	2.44	3.52	1.2828	1.3298	2.353
Bulb	0.3779	1.44	1.89	0.2252	0.972	1.222

4 Implementation Result

As shown in Fig. 5, the implementation setup explains automatic control of home appliances based on PIR sensor commands, which is used when user enters into the room.

Fig. 5 Output of proposed work

Table 2 Power consumption

Home appliances	Without smart system		With smart system		Differences (kWh)	Cost saving
	Power used in kWh	Price	Power used in kWh	Price		
Light and fan	11,027.364 11.027364 Units	99.246 Rs	7953.12 7.95312 Units	71.578 Rs	3074.244	27.668 Rs

The control board with input power supply is given 230 V, which is designed and implemented along with GSM module and current sensor. LCD initialization has to be done initially. When the power is ON, PIR sensor senses the information and the connected electronic devices will turn ON, the LCD displays the present room temperature and how much power is drawn by the appliance. The fan speed increases as temperature increases and decreases as temperature drops.

Here, GSM is used to send the message, for that we use GSM 900 hand set. It can be controlled by AT command. Through this system, the GSM module is controlled to generate and transmit text message to the number specified in the program. The text messages are interpreted by the microcontroller which controls and monitors the home appliances.

Table 2 shows the energy consumption [10] at home without smart system control and with smart system controls. The experiment is conducted for 72 h and overall energy consumed is found to be reduced by 20% less and also reduction in cost is achieved (Fig. 6).

The graph shows the comparison energy levels of with smart system and without smart system. The smart system can reduce the power consumption.

Figure 7 shows remote access of home appliances with GSM communication [7, 11].

- The #a1 is input of led1 ON (led1 = LIGHT)
- The #a0 is input of led1 OFF
- The #b1 is input of led2 ON (led2 = FAN)
- The #b2 is input of led2 OFF.

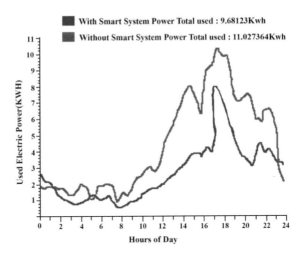

Fig. 6 Comparison of energy consumption

Fig. 7 GSM-based communication for human detection

5 Conclusion

In this paper, we proposed power consumption scheme by using automatic controlling system which saves one unit of current at consumer side and two units of current at the power station. Such intelligent power saving system is suggested to save power and enlarge the console level for the consumer with smallest amount of expenses. We implemented the automatic monitoring and control of connected appliances that increase the user comfort level especially for the disabled and elderly people.

References

1. M.S. Pan, C.J. Chen, Intuitive control on electric devices by smartphone's for smart home environments. IEEE Sens. J. **16**(11), 4281–4294 (2016)
2. A.A. Arriany, M.S. Musbah, Applying voice recognition technology for smart home networks, in *IEEE International Conference on Engineering & MIS (ICEMIS)*, Sept 2016, pp. 1–6
3. S. Kumar, S.S. Solanki, Voice and touch control home automation, in *2016 IEEE 3rd International Conference on Recent Advances in Information Technology (RAIT)*, Mar 2016, pp. 495–498
4. A.C. Jose, R. Malekian, N. Ye, Improving home automation security; integrating device fingerprinting into smart home. IEEE Access **4**, 5776–5787 (2016)
5. D. Sunehra, M. Veena, Implementation of interactive home automation systems based on email and Bluetooth technologies, in *2015 IEEE International Conference on Information Processing (ICIP)*, Dec 2015, pp. 458–463
6. D. Pavithra, R. Balakrishnan, IoT based monitoring and control system for home automation, in *2015 IEEE Global Conference on Communication Technologies (GCCT)*, Apr 2015, pp. 169–173
7. S. Jain, A. Vaibhav, L. Goyal, Raspberry Pi based interactive home automation system through E-mail, in *2014 IEEE International Conference on Optimization, Reliability, and Information Technology (ICROIT)*, Feb 2014, pp. 277–280
8. T. Perumal, S.K. Datta, C. Bonnet, IoT device management framework for smart home scenarios, in *2015 IEEE 4th Global Conference on Consumer Electronics (GCCE)*, Oct 2015, pp. 54–55
9. N. Singh, S.S. Bharti, R. Singh, D.K. Singh, Remotely controlled home automation system, in *2014 International Conference on Advances in Engineering and Technology Research (ICAETR)*, IEEE, Aug 2014, pp. 1–5
10. R. Piyare, Internet of things: ubiquitous home control and monitoring system using android based smart phone. Int. J. Internet Things **2**(1), 5–11 (2013)
11. D. Javale, M. Mohsin, S. Nandanwar, M. Shingate, Home automation and security system using Android ADK. Int. J. Electron. Commun. Comput. Technol. (IJECCT) **3**(2), 382–385 (2013)
12. I. Papp, G. Velikic, N. Lukac, I. Horvat, Uniform representation and control of Bluetooth Low Energy devices in home automation software, in *2015 IEEE 5th International Conference on Consumer Electronics-Berlin (ICCE-Berlin)*, Sept 2015, pp. 366–368

An Optimal Rule Set Generation Algorithm for Uncertain Data

S. Surekha

Abstract Nowadays, mining of knowledge from large volumes of datasets with uncertainties is a challenging issue. Rough Set Theory (RST) is the most promising mathematical approach for dealing with uncertainties. This paper proposes an RST-based optimal rule set generation (ORSG) algorithm for generating the optimal set of rules from uncertain data. At first, the ORSG approach applies the concepts of RST for identifying inconsistencies and then from the preprocessed consistent data, the RST-based Improved Quick Reduct Algorithm is used to find the most promising feature set. Finally, from the obtained prominent features, the Reduct-based Rule Generation algorithm generates the optimal set of rules. The performance of the ORSG approach is 10-fold cross-validated by conducting experiments on the UCI machine learning repository's Thyroid disease dataset, and the results revealed the effectiveness of the ORSG approach.

Keywords Rough set theory · Improved quick reduct algorithm
Optimal rule set generation · Uncertain data

1 Introduction

Data mining is the way of extracting useful information and discovering knowledge patterns that may be used for decision making [1]. The decision rules are the prescribed standards on the basis of which decisions are made for specific purpose [2]. The most important data mining techniques for generating the rules are decision tree classification, rule-based classification and association rule mining, etc. The factors that affect the performance of a data mining technique mainly includes the quality of the data and the features that are used to represent the data. But in real

S. Surekha (✉)
Jawaharlal Nehru Technological Univeristy, Kakinada, Andhra Pradesh, India
e-mail: surekha.cse@jntukucev.ac.in

© Springer Nature Singapore Pte Ltd. 2018
S. S. Dash et al. (eds.), *Artificial Intelligence and Evolutionary Computations in Engineering Systems*, Advances in Intelligent Systems and Computing 668,
https://doi.org/10.1007/978-981-10-7868-2_13

world, data is represented with so many features and all these features are not relevant to a specific task and the existence of such irrelevant features in the given data increases the computational complexity of the mining algorithm, which further decreases its performance. Hence, the features that are relevant for a particular task are to be identified, and the process of selecting the most relevant set of features is called as feature selection [3]. Rough Sets [4] have many applications in the process of knowledge discovery such as inconsistent removal, feature selection, discretization, etc. In this paper, RST-based Improved Quick Reduct Algorithm is applied to obtain the most promising features, and then, these selected features are submitted to the reduct-based rule generation algorithm to generate the optimal set of rules.

1.1 Paper Organization

Section 2 gives the methodology of the ORSG approach, Sect. 3 illustrates the RST-based inconsistent removal, Sect. 4 presents the improved quick reduct algorithm, and the reduct-based rule generation is explained in Sect. 5. Section 6 represents the result analysis, and finally, Sect. 7 concludes the paper.

2 Methodology

The methodology of the ORSG approach is given in Fig. 1. First, the Rough Set concepts are applied to eliminate inconsistencies in the given data and then from the preprocessed data, identify the most relevant features by using Improved Quick Reduct Algorithm and then submit the reduced dataset to Rule generation algorithm. The efficiency of the ORSG approach is evaluated using 10-fold cross validation test [5].

3 RST-based Approach for Dealing with Inconsistent Data

As a first step of the proposed approach, existence of inconsistencies in the data is to be identified. In this section, the concepts of RST [6] are applied to identify inconsistencies in the training data and such inconsistencies are to be removed to make the training data consistent [7].

Rough Set Theory (RST), introduced by Zdzisław I. Pawlak, is a mathematical approach to deal with vagueness and uncertainty in data, without requiring any additional knowledge about data. The basic concepts of RST can be found in [3, 4, 6–8].

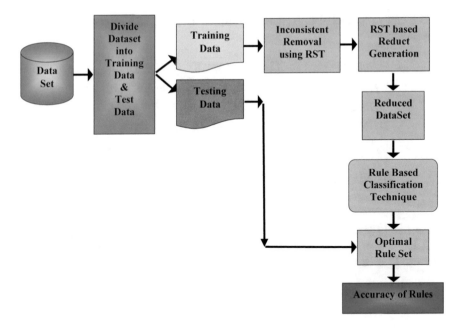

Fig. 1 Methodology of ORSG approach

To explain the concepts and approaches given in this paper, a sample dataset consisting of four conditional attributes and one decision attribute with ten objects is taken into consideration and is given in Table 1. The four conditional attributes of the sample data can be represented as C = {Degree(D), Experience(Exp), TechnicalSkills (TS), English(E)}. The decision attribute with three possible classes, namely Good, Satisfactory, and Unsatisfactory, is represented as D = {Appraisal}.

The sequence of steps for identifying inconsistencies using RST is explained below.

Table 1 Sample dataset with inconsistencies

U	Degree	Experience	TechnicalSkills	English	Appraisal
1	M.Tech	High	Excellent	Low	Good
2	M.Tech	High	Good	Fluent	Good
3	M.C.A	High	Excellent	Fluent	Good
4	M.C.A	Medium	Excellent	Low	Satisfactory
5	M.Tech	Medium	Excellent	Moderate	Satisfactory
6	M.C.A	Medium	Average	Low	Unsatisfactory
7	M.Sc	Medium	Average	Low	Unsatisfactory
8	M.Sc	Low	Average	Low	Unsatisfactory
9	M.Sc	Low	Good	Moderate	Satisfactory
10	M.Sc	Low	Good	Moderate	Good

The equivalence classes [7] generated for the conditional attribute set is obtained as follows:

$$U/\text{IND(Degree)} = \{\{1, 2, 5\}, \{3, 4, 6\}, \{7, 8, 9, 10\}\}$$
$$U/\text{IND(Experience)} = \{\{1, 2, 3\}, \{4, 5, 6, 7\}, \{8, 9, 10\}\}$$
$$U/\text{IND(TechnicalSkills)} = \{\{1, 3, 4, 5\}, \{2, 9, 10\}, \{6, 7, 8\}\}$$
$$U/\text{IND(English)} = \{\{1, 4, 6, 7, 8\}, \{5, 9, 10\}, \{2, 3\}\}$$
$$U/\text{IND(Appraisal)} = \{\{1, 2, 3, 10\}, \{4, 5, 9\}, \{6, 7, 8\}\}$$

$$U/\text{IND(C)} = (U/D) \otimes (U/\text{Exp}) \otimes (U/\text{TS}) \otimes (U/E)$$
$$= \{\{1\}, \{2\}, \{3\}, \{4\}, \{5\}, \{6\}, \{7\}, \{8\}, \{9, 10\}\}$$

As per [7], any inconsistencies in the data can be identified by finding the boundary region for the set of conditional attributes C with respect to the decision attribute D is as follows:

$$\text{C-BoundaryRegion(D)} = \text{C-UpperApproximation(D)}$$
$$- \text{C-LowerApproximation(D)}$$

The C-lower approximation [7] with respect to D is obtained as follows:
The target set D is consisting of three subsets, that is,

$$D_1 = \{1, 2, 3, 10\}, D_2 = \{4, 5, 9\} \quad \text{and} \quad D_3 = \{6, 7, 8\}$$
$$\underline{C}D_1 = \{\{1\}, \{2\}, \{3\}\}, \underline{C}D_2 = \{\{4\}, \{5\}\} \quad \text{and} \quad \underline{C}D_3 = \{\{6\}, \{7\}, \{8\}\}$$

Therefore, $\underline{C}D_1 \cup \underline{C}D_2 \cup \underline{C}D_3 = \{\{1\}, \{2\}, \{3\}, \{4\}, \{5\}, \{6\}, \{7\}, \{8\} \}$
The C-upper approximation [7] with respect to D is obtained as follows:

$\overline{C}D_1 = \{\{1\}, \{2\}, \{3\}, \{9, 10\}\}$ Since, $\{1\} \cap \{1, 2, 3, 10\} = \{1\} \neq \Phi$ and
$\{9, 10\} \cap \{1, 2, 3, 10\} = \{9, 10\} \neq \Phi$.
Similarly, $\overline{C}D_2 = \{\{4\}, \{5\}, \{9, 10\}\}, \overline{C}D_3 = \{\{6\}, \{7\}, \{8\}\}$.
Therefore,
$\overline{C}D_1 \cup \overline{C}D_2 \cup \overline{C}D_3 = \{\{1\}, \{2\}, \{3\}, \{4\}, \{5\}, \{6\}, \{7\}, \{8\}, \{9, 10\}\}$.
C-boundary region [7] of D can be calculated as follows:

$$\text{BND}_C(D) = \{\overline{C}D_1 \cup \overline{C}D_2 \cup \overline{C}D_3\} - \{\underline{C}D_1 \cup \underline{C}D_2 \cup \underline{C}D_3\} = \{9, 10\}$$

That is, objects 9 and 10 are the inconsistent objects. So, the elimination of objects 9 and 10 makes the dataset consistent.

4 Improved Quick Reduct Algorithm

The RST-based Improved Quick Reduct Algorithm [9] generates the most promising set of features called as Reduct. The Improved Quick Reduct Generation algorithm is based on the dependency degree of attributes and is given below.

Pseudo code for Improved QuickReduct Generation (C_a, D_a)

Input: C_a, the set of conditional attributes;
     D_a, the set of decision attributes.
Output: RS, minimal subset of conditional attributes.

---

(1)	RS ← { }	// (Initially the Reduct set is empty)
(2)	$\gamma_{Best} = \gamma_{Prev} = 0$ // γ_{Best} is the best γ , γ_{Prev} is the previous γ dependency	
(3)	Do	
(4)	TS ← RS	// TS is the temporary set to hold the reduct
(5)	$\gamma_{Prev} = \gamma_{Best}$	
(6)	∀ X∈(C_a −RS)	
(7)	If Max($\gamma_{\{RSU X\}}$ (D_a)) > γ_{Prev} then	
(8)	TS ←RSU{X}	
(9)	$\gamma_{Best} = \gamma_{TS}$	
(10)	RS←TS	
(11)	Until $\gamma_{Best} = \gamma_{Ca}(D_a)$	
(12)	Return RS	

The following steps explain the reduct generation using Improved Quick Reduct Generation algorithm for the data in Table 1.

4.1 *Worked Out Example*

For the first eight consistent objects in Table 1, the dependency degree of all attributes is obtained as follows:

$$\gamma_{Degree}(\text{Appraisal}) = \left|POS_{Degree}(\text{Appraisal})\right|/|U| = |\{7, 8\}|/8 = 2/8 = 0.25$$

$$\gamma_{Experience}(\text{Appraisal}) = |\{1, 2, 3, 8\}|/8 = 4/8 = 0.5$$

$$\gamma_{TechnicalSkills}(\text{Appraisal}) = |\{2, 6, 7, 8\}|/8 = 4/8 = 0.5$$

$$\gamma_{English}(\text{Appraisal}) = |\{2, 3, 5\}|/8 = 3/8 = 0.375.$$

The dependency degree of the entire set of conditional attributes C is obtained as,

$$\gamma_C(\text{Appraisal}) = |\{1,2,3,4,5,6,7,8\}|/8 = 8/8 = 1$$

Initially, RS = TS = Φ and $\gamma_{\text{Prev}} = \gamma_{\text{Best}} = 0$.

From the above, the attributes with the same maximum dependency degree are *Experience* and *TechnicalSkills*.

So, first select *Experience*, TS = {Experience} and $\gamma_{\text{TS}} = 0.5$ and $\gamma_{\text{Prev}} = 0$.

The condition $\gamma_{\text{TS}} > \gamma_{\text{Prev}}$ is satisfied. Hence, add the attribute *Experience* to the reduct set and the reduct set RS becomes, RS = TS = {Experience}.

Next, the best attribute from the remaining list is to be selected, so that the addition of which, should improve the dependency degree.

So, select the attribute TechnicalSkills, that is,

$$\text{TS} = \text{TS} \cup \{\text{TechnicalSkills}\}$$
$$= \{\text{Experience}, \text{TechnicalSkills}\}$$

Then, U/IND(Experience,TechnicalSkills) = {{1, 3}, {2}, {4, 5}, {6, 7}, {8}} and $\text{POS}_{\{\text{Experience, TechnicalSkills}\}}(\text{Appraisal}) = \{1, 2, 3, 4, 5, 6, 7, 8\}$

Therefore, $\gamma_{\text{TS}} = \gamma_{\{\text{RS} \cup \text{TechnicalSkills}\}} = |\{1,2,3,4,5,6,7,8\}|/8 = 1$.

Similarly, calculating the degree of dependency by including the remaining attributes *Degree* and *English*, the dependencies are obtained as,

$$\gamma_{\{\text{RS} \cup \text{Degree}\}} = 0.75 \quad \text{and} \quad \gamma_{\{\text{RS} \cup \text{English}\}} = 0.625$$

So, add the attribute *TechnicalSkills* to the existing reduct set and TS as,

$$\text{TS} = \text{TS} \cup \{\text{TechnicalSkills}\}$$
$$= \{\text{Experience}, \text{TechnicalSkills}\}.$$

The dependency degree of the TS is equal to the dependency degree of the full set. Hence, stop the process and terminate with RS as the reduct.

That is, Reduct = RS = {Experience,TechnicalSkills}.

5   Optimal Rule Set Generation

Rule-based classification [10] classifies the data objects based on the strength of the rules. The strength of an association rule A $\rightarrow$ D is the number of samples that contain both A and D to the total samples that contain only A.

A rule from Table 1 is given below.

$$R_1 : \text{If}(\text{Degree} = \text{M.Tech}) \text{ then} \quad \text{Appraisal} = \text{Good}$$

The strength [10] of the above rule can be calculated as follows:

$$\text{Strength}(R_1) = |\{1,2\}|/|\{1,2,5\}| = 2/3 = 66\%$$

The steps for the generating rules using rule set generation algorithm are given below.

Step 1 *Determine the Strength of each possible Rule*

For the attribute *Degree,* the possible rule set is as follows:

R_1: (Degree = M.Tech $\rightarrow$ Appraisal = Good) = 2/3 = 66%
R_2: (Degree = M.Tech $\rightarrow$ Appraisal = Satisfactory) = 1/3 = 33%
R_3: (Degree = M.C.A $\rightarrow$ Appraisal = Good) = 1/3 = 33%
R_4: (Degree = M.C.A $\rightarrow$ Appraisal = Satisfactory) = 1/3 = 33%
R_5: (Degree = M.C.A $\rightarrow$ Appraisal = UnSatisfactory) = 1/3 = 33%
R_6: (Degree = M.Sc $\rightarrow$ Appraisal = Satisfactory) = 1/4 = 25%
R_7: (Degree = M.Sc $\rightarrow$ Appraisal = UnSatisfactory) = 2/4 = 25%
Similarly, for the attribute *Experience*
R_8: (Experience = Low $\rightarrow$ Appraisal = Unsatisfactory) = 1/1 = 100%
R_9: (Experience = Medium $\rightarrow$ Appraisal = Unsatisfactory) = 2/4 = 50%
R_{10}: (Experience = Medium $\rightarrow$ Appraisal = Satisfactory) = 2/4 = 50%
R_{11}: (Experience = High $\rightarrow$ Appraisal = Good) = 3/3 = 100%
For the attribute *TechnicalSkills*,
R_{12}: (TechnicalSkills = Good $\rightarrow$ Appraisal = Good) = 1/1 = 100%
R_{13}: (TechnicalSkills = Excellent $\rightarrow$ Appraisal = Satisfactory) = 2/4 = 50%
R_{14}: (TechnicalSkills = Average $\rightarrow$ Appraisal = Unsatisfactory) = 3/3 = 100%
For the attribute *English*,
R_{15}: (English = Low $\rightarrow$ Appraisal = Good) = 1/5 = 20%
R_{16}: (English = Low $\rightarrow$ Appraisal = Satisfactory) = 1/5 = 20%
R_{17}: (English = Low $\rightarrow$ Appraisal = UnSatisfactory) = 3/5 = 60%
R_{18}: (English = Moderate $\rightarrow$ Appraisal = Satisfactory) = 1/1 = 100%
R_{19}: (English = Fluent $\rightarrow$ Appraisal = Good) = 2/2 = 100%

Step 2 *Elimination of Rules with Less Support Count*

Suppose, if the allowable rule strength is defined as *100%*, then the list of qualified rules are R_8, R_{11}, R_{12}, R_{14}, R_{18}, and R_{19}.

R_1: (Experience = Low $\rightarrow$ Appraisal = Unsatisfactory) = 1/1 = 100%
R_2: (Experience = High $\rightarrow$ Appraisal = Good) = 3/3 = 100%
R_3: (TechnicalSkills = Good $\rightarrow$ Appraisal = Good) = 1/1 = 100%
R_4: (TechnicalSkills = Average $\rightarrow$ Appraisal = Unsatisfactory) = 3/3 = 100%
R_5: (English = Moderate $\rightarrow$ Appraisal = Satisfactory) = 1/1 = 100%
R_6: (English = Fluent $\rightarrow$ Appraisal = Good) = 2/2 = 100%

Table 2 Consistent dataset
with relevant attributes

U	Experience	TechnicalSkills	English	Appraisal
1	High	Excellent	Low	Good
2	High	Good	Fluent	Good
3	High	Excellent	Fluent	Good
4	Medium	Excellent	Low	Satisfactory
5	Medium	Excellent	Moderate	Satisfactory
6	Medium	Average	Low	Unsatisfactory
7	Low	Average	Low	Unsatisfactory

The set of attributes that are contributing to the above rule set are {*Experience*, *TechnicalSkills*, and *English*}.

Step 3 *Elimination of Redundant Objects and Irrelevant Attributes*

Now, reduce the dataset by keeping only the relevant attributes and also eliminate the duplicates. The reduced dataset is given in Table 2.

Step 4 *Eliminate Irrelevant Attributes for each Decision*

From Table 2, it can be observed that for the decision *Good*, the attribute Experience alone is sufficient. Similarly, for the decision *Satisfactory*, the attributes Experience and TechnicalSkills are sufficient. For the decision *Unsatisfactory*, the attributes TechnicalSkills and English are important. Then, keep only the attribute values that are contributing to the respective decisions. Table 2 can be simplified as given in Table 3.

Step 5 *Rule Set Generation*

From Table 3, the rule set generated for each decision is shown in Fig. 2.

Figure 2 shows the optimal rule set generated for the full set of attributes. The generated rule set is consisting of three rules and six conditions. From the above steps, it can be observed that all the attributes are not contributing to a particular decision. Hence, calculating the rule strength for all the attributes increases the computational complexity of the algorithm. The computational complexity can be reduced by identifying only the relevant set of attributes, and also submitting the dataset with the most promising features to the rule-based classification technique generates the optimal rule set.

Table 3 Dataset with
relevant attribute values

U	Experience	TechnicalSkills	English	Appraisal
1	High	*	*	Good
2	Medium	Excellent	*	Satisfactory
3	*	Average	Low	Unsatisfactory

> **Rule 1:** If (*Experience = High*) Then *Appraisal = Good*
> **Rule 2:** If (*Experience = Medium*) and (*TechnicalSkills = Excellent*)
>   Then *Appraisal = Satisfactory*
> **Rule 3:** If (*TechnicalSkills = Average*) and   (*English = Low*)
>   Then *Appraisal = Unsatisfactory*

Fig. 2 Rule set obtained for the raw data

Table 4 Reduced dataset with relevant attributes only

U	Experience	TechnicalSkills	Appraisal
1	High	*	Good
2	High	*	Good
3	Medium	Excellent	Satisfactory
4	*	Average	Unsatisfactory
5	*	Average	Unsatisfactory

Table 5 Dataset with relevant attribute values

U	Experience	TechnicalSkills	Appraisal
1	High	*	Good
2	Medium	Excellent	Satisfactory
3	*	Average	Unsatisfactory

From Sect. 4.2, the reduct generated by the Improved Quick Reduct Algorithm for the consistent part of Table 1 is consisting of only two attributes and the reduct set is {*Experience, TechnicalSkills*}, that is, among the four conditional attributes, only these two attributes are the most relevant attributes. The following steps explain the rule generation for the reduced dataset. The reduced data of Table 1 with relevant attributes and without inconsistencies is given in Table 4.

The dataset with relevant attribute values for a decision is given in Table 5.

The rule set can be generated from Table 5 as shown in Fig. 3.

The optimal rule set generated from the reduced dataset is consisting of three rules and four conditions only. The number of conditions to be verified to arrive at a decision has been reduced to four. To know the efficiency of the ORSG approach, the rule set generated by the ORSG approach is compared with that of the rule set obtained by ID3 tree classification [5, 11]. The decision trees generated by the ID3 algorithm on both inconsistent and consistent dataset given in Table 1 are shown in Figs. 4 and 5, respectively. In Fig. 4, the leaf nodes are labeled as G, S, and US for the actual decision classes *Good, Satisfactory,* and *Unsatisfactory,* respectively. Similarly, the branches are labeled with the first letter of the respective attribute values, that is, for the attribute *TechnicalSkills,* the three possible values are Excellent(E), Good(G), and Average(A). But for the attribute *Degree,* the attribute values M.Tech, M.C.A, and M.Sc are labeled as T, C, and S, respectively. From

> **Rule 1:** If (*Experience = High*) Then *Appraisal = Good*
> **Rule 2:** If (*Experience = Medium*) and (*TechnicalSkills = Excellent*)
>   Then *Appraisal = Satisfactory*
> **Rule 3:** If (*TechnicalSkills = Average*) Then *Appraisal = Unsatisfactory*

Fig. 3 Rule set obtained by ORSG approach

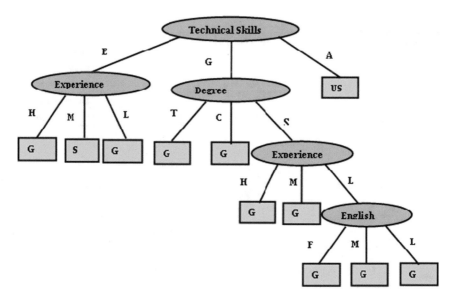

Fig. 4 ID3 decision tree for inconsistent data

Figs. 4 and 5, the number of rules and conditions generated by ID3 decision tree classification on inconsistent data is 11 and 15, respectively. But, for the consistent and reduced data, the number of rules and conditions generated by the same ID3 decision tree has been reduced to five and six, respectively. The comparison of the performance of ID3 and ORSG on both inconsistent and consistent data is given in Fig. 6.

From Fig. 6, it can be observed that the size of the rule set generated by ID3 on inconsistent dataset is very large and the rule set generated on consistent and reduced dataset by the same algorithm is reduced drastically. The rule set generated by the proposed approach ORSG on the same reduced consistent dataset is optimal with very few rules having less number of conditions when compared to the ID3 classification technique.

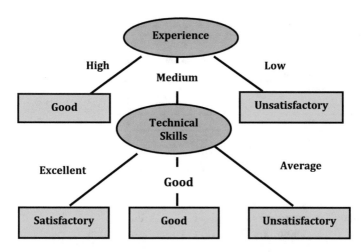

Fig. 5 ID3 decision tree for consistent data

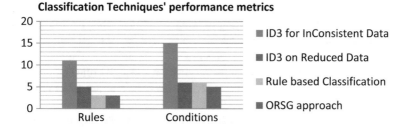

Fig. 6 Comparison of performance of ORSG and ID3

Table 6 Description of Thyroid dataset

Dataset	Feature set	Classes
Thyroid	Age; Gender; On thyroxine; Pregnant; Query on thyroxine; Sick; Thyroid Surgery; On antithyroid medication; Goitre; Tumor; Lithium; Psych; T131 measurement; Query hypothyroid; Query hyperthyroid; Hypopituitary; TSH, TSH measured; T3, T3 measured; T4U, T4U measured; TT4, TT4 measured; FTI, FTI measured	HypoThyroid, HyperThyroid, Negative

6 Experimental Analysis

Experiments were conducted on the Thyroid dataset taken from the UCI machine learning repository [12] consisting of 500 instances and 27 attributes including one decision attribute. The description of the dataset is given in Table 6.

Table 7 Prediction accuracies of Optimal Rule Set Generation Approach

Rule strength	Original data		Reduced data	
	Number of conditions	Accuracy %	Number of conditions	Accuracy %
75	14	97.14	10	97.14
80	12	97.14	10	97.14
85	10	92.85	9	92.85
90	8	84.28	7	84.28
95	8	84.28	8	84.28
100	7	74.20	6	74.20

Applying the Improved Quick Reduct Algorithm on Thyroid dataset generated only 8 attributes among the 26 as the most relevant attributes, and the reduct is as follows:

Reduct = {Gender, Onthyroxine, Query on Thyroxine, Tumor, TSH, T3, TT4, FTI}.

As the Thyroid dataset is consisting of three decision classes, the rule generation algorithm generates rules for each decision class and the optimal rule set contains only three rules. The performance of the ORSG approach on the original and reduced Thyroid dataset is measured by varying the allowable rule strength, and the average prediction accuracies along with the average number of conditions are given in Table 7.

The comparison of performances of the ORSG approach with the ID3 classification technique [5, 11] is given in Table 8.

The performance of Rule-based classification has been improved for reduced dataset. Figure 7 depicts the comparison of the results of the proposed ORSG approach with other techniques.

The results clearly depict the importance of inconsistent removal. From Fig. 7, it is very clear that the ORSG approach identifies and eliminates the inconsistencies and irrelevant features in the data. The proposed ORSG approach is better when compared with other rule generation techniques in all aspects by generating optimal results in terms of the increased rate of accuracy and less number of rules generated, and the number of conditions required to arrive at a decision is also very less.

Table 8 Comparison of accuracies of ID3 and ORSG approaches

Classification technique	No. of attributes	No. of rules	No. of conditions	Accuracy (%)
ID3 inconsistent data	26	17	156	95.50
ID3 reduced data	8	8	26	96.05
Rule set for raw data	26	3	12	97.14
ORSG approach	8	3	10	97.14

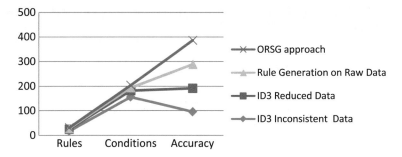

Fig. 7 Comparison of ID3 and ORSG approaches

7 Conclusion

This paper presented an efficient rule generation approach for generating optimal rule set from the large datasets with uncertainties. The ORSG approach uses the RST concepts to identify uncertainties and eliminates irrelevant features using the Improved Quick Reduct Algorithm thereby reduces the search space and generates optimal set of rules. The RST-based ORSG approach generates an optimal rule set with a very less number of conditions.

References

1. J. Han, M. Kamber, *Data Mining Concepts and Techniques* (Elsevier, San Francisco, CA, 2007)
2. P. Gogoi, D.K. Bhattacharyya, J.K. Kalita, A rough set-based effective rule generation method for classification with an application in intrusion detection. Int. J. Secur. Network **8** (2), 61–71 (2013)
3. S. Surekha, G. Jaya Suma, in Comparison of feature selection techniques for thyroid disease. *Proceedings of ICICMT'2015* (2015), pp. 20–26
4. L. Jian, S. Liu, Y. Lin, *Hybrid Rough Sets and Applications in Uncertain Decision-Making* (CRC Press, Boca Raton, 2011)
5. R. Kohavi, in *A study of cross-validation and bootstrap for accuracy estimation and model selection*. International Joint Conference on Artificial Intelligence (IJCAI), vol 14(2) (1995), pp. 1137–1145
6. Z. Pawlak, Rough sets. Int. J. Comput. Inform. Sci. **11**(5), 341–356 (1982)
7. S. Surekha, in An RST based efficient preprocessing technique for handling inconsistent data. *Proceedings of 2016 IEEE ICCIC* (2016), pp. 298–305
8. Z. Pawlak, Rough set approach to knowledge based decision-support. Eur. J. Oper. Res. **99** (1), 48–57 (1997)
9. P. Jaganathan, K. Thangavel, A. Pethalakshmi, M. Karnan, Classification rule discovery with ant colony optimization and improved quick reduct algorithm. IAENG Int. J. Comput. Sci. **33** (1), 50–55 (2007)

10. R. Vashist, M.L. Garg, Rule Generation based on reduct and core: a rough set approach. Int. J. Comput. Appl. **29**(9), 0975–8887 (2011)
11. M. Tom, *Mitchell: Machine Learning* (McGraw-Hill, Singapore, 1997)
12. A. Asuncion, D. Newman, UCI machine learning repository (2007), https://archive.ics.uci.edu/ml/datasets/Thyroid+Disease

Extended Security Solution for Women Based on Raspberry Pi

M. V. Marathe, Saurabh Kumar, Sejal Ghodke, PrashantKumar Sharma and Nupura Potdar

Abstract In today's highly developing era, cities are getting smarter solutions for everyday problems. One such problem is about women's safety. Recently, there has been a lot of incidence which could have been avoided if there was a reliable safety solution for women. There are many applications based on women's safety most of which are cell phone based. The problem with these applications is that they depend on the phone's capabilities and constraints like a battery, network, and connectivity. Thus, they somehow become unreliable in real-world scenarios. This paper focuses on creating a safety module that provides extended security to women. This paper proposes a smart solution for enhancing women safety. The proposed system is a Raspberry Pi-based module which will be a roadside unit (RSU) deployed across the city where police stations are not easily available. This module uses RFID technology to identify individuals in case of emergency. When activated, the module sends out the location of emergency to the nearest police station and to the emergency number extracted from RFID tag. It contains GPS module for location information and GSM module for sending out an alert.

Keywords Raspberry Pi · Smart city · Women's security · GPS
GSM · RFID

M. V. Marathe · S. Kumar (✉) · S. Ghodke · P. Sharma · N. Potdar
Pune Vidyarthi Griha's College of Engineering and Technology, Computer Engineering
Department, 44 Vidyanagari, Parvati, Pune 411009, Maharashtra, India
e-mail: sksaurabhsky06@gmail.com

M. V. Marathe
e-mail: mvm_comp@pvgcoet.ac.in

S. Ghodke
e-mail: sejalghodke@gmail.com

P. Sharma
e-mail: prashant.240496@gmail.com

N. Potdar
e-mail: nupurapotdar@gmail.com

© Springer Nature Singapore Pte Ltd. 2018
S. S. Dash et al. (eds.), *Artificial Intelligence and Evolutionary Computations
in Engineering Systems*, Advances in Intelligent Systems and Computing 668,
https://doi.org/10.1007/978-981-10-7868-2_14

1 Introduction

A lack of women's security has been observed in today's developing cities. Crimes against women are increasing day by day. Although there are security solutions available, there is a need for a more reliable security solution to ensure women's safety. Nowadays, many organizations are coming up with different ideas to tackle this problem, and they are using various technologies to do so, but many of these ideas are under development.

This paper focuses on a security system that will help women in dangerous situations, places where they find themselves alone or in case of medical emergency. It is also a helping aid for women who cannot afford devices like smartphones or any other expensive solutions for their security. This system will be able to detect the location of the emergency as soon as it happens and that will enable us to take actions accordingly. It is based on technologies like GPS receiver, GSM, RFID tag and reader. Since most of the applications are cell phone-based which are handy and easy to use, they still have few drawbacks such as battery constraint or cases where the phone is lost or damaged. Our paper proposes better solutions to overcome these drawbacks. We are focusing on building a roadside unit instead of a phone-based application which will use Raspberry Pi as the main processing unit. We are using technologies like RFID, GPS, and GSM to build a robust system which can be used to provide extended security to women.

2 Related Work

Some of the applications developed for women safety are as follows:

- SOS-Stay Safe: This app is freely available in play store. Unlocking the phone is not required, just keeping the app activated works. The user needs to shake the phone to send the alert. Along with shaking, tapping the home button option is also available, which will also send a message having the location, even battery status of the phone and an audio recording.
- Eyewatch Women: This app works a little different, and it not only sends the alert message but also sends an audio/video recording to the predefined contact number. It is highly appreciated for its accuracy and functioning without the need for GPS. By pressing "I am safe" button, the user can let the concerned people know of their safety.
- VithU: Channel V had this initiative of providing an interactive safety application, and they came up with this application. By pressing the power button twice, your location will be sent to the emergency contacts you have saved. This app sends location every 2 min so that tracking is made easy.
- Smart Belt: This system resembles a normal belt design wise. It has an Arduino board for activating the alarm, screaming alarm which can be heard from a considerable distance and pressure sensors. When the pressure sensor is

activated, the device will start working automatically. The screaming alarm unit will be activated, and a siren will start [1].

- Police Nearby: The developer of this application is Big Systems. The police nearby scanner is an android application that helps students and citizens to connect to their nearest police station in case of emergency. This app connects citizen to their respective city police, and it is an initiative to involve as many people in public safety with the help of their android smartphones. No signup is required to download this application.
- Sonata Act: Sonata launched this watch that works with an android application. This watch connects to the phone using Bluetooth and sends alerts. The alert contains user's location, and it is sent via the in-app alert system and by SMS.

3 Proposed System

As shown in Fig. 1, the system consists of central processing unit, preferably a microcontroller which will be responsible for interfacing and activating other components of the system. This system will be installed at strategic locations which are easily recognizable by any individual. The system will be a roadside unit consisting of Raspberry Pi as the main processing and controlling unit. All the other modules will be interfaced with this Raspberry Pi. Individuals will be provided with a RFID tag which will contain identification information along with an emergency contact number. As soon as the unit is activated by the individual using their tag, the RFID reader will get the information from the tag. The reader will then pass this information to Raspberry Pi. The alert will be sent using GSM module to the nearest police station and the emergency contact number fed in the tag. Along with sending the alert, this module will be linked to a camera which will record the event and the data will be saved in the database. All this will happen in real time. The system includes the following modules:

1. **Processing module**: This module is responsible for controlling and activating all other modules. We will be using Raspberry Pi for this purpose. It is a credit card-sized computer which is easy to use and maintain [2, 3]. Raspberry Pi is interfaced with all the other modules and is responsible for handling and redirecting data from one module to another. Performance wise Raspberry Pi is slower than a laptop or desktop; it is a complete computer but uses much less power. It is also a Linux-based computer which is perfect for the kind of system we want to build.

All of the Raspberry Pi is not an open hardware just the primary chip that runs many of the components of the board. CPU, graphics, memory, USB controller, etc., is not open source. All the projects that can be made using Raspberry Pi are very well documented, and there are many forums that support the development of new projects and applications using the Raspberry Pi. The Raspberry Pi can also be

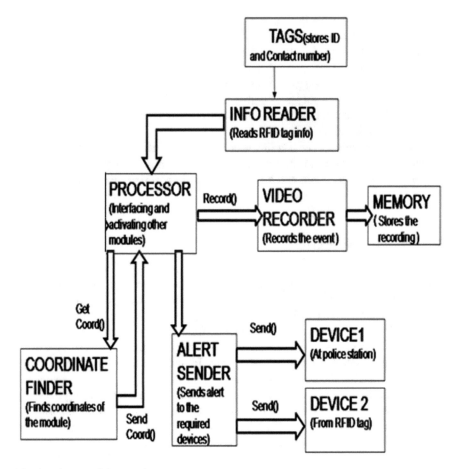

Fig. 1 Diagram of the security system

easily expanded by adding different modules like the camera module and touch screen module and greatly increased the scope of the module (Fig. 2).

The Raspberry Pi 2 comes with 40 GPIO pin which we can use to connect other modules. For easier connection, the signals can be converted and transmitted using USB ports available on Raspberry Pi.

The GPIO pin layout is given in Fig. 3.

2. **Identification module**: This module consists of a RFID tag and reader [4, 5]. RFID tags are either passive, active, or battery-assisted passive. Passive tags are smaller and cheaper as they are not equipped with a battery. Passive tags work by collecting energy from a nearby RFID reader's radio waves, whereas active tags have their own local power source (battery). As active tags are equipped with batteries, they may operate at a range of hundreds of meters from the RFID reader. Our tag will be a passive tag with enough memory to store unique

Fig. 2 Top view of Raspberry Pi 2 Model B

identification number and contact number of the user's kin. The RFID reader/writer that can be used is MFRC522; it is a highly integrated reader/writer IC for contactless communication at 13.56 MHz. Every user will be provided with one RFID tag, and they can use it to activate any unit nearby. The reader will then send the information to the Raspberry Pi.

The tag which will be used is a passive tag which means it does not have a power source to transmit signals to a long distance. Therefore, the user needs to bring the tag as near as 2–3 cm to the reader. Reading and writing on RFID tags is done using electromagnetic waves transmitted from the antenna of the transponder. The MIFARE library is available which contains source code for reading/writing on RFID tags (Fig. 4).

The RFID reader/writer is connected to Raspberry Pi using the GPIO pins. The required connection is as in Table 1.

3. **Location module**: This module consists of a GPS modem which is able to determine the coordinates of the system using MEO (Medium Earth Orbit) satellites. The longitude and latitude of the receiver are calculated from the time difference between the signals of various satellites to reach that receiver [6, 7]. The GPS module antenna requires clear sky view to receive NMEA (National Marine Electronics Association) sentences. The data by GPS receivers consist of a lot of information like latitudinal and longitudinal position, speed, and altitude.

Fig. 3 Raspberry Pi 2 GPIO pin layout

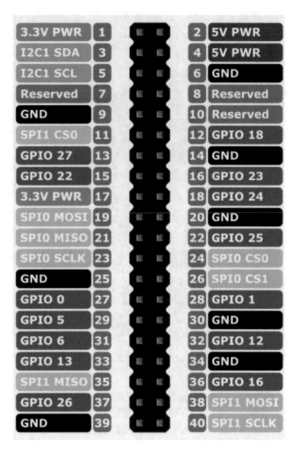

The NMEA standards send a series of independent and self-reliant sentence that contains all these data. A particular sentence begins with a "$", and it is about 80 characters in length with the data items separated by a comma.

GPS receivers are made compatible with NMEA standards and are also compatible with serial ports using RS232 protocols. This system uses the SIM 28 M GPS receiver which is a stand-alone receiver. The coordinates acquired will be incorporated in a Web link that will be sent using the alert module (sample link in Fig. 5).

The SIM 28 M can be easily interfaced with the Raspberry Pi with the help of the Tx (transmit) and Rx (receive) pins along with 5 V supply and ground (Table 2).

Fig. 4 RFID reader/writer with RFID tag

Table 1 RFID transponder to Raspberry Pi connection

RFID-RC522 pin	Raspberry Pi pin	Raspberry Pi pin description
SDA	24	GPIO8
SCK	23	GPIO11
MOSI	19	GPIO10
MISO	21	GPIO9
IRQ	None	None
GND	Any	Any ground
RST	22	GPIO25
3.3 V	1	3V3

4. **Alert module**: Once the system is activated, Raspberry Pi will send an alert message (Fig. 6) having the location of the system to the nearby police station and to the contact hard-coded in the RFID tag using the GSM modem. The GSM module we will use is SIM 800. The GSM modem can accept any SIM card and act just like a mobile phone. AT commands are used to control the

Fig. 5 GPS receiver with external antenna

Table 2 GPS to Raspberry Pi interface

GPS SIM28M pin	Raspberry Pi pin	Raspberry Pi pin description
VIN/VCC	2	5 V
GND	6	GND
RX	8	TX
TX	10	RX

Fig. 6 Sample message to be sent

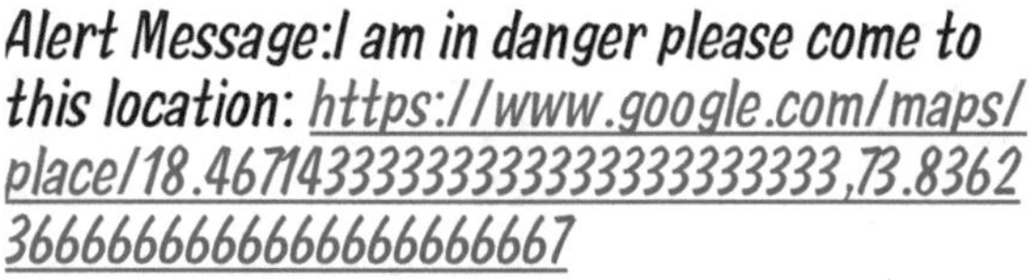

Alert Message:I am in danger please come to this location: https://www.google.com/maps/place/18.4671433333333333333333333,73.8362 366666666666666666667

GSM module. This system uses AT commands to send messages. Moreover, more than one GSM modem can be used and with slight changes in software code, these modems can work simultaneously for a more reliable alert mechanism (Fig. 7) [8].

Fig. 7 SIM800 GSM module with antenna

SAMPLE PYTHON CODE TO SEND A MESSAGE:

```
port = serial.Serial('/dev/ttyUSB0',9600,timeout=1)
port.flush()
port.write('AT+CMGF=1'+'\r\n')
rcv =port.read(10)
time.sleep(1)
port.write('AT+CMGS="{CELLPHONE NUMBER WITHOUT BRACKET}"\r')
time.sleep(0.5)
port.write('Message'+'\r\n')
port.write("\x1A")
for i in range(10):
rcv=port.read(10)
port.close()
```

5. **Camera module**: Since the system is located at one position, a camera can be
   used covering the area of the system so as when anyone activates the system it
   can record the event in case police needs video evidence to produce in court or
   for investigation. In case someone tries to damage the system, the further
   enhancement will be a pressure sensor that will activate the system and the
   camera will capture the recording.

4   Conclusion and Future Scope

The proposed system is an effective application toward extending security solution for women. This system can be adopted by smart city planners for providing a safer environment for women. Since this system uses RFID technology, it is cheaper than cell phones and can be used by anyone. It can be made easily available to everyone and can be easily installed and maintained.

The proposed security system for women can be effectively deployed in cases where help is not easily available. It not only provides extended security to women but also provides evidence to support further investigations with the help of the feed obtained from the camera module. The same system could also be used for providing extended security for senior citizens, children, etc., and for other applications like police surveillance system. With slight changes in the source code, this system can also be installed in buses and trains. With other slight modifications, the proposed system could be deployed as a pick-and-drop cab service across various locations.

References

1. B. Chougula, Smart girls security system. Int. J. Appl. Innov. Eng. Manage. **3**(4) (2014)
2. W. Anwaar, M. Ali Shah, Energy efficient computing: a comparison of raspberry PI with modern devices. Int. J. Comput. Inf. Technol. **04**(02). ISSN: 2279-0764 (2015)
3. Adafruit Learning System, in Introducing the Raspberry Pi Model B+, https://learn.adafruit. com. Mar 2015
4. D.L. Wu, W.W.Y Ng, D.S. Yeung, H.L. Ding, A brief survey on current 2 RFID applications, in *Proceedings of International Conference on Machine Learning and Cybernetics*, pp. 2330–2334, 12–15 July 2009
5. The History of RFID Technology—RFID J. [ONLINE] Available at http://www.rfidjournal. com/article/articleview/1338/1/129/
6. H.D. Pham, M. Drieberg, C.C. Nguyen, in *Development of vehicle tracking system using GPS and GSM modem*, Open Systems(ICOS), 2013 IEEE Conference, pp. 89–94, 2–4 Dec 2013
7. S.J. Lee, G. Tewolde, J. Kwon, in Design and implementation of vehicle tracking system using GPS/GSM/GPRS technology and smartphone application, IEEE World Forum on Internet of Things (WF-IoT), Seoul, pp. 353–358, 6–8 Mar 2014
8. R. Kumar, H. Kumar, in *Availability and handling of data received through GPS device*, Tracking a Vehicle, Advance Computing Conference (IACC), IEEE International, pp. 245–249, 21–22 Feb 2014

Recommendation System Based on Generalized-Weighted Tree Similarity Algorithm

D. Pramodh Krishna and K. Venu Gopal Rao

Abstract Weighted tree representation and its similarities are widely used in e-learning and e-business where items are represented using ontologies and user preferences are given by weights. However, there have been some challenges to develop recommendation systems where items are represented using weighted trees. Recommendation systems consist of content-based where similarity between items plays a major role, collaborative filtering where user rating plays a major role and hybrid approaches. This paper proposes a new hybrid recommendation system which uses generalized-weighted tree similarity to find similarity between items that also includes user rating to integrate with collaborative filtering. This new hybrid recommendation system has flexibility to include preferences for different aspects of items using weighted trees and user ratings as well. This paper addresses the challenge of using recursive weighted tree similarity in hybrid recommendation system. We established theoretical and experimental evaluation among a few example trees using our proposed recommendation systems.

Keywords Tree similarity measure · Match value · Miss value
Recommendation system · Generalized measure

1 Introduction

Recommendation system has attracted a great deal of attention in many applications like e-learning [1], e-business [2], e-commerce [3], and telecommunications. In many applications, these data are represented using tree structure. For example, telecom companies offer number of products and services. These products and

D. P. Krishna (✉) · K. Venu Gopal Rao
Department of CSE, JNTUH, Hyderabad, Telangana, India
e-mail: Pramodhkrishna.d@gmail.com

K. Venu Gopal Rao
e-mail: kvgrao1234@gmail.com

© Springer Nature Singapore Pte Ltd. 2018
S. S. Dash et al. (eds.), *Artificial Intelligence and Evolutionary Computations in Engineering Systems*, Advances in Intelligent Systems and Computing 668,
https://doi.org/10.1007/978-981-10-7868-2_15

services have complex structures and features. So finding their favorite products quickly and accurately is difficult. Recommendation systems were developed to solve this issue. Recommendation system is designed to filter out uninteresting items and suggest the items based on user's preferences.

In this paper, a tree-based recommendation system is proposed for making recommendations based on item description and user's interest. The items are represented using weighted trees; here, each node represents the concept, and each edge is associated with weight. The weight represents the importance or user preference of the concept. A generalized-weighted tree similarity technique is proposed to find the similarity between the tree structured items, and it is integrated with collaborative filtering approach for making recommendations.

This paper has theoretical aspects and practical aspects. At theoretical level, a generalized-weighted tree similarity technique is developed, and at practical level, recommendation approach is developed. Related work is presented in Sect. 2, generalized-weighted tree similarity technique is described in Sect. 3, a new hybrid recommendation method is described in Sect. 4, Sect. 5 describes case study, and the conclusion is presented in Sect. 6.

2 Related Works

Many approaches for recommendation systems were developed. Collaborative filtering method is successful and used in many applications [4]. CF method recommends item based on user's rating and rating of other like-minded users [4]. CF method contains user- and item-based CF methods [5]. Item-based collaborative filtering method was developed [6] and uses rating matrix for predicting recommendations. It uses the memory- and model-based approaches. Model-based approach uses Bayesian network, clustering, and rule-based approaches. Memory-based technique employs KNN approach. It suffers from cold start problem and data sparsity problem, and it is not scalable. CF based on semantic information was proposed [7], and it uses both the semantic similarity and rating similarity for the prediction.

In user-based CF approach, it finds k-similar users then combines user's rating on item for prediction. It also suffers from cold start and data sparsity problems. To resolve data sparsity problem, some dimensionality reduction techniques are introduced [8–10], but it leads to loss of some useful information during dimensionality reduction. Data sparsity problem can be solved by mapping users and items to a domain ontology [11]. Content-based system recommends the items, which are similar to previously preferred items by target user [12], CB approach suffers from contend dependence problem and new user problem [13, 14]. Fuzzy tree-based recommendation system [15] was proposed, and it depends on the semantics and properties of items and users. To generate the predictions, the semantic information contains the features of items and association between items. User- and item-based CF techniques are combined to get good performance.

2.1 Tree Similarity

In many applications like e-learning, e-business, e-commerce, and case-based reasoning, bioinformatics data are represented as tree structures. In previous research, trees are compared using many techniques. In literature, edit distance method is used to compare ordered and unordered trees. Similarity is measured as minimum cost of edit operation sequences. Hamming distance is used in some methods to compare trees. Hill climbing and bipartite matching techniques are used for comparing trees. In this paper, a generalized-weighted tree similarity technique is used to find the similarity between the trees [10]. Here data are represented by means of tree structure, in which each node denotes a concept and child nodes denotes the sub-concepts. The association between concepts is represented using tree. Each edge is associated with weight that represents importance of the concept or user preferences for the concept.

3 Generalized-Weighted Tree Similarity Technique (GWT-Similarity Technique)

A GWT-similarity technique is presented in this section. It contains match and miss algorithms. In this technique, the match and miss values between two trees are computed. To obtain similarity between the trees, the match and miss values are merged using generalized measure. Miss value is used to differentiate the similarity of two or more trees that have the same matching value.

Definition Weighted tree is a 5-tuple (N, E, V, A, W). Here, N is set of nodes; E is set of edges; V is set of node labels; A is set of edge labels; W is set of edge weights, and it represents the importance of the concepts or user preference on the concept.

3.1 Match Algorithm

Match algorithm determines the match value between two trees. It proceeds in top down and calculates match value in bottom up. It compares the roots of two trees. If the roots are not matched, then it returns similarity as 0. If the roots are same, then it considers the following cases.

Case 1: the roots are leaf nodes.
Case 2: the roots are non-leaf nodes.
Case 3: one root is leaf node, the other is non-leaf node and vice versa.

For case 1 and case 3, it returns match value as 1.0. In case 2, for every pair of the matching edges, it computes the average edge weights and multiplied with match algorithm at their sub-trees recursively. It sums the match value for every pair of the matching edges. Node equality factor (ϵ) is used to avoid zero match value if the roots are same and their child nodes are not matched. Match algorithm is given in Algorithm 1.

Algorithm 1: Match (T, T')

Input: T, T' are trees.
m=0, 0 <ϵ<0.5   // m is match value
If root (T) =root (T') then
                    {Case 1, Case 3}
If (root(T)&root(T') are leaf) or (root(T) is non leaf and root(T') is leaf) or (root(T) is leaf and root(T') is non leaf)
                            Return 1;
Else
{Case 2}
If (root (T) & root (T') are non leaf) then
for every pair of the matching edges $l_i \in$ T and $l'_j \in$ T' do
$$m = m + \frac{w(l_i) + w(l'_j)}{2} \times \text{Match} (T_i, T'_j);$$

 End for
Return (ϵ + (1- ϵ) $\times$m));
End If
End If
else

Return 0;

3.2   *Miss Algorithm*

Miss algorithm computes the miss value between two trees when any miss tree is appear in one of the two trees. This algorithm proceeds in top down and computes miss value in bottom up. Miss value is used to differentiate the similarity of two or more trees that have the same match value. Miss algorithm is given in Algorithm 2.

Algorithm 2: Algorithm Miss (T, T', match)

Input: roots of trees T, T'   //m is match value and λ is miss value.
λ=0.0, c=0.0, 0<ϵ<=0.5;
If root (T) = root (T') then {Case 2}
If T & T' are internal nodes then
For every pair of the matching edges $l_i \in$ T and $l'_j \in$ T' do
Miss (Ti, T'j);
End for
For every edge l_i which appears in T and absents in T' do
λ=λ+w(l_i) ×simplicity(m ,T_i); c=c+1;
End for
For every edge l'_j which appears in T' and absents in T do
λ=λ+w(l'_j)×simplicity(m , T'_j); c=c+1;
End for
Else
{Case 3}
If T is a internal node & T' is a leaf then λ=λ+simplicity (m, T);
c = c + 1;
Else
{Case 4}
If T is a leaf & T' is internal node then λ=λ+simplicity (m, T');
c=c+1;
End If
End If
End If
Return (λ/c);
Else
Return m;
End If

The GWT-similarity technique computes match and miss values and combines them to compute the similarity between trees using a generalized formula. This similarity includes both structural and semantic similarity between trees. The semantic similarity (S_1) between trees T and T' is given as

$$S_1 = \frac{((\beta^2 + 1) * m * \lambda)}{(\beta^2 * m + \lambda)} \tag{1}$$

The variable β provides the significance between m and λ values. If $\beta < 1$, then m value is more significant. If $\beta > 1$, then the λ value is more significant. If $\beta = 1$, then it provides a harmonic mean between λ and m values.

4   A Tree-Based Recommendation Technique

This tree-based recommendation technique takes the list of items rated by the target user and computes the similarity between target item i and then selects the top-k similar items. To find the similarity between two items, our method first computes the item-based CF similarity and then computes semantic similarity using GWT-similarity technique and merges them to obtain total similarity. Once the top-k similar items are determined, then prediction is made by taking weighted average of rating by target user on these similar items.

Step 1   Computing rating-based similarity

This step calculates the similarity between two items based on user's rating on items. In our approach, we used adjusted cosine similarity measure to find the similarity between items i and j. Let U_{ij} be the list of users who rated the items both i and j. The rating-based similarity (S_2) between items i and j is given in following formula.

$$S_2 = \frac{\sum_{u \in U_{ij}} \left(r_{u,i-\bar{r}_u}\right) \times \left(r_{u,j-\bar{r}_u}\right)}{\sqrt{\sum_{u \in U_{ij}} \left(r_{u,i-\bar{r}_u}\right)^2} \times \sqrt{\sum_{u \in U_{ij}} \left(r_{u,j-\bar{r}_u}\right)^2}} \tag{2}$$

where $r_{u,i}$ and $r_{u,j}$ are user u's rating on items i and j. $\bar{r}_u$ is the average rating of all items which is rated by user u. The range of S_2 is $[-1, 1]$. So it is normalized to get the range $[0, 1]$ using the formula

$$S_2 = \frac{(1 + S_2)}{2}.$$

Step 2   Computing Semantic similarity.

This step computes semantic similarity $S_1(i, j)$ between items i and j using (1) generalized-weighted tree similarity technique.

Step 3   Integrate the semantic and rating-based similarities.

In this step, semantic and rating similarities are merged to find total similarity between items i and j. Total similarity is calculated using the following formula (3).

$$\text{Total similarity}(i, j) = \alpha \times S_2 + (1 - \alpha) \times S_1. \tag{3}$$

where α is a semantic combination parameter, and its value is in the range $[0, 1]$.

Step 4   Recommendation Generation

After finding the similarity between the items, we select top-k similar items to the target item, then we can make prediction $P_{u,i}$ for target item i. $P_{u,i}$ is the prediction on item i for the target user u.

$$P_{u,i} = \frac{\sum_{n \in S_k} r_{u,n} \times \text{TotalSim}(i, n)}{\sum_{n \in S_k} \text{TotalSim}(i, n)} \qquad (4)$$

S_k denotes top-k similar items to the target item i. $r_{u,n}$ denotes user u's rating on item n.

5 Case Study

In this section, we have illustrated our proposed technique with example. Here, we are taking five users and five telecom products I1, I2, I3, I4, and I5 shown in Fig. 1. The user's opinion is indicated by rating from 1 to 5. A user-item rating matrix is given in Table 1. Here, package 5 is the new item. Therefore, rating-based similarity between I5 and others cannot be calculated.

Step 1 This step determines the rating-based similarity between items using Eq. (2). We have calculated rating-based similarity between items, and it is given in Table 2. Here, I5 is the new item; therefore, rating-based similarity between I5 and others cannot be computed.

Step 2 This step determines semantic similarity between items using generalized formula (1). Here, we assumed node equality factor (ϵ) value as 0.4. We have computed semantic similarity between items using GWT-similarity technique, and it is given in Table 3.

Step 3 This step computes overall similarity between items using Eq. (3). Let α be 0.4, total similarity between items computed which is shown in Table 4.

Step 4 This step generates the predictions using the overall similarity between items. The predictions are calculated as

$P_{1,3} = 2.23$, $P_{1,5} = 1.9$, $P_{2,3} = 1.81$, $P_{2,5} = 1.91$, $P_{3,1} = 1.19$, $P_{3,5} = 0.95$, $P_{4,2} = 0.90$, $P_{4,4} = 0.70$, $P_{4,5} = 0.98$, $P_{5,1} = 0.85$, $P_{5,4} = 0.83$, $P_{5,5} = 0.87$. Where P_{ij} denotes the prediction on the item j for user i.

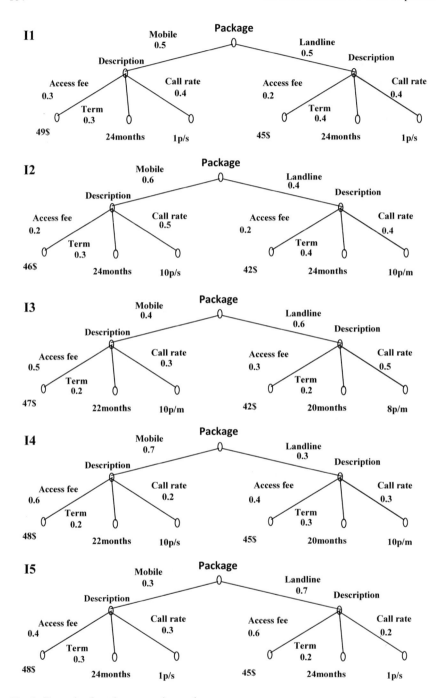

Fig. 1 Examples for telecom service packages

Table 1 Rating matrix of user and items

Telecom products					
Users	I1	I2	I3	I4	I5
U1	4	5		4	
U2	5	4		3	
U3		2	3	3	
U4	1		5		
U5		2	4		

Table 2 Rating-based similarity between items

	I2	I3	I4	I5
I1	0.70	0.29	0.66	
I2	1.0	0.25	0.80	
I3	0.25	1	0.37	
I4	0.80	0.37	1.0	

Table 3 Semantic similarities between items

	I2	I3	I4	I5
I1	0.30	0.42	0.30	0.86
I2	1.0	0.42	0.36	0.23
I3	0.42	1.0	0.22	0.40
I4	0.36	0.22	1.0	0.62

Table 4 Total similarities between items

	I2	I3	I4	I5
I1	0.46	0.36	0.44	0.51
I2	1.0	0.35	0.53	0.13
I3	0.35	1.0	0.28	0.24
I4	0.53	0.28	1.0	0.37

6 Conclusion

This study proposes a new recommendation technique using tree structured data. In this method, a generalized-weighted tree similarity technique is developed to find similarity between tree structured items, and it is combined with item-based collaborative filtering approach for generating predictions. The case study shows that the proposed method recommends the telecom service package, in which the rating for new item can be predicted. The proposed approach solves the new item problem and data sparsity problems in the traditional collaborative filtering approach.

References

1. J. Lu, in A personalized e-learning material recommender system, International Conference on Information Technology and Applications. Macquarie Scientific Publishing (2004), pp. 374–379
2. S. Alter, Moving toward E-business as usual, in *Information Systems: Foundation of E-Business*, 4th edn. (Prentice Hall, Upper Saddle River, NJ, 2001), pp. 30–67
3. J.B. Schafer, J.A. Konstan, J. Riedl, in E-commerce recommendation applications, Applications of Data Mining to Electronic Commerce (Springer, New York, 2001), pp. 115–153
4. Z. Huang, D. Zeng, H. Chen, A comparison of collaborative-filtering recommendation algorithms for e-commerce. IEEE Intell. Syst. **22**(5) (2007)
5. X. Guo, L. Jie, Intelligent e-government services with personalized recommendation techniques. Int. J. Intell. Syst. **22**(5), 401–417 (2007)
6. P. Jagalingam, A.V. Hegde, A review of quality metrics for fused image. Aquatic Proc. **4**, 113–142 (2015)
7. B. Mobasher, J. Xin, Y. Zhou, Semantically enhanced collaborative filtering on the web, in *Web Mining: From Web to Semantic Web* (Springer, Berlin, 2004), pp. 57–76
8. K. Goldberg, et al., Eigentaste: a constant time collaborative filtering algorithm. Inf. Retr. **4** (2), 133–151 (2001)
9. J.D.M. Rennie, N. Srebro, in Fast maximum margin matrix factorization for collaborative prediction, *Proceedings of the 22nd International Conference on Machine Learning*. ACM (2005)
10. D.P. Krishna, K. Venu Gopal Rao, Generalized weighted tree similarity algorithms for taxonomy trees. EURASIP J. Inf. Secur. **2016**(1), 35 (2016)
11. C.C. Aggarwal, S. Yu Philip, in Semantic based collaborative filtering. U.S. Patent No. 6,487,539, 26 Nov 2002
12. M.J. Pazzani, D. Billsus, Content-based recommendation systems, in *The Adaptive Web* (Springer, Berlin, 2007), pp. 325–341
13. G. Adomavicius, A. Tuzhilin, Context-aware recommender systems, in *Recommender Systems Handbook* (Springer, New York, 2015), pp. 191–226
14. Y. Dou, H. Yang, X. Deng, in A survey of collaborative filtering algorithms for social recommender systems, 12th International Conference on Semantics, Knowledge and Grids (SKG). IEEE (2016)
15. D. Wu, L. Jie, G. Zhang, A fuzzy tree matching-based personalized e-learning recommender system. IEEE Trans. Fuzzy Syst. **23**(6), 2412–2426 (2015)

Product Recommendation System Using Priority Ranking

S. Vasundra, D. Raghava Raju and D. Venkatesh

Abstract Numerous suggestion procedures have been produced over the previous decade, and significant endeavors in both scholarly community and industry have been made to enhance proposal precision. Recommender frameworks are fundamentally utilized as a part of e-business frameworks. In any case, it has been progressively noticed that it is not adequate to have exactness as the sole criteria in measuring suggestion quality and consider other vital measurements, for example, differing qualities, certainty, and trust, to produce proposals that are precise as well as helpful to clients. More different suggestions, probably prompting to more offers of long-tail things, could be advantageous for both individual clients and some plans of action. Existing frameworks like community collaborative filtering (CF) are utilized as a part of different headings. The primary point of the paper is to locate the best top-N proposal records for all clients as per two measures—they are precision and assorted qualities. To get greatest differences, optimization-based methodologies are utilized while keeping up satisfactory levels of exactness in the proposed technique.

Keywords Frameworks · e-business · Community collaborative filtering
Precision · Assorted qualities

S. Vasundra (✉)
Department of Computer Science and Engineering, JNTUA CEA,
Anantapur, Andhra Pradesh, India
e-mail: vasundaras@rediffmail.com

D. Raghava Raju
Department of Computer Science and Engineering, Sri Venkateswara
Institute of Technology, Anantapur, Andhra Pradesh, India
e-mail: raghava.digala@gmail.com

D. Venkatesh
Department of Computer Science and Engineering, GATES Institute
of Technology, Gooty, Anantapur, Andhra Pradesh, India
e-mail: dvvenkatesh@yahoo.co.in

© Springer Nature Singapore Pte Ltd. 2018   167
S. S. Dash et al. (eds.), *Artificial Intelligence and Evolutionary Computations
in Engineering Systems*, Advances in Intelligent Systems and Computing 668,
https://doi.org/10.1007/978-981-10-7868-2_16

1 Introduction

With the extension of e-exchange, recommender systems have transformed into a key portion of modified e-business benefits and are pivotal for e-business providers to remain reasonable. Group filtering is one of the proposition techniques, whose execution has been shown in various e-exchange applications. Collaborative filtering (CF) automates the "casual" process. It structures a pointer mass that fills in as an information hot spot for recommendations.

On the other hand, customary CF frameworks encounter the evil impacts of two or three fundamental repressions, for instance, the cool start issue, data sparsity issue, and recommender trustworthiness problem. Thus, they encounter trouble overseeing high-affiliation, learning concentrated spaces, for instance, e-learning highlight on premium. To beat these issues, examiners have proposed proposition techniques, for instance, a hybrid system joining CF with substance-based sifting. Since e-business Web districts for re-adjusting consistently have distinctive thing classes, isolating the various characteristics of these orders for substance-based moving is incredibly troublesome. In this way, it might be helpful to overcome these obstructions by improving the CF framework itself. The auxiliary designing of recommender systems is as shown in Fig. 1. Conventional CF techniques construct their recommendations in light of a singular recommender pack. Our CF technique structures twofold recommender groups an equivalent customers' social event and an expert customers' get-together as sound information sources. By then, it separates every cluster's effect on the objective customers for the objective thing classes.

2 Literature Survey

As in [1] developed a technique that joins an exactness focused count and grouped qualities focused figuring. As demonstrated by maker, such joint endeavors can yield best outcomes changing both precision and contrasts, without relying upon any semantic or setting specific information. They utilized averaging process as a part of their calculation that backings differing qualities improvements.

Fig. 1 Recommended architecture for proposed system

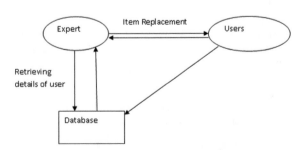

As in [2] developed a recommender count that works for compound things and organizations instead of just solitary things. For this they completed CARD structure which on a very basic level separates the utility space and grouped qualities space to sidestep the trade-off amid closeness and contrasts. The figuring he laid out is computationally beneficial and beats similar to contrasting qualities.

As in [3] proposed an approach where the position of things in proposition rundown has given for the most part essentialness. Makers said that however contrasts atmosphere individual or aggregate is expert, it is excessive that customer will see its purposes of intrigue. A well widens once-over may demonstrate the practically identical things at the most noteworthy need on the rundown and grouped things might be at the base which is not alluring reliably. Especially it is significantly fundamental if there ought to be an event of little screen contraptions where the customer is enthusiastic about quite recently beat couple of recommendations.

As in [4] said that customer's leeway are always overflowing with a helplessness which could not be tended to by top-N once-over of recommendations easily. As opposed to this, maker proposed a cloud model which is powerful at enlightening learning hazards.

The review consolidates a watchful depiction of both substances based and group situated frameworks for addressing things and customer profiles. By then, we inspected about various attempts by means of some expertise in individual and what's more aggregate contrasts. Finally, it discusses examples and future investigation which may lead toward the best in class period of recommender structures.

Recommender frameworks have expected a colossal part in helping clients discover material things from endless, especially in e-business applications, for example, Amazon and Netflix. Recommender structures are in like way great to online substance suppliers. For instance, as per Forrester Research, it is assessed that recommender frameworks address 10–30% of an online retailer's courses of action. Netflix in addition nitty-gritty that around 66% of their films leased were ones that clients may never have considered something else yet was suggested by their recommender framework. With the making eagerness for changed recommendation, much work has been finished over the traverse of the most recent decade on growing new suggestion strategies, both in the business and the adroit world. The Netflix Prize competition1 is an OK case for the making thought for these proposal strategies. In any case, paying little regard to fundamental movement, current recommendation frameworks still have diverse important inconveniences to be tended to.

As in [5] proposed to characterize past strategies as takes after. Customary recommender frameworks utilize the client's most loved weights. These weights are considered for each item qualities are measured in light of the in general evaluations for the results of clients. Learning-based techniques in which clients straightforwardly condition their basic top picks in an interactive recommendation interface. Frameworks in which the clients can distinguish their gauge individual items. As in [6] evaluated the improvements on two different data sets, one again in view of data from motion pictures' site and one from the antivirus. As in [7] proposed different strategies to predict the item ratings.

There are number of suggestions proposed for the recommender frameworks to update accuracy. The precision which the recommender structure predicts clients' examinations for the items that they have not yet rated. While suggestion exactness is certainly fundamental, there is a making understanding that precision does not all around prescribe steadiness to clients, and depending upon the exactness of proposals alone may not be satisfactory to locate the most critical things for each. Along these lines, in spite of the precision of suggestions, this article investigates the shifting attributes of recommendation that can mirror the farthest point of recommender frameworks to go past the doubtlessly self-evident, untouchable things, and make more fanciful, revamp, and long-tail proposition [8–10]. A conventional recommender framework offers suggestions to a client by surveying the appraisals of things yet to be eaten up by the client, in context of the assessments of things enough expended. Recommendations to clients are made considering the anticipated appraisals of everything for every client (i.e., the things with the most astoundingly expected examinations are the ones embraced to the client). While the greater part of current recommender structures utilize a solitary numerical rating to address each client's inclination for a given item. Recommender frameworks in some e-business settings have tested multi-criteria examinations that catch more right data about client incline concerning unmistakable parts of a thing (e.g., getting client evaluations for the story and acting bits of every motion picture in a film recommender framework). Some substance-based recommender frameworks utilize differing substance properties for thing examinations and equivalence figuring, the subjective client incline for a thing are still gotten by an essential general rating. On the other hand, my paper concentrates on multi-criteria rating frameworks that permit clients to rate distinctive different parts of a thing.

3 Proposed System

A suggestion framework gives an answer when a great deal of valuable substance turns out to be a lot of something to be thankful for. A proposal motor can help clients find data of enthusiasm by investigating recorded practices. An ever-increasing number of online organizations including Netflix, Google, Facebook, and numerous others are coordinating a proposal framework into their administrations to help clients find and select data that might be specifically compelling to them. As in [1] proposed to re-rank the rundown of hopeful things for a client to enhance the total differences. Initial, a requested rundown of suggestions is computed utilizing any separating procedure. Second, for all things having a superior expected rating than a given limit, extra components are ascertained, for example, the supreme and relative amiability of a thing (what number of clients preferred the thing among all clients or among all clients who appraised that thing, individually) and the thing's evaluating change. As per these components, the applicant things are re-positioned and just the top-N things are prescribed. Along these lines, specialty things are pushed to the proposal records and extremely mainstream things are rejected. While this re-positioning strategy can enhance the total differences, it comes to the detriment of exactness. The re-positioning methodology, quickly talked about above, can enhance suggestion differing qualities by prescribing those things that have bring down anticipated evaluations among the things anticipated to be pertinent, by changing positioning edge TR, yet it does not give coordinate control on how much differences change can be gotten. To address this confinement, Item Replacement Technique endeavors to specifically build the quantity of particular things suggested over all clients (i.e., enhance the assorted qualities in-top-N measure). The fundamental thought behind this iterative approach is as per the following. To start with, the standard positioning method-ology is connected to every client, to acquire the underlying top-N proposals, commonly with the best precision. By then, iteratively, one of the starting at now recommended things is supplanted by another confident thing that has not yet been endorsed to anyone, thusly growing the varying qualities by one unit, until the distinction augmentations to the looked for level, or until there are no more new things available for substitution. Since thing substitution is made just when it brings about a prompt change of assorted qualities by on unit, we allude to this approach henceforth as an Item Substitution Approach. The most every now and again suggested thing i_{old} is supplanted by one of the never-prescribed things i_{new} for a similar client. Among every one of the clients who got suggested thing i_{old}, a trade happens for client u_{max}, who is anticipated to rate thing i_{old} most exceptionally, taking into consideration a perhaps higher anticipated rating an incentive for the substitution thing i_{new} (and, in this way, for better exactness). At the end of the day, since any new hopeful thing for substitution i_{new} is anticipated to be lower than thing i_{old} for the picked client, the higher the forecast of thing i_{old}, the higher the likelihood of acquiring a high expectation of the new thing i_{new}.

As shown in Fig. 1, the architecture of proposed system is there. First there is a database consisting of the all products including different types like sports and movies. All the users are registered with the recommender system. All the details of the users are saved in the database such as user id, password, qualification, interests. While recommending the items to the users, the proposed system will follow the Item Replacement Technique as shown in the following algorithm.

Algorithm 1

  Step 1: Initializing top-N recommendation list
  Step 2: Replacing already recommended item i_{old} with the never recommended item i_{new}
  Step 3: Repeat the process until maximum diversity is reached.

As shown above, the process of the proposed system works. First top-N recommendation list is generated by using the standard ranking. And items will be replaced by the never recommended items.

4 Improved Re-ranking Technique

In this mechanism, the items are get prioritized according to their ratings as

$$\text{Rerank}(i) = \text{prior}(R * (u, i)^{-1} \tag{1}$$

where R represents ratings, u user, i item.

In the traditional ways, the unknown ratings of the items are given by neighboring ratings of products as in CF. Here the ratings of the unknown items are useful to the customer who are willing to get products. We use the improved re-rank technique in our project. Here we use (1) for the approach. In this technique, the items are first selected from database as new list. Then we use traditional technique to rate the items. Hence we are giving priorities to the each product based on the rating. Then we apply normal re-rank technique. The main advantage of this technique is to improve the accuracy of the recommender system.

We are defining the quality of the proposed system by using following metrics: Diversity and Accuracy

Diversity can be measured by using

$$\text{Diversity} = \bigcup_{u \in U} L_N(u) \tag{2}$$

where $L_N(u)$ represents the list of all recommended items.

Accuracy can be measured by using

$$\text{Accuracy} = \sum_{u \in U} |\text{correct}(L_N(u))| / \sum_{u \in U} |L_N(u)| \tag{3}$$

where $L_N(u)$ represents the list of all recommended items. The metric is estimated with the correct items to the all the items of all users. Table 1 presents the clear information how the user can choose the rating from 1 to 5 for the particular product.

Table 1 Rating description

Rating	Description
5	Excellent
4	Best
3	Good
2	Average
1	Bad

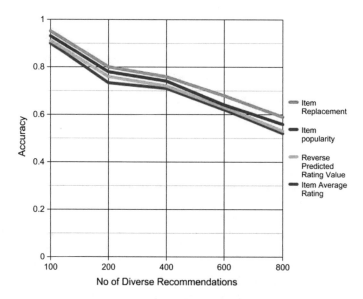

Fig. 2 Comparison of accuracy and diverse recommendations

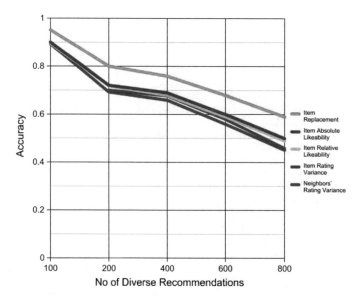

Fig. 3 Comparison of accuracy and the diverse recommendations

5 Results

Experimental results show that the proposed technique gives high accuracy than existing technique (as shown in Figs. 2 and 3).

6 Conclusion

Suggestion differences have as of late pulled in extensive consideration as an imperative viewpoint in assessing the nature of proposals. Customary recommender frameworks normally prescribe the top-N most exceedingly anticipated things for every client, subsequently giving great prescient precision, however performing ineffectively as for suggestion assorted qualities. Thusly, the proposed technique, "thing swap procedure for high total differing qualities in recommender framework," improves the differences.

The proposed procedure has a few points of interest over the suggestion re-positioning methodologies from former writing: (1) acquiring further changes in differing qualities parallel to the level of precision. The proposed enhancement methodology has been composed particularly for the differing qualities in-top-N metric, which measures the quantity of different things among the top-N proposals.

References

1. C.-N. Ziegler, S.M. McNee, J.A. Konstan, G. Lausen, in Improving recommendation lists through topic diversification, Proceedings of the 14th international conference on World Wide Web (2005), pp. 22–32
2. M.J. Pazzani, A framework for collaborative, content-based and demographic filtering. Artif. Intell. Rev. **13**(5–6), 393–408 (1999)
3. P. Melville, R.J. Mooney, R. Nagarajan, in Content boosted collaborative filtering for improved recommendations, *Proceedings of the Eighteenth National Conference on Artificial Intelligence (AAAI-02)* (2002), pp. 187–192
4. S.K. Lam, J.R. Shilling, in Recommender systems for fun and profit, 13th international conference on World Wide Web (2004), pp. 393–402
5. G. Adomavicius, N. Manouselis, Y. Kwon, *Multi-criteria Recommender Systems. Recommender Systems Handbook* (Springer, New York, 2011), pp. 769–803
6. D. Jannach, Z. Karakaya, F. Gedikli, in Accuracy improvements for multicriteria recommender systems, *Proceedings of the 13th ACM Conference on Electronic Commerce (EC 2012)* (2012), pp. 674–689
7. M. Nilashi, D. Jannach, O. bin Ibrahim, N. Ithnin, in Clustering and regression-based multi-criteria collaborative filtering with incremental updates. Inf. Sci. **293**, 235–250 (2015)

8. B. Sarwar, G. Karypis, J. Konstan, J. Riedl, in *Application of dimensionality reduction in recommender system-a case study*. Technical report, DTIC Document
9. A.I. Schein, A. Popescul, L.H. Ungar, D.M. Pennock, in Methods and metrics for cold-start recommendations, International ACM SIGIR Conference on Research and Development in Information Retrieval (2002), pp. 253–260
10. R. Bell, Y. Koren, C. Volinsky, Matrix factorization techniques for recommender systems. IEEE Comput. **42**(8), 30–37 (2009)

Managing and Control of Data Transmission in a High Mobility-based Wireless Networks

Giri M., Seethalakshmi R. and Jyothi S.

Abstract Wireless network is a communication network for voice, data, video, graphics, images, and other signals. Managing and control of such data transmission is an art in such networks. A new protocol for data transmission in high mobility context deals with formulating a new dynamic topology and reconstructing a data path from source node to the destination node without packet loss. The new packets are redirected through the discovered topology which encompasses a new message exchange protocol and retained in the network alive. The new protocol controls the data transmission in a novel way and prevents packet loss in the network by quickly formulating a new wireless network pertaining to current context. The new protocol emphasizes the stability of cluster heads and allows mobility of other nodes. It discovers the new route with the stabilized cluster heads by passing messages. Thus, the discovered route prevents data loss.

Keywords Wireless networks · Cluster · Cluster heads · Mobility
Topology · Cluster_regen_mobil_protocol · Cooperative networks

Giri M. (✉)
Department of Computer Science and Engineering, Rayalaseema University,
PP. Comp. Sci & Engg. 0485, Kurnool, Andhra Pradesh, India
e-mail: prof.m.giri@gmail.com

Seethalakshmi R.
Department of Computer Science and Engineering, Vel Tech High
Tech Dr. Rangarajan & Dr. Sakunthala Engineering College, Chennai, Tamil Nadu, India

Jyothi S.
Department of Computer Science, SPMVV University, Tirupati, Andhra Pradesh, India
e-mail: jyothi.spmvv@gmail.com

© Springer Nature Singapore Pte Ltd. 2018   177
S. S. Dash et al. (eds.), *Artificial Intelligence and Evolutionary Computations
in Engineering Systems*, Advances in Intelligent Systems and Computing 668,
https://doi.org/10.1007/978-981-10-7868-2_17

1 Introduction

Wireless networks are very popular these days in the context of communication networks. Wireless media is flourishing very well with the advent of many wireless communication strategies. Digital network is another area which reinvents new techniques for data transmission. Architecturally, a wireless network consists of nodes with wireless links. The links are categorized into strong and weak links. Strong links lead to less mobility and reliable data transmission. The weak link is leading to more mobility in the network and less reliable data transmission. Generally, the wireless networks lead to large clusters during peak transmission time. The clusters encounter a cluster head which is responsible for communicating with other clusters and intra-cluster communication. The internetworking is achieved by cluster gateway nodes. The cluster heads play a major role in the wireless network communication of majority of network category.

Cluster-based wireless networks are called cooperative networks. The cluster networks are necessitated in the context of scalable networks. There are two types of cluster communication. They are intra-cluster communication and inter-cluster communication. When massive parallel processing is encountered, it is worked out with cluster communication networks. Parallelism involves dependency processing and granularity checking. A high granular and high parallelized code can be executed on cluster networks. In such networks, data transmission takes place as requested by the code. Every node has some processing which involves some data. There are code distribution and data distribution in such networks. The cluster heads decide the data and code to be distributed across the clusters and within the cluster. After processing, the results are gathered and distributed to the requested nodes. In this context, high mobility is playing a major role. A novel algorithm for handling data transmission in such high mobility-based networks is dealt with here.

2 Related Work

In [1], Ian et al. have done work on a Survey of Mobility Management in Next-Generation-All-IP-Based Wireless Systems in 2004. This paper discusses clearly about the various techniques used in general. In [2], Yiche et al. have discussed about the Strategies of Developing Deep Ocean Water Industry Cluster and Value Network Views in 2007. This paper deals with deep ocean water industry cluster which is different from current work, but protocols are referred. In [3], Mathivaruni et al. have proceeded with a work on An Activity-Based Mobility Prediction Strategy using Markov Modeling for Wireless Networks in 2008. But Markov Modeling is very old method. Hence, we adapt a new strategy in the current work. In [4], Mehdi et al. have worked on Distributed Multiple Target-SINRs Tracking Power Control in Wireless Multi-rate Data Networks in 2013. This is different from our current work but referred to the context.

In [5], Yalcin et al. have proceeded with Minimum Energy Data Transmission for Wireless Networked Control Systems in 2014 and this method is a very old method. In [6], Henrik et al. have proceeded with Adaptive Admission Control in Interference-Coupled Wireless Data Networks: A Planning and Optimization Tool Set in 2014, and this method is encountered in the control part of the current context. In [7], Kalyani has worked on Mobility-based Clustering Algorithm for Ad hoc Network: MBCA in 2014, and this algorithm is referred to but it is not a novel algorithm. In [8], Santhi et al. have done a work on Mobility-Based Tree Construction for ZigBee Wireless Networks in 2014 but here cluster is dealt with.

In [9], Mahathir et al. have conducted a research on Enhanced Availability in Content Delivery Networks for Mobile Platforms in 2015. This work deals with content delivery networks, and the protocol is very much existing. In [10], Xiaobo et al. have done a work on Distributed Resource Reservation in Hybrid MAC with Admission Control for Wireless Mesh Networks in 2015. This is also an existing technique. In [11], Zhiyang et al. have worked on An Effective Approach to 5G: Wireless Network Virtualization in 2015, and it is a technique on virtualization. In [12], Li et al. have done a work on Control and Data Signaling Decoupled Architecture for Railway Wireless Networks in 2015, and it deals with railway networks.

In [13], Seyhan et al. have done a work on Multi-hop Cluster-based IEEE 802.11p and LTE Hybrid Architecture for VANET Safety Message Dissemination in 2016. This is dealing with VANET and referred to as an existing technique. In [14], Hung et al. have worked on Minimizing Radio Resource Usage for Machine-to-Machine Communications through Data-Centric Clustering in 2016. Data-centric clustering is useful, but here the mobility is also encountered. In [15], Seiseki et al. have conducted a research on Network Coding for Distributed Quantum Computation over Cluster and Butterfly Networks in 2016. This work deals with quantum computation in butterfly networks and in the current context deals with cluster networks with mobility. All the existing techniques have some drawbacks or the other when dealing with mobility in the cluster networks. Hence, in this work, the Cluster_regen_mobil_protocol has been developed as a novel and a new technique which is totally different from the existing technique.

3   Intra- and Inter-Cluster Cooperative Paradigm

Clusters are networks which work in unison having some commonality. Cluster is identified by the cluster head. Each node in a wireless cluster network has a unique identifier which is recognized by the network for communication. In a wireless network, there may be many clusters which may be working in cooperation. The clusters communicate using the cluster heads. Every cluster discovers and registers and records the nodes in its own network. The topology, structure, and neighborhood connectivity are well defined under stable condition. We assume the cluster

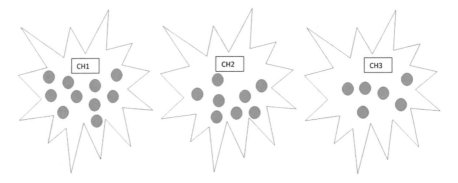

Fig. 1 Intra- and inter-cluster cooperative paradigm with cluster heads defined

heads are immobile during strong mobility-based condition. The clusters reorganize themselves once the mobility occurs.

A new registration of nodes is done with the cluster head, and the network topology is redefined. The number and nature of the clusters keep varying, and interaction in such context is also dynamic. The nodes cooperatively interact with each other. Within a cluster, when computation is needed, a node will use the preexisting path and tries to compute and returns. If communication is needed across clusters, one cluster head which is the initiator distributes the data to the other clusters via cluster head and returns. The results are fetched back. Figure 1 shows the cluster paradigm in detail. The figure shows the wireless network organized as clusters with each cluster having a cluster head. The communication within a cluster is via the cluster head, and across clusters the communication takes place via the corresponding route explored.

4 Processing in Wireless Cluster Networks

Whenever the processing of data is to take place, the data or code is pre-analyzed by the cluster network. The number and nature of nodes in the cluster are decided. The number of clusters required for processing is also designed. The master cluster then constructs a new network for processing. The data is distributed to various nodes in the cluster, and processing takes place for a given configuration. This is as for as the stable network is concerned. When node mobility occurs, there occurs an assumption that the cluster heads do not move (Fig. 2).

During mobility, the new protocol named Cluster_regen_mobil_Protocol is executed and this protocol constructs a new network for computation. The data is redistributed, and processing continues. The results are gathered and output to the user in the form of GUI or data.

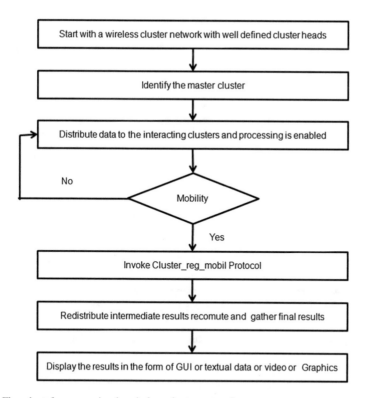

Fig. 2 Flowchart for processing in wireless cluster networks

5   Cluster_regen_mobil_Protocol

Step 1   Define the initial clusters in the current context like the one shown in Fig. 1.

Step 2   Identify the identity of the cluster heads.

Step 3   Process the code or data and determine the dependencies.

Step 4   Distribute the data via wireless links and initiate computing.

Step 5   Check for mobility of nodes by the master cluster.

Step 6   The master cluster identifies the node mobility in each cluster to which it is linked to.

Step 7   If there is mobility, the data is collected via the cluster heads of all clusters at the master node. It gathers the cluster heads and checks for the incoming and outgoing nodes during mobility.

Step 8   Each cluster head passes the bit patterns and polls for the cluster tree in the network.

Step 9   The polling in each cluster starts from the cluster head, and it records whether a node under consideration is the ascendant or the descendent node in the cluster.

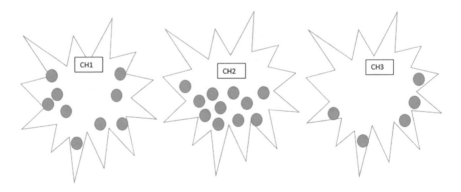

Fig. 3 Reforming of clusters in the wireless cluster network

Step 10   A new cluster tree for the cluster under consideration is constructed by the cluster head.

Step 11   After each cluster rediscovers the network, the new distribution of data is processed at the master node based on the details of the new clusters rediscovered.

Step 12   The master cluster again redistributes the data to the new clusters via the cluster heads.

Step 13   Processing proceeds and the lifetime of the nodes is also monitored.

Step 14   If any node has less power, it is highly powered from the cluster head and all nodes are made active.

Step 15   The master cluster continuously monitors for node mobility and restructuring.

Step 16   If there is node mobility, the steps 6–15 are repeated.

Step 17   If there is no node mobility, the processing continues and proceeds and output is gathered by master cluster.

Step 18   At the master cluster, the result is processed for GUI, voice, video, graphics, image, or text and displayed to the user (Fig. 3).

5.1   Basic Assumptions on the Wireless Cluster Network

The number of master cluster for the current context 1.

The initial number of clusters with cluster heads q.

Initial distribution of data on one cluster with probability:

$$y = P_i(1, r)/P_m(1, 1) \qquad (1)$$

P_i is ith node probability, and P_m is the master node probability; r is the number of nodes in the given cluster, and 1 is the number of nodes in the master cluster.

For q clusters, the probability is $q * y$.

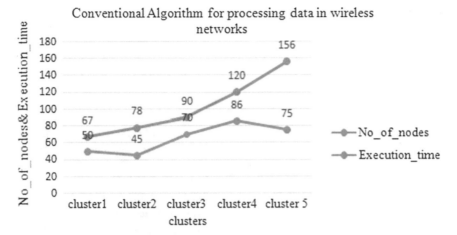

Fig. 4 Performance of the existing cluster data processing

5.2 For Node Mobility

The number of clusters with cluster heads g.

The redistribution of data on one cluster with probability:

$$y = P_i(1,j)/P_m(1,1) \qquad (2)$$

P_i is ith node probability, and P_m is the master node probability; j is the number of nodes in the reformed cluster, and l is the number of nodes in the master cluster. For g clusters, the probability is $g * y$.

5.3 Display of Results

The master cluster displays the output with the probability:

$$o = P_m(1,1) * K_i/Pg(1,v) \qquad (3)$$

P_i is ith node probability, and P_m is the master node probability; v is the number of nodes in the reformed cluster, and l is the number of nodes in the master cluster, and K_i is the random number of data displayed at the various clusters.

6   Results and Experimentation

The algorithm is simulated in NetSim package. Initial clusters are formulated with one master cluster. The other clusters are randomly chosen for the number of nodes and dynamic topology and connected. The data is initially distributed for the formed topology and the dependencies, and the processing is carried out. The diagrams below depict this. Then, when mobility occurs, the Cluster_regen_mobil_protocol is invoked and the experimentation is carried out. The results are recorded. Later, the nature of the networks and the number of nodes are varied due to mobility, and the experimentation is carried out, and results are recorded and displayed. Figure 5 depicts the initial configuration of the network, and Fig. 6 shows the configuration during network mobility, and Fig. 7 shows the configuration during the mobility later. Thus, this protocol executes the data at various situations. This is compared with the conventional algorithm which already exists. It is shown in Fig. 4, and it is also interpreted that the proposed algorithm is working better than the existing algorithm in terms of the execution time and mobility because there is no organization of the network during mobility in the existing algorithm.

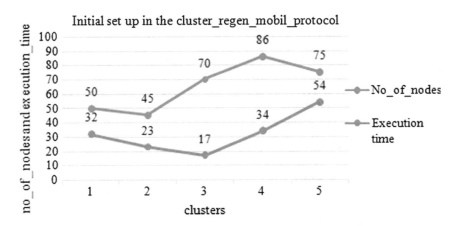

Fig. 5 Initial configuration of the Cluster_regen_mobil_protocol

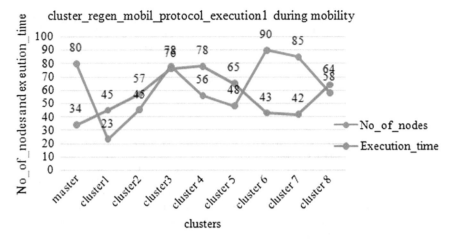

Fig. 6 Sample run1 of a Cluster_regen_mobil_protocol during mobility

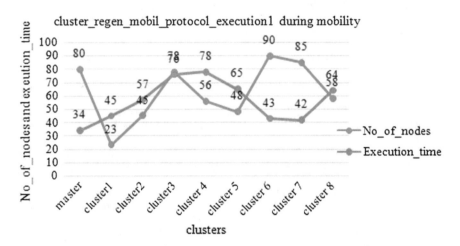

Fig. 7 Sample run2 of the Cluster_regen_mobil_protocol during mobility

7 Conclusion

The wireless cluster networks are prominent networks for processing data in the computing world. The efficiency of the network depends on the performance of the network during mobility. In the current context, the existing protocol is compared with the proposed Cluster_regen_mobil_protocol. It is interpreted from the experimentation that the Cluster_regen_mobil_protocol executes better in the cluster network during mobility than the existing protocol in terms of the time of execution and the variation in the number of nodes. The existing protocol does not emphasize a master and the cluster head nodes which do not move during mobility. This is the basis of the new protocol, and the performance is enhanced by 40% than the existing method. Thus, the Cluster_regen_mobil_protocol is evaluated and found to be better in terms of the performance execution parameters.

References

1. F.A. Ian, X. Jiang, M. Shantidev, in A survey of mobility management in next generation-All-IP based wireless systems. IEEE Wireless Commun 16–28, (2004)
2. C. Yiche, R.L. Yan, F.H. Pi, S.L. Chung, in Strategies of developing deep ocean water industry cluster and value network views, in *PICMET 2007 Proceedings* (2007), pp. 351–357
3. R.V. Mathivaruni, V. Vaidehi, V, in An activity based mobility prediction strategy using markov modeling for wireless networks, *WCECS*, Oct 2008
4. M. Mehdi, R.Z.A. ALi, G.H. Shapoor, R. Mehdi, in Distributed multiple target-SINRs tracking power control in wireless multi rate data networks. IEEE Trans. Wireless Commun. **12**(4), pp. 1850–1859 (2013)
5. S. Yalcin, C.E. Sinem, P. Pangun, Minimum energy data transmission for wireless networked control systems. IEEE Trans. Wireless Commun. **13**(4), 2163–2175 (2014)
6. K. Henrik, F. Gerhard, in Adaptive admission control in interference-coupled wireless data networks: a planning and optimization tool set, IEEE Mobile and Wireless Network Symposium. ICC (2014), pp. 2375–2380
7. Kalyani, Mobility based clustering algorithm for ad-hoc network. MBCA. IJARCSSE. **4**(12), 597–602 (2014)
8. S.G. Santhi, K. Chitralakshmi, Mobility based tree construction for ZigBee. Wireless Netw. **5**(1), 22–25 (2014)
9. A. Mahathir, K. Ibrahim, T. Zahir, Y.Z. Albert, S. Sartaj, Enhanced availability in content delivery networks for mobile platforms. IEEE Trans. Parallel Distrib. Syst. **26**(8), 2247–2257 (2015)
10. Y. Xiaobo, N. Pirabakaran, M. Klaus, C. Haitham, Distributed resource reservation in hybrid MAC with admission control for wireless mesh networks. IEEE Trans. Veh. Technol. **64**(12), 5891–5903 (2015)
11. F. Zhiyong, Q. Chen, F. Zebing, W. Zhiqing, L. Wei, Z. Ping, in *An effective approach to 5G: wireless network virtualization.* IEEE Communications Magazine, IEEE Communications Magazine—Communications Standards Supplement (2015), pp. 53–59
12. Y. Li, F. Xuming, F. Yuguang, Control and data signaling decoupled architecture for railway wireless networks. IEEE Wireless Commun. 103–111 (2015)
13. U. Seyhan, C.E. Sinem, O. Oznur, Multi hop cluster based IEEE 802.11p and LTE hybrid architecture for VANET safety message dissemination. IEEE Trans. Veh. Technol. **65**(4), 2621–2636 (2016)
14. Y.H. Hung, C.J. Tzu, D.T. Yun, C.H. Hong, Minimizing radio resource usage for machine to machine communications through data centric clustering. IEEE Trans. Mob. Comput. **15**(12), 3072–3086 (2016)
15. A. Seiseki, M. Mio, Network coding for distributed quantum computation over cluster and butterfly networks. IEEE Trans. Inf. Theory **62**(11), 6620–6637 (2016)

Optimization Technique for Flowshop Scheduling Problem

Amar Jukuntla

Abstract This paper presents a novel approach for solving flowshop scheduling problem with the objective of minimizing the makespan and maximizes the machine utilization with the help of optimization techniques. Flowshop is used for allocation of resources among the tasks to complete their scheduling process with optimization technique to get a feasible solution. This paper illustrates a proposed method with example and compared with traditional algorithms.

Keywords Optimization · Flowshop scheduling · Shop scheduling
Optimization techniques · Artificial bee colony · Genetic algorithm

1 Introduction

Scheduling is a process of optimizing work and workloads and generating reports that show a plan for the timing of certain activity to be done [1]. This scheduling can be done in two ways; in first, the sequence is planned and executed. In second, designing planning time and completion time of each task can be performed.

In scheduling, the amount of resource utilization for each task should be known so that tasks can be determined in feasibly. In addition, each task can be described in terms of resource requirement, duration, start time, and termination time. Any technological constraints can be described during execution of a task. In manufacturing industries, resources are usually called machines; task is pronounced as jobs. These jobs contain several operations, and the environment is called job shop. The scheduling problem raised due to a limited number of resources available complete task and the time determination of given resources to complete tasks. This situation arises for decision-making of scheduling [2] in resources. In industries,

A. Jukuntla (✉)
Department of CSE, Vignan's Foundation for Science, Technology & Research,
Vadlamudi, Andhra Pradesh, India
e-mail: amar.jukuntla@gmail.com

© Springer Nature Singapore Pte Ltd. 2018   187
S. S. Dash et al. (eds.), *Artificial Intelligence and Evolutionary Computations
in Engineering Systems*, Advances in Intelligent Systems and Computing 668,
https://doi.org/10.1007/978-981-10-7868-2_18

decisions are made with planning function and describe the design of a product, for making and testing the requirements to be produced. In other, planning function gives the information about resources available for production and tasks to be scheduled.

In present industrial environment, facing competition from customer requirements and expectations is very high in terms of quality, cost, and delivery times. For solving this, industries introduced shop scheduling. Scheduling the task on machines through job sequence flowshop scheduling [3] gives better throughput in terms of machine utilization and satisfies the customer needs highly. This paper concentrates only job sequence with optimal time, and different types of optimization techniques are used for getting a feasible solution.

2 System Overview

This section describes an overview of the proposed method in flowshop scheduling for manufacturing industries. Comprising this proposed method into three modules.

- System initialization, the combinatorial auction is conducted between the RA (resources) and JA's (jobs) until stopping criteria is reached. Combinatorial auction mechanism is described in [3].
- After completion of combinatorial auction, JAs are ready to schedule their tasks in machine resources. Job agents are scheduling their tasks in machine resources based on the availability of the resource. In scheduling process, machines are blocked, because all jobs on machine M_i then only next machine M_j will be available. Job agents are waiting in M_i to complete their task on M_j, this is called blocking flowshop scheduling problem is described in [4].
- For solving this, a novel approach artificial bee colony algorithm is used and scheduling jobs. Artificial bee colony algorithm is used in next forthcoming section. Overview of the entire system is shown in Fig. 1.

3 Evolutionary Algorithms

Evolutionary algorithms are population-based on stochastic optimization algorithms, and the operations are based biological evolution [3]. Recombination, mutation, and selection operations are the example for biological evolution. These algorithms are common to search randomly over the solution based on some heuristics to discover a near-optimal solution. In this only three, such algorithms are considered, namely genetic algorithm and artificial bee colony algorithm.

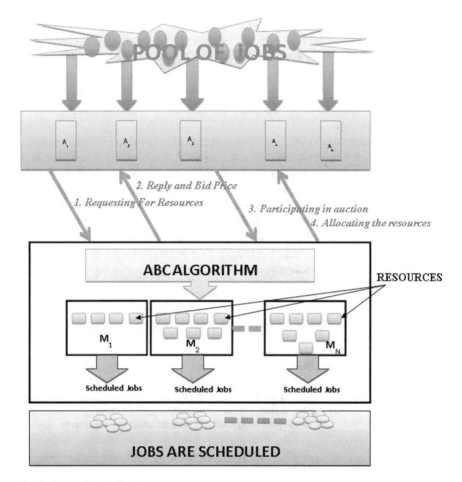

Fig. 1 System block diagram

3.1 Genetic Algorithm

The genetic algorithm [4] concept was developed by Holland in 1975. Genetic is a biological word. Applying genetic algorithms in flowshop scheduling is different in practical and needs to use different string representation and genetic operators. In the traditional representation of GA which uses 0s and 1s and it does not work for scheduling. In order to apply GA to scheduling, the structure can be described as a sequence of jobs in scheduling. In this algorithm, each and everything will be cloned with genetic gene operations like crossover, reproduction, and mutation. These operations will not give effective solution for industries, and this may increase machine idle time and dissatisfies the client's requirements.

3.2   *Artificial Bee Colony Algorithm*

The artificial bee colony (ABC) algorithm is introduced by Karaboga in 2005. This is an optimization technique; ABC algorithm is copied the behavior of honey bee colony [5]. This algorithm contains three bees. They are employee bees, onlooker bees, and scout bees. Behavior of these three bees is foraging and using them in solving the flowshop scheduling problem. Employee bees are responsible to search food sources (resources available in the machine); this searching is done randomly and shares the information with other bees/neighborhood bees called onlooker bees. Onlooker bees are used to select the best path for selecting the food source and sharing this other bee. After completion of decision-making for a food source, employee bees are foraging that path and occupying that food source and becoming the scout bees and searches the new food source randomly, because to get better food source. Considering the food sources as number of machines and bees are the scheduler agents. This procedure is repeated till it gets the optimal solution. The procedure of this algorithm is given in below.

> **Begin**
>
> Initialize the control parameters, i.e., colony size; number of food sources (resources in a machine)
>
> **Do**
>
> **Step:1.** Send Employee bees for searching food sources (resources) and share information with other bees.
>
> **Step:2.** Onlooker bees getting information from Employee bees and make decision for selecting the best food source and share best food source information with other bees.
>
> **Step:3.** Scout bees getting decision form onlooker bees and occupying the food source and search new food source for best food source.
>
> **Step:4.** Repeat all steps till termination condition is met.
>
> **End Do**
>
> **End**

The entire detailed description of this optimized method is given in the following section.

4   Proposed Methodology

The ABC algorithm is an iteration process [6–9]. At first, initialize the control parameters, population of the colony is a randomly generated resource, and this process is continued until a termination condition met. The following gives the full detailed information about ABC algorithm implementation as follows.

1. Objective of the ABC algorithm is to reduce the makespan time of the machines. Optimization of the function is as follows

$$f = \sum_{i=1}^{\text{Number of Resources}} C_{\max}^i \tag{1}$$

2. Initialize the control parameters, after sending employee bees for finding the resources in a machine by exploring the environment randomly and measure their capacity of resources.
3. The onlooker bee selects the resources and sharing the information with employee bees and determines the capacity of the resources.
4. Determine the scout bees and send them to occupy the resources.

The computational procedure of ABC algorithm in flowshop scheduling is described below:

Step:1. Control parameters of ABC algorithm are set as:

    (a) Colony size CS (total number of bees, i.e., total number of job agents).
    (b) Limit for scout bee, $\text{Limit} = \dfrac{\text{Colony Size} * D}{2}$.
    (c) where D is Dimension of the problem, D = number of jobs.
    (d) Maximum cycle is used for foraging, i.e., stopping criteria.
    (e) Number of employed bees and number of onlooker bees $= \dfrac{\text{Colony Size}}{2}$.
    (f) Number of scout bees $= \dfrac{\text{Colony Size} * D}{2}$.

Step:2. First, **initialize** the positions of resources in a machine (Colony Size/2) of employed bees, randomly using uniform distribution in the range $(-1, 1)$.

Calculating Fitness Function

$$\text{fitness}_i = \begin{cases} \frac{1}{1+f_i} \text{ if } & f_i \geq 0 \\ 1 + \text{abs}(f_i) \text{ if } & f_i < 0 \end{cases} \tag{2}$$

Take maximum fitness value for getting the best resource.

Step:3. **Employed Bees Stage** (Use maximum number of cycles for foraging resource path or stopping criteria):

    (a) Produce new solutions (resource positions) $v_{i,j}$ in the neighborhood of $x_{i,j}$ for the employed bees using the formula.

$$v_{i,j} = x_{i,j} + \Theta_{i,j}\left(x_{i,j} - x_{k,j}\right) \tag{3}$$

$$x_{i,j} = \min_{i,j} + \text{rand}(0,\ 1) * \left(\max_{i,j} - \min_{i,j}\right) \tag{4}$$

where k is a solution in the neighborhood of i, Θ is a random number in the range $[-1, 1]$, and j is the randomly selected index.

(b) Apply the greedy selection process between x_i and v_i.
(c) Calculating the probability function values P_i for the solutions x_i by means of their fitness values using the following Eq. (5).

$$P_i = \frac{\text{fitness}_i}{\sum_{i=1}^{\text{Number of Resources}} \text{fitness}_i} \tag{5}$$

Step:4. **Onlooker Bee Stage**
Chooses the resource-based on the probability P_i obtained from the employed bee stage.

(a) Compute fitness of individual resources.
(b) Fitness of new solution is better than the existing solution replaced with older solution otherwise using new solution.

Step:5. **Scout Bee Stage** Memorizes the best solution. If there is an abandoned solution, generate a new solution randomly to replace with the abandoned one. For example, Limit L = number of scout bees is given in initial condition, occurred number of scout bees are equal to L then there is no abandoned solution, if the trial counter is higher than L, there is an abandoned solution.

This procedure is continued till the objective is reached.

5 Results

Flowshop scheduling with ABC optimization is tested with existing traditional techniques of input data with two machines and each machine varying number of jobs with various resources. The computational result is compared with traditional existing systems like SPT, FCFS, CR, LPT, and WSPT and shown in Fig. 2.

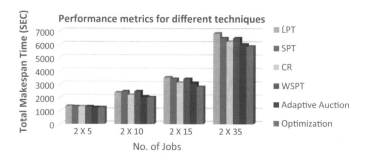

Fig. 2 Performance for different techniques

6 Conclusion

Although discussion centered on the flowshop scheduling problem, this hybrid mechanism is generic in the sense that solving the resource allocation in flowshop scheduling to get minimum makespan time. This optimization technique reduces the makespan by 8.11% compared with traditional algorithms like SPT, CR, and WSPT. For further enhancement, other evolutionary techniques like PSO algorithms can be implemented for improving the system performance.

References

1. J. Andro-Vasko, *Shop problems in scheduling* (University of Nevada, Las Vegas, 2011)
2. M. Drozdowski, Scheduling multiprocessor tasks- an overview. Eur. J. Oper. Res. 94 (1996)
3. A. Jukuntla, G.M. Kanaga, Flowshop Scheduling using multiagents with adaptive auction. Int. J. Comput. Eng. Res. **3**(3) (2013)
4. H. Ishibuchi, T Murata, A multi-objective genetic local search algorithm and its application to flowshop scheduling. IEEE Transactions on Systems, Man, and Cybernetics Society, vol. 28, no. 3 (1998)
5. D. Karaboga, An idea based on honey bee swarm for numerical optimization. Technical Report-TR06, Erciyes University, vol. 200 (2010)
6. D. Karaboga, B. Akay, A comparative study of artificial bee colony algorithn. Appl. Math. Comput. **214**(1), 2009 (2009)
7. E. Zitzler, L. Thiele, in *Multiobjective Evolutionary Algorithms: A comparative Case Study and the Strength Pareto Approach.* IEEE Transactions on Evolutionary Computation, Vol. 3, No. 4, November 1999
8. J. Amar, P. Keerthi, N. Arjun, in *SAWR: Scheduling Algorithm for WSN with Rendezvous Nodes.* 2015 IEEE International Conference on Computational Intelligence and Computing Research, 10–12 Dec 2015
9. H.C. Lau, Multi-period combinatorial auction mechanism for distributed resource allocation and scheduling. in *Proceedings International Conference Intelligent Agent Technology,* pp. 407–411

Optimal Active Node Selection, Neighborhood Discovery, and Reliability in Wireless Sensor Networks

Giri M., Seethalakshmi R. and Jyothi S.

Abstract Recently, wireless sensor networks (WSNs) are getting popular in the computing arena and embedded system category. In WSNs, sensors are deployed randomly and in a widespread manner. These sensors transmit data to the controller at periodic intervals. The number and nature of sensors vary randomly and it is intended to detect the active sensor node selection, which neglects the idle nodes from active nodes. In WSN which is a random widespread one, it is intended to discover the neighbors in the proximity and coordinate data acquisition at the controller. In any network, reliability is an important issue which deals with stability under mobility, low delay, more lifetime, optimal route, and other optimizations.

Keywords WSN · Mobility · Reliability · Neighborhood discovery
Optimization · Active nodes

1 Introduction

Wireless sensor networks encompass a wide variety of sensors which cooperatively and cumulatively lead to control and monitoring of various applications like surveillance, tracking and detection, transportation, military, and other applications. The sensors range from acceleration sensors, temperature sensors, pressure sensors,

Giri M. (✉)
Department of Computer Science and Engineering, Rayalaseema University,
PP. Comp. Sci. & Engg., Kurnool 0485, Andhra Pradesh, India
e-mail: prof.m.giri@gmail.com

Seethalakshmi R.
Department of Computer Science and Engineering, Vel Tech High Tech
Dr. Rangarajan & Dr. Sakunthala Engineering College, Chennai, Tamil Nadu, India

Jyothi S.
Department of Computer Science, SPMVV University, Tirupati, Andhra Pradesh, India
e-mail: jyothi.spmvv@gmail.com

© Springer Nature Singapore Pte Ltd. 2018   195
S. S. Dash et al. (eds.), *Artificial Intelligence and Evolutionary Computations
in Engineering Systems*, Advances in Intelligent Systems and Computing 668,
https://doi.org/10.1007/978-981-10-7868-2_19

gas sensors, biomedical sensors, water level detection sensors, Webcameras, microphones, mobile phones, and other sensors. These sensors are deployed in the field of acquisition zone which depends on the application context. The sensors properties include sensing, formatting, compressing, sending data, and lifetime. This is done periodically by the sensors. The lifetime depends on the energy stored in the sensors. If the battery gets discharged, it has to be recharged. Then, the lifetime of the sensors comes up. It will interactively participate in communication after getting recharged.

Active node selection is one of the important issues in such domains. The sensors transmit data periodically. Many nodes transmit data at a time. These data are sent one by one. This is achieved by time-division multiple access. The nodes interactively take turns to transmit data. This data is not always transmitted by all nodes. The nodes have predefined lifetime of processing. This is determined by the battery lifetime. So the nodes that do not actively participate in transmission become idle because of want of charge. It recharges then and continues. Neighborhood discovery is an important aspect in the WSN. For particular nodes, it is intended to determine who the neighbor node is. This is because with the neighbor node only it has to negotiate to transmit data. The third issue is the reliability of WSN. The reliability is the probability that there occurs no node failure. The delay is reduced. The lifetime becomes high. All these things lead to high reliability. In this context, these issues are discussed. The various protocols and algorithms are discussed.

2 Related Work

In Joanna et al. [1] has done a work on negotiation-based protocols for disseminating information in wireless sensor networks. Their method is not novel and has some drawbacks. Hence, new methods are adapted. In Eylem et al. [2] has done an experiment on mobility-based communication in wireless sensor networks. This is an existing technique. In Yulong et al. [3] has worked on intercept behavior analysis of industrial wireless sensor networks in the presence of eavesdropping attack which is again an existing technique. In Wei et al. [4] has worked on optimized node selection for compressive sleeping wireless sensor networks, and their method is very old, and hence, we adapt to the new protocol discussed in this context.

In Pouya et al. [5] has worked on scalable video streaming with helper nodes using random linear network coding. The linear coding technique is not to our context but referred to. In Heng et al. [6] has worked on optimal DoS attack scheduling in wireless networked control system and it deals with the security issue which may be adapted in future work and their control strategy is different and old. In Dong et al. [7] has done a research on design and evaluation of an open source wireless mesh networking module for environmental monitoring which is considered for reference but adapted in a novel way which is our original work.

In Haung et al. [8] proposed a new method of design and analysis of wireless mesh network for environmental monitoring application. In Dongyao et al. [9] has conducted research on dynamic cluster head selection method for wireless sensor network which has the context in current work but dealt differently.

In Pushpendu and Sudip [10] has done a work on reliable and efficient data acquisition in wireless sensor networks in the presence of transfaulty nodes which deals with reliability issues. But we incorporated our own protocol for reliability detection model. In Ju et al. [11] has done a work on adaptive and channel-aware detection of selective forwarding attacks in wireless sensor networks which deals with security of WSN, this is referred but our own protocol is considered for discussion. In Yuxin et al. [12] has done a work on active trust: secure and trustable routing in wireless sensor networks where the routing protocol is discussed and the novel and new technique is adapted in our design in the current context. In Zhen et al. [13] has conducted a work on a clustering-tree topology control based on the energy forecast for heterogeneous wireless sensor networks deals with clusters. The current work encompasses a random and dynamic distribution of sensors.

In Mianxiong et al. [14] has worked on joint optimization of lifetime and transport delay under reliability constraint wireless sensor networks. The current discussion deals in a different way the reliability problem which is a new one. In Tong et al. [15] has done a work on code-based neighbor discovery protocols in mobile wireless networks. The neighborhood discovery of proposed method attacks the problem with user-generated packet structures and packet lifetime. Thus, the existing techniques are thoroughly studied and their drawbacks are discussed. The proposed method eliminates these drawbacks and solves the problem of active node selection, neighborhood discovery, and reliability detection of WSN in a quite different and perfect manner.

3 Protocols for WSN

WSNs are the important communication network in the computer science and engineering field. The WSNs network encompasses wide variety of sensors which may be deployed in random nature. The nodes get self-organized and participate in communication. Sometimes, the nodes move in the environment and negotiate in communication. In this work, it is intended to discuss these protocols anew.

3.1 Optimal Node Selection Protocol

Figure 1 shows the typical WSN. The sensor nodes get self-organized in the network and actively transmit wireless data to the controller. The number and nature of data received vary. The data is transmitted in the form of packets. These data packets are transmitted as such using the network transmission protocol to the

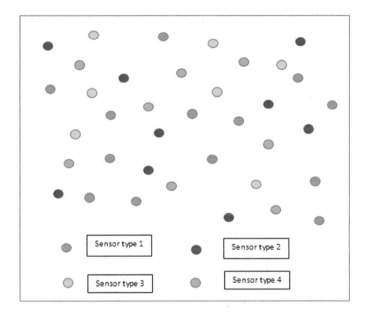

Fig. 1 Typical WSN for a specific application

Fig. 2 WSN transmission and processing

controller. These nodes transmit data periodically. This period is designed by the protocol. Figure 2 below depicts the transmission of data in WSN and processing of data in WSN.

A typical WSN is deployed in the field area and it is sensed using the wireless transmission network like Zigbee. The Zigbee is connected at the receiving side and it transmits the data to the controller. The controller processes the data and takes decision. The decision is then incorporated in the output side. This is the basis of WSN. The active node selection protocol is explained in the algorithm stated below.

3.1.1 WSN_ActiveNode_Selection_Protocol

Step 1: Select the application for the current context.
Step 2: Determine the various wireless sensors for the deployment.
Step 3: Choose the strategy for deployment like random spreading, spraying, dropping, and so on.

Step 4: Activate the various wireless sensors and assign node id to the sensors and allow to transmit data in a multiplexed way.

Step 5: Receive the transmitted data at the controller end.

Step 6: Store the big data in the specified database.

Step 7: Determine the wireless node from which the data is received.

Step 8: Step 7 is done by the node identifier pre-specified by the protocol.

Step 9: Check if there are idle nodes by processing the node identifiers from which the data is received.

Step 10: If any node is found idle allow the nodes to recharge using batteries.

Step 11: Check if the nodes are alive again periodically and energize if found idle.

Step 12: Repeat Steps 1 to 11 for active node selection protocol.

3.1.2 WSN Active Node Selection Model

A typical WSN encompasses a variety of sensors. The sensors are identified for the type and the facilities it offers. The sensors of the corresponding type are grouped and they are recorded. These sensors are deployed in a random manner. All the sensors are initially filled with full energy. They transmit the data sensed in the environment as they perceive. This is received at the controller center. Later the nodes become idle for want of energy. We identify such nodes using the WSN_ActiveNode_Selection_Protocol, and this protocol helps in managing and controlling the WSN from the required perspective.

3.1.3 WSN Active Node Selection Mathematical Model

Let the number of category of sensors is m. The number of sensors in each category is l. The initial total energy empowered for each sensor is Einit. Energy consumed during transmission is

$$Econs = Einit - Etrans \qquad (1)$$

The initial total energy in the network is l * Einit. The energy during transmission is l * Econs. Energy loss due to transmission is

$$Eloss = Einit - Econs \qquad (2)$$

The condition for node deactivation is Eloss < Ethresh where Ethresh is the threshold value of energy for the specified sensor. The condition of active node selection is Eloss > Ethresh. If the node is deactivated then allow the node to recharge. The recharge time is

$$Time\,taken(Erecharge) > \ = Time\,taken(Einit) \qquad (3)$$

3.2   WSN Neighborhood Discovery

After all nodes become active, the protocol explores the neighbor of the particular sensor in the network. In WSN, the nodes have the capability to move in the current context. This depends on the nature and property of the WSN chosen. The application-specific sensors are critical to the application chosen. These sensors interact amongst themselves for something or the other. Some sensors have common communication and data digitization circuits. In this context, the neighborhood discovery is very important. The protocol for neighborhood discovery is dealt with here. Figure 3 shows the model for neighborhood discovery in the WSN.

3.2.1   WSN_Neighborhood_Discovery_Protocol

Step 1: Formulate the WSN.
Step 2: Select the sensors which are dependent and have commonalities.
Step 3: Choose the strategy for deployment like random spreading, spraying, dropping, and so on.
Step 4: Assign the identifiers and identify the initial configuration in the WSN.
Step 5: Allow to transmit in cooperation with each other.
Step 6: Receive the transmitted data, store and process the data at the center.
Step 7: Check for any WSN node mobility.
Step 8: If there is node mobility, the new network with different topology and architecture is formulated.
Step 9: Any node arbitrarily transmits the are_u_my_neighbor packet to the nodes.
Step 10: The nodes which receive this packet will respond with I_am_your_neighbor packet to the nodes from which it received the are_u_my_neighbor packet.
Step 11: The nodes wait for all such packets and after receiving all the replies it comes to know about the entire network.
Step 12: Every node in the WSN executes this protocol and every node knows about every other node.
Step 13: This emphasizes the neighborhood connectivity of every node and thus the protocol.
Step 14: If there is node mobility, then the steps 7–13 are executed.

Node identifier	Packet Reception	Start Time	Finish Time	Life Time	No_of_hops	Next hop	Error bit	data

Fig. 3 Neighborhood discovery format in the packet

3.2.2 Neighborhood Discovery Mathematical Model

The number of category of sensors is m. The number of sensors in each category is l. Next_hop bit pattern will receive the id of next hop. Optimal route is if Next_hop <= Hopmin.

3.3 WSN Reliability Detection Model

In any network reliability is an important issue. Reliability is the probability that there should not be any node failure. If there occurs any node failure, then the node is repaired and added to the network. The packet carries the start time, and at the other end, the finish time is marked. The total time for transmission is calculated by the protocol. The optimal mean time for transmission is encountered using this field. Every packet carries the packet lifetime. If the lifetime expires, the packet is discarded. Optimal route is identified by the number of hops field. There should be minimum number of hops for transmission. The packet error field determines if the packet or data is lost or erroneous. Then, the data is ignored and next packet is encountered. Thus, the various parameters for reliability are encountered. Figure 4 below shows the packet reliability model in WSN.

3.3.1 WSN_Reliability_Detection_Protocol

Step 1: Formulate a WSN.

Step 2: Initiate all sensors and allow the data transfer to the controller.

Step 3: Set various fields in the reception side like packet reception bit, total time taken for transmission, packet lifetime and number of hops that the packet has traversed, error packet or not in the controller for each packet that is received.

Step 4: The data are filled in the respective fields and control decision is taken on various parameters.

Step 5: If any packet is not received, then the stability field is marked incorrect. If the transmission delay is more, then the congestion in the path is checked. If the number of hops is more than the route optimization, decision is taken. If the received packet is error, then next data is considered. The current packet is ignored.

Step 6: This reliability checking is done for every packet, and stability in the network is emphasized.

Node identifier	Packet Reception	Start Time	Finish Time	Life Time	No_of_hops	Error bit	data

Fig. 4 Packet reliability model

3.3.2 Reliability Detection Mathematical Model

The number of category of sensors is m. The number of sensors in each category is l. Sensor node is identified by using its ID. The start time of the transmission is Tstart. The finish time of the transmission is Tfinish. The total time for transmission is

$$Ttime\_consumed = Tfinish - Tstart \qquad (4)$$

Lifetime of the packet is Tlife. Packet discard condition is Tlife > Texpiry. The number of hops is No_of_hops. Optimal path strategy is No_of_hops <= Nmin. Error bit is set then the packet is Perror.

4 Results and Analysis

The WSN is constructed using standard strategy using NetSIM package. The nodes are deployed and activated. The reception side is perfectly organized. The nodes start transmitting data packets. These packets are received and processed, and then control decisions are taken. The various parameters are recorded and analyzed. The resultant statistics are gathered plotted in the form of graph and analyzed. Figure 5 deals with the plot of active node selection for initial energy, consumed energy, and node deactivation against number of sensors. It is observed that if the consumed energy is minimal then the node deactivation is not present in WSN. Figure 6 depicts the plot of total energy, energy loss, and energy threshold against the number of sensors in the network. Energy loss is minimal for a high-class WSN. Figure 7 displays the plot of neighborhood discovery protocol. It shows the graph

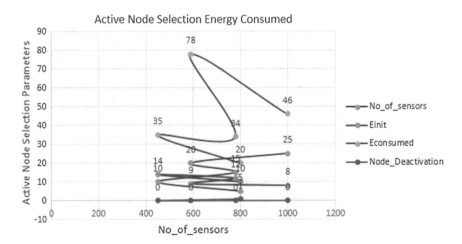

Fig. 5 Active node selection plot for energy consumed

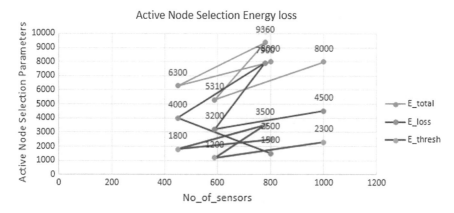

Fig. 6 Active node selection plot for energy loss

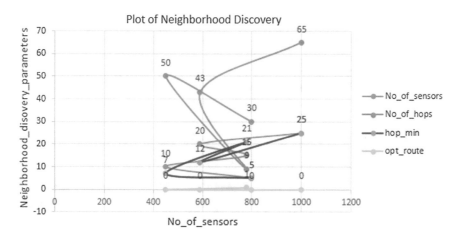

Fig. 7 Neighborhood discovery plot

of the number of hops, minimum number of hops against the number of sensors. It determines the optimal route if the number of hops is less than the threshold value. Figure 8 depicts the plot of WSN reliability analysis protocol parameters of some category. It mainly displays the total time consumed for the sensors deployed. Figure 9 shows the WSN reliability protocol on the optimal route based on the hops. It determines whether the packet has been successfully received or discarded due to expiration time. Thus, the behavior of various models like active node selection model, neighborhood discovery model, and WSN reliability detection model is clearly discussed.

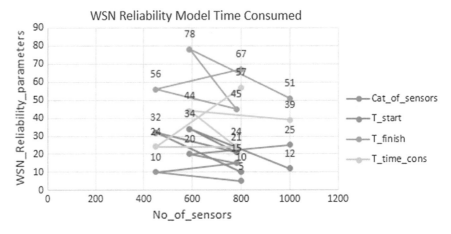

Fig. 8 Plot of WSN reliability analysis for time consumption

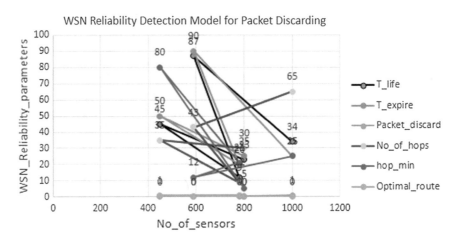

Fig. 9 Plot of WSN reliability analysis for packet discarding

5 Conclusion

Wireless sensor network is a classical network which is gaining world-class popularity because of its increased usage in various fields. In such networks, the node
activation plays a major role. The active node selection is an important issue to be
discussed, and hence in this context, it is discussed that the active nodes are polled
for energy content. If the energy expires, it becomes inactive and has to recharge for
becoming active. The next issue is the neighborhood discovery which is very
important in the mobility-based sensor network. The neighbors play a major role in
active node participation of packet transmission. By means of neighborhood

detection, an optimal route is explored. The next issue that is considered for discussion in the current context is the reliability detection model. The analysis implies that the sensors total time taken for data transmission is the difference between finish time and start time. This time depends on the number and nature of sensors. The lifetimes of the packets are recorded, and packet discarding is correlated with the lifetime. An analysis is done on this. On an average rate, packet discarding is reduced. Thus, the discussion yields better performance and suggestions for improvement in the performance of a typical WSN.

References

1. K. Joanna, H. Wendi, B. Hari, Negotiation based protocols for disseminating information in wireless sensor networks. Wireless Netw. **8**, 169–185 (2002)
2. E. Eylem, G. Yaoyao, B. Doruk, Mobility based communication in wireless sensor networks. IEEE Commun. Mag. 56–62 (2006)
3. Z. Yulong, W. Gongpu, Intercept behavior analysis of industrial wireless sensor networks in the presence of eavesdropping attack. IEEE Trans. Industr. Inf. **12**(2), 780–787 (2016)
4. C. Wei, J.W. Ian, Optimized node selection for compressive sleeping wireless sensor networks. IEEE Trans. Veh. Technol. **65**(2), 827–836 (2016)
5. O. Pouya, W. Jie, K. Abdallah, B.S. Ness, Scalable video streaming with helper nodes using random linear network coding. IEEE/ACM Trans. Networking **24**(3), 1574–1587 (2016)
6. Z. Heng, C. Peng, S. Ling, C. Jiming, Optimal dos attack scheduling in wireless networked control system. IEEE Trans. Control Syst. Technol. **24**(3), 843–852 (2016)
7. H.S. Dong, B. Saurabh, C.C. Wang, Toward optimal distributed monitoring of multi-channel wireless networks. IEEE Trans. Mob. Comput. **15**(7), 1826–1838 (2016)
8. C.L. Huang, H.L. Hsiao, Design and evaluation of an open source wireless mesh networking module for environmental monitoring. IEEE Sens. J. **16**(7), 2162–2171 (2016)
9. J. Dongyao, Z. Huaihua, Z. Shengxiong, H. Po, Dynamic cluster head selection method for wireless sensor network. IEEE Sens. J. **16**(8), 2746–2754 (2016)
10. K. Pushpendu, M. Sudip, Reliable and efficient data acquisition in wireless sensor networks in the presence of trans faulty nodes. IEEE Trans. Netw. Serv. Manage. **13**(1), 99–112 (2016)
11. R. Ju, Z. Yaoxue, Z. Kuan, S. Xuemin, Adaptive and channel-aware detection of selective forwarding attacks in wireless sensor networks. IEEE Trans. Wireless Commun. **15**(5), 3718–3731 (2016)
12. L. Yuxin, D. Mianxiong, O. Kaoru, L. Anfeng, Active trust: secure and trustable routing in wireless sensor networks. IEEE Trans. Inf. Forensics Secur. **11**(9), 2013–2027 (2016)
13. H. Zhen, W. Rui, L. Xile, A clustering-tree topology control based on the energy forecast for heterogeneous wireless sensor networks. IEEE/CAA J. Automatica Sinica **3**(1), 68–77 (2016)
14. D. Mianxiong, O. Kaoru, L. Anfeng, G. Minyi, Joint optimization of lifetime and transport delay under reliability constraint wireless sensor networks. IEEE Trans. Parallel Distrib. Syst. **27**(1), 225–236 (2016)
15. M. Tong, W. Fan, C. Guihai, Code-based neighbor discovery protocols in mobile wireless networks. IEEE/ACM Trans. Networking **24**(2), 806–819 (2016)

A Review on Security Issue in Security Model of Cloud Computing Environment

G. Venkatakotireddy, B. Thirumala Rao and Naresh Vurukonda

Abstract Cloud computing is recently everyone uses increasing tendency in distributed and parallel computing environment changes. Cloud is an Internet. It works the largest group of interconnected individual desktop computers or network servers; these are used public cloud or private cloud or hybrid. It's an active implementation by IT industry resource and remote provisioning scalable and measured documents and decentralized its resources. It is different security issues facing cloud environmental. Cloud computing consists of different services providing. One of the Software as a Service (SaaS) has on many business orientated application. It is shared software and resources and information are available on demand device access in public it allows cost and complexity of service providers to access and operational costs anytime and anywhere uses provide cloud services. Another service is a IaaS is provides protect security customer's knowledge, functions also secured. Other models Platform as a Service (PaaS) clouds. Security of PaaS clouds forms multitenant access control. Its privacy protected together the service provides and user security. problems., security and threats.

Keywords Cloud computing · Deployment model · Service level management utility · SaaS · Security issues keys · IaaS · PaaS

G. Venkatakotireddy (✉) · B. Thirumala Rao · N. Vurukonda
Department of CSE, Koneru Lakshmaiah Education Foundation,
Vaddeswaram, Guntur 522502, Andhra Pradesh, India
e-mail: gvkotireddy@gmail.com

B. Thirumala Rao
e-mail: drbtrao@kluniversity.in

N. Vurukonda
e-mail: naresh.vurukonda@gmail.com

© Springer Nature Singapore Pte Ltd. 2018
S. S. Dash et al. (eds.), *Artificial Intelligence and Evolutionary Computations in Engineering Systems*, Advances in Intelligent Systems and Computing 668,
https://doi.org/10.1007/978-981-10-7868-2_20

1 Introduction

Cloud computing is a major broad range of service. Cloud Computing is a standard being acceptable for makeable about request interconnected system connection by on share configuration form funds by authority communication, documents, the skill talent as point top user to one to one exploit part about funds acquired quickly also easily. Cloud computing in order self serve. Last user (ultimate consumer facing into agree along with accept work [1]. After late, effective application service to end user internet the cloud service termed as SaaS, PaaS, Iaas) and public the cloud the service networks, servers, storage, applications, and services. That can be supported and released with least management interaction. The ability for end user to users to utilize part of the resources acquired quickly and easily. The cloud service termed as Software as a Service (SaaS), Platform as a Service (PaaS) and Infrastructure as a Service (IaaS) and deployment models (public and private and Hybrid and Community cloud) [2–6]. The cloud provider everywhere the software service mentioned to cloud service. for Example Microsoft Azure, Amazon and Google, IBM, Google Apps Engine Etc .it is Cloud is a pool of virtualized computer resource networks, it can host a different workload and its style back-end jobs, interactive facing application workload with physical machines. It self recovery and highly scaled deployed workload and recovery from any software and hardware failures. It is real time to enable rebalancing of allocations A number of Characteristics consists of cloud data application services and infrastructure computing consists different ay-use-on demand self service-it is demand of market, Remotes hosted- is consists data are remote infrastructure, Ubiquitous—anywhere data is available, Geographic Distribution, Virtualization, Advance Security, Broad network access [7–9].

2 Threats Now in Cloud Computing

2.1 Threats

Cloud computing facing different security threats presently initiate now extant stage, data centers, on line network new trade, networks. Those hazards are exposure risk problems occur many models. The Cloud preservation collusion 2016 created by industry wide act an analysis in risk identified the following major threats [10–13].

a. Attacks by other consumer/customer.
b. Shared new technologies security vulnerabilities.
c. Application failures in provider security.
d. Insecure interface and application programming interfaces (APls).

e. Advance persistent threats (APTs).
f. Customer account traffic hijacking and service.
g. Data leakage/loss.
h. Availability and reliability service issues.
i. Weak identity, authorization, and access management.
j. Malicious insiders and denial of service.
k. Legal and regulatory authority service security issues.
l. Unknown risk profile and customer security systems integrating.

3 Cloud Computation Implementation

3.1 Steps to Cloud Prevention Issues

Preservation navigation as critic field was finding now Cloud gauge begins in the month of April-2009. Started that Preservation exposures and susceptibility in the trends cloud gauge. It is extensive a intent protocols structure action mange set up care for info, application also exposed enterprises that want to ensure the cloud customers to evaluate the and manage the security risk and delivering support of following details blow [14].

1. **Dispute Follow**: The cloud uniquely how to lose structure affects the security of data send into Consuming data in in-death understanding how to transmit data in cloud computing.
2. **Trade Clarity**: Cloud worker on detail data preservation also accept ensure effective governance of daily preservation repot, operational appraisal from any separate frame or centralized firm.
3. **Interior Preservation**: The cloud grant interior preservation data also apply security firewall also authorized control path is actual tough heavy fit through effective the cloud measure protected.
4. **Fair and Precision Service Intimation**: The serious to apply affective order also regulation intention involve why need toward cloud security needs?
5. **Scale Terms Service Adoption**: The cloud observing all improvement now effective cloud high tech is updated process impression of yours manages documents protects service agreement.

3.2 Information Security Principles

It consists of security principles: availability, integrity, and confidentiality.

1. **Availability**: Any information must be available when it is needed (store and process the information).

2. **Integrity**: Maintaining and assuring the assuming data.it is message integrity in addition to data confidentiality.
3. **Confidentiality**: This information is not made available to unauthorized individual process.
   Client computing devices consist of integrity, confidentiality, and availability.

3.3 Analyze Resources and Basis

Patron data consist of purity, confidentiality, and vacancy.
   Patron value consist of confidentiality, integrity, and availability.

4 Issues of Security to Clarify Before and After Adopting Cloud Computing

Now a days the information technology, different companies observing research area and has identified some of the security issues that an enterprise cloud computing user discourse with computing providers (Edwards, April 2009) before approving [15, 10].

1. **Act and Administrate Rule Assent**: Create the customer service provides willing to audits internal or external security authorized certification submitted.
2. **Facts Customer Access**: Leading group must order also administer their particular ideas and principles following cloud environment. There can provide administrators and control information privileged access users.
3. **Facts Confine**: Realize what is fixed to isolate your documents? Proved facts the encoding schemas are extend also sufficient.
4. **Fact Area**: Enterprise is support cloud computing provides store and process data are law and jurisdiction privacy rules and regulation applied.
5. **Distress Recovery**: Provider commitment of structure clear-cut forms of investigation corresponding probe ideas and disclosure step from litigation also prove complete support such activities previews days.
6. **Distress Recovery Verification**: Whether the data provider will be capable of storing your data and service. How much long time finds it?
7. **Data Service Warranty**: Provider support recovery data would not your data back or fail it? It is not possible condition. Apply to rules and regulation agreement condition. It supports easy replacement of application applicable.
8. **Lack of constrain**
9. **Lack of poise**
10. **Multi-activity/work problems**
11. **Hazard authority**

5 Solution of Security Issues

1. **Inventionkey**: The Discovery key cloud security provides visible security and monitoring controls its security and data to enable to adopt the cloud protecting data complying government regulations.
2. **Clear Record**: Cloud provider consists of vendor clear. If vendor early closes the commitment, apply to trade allegation.
3. **Reform Setup**: Cloud merchant trust facing grant reform service such case documents are misplaced or because of assured matter these are recovery also documents continue handled.
4. **Raise Trade Infrastructure Security**: Cloud trade has base enable to the software installation shape fashion fixtures elements like OS, firewall, route, services etc. clients it avoid events against way virtual kills.
5. **Benefit of Data Encrypt for Security Plan**: Cloud builder provide encrypted data for the security. IT leaders must approach key security, Components are needed.
6. **Authentication**: Cloud security provides user authentication for security.
7. **Recognized and Verification**: Clients provide security encrypt data that can recognize a person to access data verification.
8. **Organization Virtual Attacks**: Cloud security issues to control organization maintain cyber attacks observing government data servers hacking and unauthorized person's leakage.

6 Conclusion

Its contains probable as time and price saving to the enterprise risks are also large. Cloud data processing as gambit to level costs also time and gain propriety security risk of cloud computing. It maintains peril security executive do tense peril further forcefully take way integrate mark. It is a new portent is set of a reform we use the internet. New technologies at an expert rate, which development and human lives in case. It must be appreciated the security risks and challenges and exploiting technologies. In this paper cloud Security issues in service model of cloud environment are currently handled emphasize. Cloud is possible turn into treasure now energizing a riskless, basic also cost-effective growth for IT solution smart tense upcoming. We are finding various problems facing security affair try to solve. We are continue work implement different cryptographic—algorithms and key-management authentications, Threat Data Protection, Cloud app, threat intelligence etc. security in cloud computing.

References

1. R. Buyya, C.S. Yeo, S. Venugopal, J. Broberg, I. Brandic, Cloud computing and emerging IT platforms: vision, hype, and reality for delivering computing as the 5th utility. Future Gener. Comput. Syst. **25**(6), 599–616 (2009)
2. D.A. Fernandes, L.F. Soares, J.V. Gomes, M.M. Freire, P.R. Inácio, Security issues in cloud environments: a survey. Int. J. Inf. Secur. **13**(2), 113–170 (2014)
3. S. Subashini, V. Kavitha, A survey on security issues in service delivery models of cloud computing. J. Netw. Comput. Appl. **34**(1), 1–11 (2011)
4. H. Takabi, J.B. Joshi, G.J. Ahn, Security and privacy challenges in cloud computing environments. IEEE Secur. Priv. **8**(6), 24–31 (2010)
5. A. Sangroya, S. Kumar, J. Dhok, V. Varma, Towards analyzing data security risks in cloud computing environments. in *International Conference on Information Systems, Technology and Management*, (Springer, Berlin, Heidelberg, 11 Mar 2010), pp. 255–265
6. G. Boss, P. Malladi, D. Quan, L. Lerge gni, H. Hall, Cloud computing. (2009) http://www.ibm.com/developerswork/webspehre/Zones/hipods/library.html
7. P.K. Chouhan, F. Yao, S. Sezer, Software as a service: understanding security issues. in *Science and Information Conference (SAI)*, (IEEE, 28 Jul 2015), pp. 162–170
8. S. Pearson, A.. Benameur, in *Privacy, Security and Trust Issues Arising from Cloud Computing*. 2010 IEEE Second International Conference on Cloud Computing Technology and Science (CloudCom), (IEEE, 30 Nov 2010), pp. 693–702
9. J. Kong, A practical approach to improve the data privacy of virtual machines. in *2010 IEEE 10th international conference on Computer and information technology (cIT)*, IEEE, 2010, pp. 936–941
10. R.M. Esteves, C. Rong, Social impact of privacy in cloud computing. In 2010 IEEE Second International Conference on Cloud Computing Technology and Science (CloudCom),). IEEE, 30 Nov 2010, pp. 593–596
11. M. Miller, *Cloud Computing: Web-Based Applications that Change the Way You Work and Collaborate Online*. Que publishing, 2008
12. G. Reese, Cloud application architectures: building applications and infrastructure in the cloud. O'Reilly Media, Inc., 1 Apr 2009
13. T. Mather, S. Kumaraswamy, S. Latif, *Cloud Security and Privacy: an Enterprise Perspective on Risks and Compliance*. O'Reilly Media, Inc., 2009
14. S.W. Smith, S. Weingart, Building a high-performance, programmable secure coprocessor. Comput. Netw. **31**(8), 831–836 (1999)
15. B.H. Krishna, S. Kiran, G. Murali, R. Pradeep Kumar Reddy, Security issues in service model of cloud computing environment. Procedia Comput. Sci. **87**, 246–251 (2016)

Dynamic Navigation of Query Results Based on Hash-Based Indexing Using Improved Distance PageRank Algorithm

L. Lakshmi, P. Bhaskara Reddy and C. Shoba Bindu

Abstract World Wide Web as we know unlimited source of data, which contains list of Internet pages and infinite links. During last ten years, the size of Web has grown, as millions of web pages are adding to Web every day. So it is the most important source of information and more popular manner of communication. The main purpose of the Web data mining is to provide the hyperlink structure for the Internet pages. User-entered queries on most of websites, such as Cloud Bigtable (Cloud Bigtable is a publicly available version of Bigtable used by Google system), most of the time result in huge number of documents and hyperlinks, but only few results are related to user query, but they may not be present at top place. Web page ranking and classification of Web queries, used in combination, overcome the problem of non-relevant results for a given query. Web page results classification and most relevant information retrieval on educational data sets is our proposed methodology. In our methodology, we propose a solution to this problem by categorization of Web queries dynamically using hash-based indexing data structure and resulting web pages are ranked by using improved distance PageRank algorithm. Using our proposed method, we reduced most of non-relevant results and most important results based on content and number of hyperlinks as top results for given query.

Keywords PageRank · Improved distance · Concept hierarchies
Dynamic navigation · Cloud Bigtable

L. Lakshmi (✉) · P. Bhaskara Reddy
MLR Institute of Technology, Dundigal, Hyderabad, India
e-mail: laxmi.slv@gmail.com

P. Bhaskara Reddy
e-mail: pbhaskarareddy@rediffmail.com

C. Shoba Bindu
JNTUA University, Anantapuramu, India
e-mail: shobabindhu@gmail.com

© Springer Nature Singapore Pte Ltd. 2018
S. S. Dash et al. (eds.), *Artificial Intelligence and Evolutionary Computations in Engineering Systems*, Advances in Intelligent Systems and Computing 668, https://doi.org/10.1007/978-981-10-7868-2_21

1 Introduction

In the past ten years, the size of Web increasing day to day, it became the most important source for storing information and also used for communication [1]. The Internet acting as a scaffold for interchanging various types of data in the form of educational data, data related to research, data related to images and videos, personal information and types of software's and hardware's [2, 3]. The main aim of web structure mining is to provide the hyperlinks between the web pages, which are based number in-links and out-links, and weights of web pages [4]. Web structure mining mainly performs classification of web pages and determination important Web documents of all data the data present in Web server [5, 6]. Using Web structure mining, we calculate number of hyperlinks of each page, and then by applying PageRank algorithm, we provide most important results to user [7].

Most of search engines use the principle of document classification and PageRank algorithm for sorting web pages [8]. It is considered as a model of patron conduct, where a user searches on connections [9] at arbitrary and not using admire towards content [10]. The arbitrary surfer visits a web page with a specific probability which gets from this page rank [5]. The chance that their regular surfer faucets on one connection is exclusively given through the amount of connections on that page [7, 11]. So the page rank of one web page is not depend on the number of in-links but it also depends on the number of out-links another web page to which it is connected [12].

Most of the Internet search queries primarily based on those three approved categories, the usage of k-means, k-medoids clustering algorithm and a feature-set illustration algorithm [4, 6]. Every keyword in the search history log incorporates appearance along with person details, sessions, date and time of usage, keywords of query and the kind of information the person is trying to retrieve from Web [12, 13]. Further, every keyword changed enhanced with the question size, used by al most all search engines and Web servers for retrieval for Web documents for given query [7]. When the user visits any web page or document or performs ant transaction using during interaction session with search engines, that data can be categorized under any one the three types of navigational queries, that is, informational queries, transactional queries and navigational queries [3, 5].

Statically retrieval techniques are mainly used to categorize Web queries based totally on their purpose and number of hyperlinks [3, 10]. This static categorization is then used to robotically classify new Web queries thru genuine phrases comparable [11]. Despite, the technique is simply too prohibitive as it suitable for most related items [6]. So that we can address this problem, using classification strategies that use mainly vector space model and Bayes classifiers [4], using a vector space model received good outcomes compared to the informational retrieval using Bayes classifiers even though it used for retrieval of information with different intension [7, 13]. Experimental results provide phrase-matching functions turn out to be keyed to understand useful Web queries, but the performance is poor in the case of navigational queries [8].

Characterization or supervised training is one of the machine learning strategies for gathering related information from the given Web servers data [1, 10]. This information mainly consists of an organization of informational queries which are sets consists of most important and desired results expected by the user [5, 8]. Though group learning alludes to a gathering of characterization strategies that take in an objective work via preparing various single classifiers and consolidating their expectations [2, 6]. The standard is that a panel choice, with individual expectations consolidated suitably, ought to have better general precision, all things considered, than any individual advisory group part [12].

The main objectives of our proposed method are mainly, to provide more relevant results to the given query fast and reduce the amount of time we spend searching for information, to reduce non-relevant results for the given query, to overcome the limitations of existing navigation and PageRank algorithm (used by Google) with dynamic navigation and improved distance rank algorithm and to reduce ads or pop-up windows or "phishing" for personal data.

2 Architecture of the Proposed System

In this system, we proposed dynamic navigation of Web queries based on hash-based indexing and improved distance PageRank algorithm. Dynamic navigation of queries can be performed by using hash-based indexing, and the algorithm for improved distance PageRank is based on calculation of the distance between the authoritative pages and hub pages with good authoritative score and hub score and with the user can reach authoritative page from hub page with minimum number of clicks and can reach a hub page from authoritative page with minimum number of clicks. A good authoritative page for a given query is pointed by many good hub pages. A good hub page for a given query is pointing to many good authoritative pages.

Naturally, the distance between two hub and authoritative pages is the weight of shortest distance between i and j denoted dist ij which mainly based on number of in links and out links for i and j. Distance PageRank algorithm recursively calculates the score of a web page and converts the value to in the scale for zero to one. Here, we use distance rank vector, which mainly consists of all distances arranged in ascending order that is lower distance web pages are assigned higher ranks. The time complexity of our algorithm will be good, as it requires less number of iterations. The complete evaluation of this distance calculation will be implemented in improved distance PageRank. By using hash-based indexing for categorization of Web queries reduces the time taken to retrieve relevant documents. This method retrieves Web documents if keyword present in query exactly matches with keywords present in the hash table, which reduces most of the non-relevant results. This method can also be used in multidimensional information retrieval for given query.

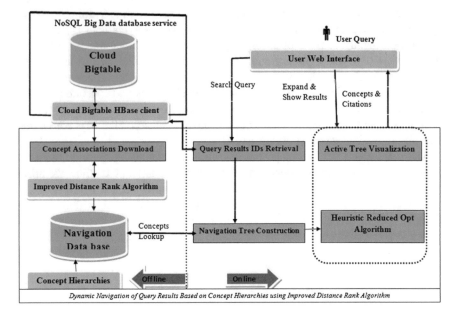

Fig. 1 Architecture of our proposed system

In addition we use a novel search interface as shown in Fig. 1, by using which the user enters a search query, all the data is present on cloud's Bigtable. First, we construct a navigation hierarchy using hash-based indexing, and the results obtained from hash-based indexing are given to improved distance PageRank algorithm to arrange the results in descending order with higher rank pages as top results. We used hash-based indexing method to selectively reveal the best concepts related to each keyword present in the given query. Using our proposed method mainly minimizes the navigation cost, and it reduces time required for navigation. So our algorithm provides efficient results even for multidimensional data.

3   Algorithm for Dynamic Navigation of Queries Based on Hash-Based Indexing Using Improved Distance PageRank

Algorithm: Dynamic Navigation and improved distance rank.
Input: Query entered by user in novel search interface.
Output: Web pages with ranks in descending order.

1. Read the query entered by user
2. List out the keywords present in the given query

3. For all keywords present in the search query, construct hash-based indexing dynamically
4. Search for keywords present in the query using extended hashing
5. Number of keywords is taken as n
6. Matching for given keywords is done by using
7. $b \leftarrow h(k) \& (2n - 1)$
8. Return b buckets matched for the given keywords
9. Apply this procedure recursively until all documents for keywords present in query are retrieved
10. All documents are arranged in descending order of their ranks
11. Calculate the count of number of incoming links and outgoing links for every document
12. Calculate unique visit count of web page
13. Calculate the distance the web pages
14. Rank is assigned to web pages from highest to lowest with minimum distance between web pages, highest unique visit count of web page and highest in links
15. Display the web pages with ranks in descending order.

4   Experimental Evaluation of the Algorithm

Evaluation of our algorithm is based on mainly two factors; firstly, numbers of relevant documents retrieved and retrieved documents are ranked using improved distance PageRank algorithm. We compared our results with graph-based Web query classification [8] and agent-based weighted PageRank algorithm [10]. In Table 1, we are representing the results that our algorithm compared with the existing algorithm.

4.1   Comparison of Relevant Web Pages Retrieval

Performance of our algorithm shows that hash-based indexing for classification of Web queries reduces most of non-relevant results and produces most relevant results for the given query. We compared our results with agent-based weighted PageRank algorithm [8]. Our algorithm produces proficient results as shown in Fig. 2.

Table 1 Number of relevant web pages retrieved by graph-based and hash-based classifications

S. no	Query no	No of keywords	Count of web pages retrieved in graph-based classification [8]	Count of web pages retrieved in our proposed method
1	Q1	6	1200	778
2	Q2	7	1500	1200
3	Q3	8	1578	1350
4	Q4	6	1879	960
5	Q5	6	2500	879
6	Q6	5	1798	676
7	Q7	4	2357	766
8	Q8	9	3500	1236
9	Q9	4	2787	673
10	Q10	6	3250	985

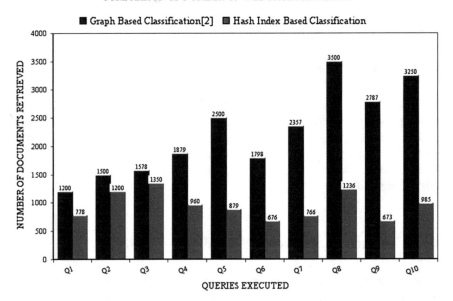

Fig. 2 Number of web pages retrieved by graph- and hash-based classifications

Table 2 Ranks calculated for web pages using agent-based and improved PageRank algorithms

S. no	URL of web page	Rank by agent-based weighted PageRank [10]	Proposed method improved distance PageRank
P1	http://sohacogroup.com.vn/index.html	567	237
P2	http://www.amicidelgiocodelponte.it/index.html	658	325
P3	http://www.uniaoparaobem.com.br/index.html	657	728
P4	http://www.hotelmajore.it/ck.html	379	297
P5	http://v-montazar.com/index.html	958	560
P6	http://www.eca.edu.au/index.html	3576	2305
P7	http://www.speyerseminar.de/index.html	875	329
P8	http://www.jsjgw.com/index.html	956	526
P9	http://www.leftoverpets.org/index.html	780	379
P10	http://www.bsc-md.de/index.html	1545	540

4.2 Comparison of PageRanks of Web Pages

After retrieval of relevant documents, we calculated ranks of each web page using our improved distance PageRank algorithm. We compared our results with agent-based weighted PageRank algorithm [10]. In Table 2, we are representing the results that our algorithm compared with the existing algorithm agent-based weighted PageRank. Results produced by our algorithm are more relevant to given query and time taken to retrieval of results also less compared to existing algorithm.

Performance of our algorithm shows that improved distance PageRank algorithm for ranking of web pages produces most relevant results for the given query in the top position in terms of both content and number of incoming and outgoing links. We compared our results with agent-based weighted PageRank algorithm [8]. Our algorithm produces proficient results as shown in Fig. 3.

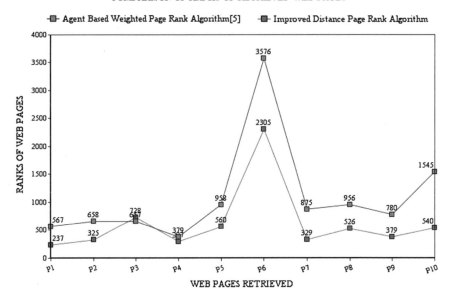

Fig. 3 Ranks of web pages using agent-based and improved distance PageRank algorithms

5 Conclusion

Retrieval of pertinent results for a given query is most complicated task; it depends on so many factors. In this paper, we proposed an efficient method to reduce non-relevant results for given query using dynamic navigation of queries using hash-based indexing and improved distance PageRank algorithm. This algorithm is based on extended hashing technique, unique visit count of web pages and distance between hubs and authorities of web pages. In this algorithm, unique visit count gives us information about the web pages which are visited more number of times, so we will get more relevant web pages; in addition to this, we are also considering the distances between hubs and authorities; if less distance between web pages, then they can be reached in less amount of time. Our proposed method reduces most of non-relevant results for given query and retrieves more relevant documents as top results based on both content and hyperlinks.

6 Future Scope

In this paper, we developed dynamic navigation of queries using hash-based indexing and improved distance PageRank algorithm to retrieve more relevant web pages, by considering unique visit count, hub scores, authority score and distance between the web pages. Our algorithm works well with database contains related data like medical data and education databases. Development of efficient search algorithm for multidisciplinary data is a future scope of our work

Acknowledgements This research was supported by department of science and technology under WOS-A. I would like to thank my supervisor Dr. P Bhaskara Reddy for his support and help through the year. Finally, I would like to thank our colleagues from MLRIT who provided insight and expertise that greatly assisted the research.

References

1. S. Amir, H. Aït-Kaci1, An efficient large scale reasoning method for the semantic web. J. Intell. Inf. Syst. **46**(135) 2–24 (Nov 2016)
2. V.K Nagappan, Dr. P. Elango, in *Agent based weighted page ranking algorithm for web content information retrieval*. Proceedings of International Conference on Computing and Communications Technologies, 2015
3. H. Dubey, B.N. Roy, An improved page rank algorithm based on optimized normalization technique. Int. J. Comput. Sci. Inf. Tech. (IJCSIT). 2183–2188 (2011)
4. A. Kashyap, V. Hristidis, M. Petropoulos, S. Tavoulari, Effective navigation of query results based on concept hierarchies. IEEE Trans. Knowl. Data Eng. **23**(10) (July 2011)
5. A. Alasiry, M. Levene, A. Poulovassilis, Extraction and evaluation of candidate named entities in search engine queries. in *WISE*, pp. 483–496, Mar 2012
6. M.Z. Bidoki, N. Yazdani, Distancerank: an intelligent ranking algorithm for web pages. Intern. J. Inf. Process. Manage. 1–16 (2007)
7. T.H. Haveliwala, Topic sensitive page rank a context-sensitive algorithm for page rank. IEEE Trans. Knowl. Data Eng. **15**(4), 784–796 (2003)
8. C. Xia, X. Wang, in *Graph-Based Web Query Classification*. IEEE Conference Publications, Feb 2015, pp. 241–244
9. S.G. Pawar, P. Natani, in *Effective Utilization of Page Ranking and HITS in Significant Information Retrieval*. Proceedings of International Conference for Convergence of Technology, 2014
10. A. Figueroa, G. Neumann, in *Exploiting User Search Sessions for the Semantic Categorization of Question-Like Informational Search Queries*. International Joint Conference on Natural-Language Processing, Jan 2013, pp. 902–906
11. L. Li, L. Zhong, G. Xu, M. Kitsuregawa, A feature free search query classification approach using semantic distance. Expert Syst. Appl. Int. J. **39**(12), 10739–10748 (2012)
12. S. Brin, L. Page, The anatomy of a large-scale hypertextual Web search engine. Page Rank Comput. Netw. ISDN Syst. **30**(1–7), 107–117 (1998)
13. Y. Liu, in *Supervised HITS Algorithm for MEDLINE citation Ranking*. IEEE 7th International Symposium on Bioinformatics and Bioengineering, 14–17 Oct 2007, pp. 1323–1327

Defending DDoS in the Insecure Internet of Things: A Survey

Manisha Malik, Kamaldeep and Maitreyee Dutta

Abstract As Internet of things (IoT) continues to entrench into our homes, offices, hospitals, and other walks of life, the stakes are too high to leave security to chance. IoT devices are loosely secured and vulnerable to a number of attacks. One of the most common attacks is the distributed denial-of-service (DDoS) attack. DDoS attacks are easiest to launch and very difficult to defend. Although DDoS is an old Internet attack and a number of defenses are available but a large number of new, constrained and always on IoT devices have increased the attack surface beyond imagination. Earlier, defending DDoS in IoT continued to torment researchers as well as enterprises. This review paper unveils how IoT offers a second chance to improve DDoS defenses and provides a survey of such mechanisms with respect to technologies that drive the IoT protocol stack like IPv6, RPL, IEEE 802.15.4, 6LoWPAN, and CoAP.

Keywords DDoS · IoT · IPv6 · RPL · Intrusion detection · IEEE 802.15.4 6LoWPAN

1 Introduction

Internet of things (IoT) is defined as the ever-growing network of physical things that are connected to the Internet and range from sensors to actuators, microwaves to pacemakers, cars to electricity meters, thermostats to light bulbs, and so on. These things or devices are characterized by extremely limited resources in terms of

M. Malik · Kamaldeep (✉) · M. Dutta
Department of Computer Science and Engineering, National Institute
of Technical Teachers Training and Research, Chandigarh, India
e-mail: kamal.katyal@yahoo.com

M. Malik
e-mail: manishamalik53@gmail.com

M. Dutta
e-mail: d_maitreyee@yahoo.co.in

© Springer Nature Singapore Pte Ltd. 2018
S. S. Dash et al. (eds.), *Artificial Intelligence and Evolutionary Computations
in Engineering Systems*, Advances in Intelligent Systems and Computing 668,
https://doi.org/10.1007/978-981-10-7868-2_22

memory, processing capacity, and battery power. In other words, IoT is the Internet extension of traditional wireless sensor networks (WSN) where not only sensors but all the things have the ability to identify, sense, process, and communicate with other things. The advent of new protocols like Internet Protocol version 6 (IPv6) over Low-power Wireless Personal Area Networks (6LoWPAN) [1], Routing over Low Power and Lossy Networks (RPL) [2], Datagram Transport Layer Security (DTLS) [3], IEEE 802.15.4 [4], Constrained Application Protocol (CoAP) [5], and so on which are developed specifically for constrained devices to ease the realization of IoT networks.

Although the seeds of the term IoT were planted in the year 1999 by British entrepreneur Kevin Ashton, it is today that IoT is becoming a commercial success. The reasons for why we are seeing the rise today and not in the early 1990s can be the mobile revolution, the rise of Internet application programming interfaces (API), fall in the cost of components and services, the rise of IoT community, and much more [6]. The IoT possibilities are limitless, and so is the number of devices that could manifest. A number of research challenges for the Internet of things elaborated in [7] include massive scaling, openness, robustness, architecture and its dependencies, big data, security, and privacy. Other challenges like lack of standards, interoperability, legal issues, and cultural impact on use of these technologies also inhibit IoT realization.

As per IoT forecast by Gartner Inc., 6.4 billion connected things will be in use in 2016, up by 30% from 2015 and by the year 2020, the number will reach 20.8 billion [8]. These heterogeneous devices form a highly interconnected network where all kinds of communications seem to be possible, even unauthorized ones. More connected devices mean more attack vectors and more possibilities for hackers to target. According to Proofpoint's report [9], more than 750,000 phishing and spam emails were launched from IoT devices (including televisions and fridge) in the period December 2013–January 2014. Such incidents highlight the need to address IoT security issues. According to Open Web Application Security Project (OWASP), the IoT vulnerabilities range from device's small memory to insecure operating system/firmware, weak web and physical interfaces to insecure network access and lack of transport encryption to insufficient authentication and authorization [10]. As a result, the security requirements for IoT devices and networks become critical.

The main motivation of this survey is to present an overview of research challenges in IoT security and provide a taxonomy of DDoS attacks in the IoT. The rest of this paper is organized as follows. Section 2 discusses the research issues associated with IoT security. Section 3 details DDoS attacks in IoT, and Sect. 4 reviews recently proposed defense solutions to combat such attacks. Finally, concluding remarks and scope for future work are provided in Sect. 5.

2 IoT Security

The widely heterogeneous and scalable nature of IoT devices raises various challenges in terms of data, communication, and network security. Also, conventional Internet security protocols are often acknowledged as unusable in this type of networks particularly due to the extremely limited resources of the devices. Traditionally, security concepts that must be provided include confidentiality, integrity, availability, authentication, authorization, and privacy. Additionally, securing the communication, data (at rest and transit, both), and network in IoT depends on multiple layers of protection including security at the end point. Several issues and challenges need to be addressed in order to ensure a secure IoT environment which are listed in Table 1 and have also been identified in [11–13].

The IoT communication stack [14] as standardized by the Internet Engineering Task Force (IETF) consists of a number of protocols explicitly designed to support Internet communication with constrained devices. These include IEEE 802.15.4 at the physical and MAC layers, RPL and IPv6 at the network layer, 6LoWPAN at the adaptation layer, and CoAP at the application layer. Figure 1 compares the IoT protocol stacks with the TCP/IP-layered model.

As far as security is concerned, IEEE 802.15.4, recently updated in 2011, supports multiple modes of security at the MAC layer to provide confidentiality, data authenticity, data integrity, access control, and protection against various replay attacks [13]. RPL and CoAP too support multiple modes of security which essentially differ on how key exchange and authentication is performed. However, the 6LoWPAN adaptation layer does not have any inherent security mechanisms of its own and relies on network layer security mechanisms like (Internet Protocol Security/Internet Key Exchange) IPsec/IKE [15, 16].

Table 1 Security challenges in IoT

Security challenge	Prospective solutions
Interoperability	Development of interoperable security solutions that do not hinder the operation of widely heterogeneous and resource-constrained IoT devices
Hardware security	Hardening the devices by leveraging hardware security features like TPM/TEE, TrustZone, and crypto acceleration
Secure bootstrapping	Ensuring authenticated and encrypted code updates to IoT devices
Secure communication	Implementing lightweight and robust security protocols like DTLS to ensure data, communication, and network security
Data protection	Encryption of data at rest and in transit and respective access control mechanisms
Scalability	Security solutions like distribution and management of cryptographic keys should be scalable to millions of IoT devices
Resilience	Ensuring fault tolerance to attacks and implementing event monitoring through IDS and firewalls

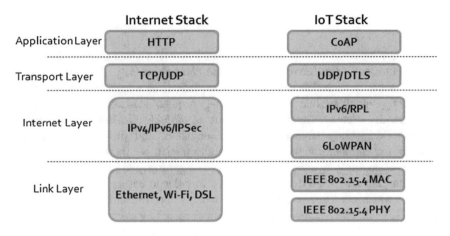

Fig. 1 Comparison of conventional Internet and IoT stack

A number of studies and researches have been conducted to protect IoT networks using these protocols. These researches mainly focus on adaptations of standard Internet security protocols like IPsec/IKE, DTLS, TLS [17], HIP-DEX [18], etc, for IoT. For instance, in [19], the authors proposed a compression of IKE header at the 6LoWPAN layer to address the problem of key management in IoT. The proposed scheme encouraged cross-layer security mechanisms by implementing lightweight IKEv2 for IPsec as well as IEEE 802.15.4 security. In [20], the authors proposed a version number and rank authentication scheme to counter internal attacks against RPL. Similarly, a number of proposals to secure CoAP using DTLS have also been proposed [21, 22]. Despite the tremendous efforts of researchers and industry organizations, a number of issues and challenges still need to be addressed to protect against various forms of attacks in order to realize a secure IoT environment.

In this paper, the focus is mainly on DDoS attacks that are very easy to trigger but difficult to prevent and defend. These attacks continue to torment researchers and network administrators even after years of its existence, and with advent of IoT, the severity and scale of DDoS attacks have drastically increased.

3   A Taxonomy of DDoS in IoT

While the IoT technology inherits the attacks of the conventional Internet, the emergence of new protocols specifically for IoT adds to the number of possible attacks. One of these attacks that has been a major concern in the conventional Internet security and common to IoT is the distributed denial-of-service (DDoS) attack. DoS attack aims to consume the resources of a remote host or network, thereby denying or degrading services to legitimate users. When a group of attackers perform this attack, it is called DDoS attack. Recently, in September 2016, a series of massive DDoS attacks propagated through compromised IoT devices and targeted Brian Kreb's website, "Krebs on Security", OVH, a known Web hosting provider, and "Dyn", a well-established DNS provider [23]. These massive attacks have highlighted the risks resulting from inadequate security mechanisms in IoT devices, together with their devastating effects on the Internet itself. Also, it is very easy to initiate a DDoS on IoT networks keeping in mind the limited bandwidth and processing capability of IoT devices compared to the attacker; that is, it is easy to exhaust IoT devices and deny resources to them.

A couple of DDoS classification in IoT has been proposed in the literature. In [24], the authors only focussed on numerous attacks on RPL like selective forwarding, sinkhole, wormhole, hello flood attacks, etc. The authors also proposed respective countermeasures including IDS and firewall to defend against these attacks. However, they did not cater to all DDoS attacks in the IoT protocol stack. In [25], the authors provided a brief classification of DDoS attacks in IoT but did not provide any prospective mechanisms to defend them. To the best of our knowledge, this is the first attempt to provide a classification based on the IoT protocol stack layers.

In this section, we present a taxonomy of DDoS attacks in IoT on the basis of layers defined in the IETF-standardized protocol stack for IoT [14]. Figure 2 summarizes the proposed taxonomy. Each classified attack has its own initiation and triggering process which decides the appropriate defense mechanism to be taken is discussed in Sect. 4.

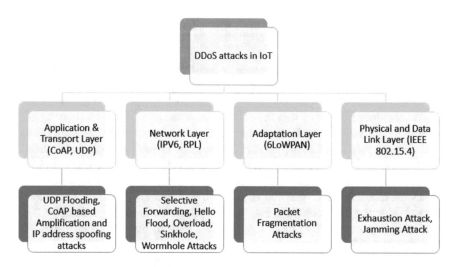

Fig. 2 Taxonomy of DDoS in IoT

3.1 Application and Transport Layer DDoS Attacks

To enable seamless transport and support of Internet applications, the IETF's Constrained RESTful Environments (CoRE) working group introduced CoAP as the de facto standardized protocol at application layer. CoAP is explicitly designed to meet the requirements of low power and lossy networks in the IoT like low overhead, simplicity, multicast support, and less energy consumption. To transport CoAP messages from one node to another, User Datagram Protocol (UDP), an unreliable datagram transport mechanism, is used. This use of unreliable and connectionless UDP protocol at the transport layer adds to DDoS risks. One such attack is known as the UDP flooding attack. In UDP-flooding attack, the attacker floods random ports on targeted host with IP packets. The target machine then checks for associated application, finds none, and sends back an ICMP destination unreachable message to the sender. As more UDP packets are received, the target machine keeps on generating ICMP messages and becomes unresponsive to legitimate requests.

CoAP, on the other hand, implements a request/response layer that includes a method code or response code, respectively. CoAP servers reply to request packet with response packet. An attacker might use CoAP nodes to turn small attack packet into a larger attack packet, a process known as amplification. By exploiting the amplifying properties of this protocol, an attacker can overload a victim with large amount of traffic. Additionally, the lack of handshake in UDP gives rise to a number of IP address spoofing attacks which may lead attackers to launch DoS attacks on IoT networks [6].

3.2 Network Layer DDoS Attacks

IoT is driven by a number of communication technologies but the foundation is provided by the IPv6 protocol because of its ability to support a large number of Internet Protocol (IP) addresses. However, adoption of IPv6 is slow, and security implementations that take it into full consideration are also lagging. Various tools to identify malicious traffic in IPv6 are immature, and the devices that perform IPv4 to IPv6 translation are quite fragile. This leaves a lot of IoT networks vulnerable to DDoS attacks. Besides, Internet Protocol Security (IPSec) is not often deployed widely due to performance considerations and in cases when attackers launch DDoS attacks with their real identity, IPSec is helpless. Protocols like Internet Control Message Protocol (ICMPv6) and Duplicate Address Detection (DAD) in IPv6 networks also suffer from vulnerabilities that enable attackers to flood the victim with falsified request/response messages [26].

DDoS attack in an IoT network can be initiated by malicious nodes from within or outside of the network. Attacks that occur due to internal malicious nodes are usually specific to routing protocol used to route packets within the network. One such standardized IPv6 routing protocol for low power and lossy networks is the RPL protocol. RPL aims to reduce energy consumption at the constrained nodes and supports a number of traffic flows like point-to-point, point-to-multipoint, or multipoint-to-point communication. However, RPL is vulnerable to attacks that already exist in WSN accompanied by those against the IoT. For example, selective forwarding attack allows launching DoS in RPL where malicious nodes selectively forward packets with intention to disrupt routing paths. Other flooding attacks in RPL include Hello Flooding attack, Black hole, Neighbor discovery, Sybil, and Clone ID attacks [24].

3.3   Adaptation Layer DDoS Attacks

The adaptation layer is basically a sub-layer that defines a way to transport IP packets over link layer. Following the path to realize an IoT world, the IETF IPv6 over low-power WPAN (6LoWPAN) working group was started in 2007 to work on specifications for transmitting IPv6 over IEEE 802.15.4 networks. Low power WPANs are often characterized by small packet sizes, low bandwidth, large number of devices, high unreliability, etc. [14]. Besides, the payload size in IEEE 802.15.4 networks is limited to 127 bytes whereas the maximum transmission unit (MTU) in IPv6 networks is at least 1280 bytes. Thus, a full IPv6 packet is not able to fit in a single IEEE 802.15.4 frame. Hence, there is a need to specify some adaptation mechanisms (via 6LoWPAN) that allow IPv6-enabled Internet communications in the IoT. These adaptation mechanisms include optimizations to headers like compression, fragmentation, and reassembly. However, these mechanisms often render buffering, forwarding, and processing of fragmented packets challenging on resource-constrained devices. This gives rise to packet fragmentation attacks in which a malicious node sends forged, duplicate, or overlapping fragments that threaten the availability and normal functioning of such devices.

3.4   Physical and Link Layer DDoS Attacks

At the physical and link layers, IEEE 802.15.4 standard defines the operation of low-rate Wireless Personal Area Networks. It can be used with Zigbee [27], WirelessHART [28], 6LoWPAN, etc., each of which further extends the standard by developing the upper layers. Since IEEE 802.15.4 is a wireless communication standard, it suffers from all the common exhaustion and jamming attacks that are possible in a wireless medium besides some specific attacks on IEEE 802.15.4. Exhaustion attacks involve exploitation of some initiation or connection procedures typically association procedures in which a node tries to associate itself with more than one coordinator [29]. Jamming attacks disrupt wireless communication by emitting radio and noise interference signals which can be constant, random, reactive, or disruptive [30]. The presence of extremely resource-constrained devices makes IoT networks an easy target of such attacks. Specific DDoS attacks on IEEE

802.15.4 include wideband, pulse, node-specific and message-specific denial including bootstrapping attacks specified in [25].

To conclude, it would be correct to say that IoT by its design is inherently insecure and the fact that IoT devices are always online makes them more susceptible to such attacks. Furthermore, hackers may also infect IoT devices and use them as an army for botnet attacks. For example, smart health band which we wear that syncs its data with mobile phone may be acting as a potential bot or a zombie in some DDoS attack. Thus, we can say that IoT and DDoS altogether is a dangerous mix. Hence, prevention and defense measures against such attacks for the IoT are extremely important and are discussed in the next section.

4 Defending DDoS in the IoT Age

There is no question that IoT is changing the way we do business, the opportunities that lie before us, and the security threats we face. At the RSA 2015, IDC analyst Chris Christiansen argued that there is no money in security and embedded security in consumer IoT devices is minimal. Although enterprises, researchers, and vendors are constantly working on solutions to protect IoT devices, there is always a trade-off between device security and market profits. Enterprises are eager to get their products to the market and in this battle of security and profit, profit always wins.

We can classify the DDoS attack defense strategies into two types:

- Proactive attack defense strategies take the precautionary steps in preventing the attacks.
- Reactive attack defense strategies work in two ways, either they try to respond to the attack after the attack situation has been detected or they aim to identify the source of attacks.

Proactive DDoS defense measures in IoT utilize a security framework that integrates security not only in the perimeter but also in the devices themselves. Some of such proactive measures are as follows:

- Hardening the IoT device for secure boot and leveraging hardware security features like the Trusted Platform Module/Trusted Execution Environment (TPM/TEE), TrustZone, crypto acceleration, and so on.
- Deploying intrusion detection and prevention system (IDPS) and adoption of robust encryption mechanisms.
- Segmenting IoT device in its own network and deploying firewalls for network access.
- Ensuring strong authentication passwords.
- Enabling visibility and audit reporting of IoT devices to enterprise management.
- Deploying intelligent gateway devices to enhance security of the network perimeter.
- Managing vulnerabilities and ensuring secure firmware updates.

Table 2 DDoS defenses in IoT

Layer in standardized IoT protocol stack	IoT technology	DDoS attack type	DDoS defense mechanism
Application and transport layer	CoAP	Amplification and IP address spoofing attacks	Mandating the implementation of security modes like pre-shared key, raw public key or certificates [5]
	UDP	UDP flooding	Hybrid IP traceback mechanisms [31], IPSec [15], Rate limiting ICMP [32]
Network layer	IPv6	Flooding and application layer attack	Cryptographically generated addresses [33], IPSec [15], Hybrid IP traceback mechanisms [31]
	RPL	Selective forwarding	Look for disjoint routing paths and implement efficient IDS [24]
		Hello flood	Bidirectional verification using shared secrets [34]
		Overload	Introduce traffic quotas each node is allowed to send and node isolation if traffic goes above certain threshold [34]
		Sinkhole	Allow only trusted communication, introduce traffic quotas each node is allowed to send and node isolation if traffic goes above certain threshold [34]
		Wormhole	Perfect wormhole attack impossible to detect and a research challenge [24, 34]
Adaptation layer	6LoWPAN	Packet fragmentation	Addition of new fields to 6LoWPAN header like timestamp and nonce to prevent fragment replays [35], message purging and ensuring per-fragment authentication [36]
Physical and link layer	IEEE 802.15.4	Jamming	Distributed fuzzy logic and anomaly-based detection [29, 30]
		Exhaustion	

A broad classification of eminent technologies that collaborate directly or indirectly to push the IoT technology, possible DDoS attacks, and corresponding defense measures is shown in Table 2.

5 Conclusions and Future Scope

The rapid growth of denial-of-service attacks in the IoT networks has led to a great number of defense solutions. Much of the work is focused on mitigating the flooding-based DDoS attack by either extending traditional Internet DDoS defenses or by using WSN-based solutions. With such points in mind, this paper provides an extensive analysis of various DDoS attacks on relatively newer IoT protocols.

The paper also addresses existing research proposals to defend such attacks. However, it should be noted that despite tremendous efforts, a lot of research still needs to be done to combat these attacks. Regarding the deployment of IDS in IoT, security gateways that act as an intermediate between IoT networks and the Internet may prove to be useful to implement heavy computations. However, such approaches are usually insufficient in terms of fast response times, real-time detection, and the ability to defend against combination of these attacks. Besides, a lot of research needs to be done to identify and thwart low volume unusual application layer DDoS attacks that may pass the 9 Gbps mark. Cloud-based web application firewalls which have the capability to administer the growing number of mobile devices are also emerging to defend against such attacks. Future research will also emphasize on detection solutions that are lightweight in nature and at the same time limit the manifestation of DDoS attacks in future.

References

1. G. Montenegro, N. Kushalnagar, J. Hui, D. Culler, Transmission of IPv6 Packets over IEEE 802.15.4 Networks. RFC 4944 (2007)
2. T. Winter et al., RPL: IPv6 Routing Protocol for Low-Power and Lossy Networks. RFC 6550 (2012)
3. E. Rescorla, N. Modadugu, Datagram transport layer security version 1.2. RFC 6347 (2012)
4. IEEE Standard for Local and metropolitan area networks–Part 15.4: Low-Rate Wireless Personal Area Networks (LR-WPANs) Amendment 1: MAC sublayer, IEEE Std 802.15.4e-2012 (Amendment to IEEE Std 802.15.4-2011) (2011)
5. Z. Shelby, K. Hartke, C. Bormann, The constrained application protocol. RFC 7252 (2014)
6. P. Corcoran, The internet of things: Why now, and what's next? IEEE Consum. Electron. Mag. 5, 63–68 (2016)
7. J.A. Stankovic, Research directions for the internet of things. IEEE Internet Things J. 1, 3–9 (2014)
8. Rob van der Meulen.: Gartner Says 6.4 Billion Connected Things Will Be in Use in 2016, Up 30 Percent From 2015 (2015)
9. Proofpoint Inc., Proofpoint uncovers Internet of things (IoT) Cyberattack (2014)
10. Internet of things Top Ten, https://www.owasp.org/images/7/71/Internet_of_Things_Top_Ten_2014-OWASP.pdf
11. O. Gracia-Morchon, S. Kumar, S. Keoh, R. Hummen, R. Struik, Security Considerations in the IP based Internet of things. IETF Internet Draft (2013)
12. B. Sarikaya, Security bootstrapping for resource-constrained devices. IETF Standards Track (2013)
13. J. Granjal, E. Monterio, J.S. Silva, Security for the internet of things: A survey of existing protocols and open research issues. IEEE Commun. Surv. Tutorials. 17, 1294–1312 (2015)
14. M. Palattella, N. Accettura, X. Vilajosana, T. Watteyne, L. Grieco, G. Boggia, M. Dohler, Standardized protocol stack for the internet of (important) things. IEEE Commun. Surv. Tutorials. 15, 1389–1406 (2013)
15. S. Kent, K. Seo, Security architecture for the internet protocol. RFC 4301 (2005)
16. C. Kauffman, Internet key exchange (IKEv2). RFC 4306 (2005)
17. T. Dierks, E. Rescorla, The Transport Layer Security (TLS) Protocol. RFC 5246 (2008)
18. R. Moskowitz, HIP Diet EXchange (DEX). IETF Internet Draft. (2012)

19. S. Raza, T. Voigt, V. Jutvik, in *Lightweight IKEv2: A Key Management Solution for both the Compressed IPsec and the IEEE 802.15.4 Security*. IETF Workshop on Smart Object Security, (Paris, France, 2012), pp. 1–2
20. A. Dvir, T. Holczer, L. Buttyan, in *VeRA- Version Number and Rank Authentication in RPL*. 8th IEEE International Conference on Mobile Adhoc and Sensor Systems (MASS), (IEEE, Spain, 2011), pp. 709–714
21. K. Hartke, Practical issues with datagram transport layer security in constrained environments. Internet Draft (2014)
22. S. Raza, T. Daniele, T. Voigt, in *6LoWPAN Compressed DTLS for CoAP*. 8th IEEE—International Conference on Distributed Computing in Sensor Systems (DCOSS), (IEEE, Hangzhou, 2012), pp. 287–289
23. European Union Agency for Network and Information Security. Major DDoS attacks involving IoT devices. (2016)
24. L. Wallgren, S. Raza, T. Voigt, Routing attacks and countermeasures in the RPL-based internet of things. Int. J. Distrib. Sensor Netw. **2013**, 1–11 (2013)
25. K. Sonar, H. Upadhyay, A survey: DDoS attack on internet of things. Int. J. Engg. Res. Develop. **10**, 58–63 (2014)
26. C.E. Caicedo, J.B.D. Joshi, S.R. Tuladhar, IPv6 security challenges. IEEE Comput. **42**, 36–42 (2009)
27. Zigbee Alliance. Zigbee Specification, 344–346. (2006)
28. A. Kim, in *When HART Goes Wireless: Understanding and Implementing the WirelessHART Standard*. IEEE International Conference on Emerging Technologies and Factory Automation (ETFA), (IEEE, Hamburg, 2008), pp. 899–907
29. C. Balarengadurai, S. Saraswathi, in *Detection of Exhaustion Attacks Over IEEE 802.15.4 MAC Layer Using Fuzzy Logic System*. 12th International Conference on Intelligent Systems Design and Applications (ISDA), (IEEE, Kochi 2012), pp. 527–532
30. C. Balarengadurai, S. Saraswathi, in *Detection of Jamming Attacks in IEEE 802.15.4 Low Rate Wireless Personal Area Network Using Fuzzy Systems*. International Conference on Emerging Trends in Science, Engineering and Technology (INCOSET), (IEEE, Tiruchirappalli, 2012), pp. 32–38
31. M.H. Yang, M.C. Yang, RIHT: a novel hybrid IP traceback scheme. IEEE Trans. Inf. Forensics Secur. **7**, 789–797 (2012)
32. F. Gont, G. Gont, C. Pignatarao, Recommendations for filtering ICMP messages. IETF Internet Informational Draft (2013)
33. T. Aura, Cryptographically generated addresses. RFC 3972 (2005)
34. T. Tsao et al., A security threat analysis for the routing protocol for low-power and lossy networks (RPLs). IETF Informational Draft. RFC 7416 (2015)
35. H. Kim, in *Protection Against Packet Fragmentation Attacks at 6lowpan Adaptation Layer*. International Conference on Convergence and Hybrid Information Technology, (IEEE Press, Daejeon (2008), pp. 796–801
36. R. Hummen, J. Hiller, H. Wirtz, M. Henze, H. Shafagh, K. Wehrle, in *6LoWPAN Fragmentation Attacks and Mitigation Mechanisms*. 6th ACM conference on Security and privacy in wireless and mobile networks (WiSec'13), (IEEE Press, Budapest (2013), pp. 55–66

Competent K-means for Smart and Effective E-commerce

Akash Gujarathi, Shubham Kawathe, Debashish Swain,
Subham Tyagi and Neeta Shirsat

Abstract The paper compares various clustering algorithms with k-means algorithm used in e-commerce. It gives a brief introduction to the e-commerce system. K-means algorithm is largely used for the clustering, so it investigates the k-means algorithm and factors out the advantages and the drawbacks of the traditional k-means approaches. For the drawbacks of the traditional approaches, the paper tries to refine the traditional algorithm. The new algorithm is expected to increase the effectiveness and the cluster quality. Paper also proposes a unique collaborative recommendation pool approach based on k-means clustering algorithm. We adopt the modified cosine similarity to figure out the similarity between users in the same clusters. Then, we produce recommendation results for the target users. By mathematical analysis, we prove that our clustering algorithm surpasses traditional k-means algorithm.

Keywords Data mining · Clustering · Marketing · K-means · Cosine similarity
Association rules

A. Gujarathi (✉) · S. Kawathe · D. Swain · S. Tyagi · N. Shirsat
Department of Information Technology, PVG's College of Engineering
and Technology, Pune, Maharashtra, India
e-mail: akash.gujarathi99@gmail.com

S. Kawathe
e-mail: skawathe@gmail.com

D. Swain
e-mail: debashishswain1996@gmail.com

S. Tyagi
e-mail: shubham11tyagi@gmail.com

N. Shirsat
e-mail: neeta.shirsat@gmail.com

© Springer Nature Singapore Pte Ltd. 2018   235
S. S. Dash et al. (eds.), *Artificial Intelligence and Evolutionary Computations*
in Engineering Systems, Advances in Intelligent Systems and Computing 668,
https://doi.org/10.1007/978-981-10-7868-2_23

1 Introduction

The era of World Wide Web has changed every traditional system with the electronic system, and even business is no excuse for it. The term e-commerce was first introduced in the late 1970s by Michael John Aldrich; today, e-commerce has overtaken the traditional marketing system, hence understating the interest of the customers is important to design the marketing strategy. The characteristics of e-commerce-oriented data mining algorithms [1] are to perform operations like association rules [2], classification, and clustering which are used to get the customers information. Through discovering, analyzing, and processing the customer's information, companies can get potential data about the customers. Companies can use this data to yield larger profits and design their business policies according to the group of customers. Clustering is used to group the similar data from the unlabeled dataset based on some metric [3]. There are basically two approaches for the clustering:

- Hierarchical clustering
- Partitioning clustering.

But for clustering, k-means algorithm is widely preferred as k-means algorithm takes in the set of unlabeled data and gives out the clusters of similar data and differentiates the dissimilar data significantly; it is a distance-based clustering model that uses exemplars for clustering, and it works better with clusters that are spherical in vector space. K-means has a linear time complexity, that is, $O(n)$. In contrast, hierarchical clustering does not use exemplars for clustering it has quadratic time complexity, that is, $O(n^2)$. It is independent of a number of clusters.

This paper gives overview of e-commerce system; in Sect. 3, it defines the traditional k-means algorithm and lists the defects of the traditional k-means. Section 4 introduces the collaborative recommendation pool approach which increases the efficiency of k-means algorithm. Finally, the conclusion is provided.

2 Related Work

In 1979, Michael Aldrich demonstrated the first online shopping system [4]. In the current era, the e-commerce has largely overtaken the traditional business system. The advantage of the e-commerce system is that it provides the global platform for the business with smaller economic investment which is totally contrary to the traditional business approach. Because of this advantage, many business companies have shifted toward the e-commerce system. As per NASSCOM, the e-commerce industry in India will be a \$100bn industry by 2020 [5]. Thus, it has become important for e-commerce business to provide a quality service in terms of the product security management, etc. E-commerce mainly focuses on 'one-to-one' customer-centric approach. Hence, understating the interest of the customer is important. For this, the companies use different clustering methods to group the similar customer. Karl Pearson developed cosine similarity function. It is used to measure the linear dependence between two variables. It gives the values between +1 and −1, where 1 is the total positive linear correlation, 0 is no linear correlation, and −1 is the total negative correlation. Formula for the Pearson correlation coefficient is [6].

$$\cos(r(A), r(B)) = \frac{\sum_{i=1}^{n} A_i B_i}{\sqrt{\sum_{i=1}^{n} A^2} \sqrt{\sum_{i=1}^{n} B^2}} \tag{1}$$

3 Cluster Analysis in E-commerce

Clustering forms a class or clusters of data instances based on information of the data instances or relationship between the instances so that instance in the same cluster or class has higher similarities and instances in the different class have higher dissimilarities. Formation of the cluster is based on the measurement of the data instances. Lesser the distance implies higher similarity, and greater distance implies higher dissimilarity.

3.1 Defects of Traditional K-means

- Traditional k-means algorithm demands the user to give k values in advance, and different k values lead to different cluster result.
- It uses large-scale sparse matrix.
- After iteration, traditional k-means will compute new average value again to obtain new cluster center. Most of its algorithm time is taken on recalculating distance between each sample and the new cluster center.

4 Collaborative Filtering Recommendation Approach

In this section, the paper describes the collaborative filtering recommendation approach. This approach analyzes the similarity and dissimilarity between the data objects to improve the quality of clusters.

Here,

A, B, C: User of the Web site

P1–P7: Web pages of Web site

In Table 1, we have considered access time of each user on a web page. Consider users A and B with time vector $r(A)$ and $r(B)$. We need a similarity metric

Table 1 Sample records

	P1	P2	P3	P4	P5	P6	P7
A	4	–	–	5	1	–	–
B	5	5	4	–	–	–	–
C	–	–	–	2	4	5	–

Table 2 Sample records

	P1	P2	P3	P4	P5	P6	P7
A	4	0	0	5	1	0	0
B	5	5	4	0	0	0	0
C	0	0	0	2	4	5	0

Table 3 Sample records

	P1	P2	P3	P4	P5	P6	P7
A	4	10/3	10/3	5	1	10/3	10/3
B	5	5	4	14/3	14/3	14/3	14/3
C	11/3	11/3	11/3	2	4	5	11/3

(A, B) to understand the similarity between users. The key to defining the similarity function is how we deal with unknown values, that is, the pages that are not accessed. One of the default options is that we replace the empty values with 0 (Table 2).

Intuitively, user A and user B look similar as compared to user A and user C. To verify similarity between the users, we use the cosine similarity function [6].

$$\text{Similarity}\,(A, B) = \cos(r(A), r(B)) \tag{2}$$

$$\text{Similarity}\,(A, B) = \cos(r(A), r(B)) = \frac{\sum_{i=1}^{n} A_i B_i}{\sqrt{\sum_{i=1}^{n} A^2}\sqrt{\sum_{i=1}^{n} B^2}} \tag{3}$$

After calculating the function, we get,

$$\text{similarity}\,(A, B) = 0.38$$
$$\text{similarity}\,(A, C) = 0.32$$

Even though it shows the difference but the dissimilarity is not significant enough. One of the other approaches of recommendation pool described [7] the pages that have not been accessed are replaced by the mean of the access time of each user (Table 3).

Table 4 Sample records

	P1	P2	P3	P4	P5	P6	P7
A	2/3	0	0	5/3	−7/3	0	0
B	1/3	1/3	−2/3	0	0	0	0
C	0	0	0	−5/3	1/3	4/3	0

After calculating the function, we get,

$$\text{similarity } (A, B) = 0.95$$
$$\text{similarity } (A, C) = 0.84$$

The above approach does not illustrate the contrast in the similarities between the user A and user B compared to user A and user C significantly. This paper tries to contrast the similarity difference significantly by introducing the centered cosine similarity which is also known as Pearson coefficient.

In this approach, the mean of the time vector is subtracted from the individual time to obtain a new time vector with the empty values being replaced by the '0'. By this approach, the mean of the time vector is subtracted from the individual time to obtain a new time vector with the empty values being replaced by the '0' (Table 4).

Here, one interesting fact that can be inferred is that positive time means that the user basically likes the specific product more than average and negative time means that the user dislikes the specific product.

Now after calculating the cosine similarity, we get,

$$\text{similarity } (A, B) = 0.009$$
$$\text{similarity } (A, C) = -0.56$$

This shows that the users A and B have more similarities compared to the users A and C. This approach is superior as it captures intuition better and treats the missing values as average (Fig. 1).

5 Competent K-means Algorithm

Using the new recommendation pool system, this paper proposes a new "Competent K-means" algorithm which is expected to increase the cluster quality and has higher search accuracy.

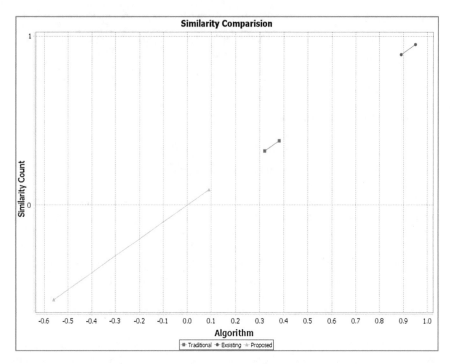

Fig. 1 This is the simulation result of all the three approaches. Here, traditional refers to the approach 1, that is, replacing empty values by 0; existing refers to approach 2, that is, replacing empty by mean; and proposed refers to approach 3, that is, using cosine similarities. '0' is considered as the lowest value of similarity, and '1' is considered as the highest value of similarity. The output of all the approaches clearly shows that the third approach, that is, the proposed approach outperforms the other two approaches. Here, the first approach shows the difference of 0.06; though the second approach shows the difference of 0.11, its similarity values are really near to 1; hence, it is difficult to differentiate between similarity of A, B and A, C. In third approach, it shows a clear difference between two points as 0.551

5.1 Algorithm

The competent k-means clustering algorithm is described below:

```
Input: Initial k_update and recommendation pool(T_r)
Output: C_set of recommendation pool
k_update=(k_update/2);
Update the recommendation pool using proposed technique;
initialize(T_r,C_set,k_update );
 while k_update ≤ K Do begin
   C_set=k-means(T_r,C_set,k_update);
   Upper=0;
   C_new = null;
```

```
for each c ⊆ Tᵣ begin
Dₙₑw = 0;
   for each c⊆Cₛₑₜ  begin
   Dₙₑw = Dₒₗd + distance(Tᵣ, t, c);
   end;
   if Dₙₑw > Upper then begin;
   Upper = Dₙₑw ;
   Cₙₑw = t;
   end;
   end;
  Cₛₑₜ = Cₛₑₜ ∪ {Center};
  K = kᵤₚdₐₜₑ + 1;
end;
return Cₛₑₜ;
end.
```

5.2 Advantage

- A clear distinguishing between users hence there is no close clustering.
- New cluster center will be from cluster members only, hence no need to store extra element.
- The largest element from the center is considered as the next cluster, hence no need of recalculation of cluster center in every iteration.
- No need to specify the cluster heads in advance.

6 Conclusion

This paper compared the traditional clustering algorithms with k-means algorithm. The examination of the traditional k-means algorithm pointed out drawbacks of the approach. The results of the unique collaborative recommendation pool proved the efficiency over the traditional method. The addition of this collaborative recommendation pool in the k-means algorithm is expected to improve the cluster quality and search accuracy of the algorithm. The experimental implementation of the algorithm is going on.

References

1. P. Baumann, D.S. Hochbaum, Y.T. Yang, A comparative study of leading machine learning techniques and two new algorithms submitted (2015). [online] Available at: https://scholar. google.ch/citations?user=v52HA-QAAAAJ&hl=en. Accessed 4 Jan 2017
2. R. Agrawal, R. Srikant, Fast algorithm for mining association rules in large databases, in *Proceedings of the 20th International Conference Very Large Databases, San Francisco: Morgan Kaufmann Publishers Inc.* pp. 487–499
3. M. Emre Celebi, H.A. Kingraci, P.A. Vela, A comparative study of efficient initialization methods for the K-means clustering algorithm. Expert Syst. Appl. **40**, 200–210 (2013)
4. E. Tkacz, A. Kapczynski, *Internet—Technical Development and Applications* (Springer, Berlin 2009), p. 255, ISBN 978-3-642-05018-3, 2011
5. Survey.nasscom.in. India to be $100bn e-commerce industry by 2020| NASSCOM. [online] Available at: http://survey.nasscom.in/india-be-100bn-ecommerce-industry-2020. Accessed 4 Jan 2017
6. J.S. Breese, D. Heckerman, C. Kadie, *Empirical Analysis of Predictive Algorithms for Collaborative Filtering,* UAI'98 Proceedings of the Fourteenth conference on Uncertainty in artificial intelligence (1998), pp. 43–52
7. L.-R. Wang, Z.-F. Wu, *Implementation on E-commerce Network Marketing Based on WEB Mining Technology,* Seventh international conference on measuring technology and mechatronics automation (2015), pp. 556–560

Installation Automation Tool

Bhanuprakash Vattikuti, N. V. K. Chaitanya and Amar Jukuntla

Abstract Installation automation tool is aimed to reduce the human effort involved in software installation and uninstallation tasks in environments where a set of software applications are needed to be installed or uninstalled over a set of systems with same specifications. The main aim is to reduce the necessity of human presence through the process. So, this tool can be used to add or remove software over multiple systems by automating the parts which needs human intervention. This can be a great help in, which involves a large number of systems being connected and treated as a whole and help will be provided across all the systems.

Keywords Installation · Automation · Authentication · Generalization
Uninstallation

1 Introduction

In environments, such as laboratories and companies, some specific softwares are needed to be added to every system to execute the specific tasks. In traditional way, a person have to manually install the software on each of the computer one by one, thus making the person expend a large amount of time, time that can be effectively used in other works. So, making any types of changes to the systems in such an environment is like a black hole taking in time and work. As all the computers concerned are of a similar configuration, with the software needed to be installed being the same, this boils down to the person doing the same set of repetitive

B. Vattikuti (✉) · N. V. K. Chaitanya · A. Jukuntla
Department of CSE, Vignan's Foundation for Science, Technology & Research,
Vadlamudi, Andhra Pradesh, India
e-mail: bhanup6663@gmail.com

N. V. K. Chaitanya
e-mail: nvk.681@gmail.com

A. Jukuntla
e-mail: amar.jukuntla@gmail.com

© Springer Nature Singapore Pte Ltd. 2018   243
S. S. Dash et al. (eds.), *Artificial Intelligence and Evolutionary Computations
in Engineering Systems*, Advances in Intelligent Systems and Computing 668,
https://doi.org/10.1007/978-981-10-7868-2_24

actions to install the software in each and every one of the systems. Sounds boring doesn't it? Not only is it incredibly boring, it is also very taxing on the workforce involved as the number of workers tasked with installation is generally very low compared to the number of systems. The solution we propose for this is installation automation tool.

The basic premise of the idea is to replace human input with automation of these tasks. Doing the same thing over and over again is something better suited to robots, rather than humans. The tool basically takes care of the redundant work required by the user by itself, thus reducing human effort and time. It still requires the basic inputs and the admin's permission to do its required work.

Installation automation tool is a form of automation which replaces human effort with automation installing and uninstalling of software applications in Linux machines [1]. This Tool is useful when there is a need to install a set of software in multiple systems. It starts the installation process by itself and proceeds to install those software onto the multiple systems automatically.

Like computer laboratories in colleges or a specific department in a company, groups of computers are widely found in all facets of society. In the absence of this tool, a user will have to spend his time installing software in sequence and wait till one process is done before starting on another one, which is a time-consuming process and needs a large manpower in the case of a huge number of systems.

Installation automation enables the installation and uninstallation of multiple software at a time with minimal manpower. The user has to give authenticated system root password to use this tool; this provides the admin access to the software for its computation. This allows the tool to configure the required human input required in the installation process of the software as well as provide it with the necessary authority required to install it into a system. Similar Tools were available in Windows platform. Some of them are still in use for the Windows 7.

2 Implementation of Installation Automation Tool

2.1 User Authentication

The first step of using installation automation tool is user authentication. In this step, the tool ensures that it can only be used by the person authorized by the admin. This step also provides the tool with admin access which is mandatory for the successful completion of installations [2].

This step provides security in a way that only the person with the user ID and password will be able to use the installation automation tool to configure his system.

If the user is not able to open the tool with admin credentials, it will close by itself after a specific number of attempts [3]. Authentication interface will provide the user with three chances to get authenticated after which, an event to terminate installation automation tool is raised.

2.2 Available Software List

When setting up a computer laboratory or a corporate workplace, it is repeatedly required to install the same set of software in multiple systems. Here, the software to be installed is the same over all the systems. The proposed installation automation tool will be configured with the software set, and it will be able to handle an alternation in system specifications like 32/64 bit and so on [4, 5].

The user is only required to give the set of software to be installed over the group of systems along with the admin password required to configure them [2]. In case of a failure or any other issues, a popup will be raised which can be taken care of by the user any time he finds suitable.

The installation [6] interface displays a set of checkboxes, with their corresponding software, along with the option for uninstallation of that particular software. This is possible since the installation automation tool has been made to handle multiple workspaces/laboratories, etc.

2.3 UnInstallation Block

Sometimes, the installation of any particular software is done due to the previous software being outdated or for better functionality. This implies that the previously present software is now redundant and serves no purpose whatsoever. To deal with this possibility, the installation automation tool is also enabled with the capability to uninstall pre-existing software through its interface [6]. This is to increase the flexibility and functionality of the tool as well as to serve the user in better ways.

Uninstallation is a similar task to perform with a different result. The total burden in uninstalling any given software is roughly the same as installing it, so this helps the flexibility of the installation automation tool, allowing it to always be there for the user's rescue.

3 Sequence Flow of Installation Automation Tool

Installation automation tool follows the below-mentioned steps.

Begin
    Initialize the Tool
    Do
        Step:1.   Enter the valid root password.
        If Password is valid **then**
            If Select Application List
                Step: 2 show the list applications to install.
                Step: 3 Select from list of applications.
                If click on install **then**
                    Install the selected applications.
            If select view installed **then**
                Step: 4 Display list of applications that are installed.
                Step: 5 Select applications to be un-install.
                If click on uninstall **then**
                    Uninstall the selected application from the
                computer.
    End Do
End

4 Installation Automation Tool

4.1 Authentication Interface

Fig. 1 shows the authentication interface screen when the user needs to input admin credentials.

The purpose of the authentication interface is to collect the user ID and password from the user [7]. The interface validates the ID and password and then authenticates the usage of the installation automation.

4.2 Application Installation Interface

Fig. 2 shows the application installation interface screen where the user can select the application for installation [7, 8].

User Name:	Enter admin username
Password:	Enter password
	Submit

Fig. 1 Authentication interface

☐ Argo UML   ☐ GCC
☐ Notepad++   ☐ Open JDK
☐ Python 3.2

Install

Fig. 2 Software applications installation interface

☐ Argo UML   ☐ GCC
☐ Notepad++   ☐ Open JDK
☐ Python 3.2

Un Install

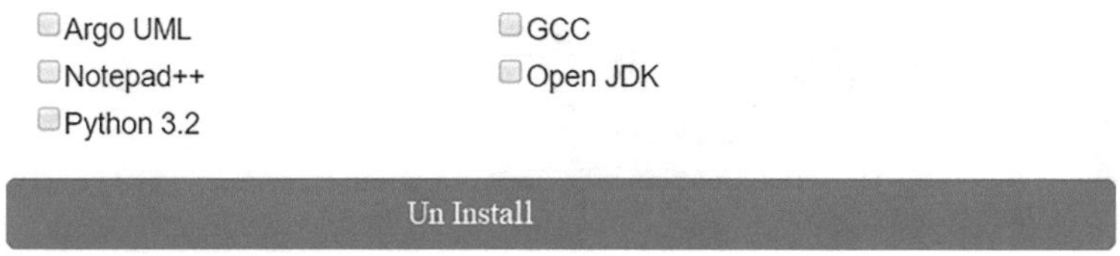

Fig. 3 Software applications uninstallation interface

4.3 Application Uninstallation Interface

Figure 3 shows the authentication interface screen where the user can select the application for uninstallation from the list of software applications already installed on the machine [7, 8].

5 System Architecture

It follows a three-tiered architecture with the operating system at the base level. The programming of the two remaining levels is done in Python with a lot of modules all of them being implemented by the "Tkinter" interface. This Tkinter interface connects the operating system to the actual implementations of the tool being contained in the package. The top-level tier consists of the user interface being implemented by the three user interaction interfaces "AUTHORIZATION, INSTALLATION, and UNINSTALLATION." This comprises of the user interface layer being the basic level at which the user interacts with the installation tool.

As shown in the Fig. 4, Tkinter acts a broker between user interfaces and the operating system. Input from the user which will be in the user interfaces is carried to the operating system with the help of the Tkinter package of Python.

OS package of Python will be used to access all the commands of the operating system.

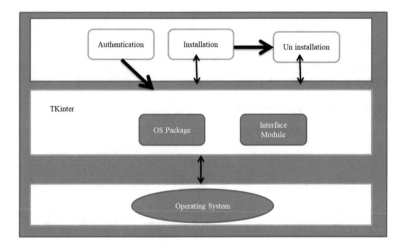

Fig. 4 System architecture

6 Results

6.1 Merits

The installation automation tool provides a simple interface to install/uninstall a set of software over multiple systems allowing even a layman to operate it, and it gives an experienced user the flexibility to freely deal with the multiple systems at the same time. We can choose the mode of installation as default or custom with the custom mode offering deeper access into the working of the tool. This also saves the time of the user.

6.2 Demerits

Installation automation tool when used in a small environment does not show a significant disadvantage as there is almost no difference between using the tool on a limited number of systems and manually installing the software on the systems one by one.

This has only been implemented as to work on Linux kernel, as the programming required will have to be custom made to tailor-fit different types of operating systems. This also points out that it can be made available to all operating systems as long as the required workforce is invested.

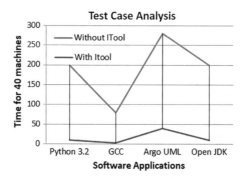

Fig. 5 Test case analysis graph

Table 1 Comparison of time taken by manual installation and automation installation

Application	Manual run time	Automated run time
Python 3.2	5 min * 40 = 200 min (3 h 20 min)	(3) * 40
GCC	2 min * 40 = 80 min (1 h 20 min)	
Argo UML	7 min * 40 = 280 min (4 h)	
Open JDK	5 min * 40 = 200 min (3 h 20 min)	

6.3 Test Cases

Installation automation tool is tested in a laboratory environment of 40 machines of four different set specifications manually on each machine, and by using this tool for four applications and considering the copy time of applications as content, the tool has been analyzed.

Below is the resultant graph that is obtained by taking software application on the X-axis and time taken by the 40 machines on the Y-axis (Fig. 5).

Observations are tabulated in Table 1.

7 Conclusion

Installation automation tool provides an easy-to-use interface to set up the given software set, over multiple systems, thus reducing the human effort involved in a typical installation process. This will be of great help in the intended target fields with an estimated efficiency of at least 70% in cutting down the manpower and time needed to be invested normally in installing or uninstalling the software in the group of systems. This saves the time of the users.

References

1. J. Alglave, A.F. Donaldson, D. Kroening, M. Tautschnig, Making Software Verification Tools Really Work, Automated Technology for Verification and Analysis, in *Proceedings 9th International Symposium, ATVA 2011, Taipei, Taiwan,* 11–14 Oct 2011
2. Executing linux commands via python programming language: https://www.cyberciti.biz/faq/python-execute-unix-linux-command-examples/
3. Linux kernal and modules, working: https://www.linux.com/
4. Working with python language: https://www.python.org/
5. Using modules and packages in python: https://docs.python.org/3.5/
6. Installing and uninstalling softwares in linux os: https://itsfoss.com/remove-install-software-ubuntu/
7. Creating user interface via python using tkinter. https://www.tutorialspoint.com/python/python_gui_programming.htm
8. Reference to functions of tkinter in python to develop gui. http://www.python-course.eu/python_tkinter.php

Missing Values Imputation Using Genetic Algorithm for the Analysis of Traffic Data

Ranjit Reddy Midde, Srinivasa K. G and Eswara Reddy B

Abstract The missing information issue stays as a trouble in transportation data framework, which genuinely confined the utilization of insightful transportation framework, for example, movement control and activity stream forecast. To take care of this issue, various ascription techniques had been proposed in the most recent decade. Notwithstanding, few existing reviews had completely utilized the spatial relationship of movement information ascription. Street mishap casualty rate relies on many elements, and it is an exceptionally difficult undertaking to examine the conditions between the qualities in view of the numerous natural and street mischance variables. Any missing information in the database could darken the revelation of essential variables and prompt to invalid conclusions. We propose a new method for missing data imputation named genetic algorithm with classification. Our results indicate that the proposed method performs significantly better than the existing algorithms.

Keywords Data mining · Missing data · Classifier · Traffic data
Genetic

Ranjit Reddy Midde (✉)
Department of CSE, JNTUA, Anantapuramu, Andhra Pradesh, India
e-mail: midderanjit@gmail.com

Srinivasa K. G
Department of IT, CBP Government Engineering College, New Delhi, India
e-mail: kgsrinivasa@gmail.com

Eswara Reddy B
Department of CSE, JNTUA, Ananthapuramu, Andra Pradesh, India
e-mail: eswarcsejntua@gmail.com

© Springer Nature Singapore Pte Ltd. 2018
S. S. Dash et al. (eds.), *Artificial Intelligence and Evolutionary Computations in Engineering Systems*, Advances in Intelligent Systems and Computing 668,
https://doi.org/10.1007/978-981-10-7868-2_25

1 Introduction

A spatiotemporal challenge can be portrayed as a question that has no short of what one spatial property and one brief property. The spatial properties are zone and geometry of the question. The transient property is timestamp or time between times for which the dissent is true blue. The spatiofleeting inquiry as a rule contains spatial, common, and topical or nonspatial properties. Instances of such inquiries are moving auto, forest fire, and earth shake. Spatiotemporal instructive accumulations fundamentally discover changing estimations of spatial and topical qualities over a time span. An event in a spatiotemporal dataset depicts a spatial and common ponder that may happen at a particular time t and region x. Instances of event sorts are earth tremor, tropical storms, road action stick, and road setbacks. In traffic data analysis, the data to be analyzed is both spatial and temporal. Henceforth, it is fundamental to perceive viably the spatial and short-lived parts of these events and their associations from tremendous spatiotemporal datasets of a given application territory. Spatiotemporal examination can be requested as common data examination, spatial data examination, dynamic spatiotemporal data examination, and static spatiotemporal data examination. The common data examination settles the spatial estimation and separates how topical qualities data change with time. Examination of precipitation, temperature, and moisture of a given region over a time period is an instance of this kind. The spatial data examination separates how topical properties data change with respect to a partition from a spatial reference at a predefined time. Examination of advance in temperature and sogginess values while moving a long way from sea float at a given time is an instance of this sort. The component spatiotemporal data examination fixes topical qualities estimation and dismembers how spatial properties change with time. Examination of moving auto data and spread of fire are instances of this class. The static spatiotemporal data examination settles the transient and topical property estimations and studies the spatial estimation. An instance of this is finding territories having same precipitation at same time. Examination of significant volume of spatiotemporal data without settling any estimation is to a great degree troublesome and complex. However, the data mining can be used to uncover darken illustrations and examples inside the data. As of late, advances in intelligent transportation systems (ITS) innovations have permitted transport experts to gather unfathomable measures of activity observation information for use in applications, for example, blockage checking, episode discovery, voyager data, and versatile movement control. Activity information, for example, movement volume and speed, is gathered through various assorted advancements, for example, inductive circle finders, GPS tests, and video picture location frameworks. Be that as it may, paying little mind to the innovation utilized, activity information is regularly deficient because of equipment or programming glitches, control disappointments, and transmission mistakes. The extent of missing information is not trifling. For instance, in Alberta, Canada, the greater part of the expressway activity checks has missing qualities [1], while in Beijing, China, the normal missing proportion of movement information is around 10% [2].

The issue of missing activity information altogether debases the execution of ITS applications. For example, movement anticipating models have a tendency to be less powerful or even pointless with an inadequate dataset. Essentially, propelled movement flag control frameworks require adequate activity stream information to create ideal control plans. Therefore, it is basic to fill the holes in the information with fitting qualities through a reasonable and viable ascription handle. The ordinary approach toward attributing missing activity information for a street portion is to depend on its authentic information or current information from neighboring street sections. This approach has some basic confinements. Ascription in view of verifiable information performs inadequately within the sight of uncommon activity conditions that veer off from recorded standards. Another disadvantage is that because of the high level of changeability in the connections between the activity parameters of neighboring street fragments, it is for the most part important to develop an expansive number of area particular models for movement information ascription.

2 Related Works

Regular ways to deal with atmosphere flag estimation from information are powerless to predispositions if the instrument records are inadequate, cover varying periods, if instruments change after some time, or if scope is poor. A technique was displayed for acquiring fair, maximum likelihood (ML) evaluations of the climatology, patterns, as well as other wanted climatic amounts given the accessible information are from a variety of settled watching stations that report sporadically. The adroitly clear strategy took after a blended model approach, making utilization of understood information examination ideas, and abstained from gridding the information. It was impervious to missing information issues, including "determination inclination," and furthermore in managing basic information heterogeneity issues and gross blunders. In literature, some strategies are proposed for quantitative mistake investigation. It has been widely studied, to use the historical and finely tuned information, to detect system malfunctioning or error in data collection, without the need of physical inspection of the equipement under consideration. These techniques have been successful in numerous practicle applications like analying the existing climate data to identify frail meridional winds in the central lower stratosphere. Creators assessed the execution of various estimation methods for the infilling of missing perceptions in outrageous everyday hydrologic arrangement. Summed up relapse neural systems were proposed for the estimation of missing perceptions with their information design decided through an enhancement approach of hereditary calculation (GA). The adequacy of the GRNN–GA procedure was obtained through near execution investigations of the proposed strategy to existing methods. In light of the after-effects of such relative investigations, particularly on account of the English River (Canada), the GRNN–GA procedure was observed to be an exceptionally aggressive strategy when contrasted with the current counterfeit neural systems strategies. Furthermore, in view of the criteria of mean square and supreme blunders, a point-by-point relative investigation including the GRNN–GA, k-nearest neighbors, and various ascription for the infilling of missing records of the Saugeen River (Canada) likewise observed the GRNN–GA system to be predominant when assessed against other contending strategies. The as often as possible utilized ascription techniques for missing movement information are recorded (neighboring) attribution strategies, spline (counting straight)/relapse ascription techniques, autoregressive incorporated moving normal (ARIMA) models, and probabilistic principal component analysis. These techniques concentrate on crediting missing information for a solitary circle identifier, which frequently uses the worldly connections, for example, day mode periodicity, week mode periodicity, and interim variety of movement information to evaluate missing information. All things considered, the activity information is spatiofleeting corresponded. Contrasted and fleeting connections, the spatial relationships of activity information have not been completely used. The most condition of-craftsmanship techniques just utilize spatial data from neighbor indicators. Be that as it may, the movement information is connected in short, separate, as well as in a vast territory particularly in a turnpike passage. Accordingly, just utilizing neighbor indicator data is not the best approach for ascription of missing activity information. As of late, a tensor (multiway cluster)-based technique has been connected to missing activity information attribution and exception movement information recuperation. The activity information is modeled by multiway grid (tensor) design, and the missing movement information is assessed by tensor finish strategy. Tensor fruition takes into account consolidating and using the multimode worldly connections to appraise the missing information, which has been ended up being a proficient device to model activity information for missing movement information ascription. In spite of the great after-effects of tensor-based technique, this work is as yet connected to single circle locator missing information ascription. Each inadequate data is replaced by some mimicked value, which will be replaced with recreated esteem substitute, in each iteration by analyzing the dataset [3]. The aim is, conceivably, to produce assesses that better demonstrate genuine difference and instability in the information than do relapse strategies. This grants master staff and programming to be utilized to make credited datasets that can be broken down by generally gullible clients outfitted with standard programming. It can be exceptionally powerful, especially for little to direct levels of missingness, where the missing information system is composed, and for datasets that are to be put in people in general area. Kuligowski and Barros [4] proposed the utilization of a back proliferation neural system for estimation of missing information by utilizing simultaneous precipitation information from neighboring gages. Brockmeier et al. [5] investigated different missing information taking care of procedures, and the creators have given exact relative examination of cancellation and ascription techniques. Abebe et al. [6] proposed the utilization of a fluffy govern-based model for substitution in missing precipitation information utilizing information from neighboring stations. The creators have given observational similar investigation of results utilizing the fluffy govern-based model and results utilizing an ANN demonstrate and a customary measurable model. The fluffy lead-based model performs marginally better. They [6] investigated the utilization of different attributions for the examination of missing information. Sinharay et al. [7] proposed cyclic alliance of information planned for maturing ANN models to evaluate missing qualities in month to month surplus datasets. Khalil et al. [8] utilized ANN models to substitute the missing estimations of wave information. Bhattacharya et al. [9] proposed the utilization of a self-sorting out guide (SOM) for attribution of information alongside the multilayer perceptron (MLP) and hot-deck strategies. Fessant and Midenet [10] gave similar investigation on rundown savvy cancellation, mean substitution, straightforward relapse, relapse with a mistake term, and the EM calculation. Musil et al. [11] investigated univariate direct, spline, and closest neighbor addition calculation, multivariate regularized expectation–maximization calculation, closest neighbor, self-sorting out guide, multilayer perceptron (MLP), and in addition cross breed strategies where joining the best components of univariate and multivariate techniques is consolidated in air quality datasets. Junninen et al. [12] proposed new attribution technique for fragmented parallel information. Subasi et al. [13] investigated attribution of missing qualities with self-sorting out guide, multilayer perceptron, multivariate closest neighbor, regularized desire boost calculation, and various ascriptions for precipitation spillover handle informational index [14].

2.1 Technique of Encoding

The chromosome (cms) is typically imparted in a progression of segments, and each part of it is recognized as an excellence. According to the subject resolves, an excellence would be portrayed with the sort of twofold, honest to goodness amount, or distinctive arrangement. One of the best techniques which is easy and traceable is *bit string encoding*, used by genetic researchers.

2.2 Fitness Function

This is utilized to center conviction level of the updated answers for issue. Conventionally, there is a health worth associated with each cms. An advanced health worth suggests that cms or result is more moved up to a topic, while a lesser opinion well-being demonstrates less propelled chromosome. Health qualities are eventual outcomes of target limit. As likelihood, $P[o|\lambda]$ is a fitting worldview used inside of the destination ability to center the way of the cms. The probability ($P[o|\lambda]$) is processed by the best likelihood procedure.

2.3 Selection

The purpose of selection is to choose cms and structure the mating pool from the population. Determination instrument mimics the survival of the fittest segment in nature. Typically, more stronger cms gets an advanced quantity compare to its successors, while the not stronger cms will go on end. A practical control is used inside of this decision framework. Each cms in the people is associated with a section in the practical controls. As shown by the health opinion of the cms, the element would have a greater area when the contrasting chromosome has a better well-being; a lesser health excellence will incite an additional unobtrusive separation.

$$p_i = \frac{F_i}{\sum_{i=1}^{M} F_i}, i = 1, 2, \ldots, M \tag{1}$$

Here (P_i) is the standardized wellness estimation of a mth cms structure in the population and (F_i) is wellness estimation of a cms in the populace. In the pre selection phase cms are chosen. In post determination based on choice utilizes a wellness estimation with cms to type cms from most elevated to least.

2.4 Crossover

This would be utilized to link subparts of the folks to convey successors. This holds a couple of parts of gatekeeper genetic objects. They picked folks without a doubt cms. It may be seen that this administrator is expected to combine the streamlined genetic materials in the folks mutually to convey additional picked up children.

2.5 Mutation

It gives overall looking for capacity to GA by self-assertively altering the measure of characteristics in the cms. Preceding the modification of a form argument, change tempo would be stood out from a self-assertively delivered probability along experiment whether the vary tempo is greater than (or) identical to the aimlessly made probability.

3 Proposed Work

The whole process of the proposed technique is shown in Fig. 1. The missing values are selected from the training attributes; each extracted value is validated with database. Here, we are taking genetic search to identify the missing values. In this work, we treat the time series data as hours, minutes, and seconds. We are considering the Td as a set of available time series data and Ma is considered to be missing value.

Algorithm 1

St1: Find the value for missing attribute
St2: Generate the new population from available training set
St3: Classify the available missing values set $C_m = \{C_1, C2,...Cn\}$
St4: Calculate the fitness function as in Eq. (1)
St5: Validate the missing value with attributes
St6: Compare the value with training time series
St7: Repeat above steps until finding the best value
St8: Store the value in database.

The proposed technique quality is calculated for the root mean squared error (RMSE). In this, we calculate between the predicted missing points and the original points.

$$\text{Rmse} = \sqrt{\left(\frac{1}{M}\sum_{m}^{M}\left(a_{\text{original}} - a_{\text{predicted}}\right)_m^2\right)} \qquad (2)$$

Missing ratio is calculated using the following formula

$$\text{mr} = m * n * 100\% \qquad (3)$$

where n is data number of test data.

Fig. 1 Proposed work

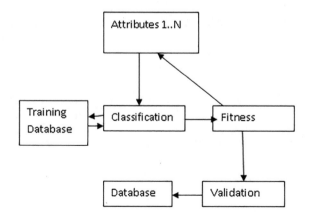

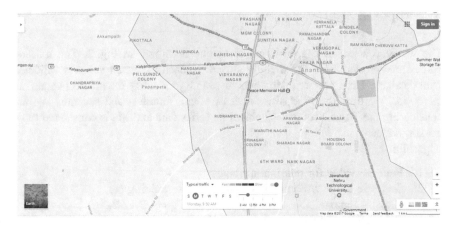

Fig. 2 Anantapur traffic map

4   Experimental Results

We tested the proposed work with Anantapur traffic dataset. The sample traffic data of Anantapur is shown in Fig. 2. We have taken this image from Google Maps.

Tables 1, 2 and 3 describe the dataset forms as full dataset, complete dataset, and missing dataset. Here why we have chosen this dataset means if there is any accident occurred at the traffic that data about the passengers and vehicles will be missing. So in order to identify the missing values, we classify the dataset and we come to one decision.

Table 1 Full dataset

Road no	Driver status	Weather condition	Number of passengers	Injury status	Time	Date
r1	Drunk	Good	3	Minor	13:20	1/1/2017
r2	Normal	Fair	4	No injury	2:32	4/1/2017
r3	Drunk	Dad	6	Major	20:30	5/2/2017
r4	Normal	Bad	5	Major	4:30	5/2/2017
r5	Drunk	Bad	6	Major	20:40	3/3/2017
r6	?	?	4	Minor	21:50	2/1/2017
r7	?	Good	4	Major	22:15	6/2/2017
r8	Drunk	Good	3	?	23:30	8/2/2017
r9	Normal	?	?	No injury	22:40	10/3/2017

Table 2 Complete dataset

Road no	Driver status	Weather condition	Number of passengers	Injury status	Time	Date
r1	Drunk	Good	3	Minor	13:20	1/1/2017
r2	Normal	Fair	4	No injury	2:32	4/1/2017
r3	Drunk	Bad	6	Major	20:30	5/2/2017
r4	Normal	Bad	5	Major	4:30	5/2/2017
r5	Drunk	Bad	6	Major	20:40	3/3/2017

Table 3 Missing dataset

Road no	Driver status	Weather condition	Number of passengers	Injury status	Time	Date
r6	?	?	4	Minor	21:50	2/1/2017
r7	?	Good	4	Major	22:15	6/2/2017
r8	Drunk	Good	3	?	23:30	8/2/2017
r9	Normal	?	?	No injury	22:40	10/3/2017

Fig. 3 RMSE values

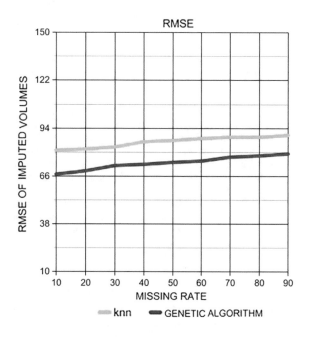

Figure 3 shows that *x*-axis represents missing rate and *y*-axis represents RMSE volumes. It is clearly shown that proposed method gives better results than existing technique.

5 Conclusions

Spatial transient packing is a methodology of accumulation articles in perspective of their spatial and common likeness. It is by and large new subfield of data mining which expanded high reputation, especially in geographic information sciences as a result of the inevitability of an extensive variety of zone-based or environmental contraptions that record position, time, and/or normal properties of a dissent or set of articles logically. Subsequently, particular sorts and a great deal of spatiotransient data got the opportunity to be unmistakably opened that familiarize new challenges with data examination and require novel approaches to manage learning disclosure. The joining of the spatiofleeting neighborhood empowers the model to conjecture successfully under missing information. The proposed display outflanks a scope of benchmark models for gauging under ordinary conditions and under different missing information scenarios. This paper proposes another system with genetic calculation with classifier. These exploratory outcomes demonstrate that proposed system gives better outcomes in contrast to the existing method.

References

1. M. Zhong, P. Lingras, S. Sharma, Estimation of missing traffic counts using factor, genetic, neural, and regression techniques. Transp. Res. Part C Emerg. Technol. **12**, 139–166 (2004)
2. L. Qu, L. Li, Y. Zhang, J. Hu, PPCA-based missing data imputation for traffic flow volume: a systematical approach. Intell. Transp. Syst. IEEE Trans. **10**, 512–522 (2009)
3. R.J. Little, D.B. Rubin, *Statistical Analysis with Missing Data* (Wiley, New York, 1987)
4. R.J. Kuligowski, A.P. Barros, Using artificial neural Networks to estimate missing rainfall data. J. AWRA **34**(6), 14 (1998)
5. L.L. Brockmeier, J.D. Kromrey, C.V. Hines, Systematically missing data and multiple regression analysis: an empirical comparison of deletion and imputation techniques. Multiple Linear Regression Viewpoints **25**, 20–39 (1998)
6. A.J. Abebe, D.P. Solomatine, R.G.W. Venneker, Application of adaptive fuzzy rule-based models for reconstruction of missing precipitation events. Hydrol. Sci. J. **45**(3), 425–436 (2000)
7. S. Sinharay, H.S. Stern, D. Russell, The use of multiple imputations for the analysis of missing data. Psychol. Methods **6**(4), 317–329 (2001)
8. K. Khalil, M. Panu, W.C. Lennox, Groups and neural networks based stream flow data infilling procedures. J. Hydrol. **241**, 153–176 (2001)
9. B. Bhattacharya, D.L. Shrestha, D.P. Solomatine, in *Neural Networks in Reconstructing Missing Wave Data in Dimentation Modeling*. The Proceedings of 30th IAHR Congress, Thessaloniki, Greece Congress, 24–29 Aug 2003 Thessaloniki, Greece

10. F. Fessant, S. Midenet, Self-organizing map for data imputation and correction in surveys. Neural Comput. Appl. **10**, 300–310 (2002)
11. C.M. Musil, C.B. Warner, P.K. Yobas, S.L. Jones, A comparison of imputation techniques for handling missing data. West. J. Nurs. Res. **24**(7), 815–829 (2002)
12. H. Junninen, H. Niska, K. Tuppurainen, J. Ruuskanen, M. Kolehmainen, Methods for imputation of missing values in air quality data sets. Atoms. Environ. **38**, 2895–2907 (2004)
13. M. Subasi, E. Subasi, P.L. Hammer. New imputation method for incomplete binary data, *Rutcor Research Report* (2009)
14. A.M. Kalteh, P. Hjorth, Imputation of missing values in precipitation-runoff process database. J. Hydrol. Res. **40**(4), 420–432 (2009)

Steganography Technique to Prevent Data Loss by Using Boolean Functions

Satya Ranjan Dash, Alo Sen, Sk. Sarif Hassan, Rahul Roy,
Chinmaya Misra and Kamakhya Narain Singh

Abstract Steganalysis technique embedded the secret message, so we have to conceal the very important data. Our main objective is undetectability, robustness and capacity of the hidden data. Existing image steganography techniques used various methods to hide data in a perfect manner. We used many Boolean functions to achieve a data (image)-hiding method or image steganography without loss of any information. Also, the data can be hidden and unhidden efficiently through the Boolean functions. Our experiments and corresponding outcome deliver that it is more secured than the existing approaches of steganography.

Keywords Image steganography · Boolean function · Data hiding
Human visual system · Pixel value differencing

S. R. Dash (✉) · A. Sen · C. Misra · K. N. Singh
School of Computer Application, Kalinga Institute of Industrial Technology,
Deemed To Be University, Bhubaneswar, India
e-mail: sdashfca@kiit.ac.in

A. Sen
e-mail: alosen10@gmail.com

C. Misra
e-mail: cmisra@yahoo.com

K. N. Singh
e-mail: kamakhya.vbhcu@gmail.com

Sk. Sarif Hassan
Department of Mathematics, University of Petroleum and Energy Studies,
Uttarakhand, India
e-mail: sarimif@gmail.com

R. Roy
Applied Electronics and Instrumentation, Asansol Engineering College,
Asansol, India

© Springer Nature Singapore Pte Ltd. 2018
S. S. Dash et al. (eds.), *Artificial Intelligence and Evolutionary Computations
in Engineering Systems*, Advances in Intelligent Systems and Computing 668,
https://doi.org/10.1007/978-981-10-7868-2_26

263

1 Introduction

Steganography is known as "invisible" medium. It means to cover up communication reality in another medium. Image steganography is the most popular technique to hide image, video, etc. It has been developed spatial domain method through which image pixel can be vague with same bits. Image can say many things what human can see with human visual system (HVS). Steganography is not a new word; from ancient ages, people were sending the secret message or information in various formats which is coming under ancient steganography. Secret message is sent through various techniques like secret letter in each word or secret word in each sentence, sometime with the different sizes of grasses which indicate different meaning.

Nowadays we are using digital steganography, as modern era has Internet and digital signal processing due to which now become digital, and here, we are using different kinds of steganography techniques like image-based steganography. Embedded hidden message is also possible in video, audio file also. We also have different kinds of steganography algorithm to embed the message. We also are using DNA cryptography to hide and unhide the data [1].

Steganography is important due to its secret communication over the Internet. In image steganography, secret communication can be sent by embedding text message into cover image, thus generating stega-image [2]. We have different kinds of steganography techniques depending on the suitability described in Fig. 1.

Different steganography measures are

- Capacity—where maximum data we can embedded into the image.
- Perceptual transparency—where subsequent information hiding into cover page, perceptual quality will be converted into stego-image.
- Robustness—where after embedding, data should stay together into stego-image goes into some alteration such as cropping, filtering etc.
- Computation complexity—where cost is computationally evaluate for embedding and extracting a hidden message.

There are different steganography techniques, and all the techniques have their pros and cons. One of these techniques is masking and filtering. In this technique, we

Fig. 1 Types of steganography

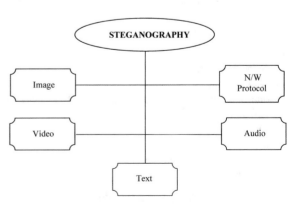

embedded information in added important areas than now hiding into the noise level, but the problem is that it can be applied into only grey scale images and limited to 24 bits. Steganography is employed in various useful applications.

The performance of steganography can be calculated from various factors, and the most important factor is the undetectability of data from an unseen communication. Further connected procedures are the steganographic capability, which is the most information that can carefully entrenched in a work lacking statistically assessable objects [3], and toughness, which refers to how well the steganographic classification resists the image out of hidden data.

2 Related Work

Steganography can be used any digital media comparing to watermarking, and encryption. In steganography, the keys are optional but in case of encryption it is necessary. At least two unless self-embedded files are required, but in case of encryption and steganography, the detection method is blind, but in case of watermarking, it is usually informative. The objective of steganography is to secret communication, but it protects the data in case of encryption. Steganography is very ancient except its digital version. The LSB-based image steganography embeds the undisclosed in the least significant bits of pixel values of the cover image (CVR). The idea of LSB embedding is easy. Also, pixel value differencing (PVD) is another technique to hide data and embedded the secret message. Transform domain technique [4] is more difficult but better than spatial domain method. It conceal information in areas of the picture that are fewer uncovered to density, cropping and processing along with many other techniques. It is also used to the stego objects [5–7] like images, audio, and video as well as file structures, and html pages. Most favourable pixel correction development is used to improve stego-image value obtained by effortless LSB replacement method. To make sure that the adaptive amount k of LSBs remains unaffected after pixel adaptation, the LSBs figure is computed by the high-order bits [8].

There are various steganography methods or techniques like in binary file techniques watermark can be embedded by making some changes in binary code, and in text technique, which is used in document to embed information inside a document we can simply alter some of its characteristics and updated thing can not visible to the human eye. We can also used wavelets to encode the whole image, sound techniques and video steganography techniques for the secret communication [9]. Al-Ataby and Al-Naima [10] had developed a customized image steganography procedure based on wavelet transform. The developed technique assures high level of security. It compresses the data with increase in the capacity or payload of the steganography process. However, this technique can produce high computational overhead. Al-Shatnawi [11] proposed an image steganography method which hides the undisclosed communication on probing about the similar kinds of bits among the secret message and the pixel value [12].

Table 1 Fifteen projection operators

	F0	F1	F2	F3	F4	F5	F6	F7	F8	F9	F10	F11	F12	F13	F14	F15
0,0	0	0	0	0	0	0	0	0	1	1	1	1	1	1	1	1
0,1	0	0	0	0	1	1	1	1	0	0	0	0	1	1	1	1
1,0	0	0	1	1	0	0	1	1	0	0	1	1	0	0	1	1
1,1	0	1	0	1	0	1	0	1	0	1	0	1	0	1	0	1

3 Proposed Work

We have aimed to create a secure data hiding technique through which we can have 100% recovery of data, meaning that we can hide the data without any loss of information and unhide the data successfully. There is always any i and j which will be treated as projection operators. We have created projection operators based on Boolean function. Table 1 illustrates the Boolean functions.

We have illustrated the data hiding algorithm as follows.

3.1 Proposed Algorithm

3.1.1 Data Hiding Method

Step 1: Read the image and calculate the size of the image.

Step 2: Get two pixels of region of interest (ROI) of square area, namely A and B by generating random pixel values. The size of A and B should be $n * n$ matrix.

Step 3: Convert the rgb image to binary image.

Step 4: For $i = 1$ to n

For $j = 1$ to n

Calculate C_i such that $C_i = A$ op B, where op is a projection operator.

Step 5: Check the values of C_i with the values of A and B, respectively, to find whether they are same or not.

Step 6: Finally, we will be getting an image with shuffled region of interest (ROI).

4 Experimental Results

We have used MATLAB for our experiment. We have taken an image in Fig. 2 having size 252 * 256. Then we have chosen two random ROI and get two cropped image Fig. 3 having size 100 * 100. Figures 4 and 6 define the corresponding

Fig. 2 Color image of
252 * 256 size

Fig. 3 First cropped image of
size 100 * 100

Fig. 4 Binary image of first
cropped image

binary images of Figs. 3 and 5 respectively. Thus, we are getting the pixel of region of interest (ROI) for square area, namely A (Fig. 4) and B (Fig. 6) matrices.

We have applied projection operator upon A and B which create C having size 100 * 100. The operation is as follows:

C [i] [j] = A [i] [j] F_0 to F_{15} B [i] [j] (Totally, 15 Boolean functions have been projected upon A and B and produced 15 different matrix). Then, each new matrix has been checked with A and B for equality, and finally, we are getting an image with shuffled region of interest (ROI).

The above experiment is based on only one sample report. We have done 500 times execution of same program to generate different matrices A and B, so that we can find out those projection operator for which most of the shuffled region of interest (ROI) will be similar to corresponding A and B matrices. Hence, data hiding

Fig. 5 Second cropped
image of size 100 * 100

Fig. 6 Binary image of
second cropped image

will be more efficient with 100% preservation of data by those 15 projection
Boolean functions.

5 Analysis

- Figure 7 illustrates those Boolean functions which are common in most of the
  iterations among total 500 iterations.
- According to the priority from Fig. 7, the Boolean functions are as follows:
- F_4 and F_6: mostly found within 101–200 iterations, where the resultant shuffled
  ROI was matched with either A or B matrix.
- $F_2,F_4,F_6,F_7,F_9,F_{13},F_{15}$: mostly found within 301–500 iterations, where the
  resultant shuffled ROI was matched with either A or B matrix.
- F_2,F_4,F_6,F_7: mostly found within 101–200 iterations, where the resultant shuffled ROI was matched with either A or B matrix.
- $F_0,F_1,F_3,F_5,F_8,F_{10},F_{11},F_{12},F_{14}$: not found in any of the iterations.

So, the Boolean functions with the higher priority can be efficiently used for
100% data recovery while unhiding the image without no information loss.

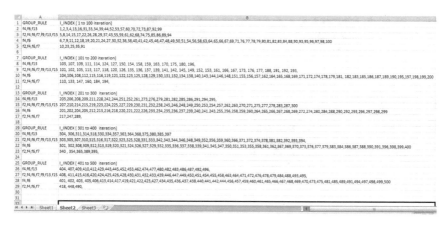

Fig. 7 Boolean functions of iterations

6 Conclusion

In this paper, a novel steganography technique was offered, implemented and analysed by using 15 Boolean functions. We have successfully hidden the image by using these functions. We have taken randomly two ROIs of same $n * n$ matrix from the whole image. Our data hiding method has used 15 boolean functions to accomplish 100% recovery of data, it means we can embedded maximum data into the image and unhide it successfully without loss of any information. We had not illustrated the unhiding methodology, but in the invert process, we can unhide the data efficiently. According to analysis, some Boolean functions got no priority over others. We will try to focus on those operators for better distribution.

References

1. A. Aieh, et al., Deoxyribonucleic acid (DNA) for a shared secret key cryptosystem with Diffie hellman key sharing technique. in *Computer, Communication, Control and Information Technology (C3IT), 2015 Third International Conference on*. IEEE, 2015
2. M. Hussain, M. Hussain, A Survey of Image Steganography Techniques. Inter. J. Adv. Sci. Tech. **54**, 113–124 (2013)
3. I. Cox et al., *Digital watermarking and steganography* (Morgan Kaufmann, Burlington, 2007)
4. S. Katzenbeisser, F. Petitcolas, *Information Hiding Techniques for Steganography and Digital Watermarking* (Artech house, Norwood, 2000)
5. H.S.M. Reddy, K.B. Raja, High capacity and security steganography using discrete wavelet transform. Int. J. Comput. Sci. Secur. (IJCSS) **3**(6), 462 (2009)
6. S. Katzenbeisser, F. Petitcolas, *Information Hiding Techniques for Steganography and Digital Watermarking* (Artech house, Norwood, 2000)
7. P. Kruus et al., A survey of steganography techniques for image files. Adv. Secur. Res. J. **5** (1), 41–52 (2003).

8. H. Yang, X. Sun, G. Sun, A high-capacity image data hiding scheme using adaptive LSB substitution. Radioengineering **18**(4), 509–516 (2009)
9. S. Channalli, J. Ajay, Steganography an Art of Hiding Data. *arXiv preprint* arXiv:0912.2319 (2009)
10. A. Al-Ataby, F. Al-Naima, A modified high capacity image steganography technique based on wavelet transform. Int. Arab J. Inf. Technol. **7**(4), 6 (2008)
11. A.M. Al-Shatnawi, A new method in image steganography with improved image quality. Appl. Math. Sci. **6**(79), 3907–3915 (2012)
12. K.B. Raja, et al., A secure image steganography using LSB, DCT and compression techniques on raw images. in *Intelligent Sensing and Information Processing, 2005. ICISIP 2005. Third International Conference on*. IEEE, 2005

Ant Colony Clusters for Fast Execution of Large Datasets

Konda Sreenu and Boddu Raja Srinivasa Reddy

Abstract Today's world is fully dependent upon computer systems. As the population is growing, the database is also growing. We are rich in information but unable to access them fast and clearly. Business experts want data within a fraction of seconds, which is not possible because of large volumes. It takes a time to view, select, or accumulate large datasets. It is important to have information fastly, clearly, and accurately. This paper focuses on those issues. Deep analytics is the application of sophisticated data processing techniques on the large collection of multi-source datasets. These datasets may be semi-structured or structured or unstructured. For data mining, there should be mining expert and a business analyst. This process may take a long day or more. We need not wait for mining expert; a business user can himself interface with the databases. Paper focuses on methods like columnar databases, run-length environment technique, and ant colony clusters. The business user can interact with database management system easily. The business user can extract domain knowledge related to the current system.

Keywords Large databases · Deep analytics · Datasets · Columnar
Run-length environment · Ant colony

1 Introduction

In today's world, computer memories are increasing with add-on data every day. Memory is growing from GBs to TBs, and in future, it may go up to PBs and exabyte. Same way, more memories are required periodically. There should be communication with all available data. Mostly, business organizations depend upon

K. Sreenu (✉)
Department of CSE, Sir C R Reddy College of Engineering, Eluru, Andhra Pradesh, India
e-mail: sreenukcupid@gmail.com

B. R. S. Reddy
Department of CSE, Ramachandra College of Engineering, Eluru, Andhra Pradesh, India
e-mail: brs_123@yahoo.com

© Springer Nature Singapore Pte Ltd. 2018
S. S. Dash et al. (eds.), *Artificial Intelligence and Evolutionary Computations in Engineering Systems*, Advances in Intelligent Systems and Computing 668,
https://doi.org/10.1007/978-981-10-7868-2_27

private and global data. Global data is useful for market research. There may be a comparison between available data (private) with global data. For every organization will have n number of relations and m number of records. The organization can maintain different databases at different locations. When we see collectively, the records or rows or tuples may be in thousands or lakhs. Nowadays, organizations are using terminals to their host computers. Large volumes of data space are consumed every day. This is trivial.

Data mining process is the solution for extracting information from large databases. For mining, there should be a business analyst and a mining expert. Both of them sit together and start mining process, which may go for a long day or more. When we integrate data in different databases, the volume also increases. When two relations are combined, then there will be result relation which occupies memory. When memory is more speed of the system falls down. Same way, obtaining a result from the various databases will take time. But we are in fast-growing sophisticated world, where users want data within a fraction of seconds.

There are different types of data we can store in databases. The data types that can be stored in secondary storage are machine and sensor data, social media data, image data, video data, enterprise content, transaction and application data, third party data, space-related data, and much more. A single computer cannot accommodate this year's long data. It will take more time to display results. Real-time analysis of such large datasets can be distributed over hundreds or thousands of computers. Data analytics is the study based on systematic computational analysis of data or statistics, or it is information resulting from systematic analysis of data or statistics. Data analytics is related to cloud computing.

A host computer which wants to search for data should contain large memory to hold data that is distributed over a group of computers or geographically place anywhere. When we combine more datasets geographically, the result of the combination may go slow or it makes take hours long to display results. The business user cannot wait for hours or days. There should be a speed process for results; the process of querying data includes columnar databases and in-memory analytics.

Carrefour is supermarket located at different places geographically. It has 11,935 outlets around the world with 381,227 employees. Supermarket operates inventory for goods, selling and buying products, daily sales, and customer data. Supermarket contains a lot of US dollars transactions every day. They maintain local databases and central databases. Every midnight, the new price should be added to items; same way, daily transactions should reach central database every day. A lot of memory will be in use for a supermarket. They have mining experts and business analysts to care about the supermarket. There may be situations where the business user wants to know domain knowledge of the sales. The business user cannot wait for hours. He requires a report within a fraction of seconds, at least minutes.

According to ER diagram, an organization will have interrelated entities. From one entity, we can get other entity details because they have a relation. This mechanism is called DBMS and RDBMS. The entity holds similar data. For each row, all columns are fixed. Each row will have different data but data of each column will be the same type. Columns have a name, but row does not have any name. Likewise, each row will have a unique key to identify it. For supermarket or hypermarket will have x cash counters for customers. Cashier users will collect information about new customers, old customers, and they will be billed. Each bill will be unique. The business user may require total sales.

$$\text{Total\_sales} = \sum \text{sales}(i), \; i = 1 \text{ to } n \text{ databases.}$$

2 Existing System

We have large databases, a lot of information to extract from various sources of databases. We use data mining techniques to extract valuable information. These techniques are related to advanced data analysis. Data warehousing and data mining are part of data analysis. Data mining can effectively work on www. Data can be stored in many different kinds of databases and information repositories. A repository is of multiple heterogeneous data sources organized under a unified schema. Data warehouse technology includes storing or collecting data from various databases, data cleaning, data integration, and OLAP. Data mining refers to extracting or mining knowledge from large databases. Sometimes it also refers to knowledge mining. Data mining process includes various steps: data cleaning and integration, selection and transformation, data mining and evaluation and presentation of the result.

Data mining system contains a database, data warehouse, World Wide Web, and other repositories as there database. A database system is also called a database management system (DBMS), known as a database. For data mining, we require two professionals. One who knows about mining but do not know about business. Another professional is the business analyst but does not know about mining techniques. Data mining professional will collect all the tables and make a unified schema which is suitable for mining. A data cube is prepared based on dimensions. The business analyst will add at this end to raise queries. They extract data in the same way by changing dimensions.

Both business analyst and mining expert should be present for mining. Mining process can carry on for a long day or more. They have to wait for the result for a long time. They have to prepare software or use open sources software for mining.

3 Literature Survey

Kusiak [1] proposed data mining approach for extracting knowledge from various datasets. The whole process is based on steps involved in mining. Data mining is a vast subject that is related to extracting data from various databases like repositories. There are data mining tools for discovery of knowledge relationships, pattern, and knowledge. The author focuses on subsets of all data, which is decomposed from large databases to datasets. Based on a collection of datasets, they go for cleaning, integration, mining, and result in elevation.

Prasad et al. [2] proposed WEBLOG preprocessing techniques for data mining. KDD is a process of extracting data from different datasets. Web usage mining is usually an automated process whereby web servers collect data from customer interestingness and report it to business analyst and mining expert. Data is collected from the server where web server applications reside. By carrying like this, we can know about customer interest. Weakness is that they are large programs for mining and result is understood only by mining expert. Mining expert will further explain the results.

Sreenu and Reddy [3] proposed customer knowledge extract based on neural network and k-means clustering algorithm that suits to large business. Every business organization will have a Web site. A Web site can register customer and access server pages from the server. Based on his mouse click on different items, parameters generate data. This data is prepared to identify the location of an item. Based on this information, business analyst can come to understanding on what customer is interested. Benefits and discount offers can be sent to the customer. This process will increase the online business.

Kumar and Ratna Kumar [4] proposed data sharing paradigm in distributed systems, such as social networks, that have been increasing demands and concerns for distribution security. One of the most challenging issues in data sharing systems is the enforcement of access policies and the support of policies update. Introduced ciphertext-policy attribute-based encryption (CP-ABE) escrow problem solution, which is constructed using the secure two-party computation between the key generation center and the data storing center, but does not focus on focus on datasets.

Ramya et al. [5] introduced new data mining technique to have customer interest over items that are available on the business web site like Flipkart, Amazon, and so on. An algorithm used here is related to preprocessing technique and k-means. No focus is on large datasets.

Rao and Ratna Kumar [6] introduced data mining techniques that are necessary to extract data from large volumes. Focus on removing the replicate or duplicate occurrences that result data is exhibited.

Venkata Siva Satish and Ratna Kumar [7] introduced performance increase between cloud users. Cloud computing is a word heard nowadays on a globe. Cloud uses a huge amount of memory and communication between users. Cloud information is shared among different users. When there are large numbers of users, then performance will go down. Freight stabilization increases the performance in between cloud users.

Manusha et al. [8] introduced genetic programming approach capable of eliminating duplicate citations. Everyday technology is growing, but we cannot access data efficiently. For this, there are various reasons for it may search and issue duplicates which make the results to be slow down. Focus on eliminating duplicates in citations.

Seekoli and Ratna Kumar [9] introduced permutation-based techniques. There will be privacy data in an organization, which should not be exhibited to the public, for example an employee salary. The paper focuses on privacy micro-data. Discuss better goals to capture the need of privacy protection for numerical sensitive attributes-based techniques are used to answer aggregate queries more accurately than generalization-based approaches.

Suresh Kumar et al. [10] introduce index and binary search algorithms searching large databases. Every day, lots of transactions are added to the databases in an organization. More transactions require more memory. Retrieving data from large databases is a challenging task. Focus on how fast data can be accessed.

4 Proposed System

In this research work, we designed and implemented a pattern to solve business user requirements without data mining expert and business analyst. The process of mining will take the time to display results. But business user wants result very fast. Data is collected from different repositories and other sources. Collected data cleaning process will carry on. Data cleaning means removing unwanted non-key attributes, tuples with null values. After cleaning, tuples are made to be ant colony cluster. Each cluster can be of 100 data records size or chunk with fixed size. If the data block is more than 100 records, then it is added to next block. Data in blocks can be compressed. Data with numeric can be compressed, but alphabets cannot be compressed. There are so many algorithms like bitmap index, look-and-say sequences, run-length limited, LZW, or run-length encoding for compression. It will be easy to accumulate large data. There should be large memory enough to hold data collected from various sources. Results are made to display for the user. Data can be displayed in reports form, picture form, text form, or any other way which can be understood by the business user.

The proposed method is called ant colony cluster method as shown in Fig. 1. The data blocks are made to available like ants row. Data blocks can be expanded to their right side. It can grow to the maximum as much memory is available. After data blocks are prepared, based on business user requirements, we can use aggregate functions like SUM, COUNT, AVG, MAX, MIN

Step 1:

Collect data from various database sources.

Step 2:

Data cleaning process, remove unwanted non-key attributes, and columns with null values.

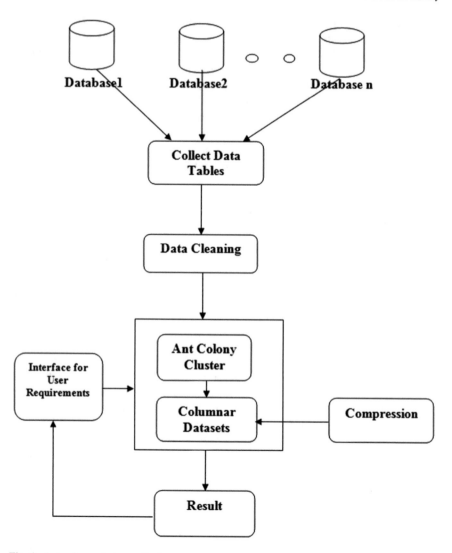

Fig. 1 Ant colony cluster method

Step 3:

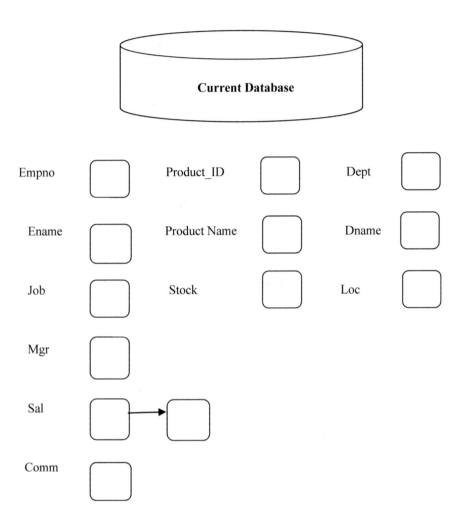

Step 4:

Arrange the data in columnar database manner.
Columnar database has data blocks for 100 record columns.
Normally, table is given in the following manner.

X	Y	Z
1	Ravi	34
2	Smith	40

Step 5:

Apply aggregate functions or any other calculation methods over the data.

5 Implementation

In this paper, we had evaluated experiments over large dataset in database management system. We had collected tables from various supermarkets and evaluated them. No need to wait for a data mining expert; after acquiring the results, the company may consult business analyst or any other third party.

$$\sum db = db1 + db2 + \dots$$
$$db1 = R(X,Y, Z), db2 = R(A, B, C, D).$$

db = db1 + db2 + ...

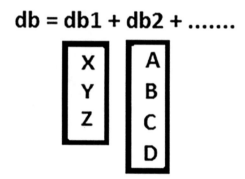

Data compression techniques are applied on X data. Suppose the X contains data in following manner X = 11010100000001, is compressed with exponents in given manner X = 11010^71

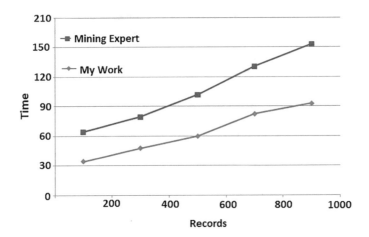

6   Conclusion and Future Scope

Nowadays computers are fast growing devices around the world. Every company depends upon computers to store their data safely and securely. There may be situations where the user wants to have knowledge about available data. This method starts from collecting the data from various repositories onwards. Data cleaning process will be next; after that dataset is handover to ant colony clusters. Each cluster will occupy fix memory. Data inside each cluster can be compressed, if there is a possibility. User can apply various methods to exhibit data. User can use this method to have domain knowledge about what is available.

References

1. A. Kusiak, Decomposition in data mining: an industrial case study. IEEE Trans. Electr. Packag. Manuf., **23**(4), ISSN (Print): 1521-334X, ISSN (Online): 1558-0822 (2000)
2. G.V.S.C.H.S.L.V. Prasad, M.S.R.L. Reddy, K.B. Rao, C.S. Kumar, Approach for developing business statistics using data web usage mining. IJACTE. **1**(2), ISSN (Print): 2319-2526 (2012)
3. V.P.S.N.K. Sreenu, B.R.S. Reddy, Approach for developing business statistics using weblog mining. IJRAET. **2**(11), ISSN (Online): 2347-2812 (2014)
4. G.R.C. Kumar, P. Ratna Kumar, Providing security and efficiently in attribute-based data sharing with novel technique. IJEECS. **3**(12), ISSN (Online): 2347-2820 (2015)
5. A. Ramya, K. Sreenu, P. Ratna Kumar, Preprocessing a unsupervised approach for web usage mining. Int J SocNet & Vircom. **1**(2), ISSN: 2252-8784 (2012)
6. A.H. Rao, P. Ratna Kumar, Data method and mining techniques for better business organization. IJEECS. **3**(12), ISSN (Online): 2347-2820 (2015)

7. S. Venkata Siva Satish, P. Ratna Kumar, A freight stabilization model based on cloud segregating for the shared cloud. IJRCCT. **3**(11), ISSN (Online): 2278-5841, ISSN (Print): 2320-5156 (2014)
8. S.D. Manusha, V. Karthik, P. Ratna Kumar, De-duplication of citation data by genetic programming approach. IJRAET. **1**(3), ISSN (Online): 2347-2812 (2013)
9. A. Seekoli, P. Ratna Kumar, Slicing: a new approach to privacy preserving high dimensional data publishing. IJSETR. **2**(15), ISSN (Online): 2319-8885, (2013)
10. S.V. Suresh Kumar, M.G. Babu, J.S.V. Gopal Krishna, S.M.B. Chowdary, Fast searching technique on sequential large databases in an organization, Unpublished

Intelligent Traffic System to Reduce Waiting Time at Traffic Signals for Vehicle Owners

Green Corridor for Emergency Service Vehicles

M. S. Satyanarayana, B. M. Murali Mohan and S. N. Raghavendra

Abstract The main objective of this research is to control the traffic issues in developing countries like India. As per the survey in metropolitan cities like Bangalore per day, there will be 4000 four wheeler vehicles are getting sold out. Just imagine the increase in traffic, which made the vehicle owners to wait at signals more than the driving time to reach their destination. And another major issue because of this is providing path to the emergency service vehicles like ambulance and fire engine. The best solution for this problem is automatic creation of Green Corridor whenever it is needed. To provide a better solution for these two problems especially in any metropolitan city like Bangalore, this system will be proposed based on the latest technology called Internet of Things (IOT). This system will give solution for waiting time at signals by using the technique called intelligent camera and wireless sensor networks. And coming to the Green Corridor it is purely a sensor-based application or hardware which is going to be fixed at signals to solve this issue. The entire sample data to carry out this process is collected from Bangalore city in order to create data sets. And the sample data sets will be treated as an input for the application which is developed to reduce waiting time at signals, whereas with respect to Green Corridor it is going to be live data, the storage of information or data not really required. In this project, the concept of big data is used especially to process live data for the Green Corridor.

Keywords IOT · Green Corridor · Emergency services · Wireless sensor networks Big data

M. S. Satyanarayana (✉) · B. M. Murali Mohan
Department of CSE, Sri Venkateshwara College of Engineering, Bangalore, India
e-mail: satyanarayanams@outlook.com

B. M. Murali Mohan
e-mail: muralimohan.bm91@gmail.com

S. N. Raghavendra
Department of ISE, Sri Venkateshwara College of Engineering, Bangalore, India
e-mail: raghavendraewit@gmail.com

© Springer Nature Singapore Pte Ltd. 2018
S. S. Dash et al. (eds.), *Artificial Intelligence and Evolutionary Computations in Engineering Systems*, Advances in Intelligent Systems and Computing 668,
https://doi.org/10.1007/978-981-10-7868-2_28

1 Introduction

The main problem in cities nowadays is traffic. Everyone is worried about traffic but nobody is worried about how to reduce this traffic problem. The manual solution to reduce this traffic in metropolitan cities' like Bangalore is by reducing the number of vehicles which are travelling day to day on roads by adopting the techniques like car pooling, high usage of public transport [1] or using of point-to-point sub-urban or metro trains as source of transport. Practically, it is not possible because of so many reasons like time, distance from pickup points [2] and drop points to office or home and coordination between car poolers. So, the only solution which will make the vehicle owners happy is reducing the journey or travel time between source and destination. As per the survey, 80% of traffic in metropolitan cities is not because of number of vehicles, it is because of congested roads and also the waiting time at signals though it is not required in some signals as the operation of standard traffic lights which are currently deployed in many junctions, are based on predetermined timing schemes, which are fixed during the installation, and remain until further resetting. Again constructing widen roads is manual process which may or may not give good results.

In this paper, the focus is more onto the second solution, i.e. reducing the waiting time at traffic signals by adopting new technologies. And not only this problem the other problem is that, as per survey in a year 15–20% people are getting died because of huge traffic at signals. If the Green Corridor is constructed automatically whenever some emergency service vehicle is coming towards the signal, it opens that direction signal, which intern really saves the life of a patient.

The main methodology adopted to develop this system is by using latest technologies like IOT [3] and wireless sensor networks. As the IT Era is moving towards smart systems, this solution will be a feasible solution to solve the above problems. This system may not solve entire problem at a stretch, but pretty sure that it will give good results over existing system or techniques.

2 Ease of Use

This system will be developed in two phases. In the first phase, the main concentration will be on reducing the waiting time at traffic signals, and in the second phase, the implementation of Green Corridor will take place.

A. Reducing the Waiting Time at Traffic Signals

As mentioned in the title, the main objective of this research is to reduce the waiting time at traffic signals, nevertheless of simulation now IOT [4] made everything possible to implement using existing boards and sensors. And of course, the best part of IOT is that the entire research project which will be carried out in the offline can be deployed directly in the real time to produce the expected results.

In this research, the methodology is to use a highly intelligent and 360° [5] rotating camera at signal location. The intelligent camera [5] will be interacting with sensor which is put up near the signals in order to control the signal. Before explaining the new methodology, the existing method should be understood.

In existing methodology, the main problem is that the time at signal lights is fixed, without considering the traffic flow [6] towards that direction. For example, there is a four direction signal light as shown in Fig. 1, the time is fixed for each direction though there is traffic or not. If one vehicle is passed in that direction, the other three direction people have to wait until unless the signal gets opened at their end.

How to solve this problem? It is the question which has risen to carry out this research. In proposed methodology, the intelligent camera will be monitoring in all the directions to calculate the length of the vehicles. Based on the length of the vehicles, the intelligent camera will interact with the sensors which are going to control the time of the signal. That means consider there is a signal where the traffic is flowing in all four directions, in the peak time, i.e. in the morning, there will be huge traffic towards south direction because of number of educational institutions located in that place. So, the intelligent camera [7] will reduce the waiting time of remaining signal and will give priority to this south signal. So that the traffic flow will be smooth and vehicle owners no need to wait long duration [3]. In the same manner, in the afternoon, the traffic flow is huge in north direction and then the

Fig. 1 Traffic at Bangalore signal

priority will be changed from south to north. To process it, the main technique followed is image processing [3] in order to find out the length of the traffic in particular direction. The priority-based algorithms are used in order to control and allocate time for each and every direction of the signal.

The proposed research though it solves the problem of waiting time at traffic signals, the only obstacle we have in developing countries like India is that vehicle length and height are not fixed. Sometime, the estimation of intelligent camera may go wrong. To avoid this, images are captured from satellite in parallel it will be compared with the images of intelligent camera. But the problem to implement this is cost will get increased. The implementation is as follows:

```
pvalRS-Priority value
TV-Traffic Volume
t1- time for Signal when TV is less.
t2-time for signal when TV is more.
t3-time for signal when Emergency Vehicle is in junction.
v-number of vehicles
G-green
Decide(PvalRS,TV)
{
   if(PvalRS)
   {
    G(t3);
   }
   else
   {
    Normal_Traffic(TV);
   }
}

Normal_Traffic(TV)
{
  max(waiting time);
  if(min(v))
  {
    G(t1);
  }
  else
  {
    G(t2);
  }
  for(i=1;i<10;i++)
  {
    G(people);
  }
}
```

B. Green Corridor

As the research on above topic continued, one more question arises or the problem found is, let us say if the signal clearance is given for south direction on high priority [8], unfortunately if ambulance or fire engine has come in north direction, then immediately the priority should be changed to north, i.e. the system has to create Green Corridor for that direction where ambulance has arrived to save the life of a patient. In order to do this, the system is fitted with an acoustic sensor which is going to read the sound frequency of ambulance or fire engine immediately the information will be passed to intelligent camera to change the priority [9]. Initially, we have trouble in communicating; now the parallel processor [10] made everything feasible to access data very easily and effectively. Figure 2 shows the need of Green Corridor in metro cities.

So once the Green Corridor is created in one direction, the other lanes may need to wait sometime but it is very less compared to existing system. This system if it gets implemented the 15–20% death cases [4] which are happening every year because of traffic can be controlled.

The only disadvantage in this Green Corridor System [11] is that if two or three ambulances coming in different directions then which one to give priority.

Fig. 2 Ambulance waiting at traffic signal

3 Advantages

1. Reduces the fuel waste at traffic signals by dynamic operation of signals.
2. Emergency vehicles can reach the destination quickly so that reducing the loss of lives.
3. Reduces the unnecessary waiting time at traffic signals.
4. Attain the safest and most efficient traffic flow.
5. Proper road lane discipline is achieved intern reducing the accidents.

4 Comparative Study

When it compared to existing system, the proposed system will give best results. The only thing which we need to take care in developing countries like India is while manufacturing vehicles [12] itself, the manufacturers should follow uniform criteria on height and width of the vehicles based on type of the vehicles like Hatchback, SUV and Terrain [8] which will also really helps in avoiding accidents.

5 Conclusion

Though the proposed system is not new to the developed countries, the developing countries have to adopt it in order to provide best services to the citizens. Automatic Traffic Solution for Emergency Vehicles such as Ambulance, Fire Engine etc. to reduce the delay in arrival of ambulance to hospital and Fire Engine to Fire Accidents with the effective traffic signal monitoring. The system operates in real time which improves traffic flow, safety, and saves costly constant human involvement. The model provides constant assistance to the emergency vehicles at each traffic intersections by creating the Green Corridor.

The main reason most of the developing countries not adopted this because of the cost. But the proposed system is cost-effective and any one can be adopted easily. And no need to demolish existing system perhaps it is an add-on for existing.

6 Future Enhancements

Technology wise probably the improvement is less, the only thing is that the automobile companies [3] should think about manufacturing vehicles with some uniform constraints like height and width.

References

1. Q.M. Ashraf, M.H. Habaebi M. Rafiqul Islam, TOPSIS-based service arbitration for autonomic internet of things. IEEE **4**, 1313–1320 (2016). http://doi.org/10.1109/ACCESS. 2016.2545741
2. I. Mashal, O. Alsaryrah T.-Y. Chung, Performance evaluation of recommendation algorithms on Internet of Things services. Phys. A: Stat. Mech. Its Appl. **451**, 646–656 (2016)
3. P. Škorput, B. Mandžuka, M. Vujić, The development of cooperative multimodal travel guide, in *Telecommunications Forum Telfor (TELFOR) 22nd, 2014*, pp. 1110–1113
4. J. Ding, A comparison of fluid approximation and stochastic simulation for evaluating content adaptation systems. Wirel. Pers. Commun. **84**(1), 231–250 (2015)
5. B. Singh, A. Gupta, Recent trends in intelligent transportation systems: a review. J. Transp. Lit. 30–34. https://doi.org/10.1590/22381031.jtl.v9n2a6 (2015)
6. M. Whaiduzzamana, M. Sookhaka, A. Gania R. Buyyab, A survey on vehicular cloud computing, J. Netw. Comput. Appl., **40**, 325–344 (2014)
7. P. Škorput, *Open Ontology Model of Cooperative Intelligent Transport Systems* (University of Zagreb, Faculty of Transport and Traffic Sciences, Zagreb, 2014)
8. S. Mandžuka, Ž. Marijan, B. Horvat, D. Bicanic, E. Mitsakis, Directives of the European Union on intelligent transport systems and their impact on the Republic of Croatia. Promet— Traffic & Transportation **25**(3), 273–283 (2013)
9. J. Hillston, Fluid flow approximation of PEPA models, in *Processing of the Second International Conference on the Quantitative Evaluation of System, IEEE Computer Society, 2015*, pp. 33–43
10. S. Mandžuka, E. Ivanjko, M. Vujić, P. Škorput, M. Gregurić, *The Use of Cooperative ITS in Urban Traffic Management, Intelligent Transport Systems: Technologies and Applications* (New York: Wiley, 2015), pp. 14.1–14.12
11. Y. Sun, Research on urban road traffic congestion charging based on sustainable development. Physics Procedia **24**, 1567–1572 (2012)
12. S. Mandžuka, *Intelligent Transport System 2 (in Croatian)—Selected Chapters* (Department for Intelligent Transport Systems, Faculty of Transport and Traffic Sciences, Zagreb, 2011)

ECG Classification Using Wavelet Transform and Wavelet Network Classifier

Dinesh D. Patil and R. P. Singh

Abstract An electrocardiogram (ECG) signal is a highly used examination in the field of cardiology; these pathologies are generally reflected by disorders of the electrical activity of the heart. In this paper, we have addressed the problem of automatic recognition of heartbeats through the development and implementation of a method combining wavelet transform with neural networks. This method consists of denoising, extraction, and classification models for robust automated ECG analysis. For the classification module, a hybrid network combining neural networks and wavelets has been proposed, implemented and evaluated for identification of the ECG classes. This technique is based on use of wavelet functions as activation function in neural networks, which allowed the wavelet network to have a better adaptability and flexibility during the learning process given the parameters of translation and expansion of the functions of wavelets. Indeed, the evaluation of the results obtained by the implemented wavelet network is satisfactory with respect to other neural networks in terms of the rate of classification of the heartbeats. The association of neural networks with the wavelet functions made it possible to extract the strengths of the two techniques (the learning capacity of the neural models and the multiresolution analysis of the wavelets). The results obtained showed that the proposed method can be considered as an effective method for the classification of cardiac arrhythmias with a very acceptable accuracy of more than 98.78%.

Keywords DWT · ECG · Neural network · Wavelet network
QRS

D. D. Patil (✉) · R. P. Singh
Department of Computer Science and Engineering, Sri Satya Sai University
of Technology & Medical Sciences, Sehore, Madhya Pradesh, India
e-mail: dineshonly@gmail.com

R. P. Singh
e-mail: prof.rpsingh@gmail.com

© Springer Nature Singapore Pte Ltd. 2018
S. S. Dash et al. (eds.), *Artificial Intelligence and Evolutionary Computations
in Engineering Systems*, Advances in Intelligent Systems and Computing 668,
https://doi.org/10.1007/978-981-10-7868-2_29

289

1   Introduction

Cardiovascular disease is the most common cause of death worldwide, according to annual statistical studies conducted at the World Health Organization (WHO) level [1]. Therefore, the diagnosis of these dangerous diseases seems a vital task. In cardiac services, at the hospital level, an electrocardiogram (ECG) signal is still one of the predominant and widely used tools for the diagnosis and analysis of cardiac arrhythmias.

The application of artificial intelligence (AI) methods on the ECG signal has become a very important trend for the recognition and classification of different types of cardiac arrhythmias. Thus, many solutions have been proposed for the development of automated systems for the analysis, recognition, and classification of the ECG.

Often, we find several techniques in the literature that apply the approaches of artificial intelligence [2], and more particularly the networks of artificial neurons for the automatic analysis of electrocardiogram (ECG) signal.

2   QRS Detection of the ECG Signal by Wavelets

In this work, the characterization of the ECG signals lies in the detection of the complex QRS of each cardiac beat, in order to extract the necessary representative parameters which will then allow recognizing well the normal beats and the pathological beats. This operation should be as efficient as possible because it reduces the size of the data and retains information that can be interpreted by specialists. It is carried out in two main stages:

1. Preprocessing step whose role is to eliminate artifacts that would hamper the correct segmentation of the beats [3].
2. Detection of the QRS complexes by the localization of the R peaks in the ECG signal [4].

In general, the characterization operation of the ECG signal can be broken down into two steps illustrated in Fig. 1.

2.1   ECG Signal Preprocessing

The different noises that stain an ECG signal are considered undesirable and can alter more or less the clinical information. In addition, the difficulties in detecting QRS complexes reside essentially in the great variability of the signal shape and the presence of these unnecessary noises of various origins in the ECG. It is therefore

Fig. 1 Steps of the segmentation of the ECG signal

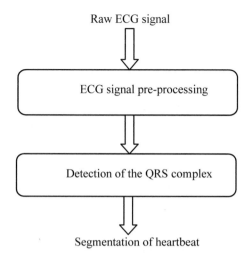

Raw ECG signal

ECG signal pre-processing

Detection of the QRS complex

Segmentation of heartbeat

important to know what types of noise can contaminate an electrocardiogram (ECG) signal.

2.2 Types of Noise Present in the ECG Signal

Technical Noises: Technical noises are noises caused by the material used during recording.

Physical Noises: Physical noises are artifacts generated by either the electrical activities of the human body such as muscular contractions or movements during breathing.

2.3 Signal Preprocessing

Assume $\{\beta_{jk}, j, k \in Z\}$ are detail coefficients of the noisy signal. In the hard thresholding, one replaces β_{jk} with $\widehat{\beta}_{jk}$ defined by:

$$\widehat{\beta}_{jk} = \begin{cases} \beta_{jk}, & \text{if } |\beta_{jk}| \geq T_r \\ 0, & \text{if } |\beta_{jk}| < T_r \end{cases} \tag{1}$$

where T_r is a certain threshold. In the soft thresholding, one replaces β_{jk} with $\widehat{\beta}_{jk}$ defined by:

$$\widehat{\beta}_{jk} = \begin{cases} \text{sign}(\beta_{jk})(|\beta_{jk}| - T_r), & \text{if } |\beta_{jk}| \geq T_r \\ 0, & \text{if } |\beta_{jk}| < T_r \end{cases} \tag{2}$$

The value of the threshold T_r is based on some prior information that may exist on the signal. A fixed threshold T_r is introduced by Donoho and Johnstone [5] in the form:

$$T_r = \sigma\sqrt{2\log(n)} \tag{3}$$

where σ is estimated using the median of the absolute deviation of the detail coefficients of the first wavelet decomposition of the signal, and n is their number:

$$\sigma = 1.483 \times \text{median}\left[(\beta_{J-1,k})k \in Z\right]$$

Let J be the highest resolution level of the signal. Steps of high-frequency noise cancellation using soft thresholding are as follows:

1. Do the first decomposition of the ECG signal $x(t)$.

$$x(t) = \sum_{k=0}^{2^{J-1}} \alpha_{Jk}\phi_{Jk}(t) \tag{4}$$

First decomposition of $x(t)$ gives:

$$x(t) = \sum_{k=0}^{2^{J-1}-1} \alpha_{J-1,k}\phi_{J-1,k}(t) + \sum_{k=0}^{2^{J-1}-1} \beta_{J-1,k}\psi_{J-1,k}(t) \tag{5}$$

Compute σ from $\beta_{J-1,k}, k \in Z$ and T_r from Eq. (3).

2. Compute the complete discrete wavelet decomposition of the signal. The expression of $x(t)$ becomes:

$$x(t) = \alpha_{00}\phi_{00}(t) + \sum_{j=0}^{J} \sum_{k=0}^{2^j-1} \beta_{jk}\psi_{jk}(t) \tag{6}$$

Replace all the detail coefficients β_{jk} with $\widehat{\beta}_{jk}$ (Eq. 2). We obtain the new signal $\widehat{x}(t)$:

$$\widehat{x}(t) = \alpha_{00}\phi_{00}(t) + \sum_{j=0}^{J} \sum_{k=0}^{2^j-1} \widehat{\beta}_{jk}\psi_{jk}(t) \tag{7}$$

3. Compute the inverse discrete wavelet transform of $\widehat{x}(t)$ to obtain the denoised signal corresponding to the original signal $x(t)$:

$$\widehat{x}(t) = \sum_{k=0}^{2^J-1} \widehat{\alpha}_{Jk}\phi_{Jk}(t) \tag{8}$$

2.4 Detection of the QRS Complex

To perform an automatic analysis of the ECG signal, detection of QRS complexes is a very important step. The detection of QRS complexes can be accomplished by a simple thresholding of the signal since the R waves are generally larger than the other waves in terms of amplitudes.

2.5 Principle of the QRS Complex Detection Algorithm

The method proposed for the delineation of QRS complexes is based on a multiresolution analysis by the wavelet transform; the general diagram of the different steps of the developed algorithm is illustrated in Fig. 2.

Thus, in the rest of this work, the Daubechies' db4 mother wavelet is chosen to be used for the decomposition of the ECG signal.

Figure 3 shows a portion of the ECG signal of the patient '100' in its filtered version from the preprocessing stage, whereas the eight details of the wavelet decomposition 'db4' of this signal are shown in Fig. 4.

2.6 The Detection of the R Peak

Details D3–D6 are retained, and all other details are deleted. The resulting signal is shown in Fig. 5b. The signal obtained is raised to the square in its positive part. This operation makes it possible to accentuate the R wave and to attenuate the other waves (as illustrated in Fig. 5c). Adaptive thresholding is performed on this signal to detect peak R.

Based on the impossibility of having two heartbeats in less than 0.25 s, a final step of eliminating the falsely detected R peaks is carried out before validating the detection results. The final result of detection of the R waves is illustrated in Fig. 6.

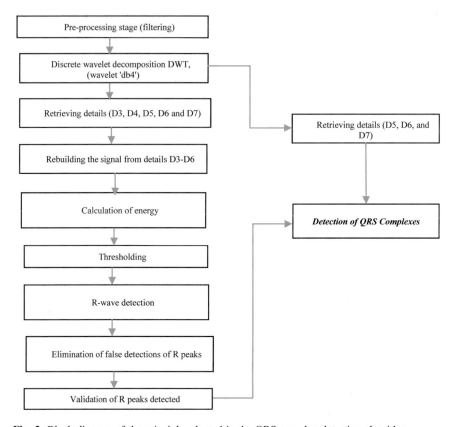

Fig. 2 Block diagram of the principle adopted in the QRS complex detection algorithm

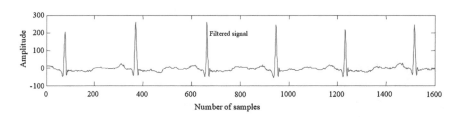

Fig. 3 Patient '100' filtered ECG signal

2.7 Detection of Q and S Waves

Once the R peaks are detected, the Q and S points must be identified to detect the QRS complex completely. In general, waves Q and S have a high frequency, they are of small amplitude, and their energies are mainly on small scale. To do this, the

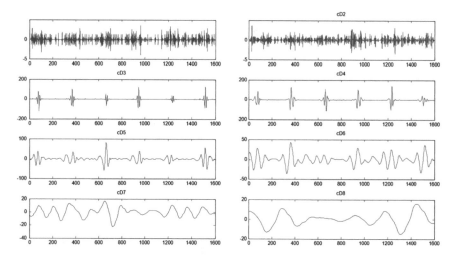

Fig. 4 Eight details of the ECG signal decomposition of the patient '100'

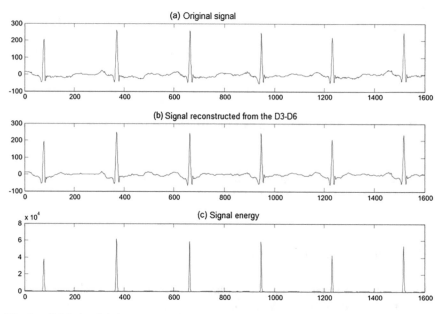

Fig. 5 **a** ECG signal, **b** its reconstruction from details D3–D6, and **c** its energy

decomposition coefficients of D5–D7 are retained to reconstruct the signal (as shown in Fig. 7a). All other details of the signal have been removed.

Moreover, the waves Q and S are negative deflections which occur on either side of the peak R over a maximum interval of 0.1 s in a usual manner. The left-hand wave noted Q is taken to be the minimum amplitude that precedes the R peak and

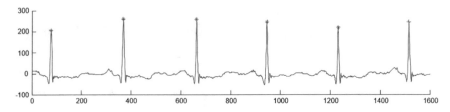

Fig. 6 R peak detection

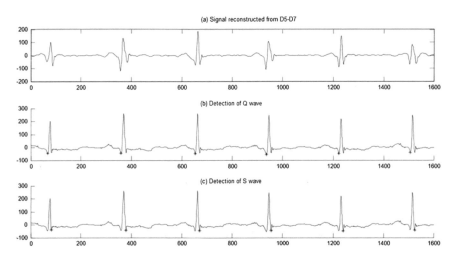

Fig. 7 **a** Reconstruction of the signal from details D5–D7, **b** detection of the Q wave and **c** the S wave

the S wave the minimum amplitude that follows it. The Q and S waves' detection result is shown in Fig. 7b, c, respectively.

2.8 Segmentation of Heartbeat

Zhang and Benveniste [6] have described a wavelet lattice structure allowing parametric modeling of multidimensional functions $\widehat{g}(x)$, according to the expression:

$$\widehat{g}_l(x) = \sum_{l=1}^{L} w_l \cdot \psi(D_l \cdot R_l \cdot (x - T_l)) + \bar{g} \tag{9}$$

where

L is the number of neurons on the input layer;

D_l, R_l, and T_l denote the affine transformations applied to the vector x before processing by the mother wavelet in the neuron of index l;
And $\bar{g}$ is the continuous component (mean value) of the approximation.
The vector p groups all the parameters of the wavelet network, namely the weights w_l, the coefficients of the affine transformations, and $\bar{g}$.
The parameter p is adjusted by an iterative optimization algorithm, which aims to minimize the residual:

$$\rho(p) = \sum_x \left[g(x) - \widehat{g}_p(x) \right]^2 \tag{10}$$

for all available values of input x.

2.9   Wavelet Network Model

When dealing with modeling or classification problems, it is often common to deal with multivariable inputs, and for this, it is necessary to use multidimensional wavelets whose expression is given by:

$$\Psi_j(x) = \prod_{k=1}^{N} \psi\left(z_{jk}\right) \tag{11}$$

where

$$z_{jk} = \frac{x_k \pm b_{jk}}{a_{jk}} \tag{12}$$

where x_k is the kth component of the input vector x, and z_{jk} is the component centered by b_{jk} and expanded by a factor a_{jk}.
A multidimensional wavelet is defined as the product of one-dimensional wavelets; it is also said that the wavelets are separable.
The hidden neurons of a wavelet array can be considered as multidimensional wavelets.

2.10 Learning Wavelet Networks

Learning of wavelet networks is supervised, using well-annotated examples (a learning base). Let a learning base consisting of D examples, each example n consisting of a vector xn (k) applied to the inputs of the wavelet network and of the vector 'dn (k)' corresponding values for the outputs. The value of 'yn (k)' corresponds to the actual output of the network which corresponds to the input 'xn (k)'.

The learning of a wavelet network is defined as an optimization problem which consists in identifying the network parameters minimizing a global error function (cost function).

The network parameters can be divided into two classes:

1. The structural parameters of the wavelet functions, that is to say the translations and the dilations.
2. The synaptic weights of the network.

For each example $x_n(k), (n \in D)$ of the learning base, a quadratic error function defined by the following equation is computed.

In a general case, it is also assumed that the wavelet network has a number 'R' of output neurons.

$$e(n) = \frac{1}{2} \sum_{r=1}^{R} [d_r(n) - y_r(n)]^2 \tag{13}$$

where

$r = 1 \ldots R$: Number of output neurons.

$n = 1 \ldots D$: Number of learning examples.

$k = 1 \ldots N$: Number of input neurons (length of the input vector).

$j = 1 \ldots L$: Number of hidden neurons.

For the whole training set D, we can define the cost function (also called the mean squared error (MSE)):

$$E(n) = \frac{1}{D} \sum_{n=1}^{D} e(n) \tag{14}$$

The optimization expression is given by the following equation:

$$\frac{\partial E}{\partial \theta} = - \sum_{n=1}^{D} e(n) \frac{\partial y(n)}{\partial \theta} \tag{15}$$

θ is the vector grouping the set of adjustable parameters:

$$\theta : \left\{ a_{jk}, b_{jk}, w_j, w_{jk} \right\}, \quad j = 1 \ldots L, \ k = 1 \ldots N$$

where $\frac{\partial y(n)}{\partial \theta}$ is the value of the gradient of the actual output of the network with respect to the parameters θ.

3 Results Analysis

3.1 Description of the MIT-BIH Database

The MIT-BIH database [7] is a universal database containing 48 half-hour two-way records (DII and V5). It has been collected by researchers to be used as a reference for the validation and comparison of algorithms on the ECG signal.

3.2 Collection of the Database and Choice of Targeted Arrhythmias

An arrhythmia corresponds to any disturbance in the regular rhythmic activity of the heart (amplitude, duration, and shape of the rhythm). From the point of view of the diagnosis of arrhythmia, the most important information is contained in the QRS complex.

There are different categories of heartbeats, which are as follows: normal beats (N), left branch blocks (LBBB), right branch blocks (RBBB), premature ventricular contractions, and premature atrial contractions (APC). As mentioned earlier, all beats used are extracted from the records in the MIT-BIH database.

3.3 Feature Input for Classifier

The characterization parameters chosen are the same parameters on which the cardiologist bases his diagnosis.

3.4 The RRp Interval

We call RRp the distance between the peak R of the present beat and the peak R of previous beat. This parameter is an indicator of prematurity of the beat.

3.5   The Interval RRs

We call RRs the distance between the peak R of the present beat and the peak R of next beat.

3.6   The Ratio of RR Intervals (RRs / RRp)

The RRs/RRp ratio is a parameter that characterizes the rhythm. In the case of a rhythm, this ratio is close to 1, but it can largely exceed this value in the case of a premature beat.

3.7   The Width of the QRS Complex

This parameter is important for the identification of pathological pulses of origin ventricular arrhythmias; these types of arrhythmias are generally characterized by a large complex QRS.

3.8   The Amplitude of the R Wave (Ramp)

This parameter is also important for the identification of pathological pulses of ventricular origin which are often characterized by either attenuation or amplification of the amplitude of the QRS complex (Fig. 8).

The algorithm of detection of the QRS complexes that we have developed in the third section allows us to calculate the values of these different parameters for the characterization of the cardiac beats.

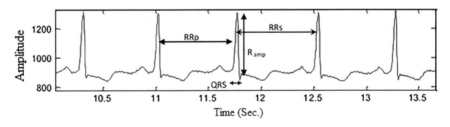

Fig. 8 Characterization parameters on a heartbeat of a healthy subject [RRp RRs RRs / RRp QRS Ramp]

3.9   Performance Evaluation Parameters

The classified heartbeat will be compared with the annotations associated with each record in the MIT database to determine the error of the classification.

1. $\text{Accuracy} = \dfrac{\text{true positive} + \text{true negative}}{\text{true positive} + \text{true negative} + \text{false positive} + \text{false negative}}$

2. $\text{Precision} = \dfrac{\text{true positive}}{\text{true positive} + \text{false positive}}$

3. $\text{Sensitivity} = \dfrac{\text{true positive}}{\text{true positive} + \text{false negative}}$

3.10   Result and Discussion

The results of the classification of the five types of heartbeats (N, PVC, APC, RBBB, and LBBB) by the wavelet network that we have implemented are illustrated in the form of a confusion matrix illustrated in Table 1.

From the results summarized in Table 1, we have to note that most unidentified normal beats are those classified by the wavelet network as beats of the APC type (five beats in total).

For the evaluation of the performance and quality of the wavelet network classification, we calculated the classification rate, specificity, and sensitivity (PVC, RBBB, LBBB, and APC). These results are summarized in Table 2.

Table 1 Confusion matrix

Confusion	N	PVC	RBBB	LBBB	APC
N	48	3	0	2	5
PVC	0	44	1	2	10
RBBB	6	2	50	3	1
LBBB	3	1	2	48	1
APC	3	2	1	3	52

Table 2 Results of wavelet network classification for different types of heartbeat

Evaluation parameters	Type of beats	Value (%)
Specificity (%)	N	99.23
Sensitivity (%)	PVC	99.05
	RBBB	98.43
	LBBB	98.51
	APC	98.51
Classification rate	Total	99.42

Table 3 Comparison of results

References	Method	Description	Classification rate (%)
	Proposed	Wavelet transform with wavelet neural network	99.78
[8]	FCM + PCA + WT + NN	Fuzzy C-means, principal component analysis and wavelet transform with neural network	99.17
[9]	NF	Neuro-fuzzy classifier	99.42

The results presented in Table 2 show satisfactory classification rates of the proposed method. Discrimination between heartbeats is very acceptable for different classes.

3.11 Comparison with Other Methods

The below table summarizes the results obtained for the classification of cardiac arrhythmias by the proposed method and other methods.

Therefore, all these factors will influence the results of the classification of the heartbeat. It is evident from Table 3 that the results of the implemented wavelet network are relatively satisfactory with respect to the results of the others work of the classification rate of the ECG beats.

4 Conclusion

Through this paper, we explored the capabilities of artificial neural networks and wavelet functions in the same hybrid system combining the two techniques to solve a classification problem of cardiac arrhythmias. The proper characterization of the ECG signal by relevant parameters is a necessity for discrimination between different targeted arrhythmias. We note that the wavelet functions can be considered as parameterized functions at the level of the neural networks whose parameters of translation and expansion are adjustable during the learning. Since the wavelets are local functions, the problem of the initialization of the dilations and translations is very important. We have used a simple initialization procedure that takes this property into consideration. Indeed, the results showed that the proposed method can be considered as a promising prototype for ECG classification with a good accuracy of more than (99.42%) even though the representative vector of the beats consists only of five parameters, which shows the effectiveness and relevance of the parameters chosen.

References

1. D. Lloyd-Jones, R.J. Adams, T.M. Brown, M. Carnethon, S. Dai, G. De Simone, T.B. Ferguson, E. Ford, K. Furie, C. Gillespie, A. Go, Heart disease and stroke statistics—2010 update. Circulation **121**(7), e46–e215 (2010)
2. L. Gang, Y. Wenyu, L. Ling, Y. Qilian, Y. Xuemin, An artificial-intelligence approach to ECG analysis. IEEE Eng. Med. Biol. Mag. **19**(2), 95–100 (2000)
3. F. Strasser, M. Muma, A.M. Zoubir, Motion artifact removal in ECG signals using multi-resolution thresholding. In *Signal Processing Conference (EUSIPCO), 2012 Proceedings of the 20th European* (IEEE, 2012), pp. 899–903
4. S.L. Joshi, R.A., Vatti, R.V., Tornekar, A survey on ECG signal denoising techniques. In *2013 International Conference on Communication Systems and Network Technologies (CSNT)* (IEEE, 2013), pp. 60–64
5. D.L. Donoho, I.M. Johnstone, Adapting to unknown smoothness via wavelet shrinkage. J. Am. Stat. Assoc. **90**(432), 1200–1224 (1995)
6. Q. Zhang, A. Benveniste, Wavelet networks. IEEE Trans. Neural Netw. **3**(6), 889–898 (1992)
7. MIT-BIH Arrhythmia Database—PhysioNet. Online available at: https://www.physionet.org/physiobank/database/mitdb/
8. J.S. Wang, W.C. Chiang, Y.L. Hsu, Y.T.C. Yang, ECG arrhythmia classification using a probabilistic neural network with a feature reduction method. Neurocomputing **116**, 38–45 (2013)
9. T.M. Nazmy, H. El-Messiry, B. Al-Bokhity, Adaptive neuro-fuzzy inference system for classification of ECG signals. In *2010 The 7th International Conference on Informatics and Systems (INFOS)* (IEEE, 2010), pp. 1–6

Developing Scans from Ground Penetrating Radar Data for Detecting Underground Target

A, B and C Scan

Siddhanth Kandul, Manish Ladkat, Radhika Chitnis, Rushikesh Sane and D. T. Varpe

Abstract Ground-penetrating radar (GPR) is a newly evolving technology which aims at detecting objects buried under earth's surface. The GPR system has a set of transmitters and receivers. The transmitters emit electromagnetic waves under earth's surface, and their reflections are recorded in the form of 3DR files. The data in 3DR files can be processed to form A-scan, B-scan, and C-scan. These graphs can be used to draw important conclusions about the presence of target. This paper focuses on generation of scans and the programming approach used to develop them.

Keywords 3DR data · GPR system · A-scan · B-scan · C-scan

1 Introduction

There are about 10,000 mines active and buried inside earth's surface, and they take a toll of more than 25,000 causalities in the world. A robust GPR system will help in detecting the underground mines and in turn reduce the number of causalities in

S. Kandul (✉) · M. Ladkat · R. Chitnis · R. Sane · D. T. Varpe
Department of Information Technology, Pune Vidyarthi Griha's College
of Engineering and Technology, Pune, Maharashtra, India
e-mail: siddhanth.kandul@gmail.com

M. Ladkat
e-mail: manishladkat@gmail.com

R. Chitnis
e-mail: radhikachitnis10@gmail.com

R. Sane
e-mail: sanerushikesh@gmail.com

D. T. Varpe
e-mail: dtv_it@pvgcoet.ac.in

© Springer Nature Singapore Pte Ltd. 2018
S. S. Dash et al. (eds.), *Artificial Intelligence and Evolutionary Computations in Engineering Systems*, Advances in Intelligent Systems and Computing 668,
https://doi.org/10.1007/978-981-10-7868-2_30

the world. In order to reach strong conclusions about the presence of mine, there is a need to generate correct A-scans, B-scans, and C-scans. We aim to develop and propose the programming approach to generate GPR scans using 3DR files as input.

GPR (ground-penetrating radar) system developed in Norway is used to detect the object beneath the surface of the earth through the signals emitted. Traditionally, metal detectors are used to determine the objects inside the earth's surface, but the probability of it being the target is extremely low. There are two main reasons for this [1]:

1. It detects any metallic object which is not our target, thus making the efforts futile and increasing the false alarm rates,
2. It detects only metallic objects, and modern mines are made up of plastic and rubber materials which are not detected by the metal detector.

The above two notable limitations of metal detectors have made people to explore in this field to minimize the efforts and time required to find the target. The cost of laying a mine in the ground is less than removing the mine as the need to be searched because the exact location of the mine is unknown and which requires a lot of human efforts as well. The electromagnetic waves emitted by the GPR will help us in determining the probability of the object being a target with maximum accuracy and without generating false alarms. The output of the GPR system is further processed and enhanced with the help of the various algorithms discussed in this paper. Many kinds of technology in the area of sensor physics, signal processing, and robotics have been studied for mine detection during the last decade. This paper will introduce some sensor technologies and signal processing, specifically the image processing method.

2 Ground-Penetrating Radar

2.1 Introduction to GPR System

The GPR uses electromagnetic waves for detecting the underground object. It emits high-frequency waves into the ground, and the received signals are detected by the receiver. The frequency and the speed of the transmitted waves are known initially, and the time required for the waves to reflect back is recorded by the GPR. This helps in determining the depth at which the object is located. The waves get reflected when the emitted waves encounter surfaces between two electrically different materials, i.e., between materials having different dielectric permittivity.

Fig. 1 depicts the model of a typical GPR System.

Penetration depth of signals depends on two factors [2]:

1. Humidity in soil—More the humidity, less the penetration depth of GPR signals,
2. Wavelength of the signal—More the wavelength, more the penetration depth.

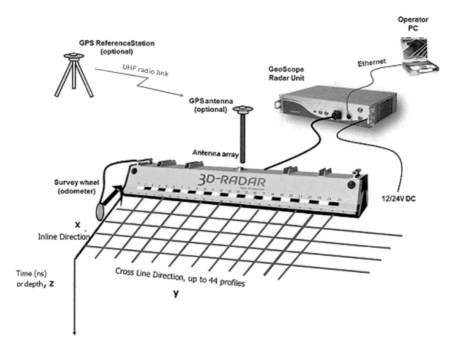

Fig. 1 GPR system

Similarly, intensity and direction of reflection depend on two factors [3]:

1. Roughness of the surface and
2. Electric properties of the surface.

Based on the above facts, we can conclude that the desired conditions are dry sand and low-frequency signals. Information provided by the GPR system is of two types:

1. Presence of the object.
2. Position of the object and the following basic formula is used.

$$R = v^* t / 2$$

where R = location of the object, v = velocity of the given medium, t = time difference between transmission and reception of signals.

GPR signals also have some limitations in itself. Low-frequency signals tend to produce low-resolution images. Also, since the EM waves cannot penetrate through water, GPR cannot locate mines located under water. The GPR outputs the data in the form of 3DR files.

3 A-scan

A-scan is obtained by a stationary measurement, emission, and collection of a signal after placing the antenna above the position of interest. The collected signal is presented as signal strength versus time delay. The information generated is only about a single point under consideration. The GPR system has a certain number of transmitter–receiver pairs (usually 31). When one transmitter emits waves, the waves are penetrated deep under earth's surface through a single specific point. When the transmitted wave penetrates earth, the waves get reflected at specific depths. There are 420 depth points from which the wave transmitted from one transmitter is reflected. The A-scan is a graph showing depth points on X-axis versus the wave intensities reflected from those depth points.

Fig. 2 depicts the generation of an A-scan formed out of readings of various intensities.

As shown in Fig. 2, we see two active peak regions. The first peak region is the air–ground reflection. Since there is a big difference in the dielectric permittivity of air and ground, we get a high-intensity region. The other undulations are due to small stones and pebbles present in the soil. These undulations are the noise and should be removed before applying image-enhancing algorithms. Before removing the noise, it is necessary to remove the portion before air–ground reflection.

Fig. 3 depicts A-scan after removing the clutter.

Figure 3 shows that the entire portion before earth's surface is encountered is removed. The intensities of the points before air–ground surface are made zero. It can either be made zero or completely removed from the storage array. Now, the next step is to remove noise, i.e., unwanted reflections from the A-scan.

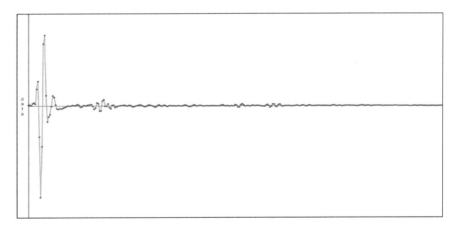

Fig. 2 A-scan

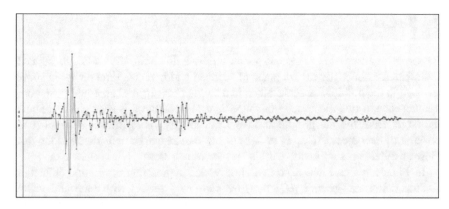

Fig. 3 A-scan with clutter removed [7]

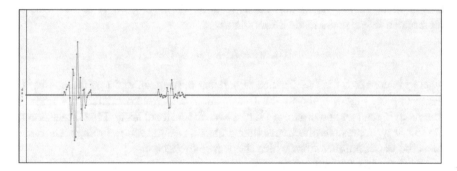

Fig. 4 A-scan with noise removed [8]

Fig. 4 shows the A-scan after removing unwanted noise.

Noise in Fig. 3 has been completely removed in Fig. 4. Figure 4 has two regions. The first region is the region having point which has maximum reflection intensity. So we have the probability of presence of object around that particular depth region. Hence, we have kept the region around that point active. Now, apart from the first region, the point having maximum reflection intensity is present in the second region. With that, if we want to analyze some other activity or object present, we may consider other region as well. It may also be possible that the object is present in the second region. It completely depends on the programmer how many regions he wants to be active and the depth which he wants to study. Also, we have made the intensities of the noisy points zero, but even this depends on the programmer. He may decrease the noise by 50% or by any desired amount. (Though noise cannot be removed 100%, in GPR Applications, the points not having very high reflection intensities can be completely ignored.)

4   B-scan

B-scan is an ensemble of A-scans, measured by a line scan. The collected signal is represented as intensity on the plane of scanned width versus time delay. In simple terms, it is the multiple A-scans taken in a single line. An A-scan could detect only the presence of two electrically distinguished objects, but a B-scan can distinguish the target from the air–ground surface and can give more information about the position of the object. An array of A-scan and B-scan can be considered as the front view of reflection sent inside earth's surface in one line.

In Fig. 5, we can see a straight line which represents air–ground reflection. A little below air–ground reflection, we can see a region with blue and yellow intensities which represents the target. With the help of B-scan, we are able to find the depth at which target is located. A hyperbola shape is observed in the region where the target is. One A-scan can only detect the presence of an object, but B-scan gives more information about location of the object. A B-scan can be represented using mathematical formula as:

$$f(x, z) = A(x_i, \ y_i, \ z_k)$$

where j is constant, i is the distance of antenna movement, and k is the number of data samples. Figure 6 shows the B-scan after background clutter, and noise is removed. B-scan can be organized into a two-dimensional array. First dimension of the 2D array represents the transmitter number, and the second dimension represents the depth number from which the wave is reflected.

Fig. 5 depicts the generation of B-scan.

Fig. 6 depicts B-scan with clutter and noise removed.

Fig. 5 B-scan

Fig. 6 B-scan with clutter and noise removed [8]

5 C-scan

C-scan is again an ensemble of multiple B-scans. That is, multiple B-scans are taken to cover a particular area of surface. The collected information is represented as intensity in a box of scanned region versus time delay. C-scan can be considered as the horizontal slice of a particular depth. It is the top view of the region at a specific depth point. A C-scan can be mathematically represented as [4]:

$$f(x, y) = A(x_i, y_j, z_k)$$

where i and j vary from 1 to size of the scanned area and k varies from 1 to the number of data samples taken. A C-scan is a three-dimensional structure, B-scan a two-dimensional structure, whereas A-scan one-dimensional structure. Figure 7 shows the C-scan implemented in our research work. Since it is difficult to imagine and display a three-dimensional structure, we show a C-scan as a horizontal slice at a certain depth.

From the programming point of view, we need a two-dimensional array to store data points for a displaying a C-scan. The first dimension represents the transmitter–receiver pair number, and second dimension represents all points in a B-scan at the same depth level. In the above figure, the yellowish spots denote the probable object. That area might be a mine or a big stone of different dielectric permittivity or might be metallic debris. C-scan helps us determine dimensions of the object present, and study of multiple C-scans will help us to know the height of the object. This C-scan can be an input for image processing algorithms which will improve and enhance the quality of image object and in turn be sure about mine probability. The image processing algorithms which can be used are KL Transformation, Hadamard Transformation, and so on [5].

Fig. 7 shows the presence of mine in a C-scan.

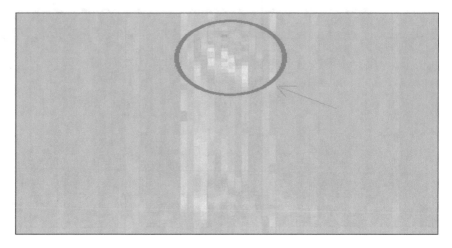

Fig. 7 C-scan

6   3DR Files

6.1   Introduction to 3DR Files

3DR stands for 3D-Radar. The output data of Geo Scope ground-penetrating radar (GPR) system is in this format. 3D-Radar AS developed this format for ground-penetrating radars (GPRs). The purpose is to provide an efficient and unambiguous file format suitable for GPRs. 3DR files are based on the HDF5 format. The nature of HDF5 makes it easy to expand and add attributes. All files saved with the 3DR file extension to keep the file format unambiguous. The Geo Scopes save GPR data to a 3DR-file, together with position information if available. It is useful for users during survey to mark the data where interesting features are visible, or at specific locations. During data acquisition, system status may change and it is also useful to mark the data with such information. In the 3DR file format, such user and system incidents are called events. The storing of GPR data, position, and events as well as configuration information and other metadata is allowed by the 3DR file structure [6].

6.2   Hierarchical Data Format

The 3DR data format is based on the Hierarchical Data Format version 5 defined by NCSA, called HDF5. For storing scientific data, HDF5 is used which is a general purpose library and file format. Two primary objects can be stored in a HDF5 file: datasets and groups. A dataset is a multidimensional array of data elements, while a group is a structure for organizing objects in a HDF5 file. Both datasets and groups can have associated metadata. Generally, HDF5 is designed for large files, i.e., larger than what can fit in memory, and can be used efficiently on parallel computing systems.

7 Conclusion

Through this work, we have developed a methodology of systematically organizing the GPR data using array data structure to generate A-scan, B-scan, and C-scan. These scans can later be an input for image enhancement and transformation algorithms like KL Transformation and Hadamard Transformation which in turn will give stronger conclusions of detecting the underground targets. If these transformations are applied successfully, it will help in reducing a huge number of causalities across the globe. This paper gives a detailed account of GPR scans with a programmer's point of view. Such an approach has never been implemented before.

References

1. Curtis Wright Company, "2010-datasheet-geoscope-3ghz"
2. C.P. Lee, Mine detection techniques using multiple sensors. (The University of Tennesse, Knoxville)
3. R. Grimberg, N. Ifitimie, G.S. Dobrescu, Ground penetrating radar and the possibility of buried objects detection
4. R.D. Dony, The Transformation and Data Compression Handbook
5. V. Kovalenko, Advanced GPR data processing algorithms for detection of anti-personal landmines
6. R.D. Dony, The transformation and data compression handbook
7. J.W. Brooks, L. van Kempenb, H. Sahlic, A primary study in adaptive clutter reduction and buried min-like target enhancement from GPR data
8. B. Vuksanovic, N. Bostanudin, H. Hidzir, H. Parchizadeh, Discarding unwanted features from gpr images using 2DPCA and ICA techniques. Int. J. Inf. Electron. Eng. 3(3), (May 2013)

Key Management Interoperability Protocol-Based Library for Android Devices

Apoorva Banubakode, Pooja Patil, Shreya Bhandare, Sneha Wattamwar and Ashutosh Muchrikar

Abstract Mobile applications generate huge amount of user's and application data which may become a serious threat to the privacy of the user if revealed. Thus, a need arises to protect such sensitive data. One of the most reliable ways to secure this data is by encrypting it. The process of data encryption squarely depends on the ability to effectively manage the cryptographic keys. Android operating system uses hardware keystore to store keys which may be vulnerable to thefts. Hence, an alternate solution is to store keys on a server. We propose to create a library which communicates with the server from the client device using Key Management Interoperability Protocol (KMIP).

Keywords Android application · API · Data security · Key management
KMIP

1 Introduction

Mobile applications have transformed the way we interact with the world in a massive way. It is used in a variety of use-cases like paying bills or shopping on-line or even finding a path back home, everything is just a click away. Mobiles

---

A. Banubakode (✉) · P. Patil · S. Bhandare · S. Wattamwar · A. Muchrikar
MKSSS's Cummins College of Engineering for Women,
Savitribai Phule Pune University, Pune, Maharashtra, India
e-mail: apoorva.banubakode@gmail.com

P. Patil
e-mail: poojapatil231995@gmail.com

S. Bhandare
e-mail: shreyabhandare25@gmail.com

S. Wattamwar
e-mail: snehaw993@gmail.com

A. Muchrikar
e-mail: ashutosh.muchrikar@cumminscollege.in

© Springer Nature Singapore Pte Ltd. 2018   315
S. S. Dash et al. (eds.), *Artificial Intelligence and Evolutionary Computations
in Engineering Systems*, Advances in Intelligent Systems and Computing 668,
https://doi.org/10.1007/978-981-10-7868-2_31

contain a huge amount of sensitive data which may become a serious threat to the privacy of the user if revealed. This has gained the interest of criminals, which in turn has forced users and application developers to protect their sensitive data.

One of the most reliable ways to protect sensitive information is by encrypting it. Encryption algorithms need cryptographic keys to convert readable data into cipher text. As the number of keys to be managed is increasing, we need key management systems that are designed to protect and manage these keys. Key management systems deal with the creation, exchange, use, and storage of keys, that is, it manages the life cycle of keys. If management of encryption keys is not carefully examined, unauthorized parties can gain access and exploit them [1].

The Android operating system uses hardware keystore to store cryptographic keys on the device. These keys can be accessed by the various APIs provided by Android operating system. Since the keys are stored on the hardware, attackers can gain access to the keys and can manipulate them. The analysis and the shortcomings of android keystore system are highlighted in [2]. Hence, an alternative solution is to store cryptographic keys on a server for centralized key management, that is, the keys are stored and managed over a key management server.

Overall paper is organized in seven sections. Section 1 gives a brief introduction of problem statement. Section 2 describes background theory which emphasizes on key management system. Section 3 gives literature survey of existing systems provided by android for data protection and key management. Section 4 gives description of the proposed system. In Sect. 5, we present the results of the system. Section 6 discusses the conclusion and future works.

2 Background Theory

2.1 Key Management System

Key management systems handle transition between phases of life cycle of cryptographic keys. Cryptographic keys undergo various stages. Figure 1 shows the life cycle of a key.

The management of keys may be handled by a server, which caters to the requests made by client devices.

A client may request for key generation, key retrieval, and deletion of keys. The client communicates with the server via a mutually agreed upon protocol. The server reciprocates to the requested operation and sends an appropriate response message. These protocols may vary depending on client and server. This non-uniformity of protocols between clients and servers increases operational, infrastructural, and implementation cost.

Hence, a single standard communication protocol is needed to communicate between cryptographic clients and servers. Key Management Interoperability Protocol (KMIP) is one such protocol.

Fig. 1 Life Cycle of
encryption key

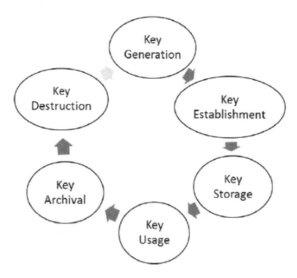

2.2 Key Management Interoperability Protocol

KMIP is governed by OASIS. KMIP addresses the problem of standardizing communication between cryptographic clients, that need to consume keys, and the key management server that create and manage those keys [3]. It is a protocol that defines a standard request and response format. Clients and key management servers use these message formats to communicate with each other.

KMIP includes three primary elements [3]:

1. Cryptographic Objects: Symmetric keys, asymmetric keys, digital certificates, split keys, and opaque objects. Operations are performed upon these objects.
2. Operations: Actions taken with respect to cryptographic objects, such as fetching an object from a key management server, modifying attributes of an object, and so on.
3. Attributes: Represents properties of the cryptographic object, such as the kind of object it is, the unique identifier for the object, and so on.

2.3 IBM's Security Key Life-Cycle Manager

For a server to communicate with a cryptographic client using KMIP it should have support for the protocol, one such server is IBM's Security Key Life-cycle Manager.

Security Key Life-cycle Manager provides centralized and automated encryption key management process in order to minimize risk and reduce operational costs of encryption key management. It offers secure and powerful key storage, key serving and key life-cycle management for IBM, and other storage solutions using KMIP [4].

3 Related Work

Android has two methods for encryption of the data: full-disk encryption and file-based encryption.

- Full-disk encryption is supported by Android 5.0 and above. Full-disk encryption uses single key protection with the device password set by user to protect user's data. The full-disk encryption considerably reduces the performance and slows down the system. In this, functioning of system applications is suspended until password is verified [5].
- File-based encryption is supported by Android 7.0 and above. In file-based encryption, different files are encrypted using different keys that can be unlocked independently.

Physical attacks, such as cold boot, against android's disk encryption are discussed in [6]. The constraint of these attacks is that attacker requires a physical access to the targeted mobile devices. The keys used in both the above methods are saved in the keystore, that is, on hardware chips [5, 7].

To implement computation and key storage on mobile platform, hardware solutions were invented.

- Secure Element: One commonly used solution for secure computation and secure key storage is the secure element [8]. It contains the hardware internals identical to smart card, which is capable of performing security-sensitive functionality such as managing cryptographic keys.
- ARM TrustZone Technology: Every smartphone irrespective of their operating systems uses ARM processors. ARM TrustZone Technology is a hardware-based solution embedded in the ARM cores where the processor cores run two execution environments, also called worlds [9]. The normal world, where any operating system runs and a special secure world, where the sensitive processes can run.

In these systems, the keys are stored on the mobile phone's hardware. The malwares can access the keystore and manipulate it. Integrity of keys is not ensured by encryption scheme used in Android keystore [2]. In a forgery attack used to breach Android keystore's security, a malware installed on the phone that has access to the keystore can break into the keystore and manipulate it by weakening the keys. If these weak keys are used by the target application to secure its data, it becomes easier for the malware to get access to this data [10].

4 Proposed System

Our proposed system consists of a key management server and multiple client applications which communicate via Key Management Interoperability Protocol (KMIP). These applications must have a KMIP client that communicates with the server to request for various key management operations. Irrespective of whether these applications are on the same android device or different devices, they need to have their KMIP client to communicate with the key management server.

The above-mentioned KMIP client program requests for KMIP operations like creation, retrieval, and deletion of keys [11]. These operations are executed and serviced by the server. To reduce this redundant task of writing the KMIP client for every application, we propose to develop a library. The library is an object-oriented API which implements the functionality of KMIP client in the form of various functions.

This reduces the application developer's effort if the library is included as a JAR (java archive) to use the functions in the android application.

4.1 Proposed Solution

Figure 2 shows the architecture of the proposed system.

The primary component of our system is CryptAPI.

CryptAPI: The API currently supports key management operations for symmetric and asymmetric keys.

This is the API for desktop application in JAR form and needs to be included in the desktop application that wants to use its functions.

There are three packages in the library, namely

1. com.ibm.kmip.impl. It contains encrypt and decrypt operations specific to the type of key that is being used.

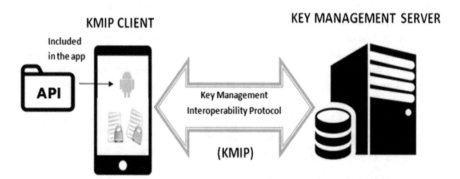

Fig. 2 Proposed system architecture

```
public static String encryptAES(byte[] strToEncrypt, String secret)

public static byte[] decryptAES(String byteToDecrypt, String secret)
```

Fig. 3 Class definition of symmetric ciphers

```
public String create_symmetric_key(String keyName)

public String locate_symmetric_key(String keyName)

public String get_symmetric_key(String UUID)

public void destroy_symmetric_key(String UUID)
```

Fig. 4 Class definition of symmetric key

2. It consists of SymmetricCiphers.java class which provides developers with functions to encrypt and decrypt data using keys fetched by functions in the 'SymmetricKey' class from objects package. Figure 3 shows class definition of symmetric ciphers.
3. com.ibm.kmip.objects. It contains the key management operations of different type of keys used. Currently, the API implements symmetric keys and asymmetric keys. It consists of a class Symmetrickey.java which implements the key management functions provided by the key management interoperability protocol. Figure 4 illustrates class definition of symmetric key.
4. com.ibm.kmip.util. This package contains utility classes that are used by other classes in this library only.

It contains a class called keyOperations.java which contains operations that help in manipulating data type of the key, the below method converts a key from string type to SecretKeySpec. Figure 5 shows the class definition of key operations.

It consists of a class called ServerRequest.java. The functions in this class are responsible for establishing a connection between the client (API) and the SKLM server. Figure 6 shows class definition of Server Request.

```
public static SecretKeySpec setKey(String myKey)
```

Fig. 5 Class definition of key operations

```
public class ServerRequest {

        public static final int SSL_PORT = 5696;
        public static String KEYSTORE_NAME = "";
        public static String KEYSTORE_PWD = "";
        public static String KMS_SERVER_ADDR = "";

        public ServerRequest(String keyStorePath, String keyStorePwd, String keyStoreIP)

                KEYSTORE_NAME = keyStorePath;
                KEYSTORE_PWD = keyStorePwd;
                KMS_SERVER_ADDR = keyStoreIP;
                initSystemProperties();
        }
}
        public void initSystemProperties()

        private static byte[] getHttpOutMessage(byte[] dataOut)

        private static String createHttpHeader(int contentLength)

        public String sendRequest(String request)

        public String readRequest(String filePath)
```

Fig. 6 Class definition of server request

It also has a ServerResponse.java class. The functions in this class are responsible for dealing with the response received from the server; they parse the response and convert it into readable messages and required values. Figure 7 illustrates class definition of Server Response.

5 Result

The API was converted into a java archive (JAR) file and imported into an android application for testing the key management functions and Cipher classes. The application used these functions to successfully create, locate, and fetch keys. These keys were used to encrypt and decrypt various files containing sensitive information. Figure 8 shows the Android Application with the entire flow of encryption and decryption of a text file.

6 Conclusion and Future Work

Though a lot of work is done on key management using Android keystore, use of KMIP for key management on mobile platforms is an unexplored area. Through this proposed system, we aim to help application developers by providing them with a library which can be included in their Android Application to provide key management, data security. The API can be easily extended according to newer versions of KMIP as and when they release.

```
public class ServerResponse {

        public static final int SSL_PORT = 5696;
        public static String KEYSTORE_NAME = "";
        public static String KEYSTORE_PWD = "";
        public static String KMS_SERVER_ADDR = "";

        public ServerResponse(String keyStorePath, String keyStorePwd, String keyStoreIP)
                KEYSTORE_NAME = keyStorePath;
                KEYSTORE_PWD = keyStorePwd;
                KMS_SERVER_ADDR = keyStoreIP;

                initSystemProperties();
        }
        public void initSystemProperties()

        public String extractField(String Final, String Path)

```

Fig. 7 Class definition of server response

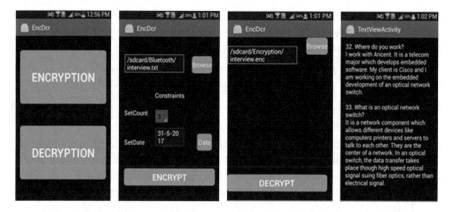

Fig. 8 Screenshots of mobile app used to test the API

The proposed system focuses on Android platform. In future, the project can be extended to include other mobile platforms also like iOS and windows. Support for other cryptographic objects like split keys and opaque objects can be added in the API.

Acknowledgements We are profoundly grateful to our mentors Mr. Mahesh Paradkar, Mr. Prashant Mestri, Mr. Chintan Thaker, Mr. Subhrojoy Roy, Ms. Neha Tirthani, and our guide Mr. Ashutosh Muchrikar for their valuable and expert guidance and continuous encouragement throughout the endeavor.

We extend our gratitude to IBM Security Labs for giving us this opportunity and invaluable guidance.

References

1. Key Management: Today's Challenge, https://www.thales-esecurity.com/solutions/by-technology-focus/key-management
2. M. Sabt, J. Traoré, in *Breaking Into the KeyStore: A Practical Forgery Attack Against Android KeyStore*. 21st European Symposium on Research in Computer Security (ESORICS) (2016)
3. Version 1.0, Key Management Interoperability Protocol (KMIP), White Paper (2009)
4. IBM Security Key Lifecyle Manager, http://www-03.ibm.com/software/products/en/key-lifecycle-manager
5. Android's full-disk encryption just got much weaker—here's why, https://arstechnica.com/security/2016/07/androids-full-disk-encryption-just-got-much-weaker-heres-why/
6. One Way and Two Way SSL and TLS, http://www.ossmentor.com/2015/03/one-way-and-two-way-ssl-and-tls.html
7. Limitations of Android N Encryption, https://blog.cryptographyengineering.com/2016/11/24/android-n-encryption/
8. Global Platform made simple guide: Secure Element. GlobalPlatform, https://www.globalplatform.org/mediaguideSE.asp
9. T.J.P.M. (Tim) Cooijmans, Secure key storage and secure computation in Android, M.S. thesis, Radboud University Nijmegen, (30 June 2014)
10. T. Cooijmans, J. de Ruiter, E. Poll, Analysis of secure key storage solutions on android. in *Proceedings of the 4th ACM Workshop on Security and Privacy in Smartphones & Mobile Devices* (SPSM'14, ACM, 2014), pp. 11–20
11. OASIS Standard, Key Management Interoperability Protocol Specification Version 1.2 (2015)

Analysis of Cloud Environment Using CloudSim

D. Asir Antony Gnana Singh, R. Priyadharshini and E. Jebamalar Leavline

Abstract Cloud computing is a thirst area of research since it provides very important services such as platform as a service (PaaS), infrastructure as a service (IaaS), mobile "backend" as a service (MBaaS), software as a service (SaaS) in the computing and communication environment. Improving the performance of these services is a major challenge being addressed by many researchers. In the recent past, the cloud environment is set up by the service providers and the service is provided to the user in on-demand basis. The quality of the services provided by cloud computing highly depends on the performance of the cloud computing setup or environment (cloud setup). Developing the real cloud environment and evaluating their performance is impractical as it requires huge investment. Hence, researchers use simulation tools to evaluate the performance of cloud computing before constructing the cloud. The CloudSim is a cloud simulation tool for modeling and simulating the cloud computing environment. This paper presents a performance analysis of cloud computing environment using CloudSim.

Keywords Cloud computing · Data center · Host · Virtual machine
CloudSim · Performance analysis

D. Asir Antony Gnana Singh (✉) · R. Priyadharshini
Department of Computer Science and Engineering, Anna University,
BIT Campus, Tiruchirappalli, India
e-mail: asirantony@gmail.com

R. Priyadharshini
e-mail: dharshinipriya245@gmail.com

E. Jebamalar Leavline
Department of Electronics and Communication Engineering,
Anna University, BIT Campus, Tiruchirappalli, India
e-mail: jebilee@gmail.com

© Springer Nature Singapore Pte Ltd. 2018
S. S. Dash et al. (eds.), *Artificial Intelligence and Evolutionary Computations in Engineering Systems*, Advances in Intelligent Systems and Computing 668,
https://doi.org/10.1007/978-981-10-7868-2_32

1 Introduction

Cloud computing is an emerging field of research. Cloud computing provides an environment to share computing resource such as primary and secondary storages, computer networks, processors and to share application and system software in the on-demand basis by different cloud service provider. The services offered by cloud computing environment include platform as a service (PaaS), infrastructure as a service (IaaS), mobile "backend" as a service (MBaaS), software as a service (SaaS), and so on. Cloud computing gives various advantages such as widened network access, more resource availability for sharing with location independence, speed elasticity to meet the sudden need of more resources. Also, the cloud services can be availed with a measurement that allows usage-based charge for services or resources. Thus, cloud computing provides various recourse sharing and services; hence, analyzing and improving the performance of the cloud computing is essential. Different cloud computing tools are used by the researchers to analyze the performance of the cloud computing environment. The CloudSim is one of the cloud computing tools to model and simulate the cloud computing environments. This paper performs an analysis of the cloud environment with different components.

CloudSim is a toolbox that is used for modeling the distributed computing such as cloud computing environment. The cloud computing environment and its applications can be modeled, simulated, and experimented using CloudSim. Hence, most of the researchers use the CloudSim framework to model, simulate, and experiment the cloud environment for cloud computing research. The researchers and developers in industry can design a specific cloud-based system using CloudSim, without concerning about the low-level details of the services and cloud-based environments. CloudSim framework is scalable in structure that allows displaying, remodeling, and experimentation on advanced distributed computing infrastructure and administrations. Its library contains the essential classes for depicting server farms, virtual machines, applications, clients, computational assets, and approaches for administration of different layers of CloudSim framework, for example, booking and administration provisioning in the online air ticket reservation system.

CloudSim contains many library functions written in Java programming language. These library functions are used to model and simulate the cloud applications and infrastructure based on the requirements of the researchers. CloudSim is more popular among researchers, analysts, and industry-based designers since it is simple to test the performance of the cloud applications developed. To set up a cloud environment using the CloudSim tool, various entities that are available in the CloudSim framework are used. The primary entities are data center, host, virtual machines, data center broker, cloudlets, and cloud coordinator [1].

1.1 Data Center

The data center is the basic and essential component used to model the core services at administrator level of the cloud infrastructures. The data center consists of user-defined set of hosts, and the hosts will maintain a set of virtual machines as shown in Fig. 1, which handles "low-level processing." The host list in a data center is fully virtualized, and it deals with virtual machines (VM) query processing instead of the cloudlet-related query processing.

1.2 Host

The host is used to assign and manage the processing capabilities (which is specified in terms of millions of instruction per second (MIPS) that the processor could perform), memory, and scheduling policy to allocate virtual machines to the processing cores which are all present in terms of the list of virtual machines managed by hosts.

1.3 Virtual Machines

Each virtual machine can be allocated to a particular host and can be scheduled to perform the processes of the system. The configuration of virtual machines depends on the hosts of the particular application and the default scheduling policy followed in the CloudSim.

Fig. 1 Representation of data center

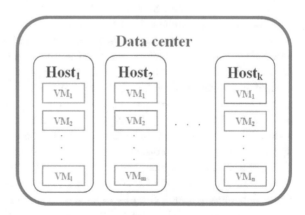

1.4   Cloudlet

The cloudlet is an application that runs in the CloudSim environment. It provides
the resources required for different applications [1, 2]. In CloudSim package, the
cloudlet is a class that contains many methods for storing the details such as
addition or deletion of the files to or from the list of files, running time of cloudlet,
length of the cloudlet.

The Features of CloudSim are,

- Modeling and simulating large-scale data centers
- Modeling and simulating virtualized server hosts
- Modeling and simulating energy-aware computational resources
- Modeling and simulating federated clouds
- Inserting simulation elements dynamically
- User-defined policies for allocation of hosts to virtual machines [1].

2   Literature Review

This section reviews the research works that are carried out by various researchers.
Cloud computing is a major research area these days, as the cloud customer can
access and utilize the information through the Web browser even with handheld
devices such as mobile phones. As the volume of the data and resource utilization in
the cloud computing increases, it is necessary to analyze and evaluate the cloud
computing performance and its environment. The modeling and simulation of cloud
computing is carried out to analyze the performance-related issues in cloud computing. This modeling and simulation is carried out using various cloud simulators
tools such as CloudSim, SPECI, CDOSIM, and DCSim.

Malhotra et al. conducted a performance analysis on the existing cloud simulators. From this analysis, the cloud simulators are roughly classified into two
categories, namely simulators based on software and simulators based on both
hardware and software [3]. Khanghahi et al. provided an overview of the cloud
evaluation strategy and emphasized the usage of simulation tools. Various performance evaluation metrics, factors, and various scenarios of the cloud computing
environment are analyzed [4]. Buyya et al. proposed a cloud simulator toolkit that is
known as CloudSim, and this extended simulation toolkit performs the modeling
and simulation of cloud computing environment to analyze the effectiveness of the
cloud performance to ensure the quality of service and other performance-related
issues. The features of the CloudSim toolkit are creation and modeling of various
CloudSim entities, simulation of federated cloud, and migration of virtual machines
(VM) to achieve reliability and scalability in cloud applications [5]. This CloudSim
can be the better choice to simulate and model the cloud environment to analyze the

performance of the services, cloud computing strategies, effective utilization of resources.

Moreover, many researchers focus their research work in the field of cloud computing toward improving the performance of cloud computing. These research areas include virtualization, resources allocation, VM scheduling, VM migration, cloud load balancing, energy-efficient computing. The virtualization is the prime concepts in the cloud computing environment. The virtualization techniques in cloud environment share the instance of the resources with many users.

Abhishek Gupta et al. conducted a performance analysis between high-performance computing (HPC) and cloud computing. From the analysis, it is identified that the cloud computing performs better than the HPC using virtualization concept in the cloud environment [6]. However, as the virtualization technique is impotent, the effective VM allocation is also an important task to improve the performance of the cloud environment. Khurana et al. presented a VM scheduling policy for the cloud model to increase the application performance under various services and demands. The scheduling policy analyzes the selection and the placement of VM to and from the host to achieve high quality of service (QoS) [7].

The cloud load balancing is another research area in cloud computing. The performance of cloud computing in terms of quality of service and optimization can be enhanced by effective distribution of workload among the computing resources. Calheiros et al. presented a provisioning technique to balance the workload in the cloud environment to improve the quality of service (QoS) of the end user. This technique can automatically balance the workload in high-scale dynamic cloud environment. Furthermore, an analytical performance evaluation is carried out to improve the workload balancing performance to automatically detect the workload changes with minimal input [8]. Lahar Singh Nishad et al. proposed a load balancing and sharing methodology to optimize the cloud data center. In order to achieve this optimization, the task allocation scheme is carried out using Round-Robin fashion [9]. The effective utilization of the resources can improve the performance of the cloud environment. Thus, Rodriguez et al. presented an algorithm with scheduling strategy to perform resource allocation to effectively utilize the cloud resources for infrastructure as a service (IaaS) in cloud environment. This algorithm uses the cloud meta-Heuristic optimization and particle swarm optimization (PSO) to minimize the execution cost. This algorithm is evaluated using the CloudSim toolkit with different size workflows. The result on this analysis shows that the algorithm produces better performance than the existing algorithms [10].

The effective dynamic resource allocation can improve the performance of the cloud environment. Xiaomin Zhu et al. developed an agent-based scheduling method in cloud computing environment to assign real-time tasks and dynamically allocate resources. The bidirectional announcement-bidding mechanism and the collaborative process consist of three phases, namely basic matching phase, forward announcement-bidding phase, and backward announcement-bidding phase.

The schedulability of this method is improved using the elasticity on dynamically added virtual machine [11].

The energy efficiency in cloud computing deals with minimizing the running cost of the data centers to provide cloud services to the customers with less cost and minimize the environmental pollution. Moreno et al. presented an energy-efficient methodology with the analysis of the workload characteristics to improve the resource and energy management [12], and Dario Bruneo et al. presented a resource allocation policy with stochastic rewards nets to improve the energy efficiency in cloud computing. The performance of this policy is compared with different resource allocation policies, and it is identified that the stochastic reward-based algorithm works better [13]. Moreover, the efficiency of the cloud computing can be improved through various ways reported in the literature as follows. Kumar et al. presented a paper that focuses on quantitative analysis of the live migration to improve cloud efficiency. They suggested various cloud parameters to be considered for the future development of the existing migration techniques to improve the live migration performance [14]. Long et al. discussed various cloud computing strategies and their features that eliminate the bottleneck effects and improve the energy efficiency of the cloud entities. The authors used CloudSim simulation toolkit to simulate and model the power utilization strategy to manage services in cloud environment [15].

From the literature, it is observed that the CloudSim can be the better choice for the analysis of the performance of the cloud computing environment. Moreover, various research areas that include virtualization, resources allocation, VM scheduling, VM migration, cloud load balancing, energy-efficient computing, etc., are focused by the researchers to improve the performance of the cloud computing environment.

3 Experimental Setup

Every CloudSim environment is created using the following steps with minor modifications in accordance with the size of the cloudlet and other cloud entities.

3.1 Steps

1. Set number of users for current simulations. The number of brokers and the number of users are directly proportional to each other.
2. Initialize the simulation by providing the current time, the number of users, and the trace flag.
3. Create the data center.

4. Create the data center broker.
5. Create virtual machines.
6. Submit the virtual machine to data center broker.
7. Create cloudlets by specifying their characteristics.
8. Submit cloudlets to data center brokers.
9. Send call to START simulations.
10. Once no more event is executed, send call to STOP simulations.
11. Finally, print the final status of the simulation.

Various examples are demonstrated using the CloudSim to illustrate the working of the CloudSim environment.

Scenario 1 In this scenario, the cloud environment is developed with single data center with one host and single cloudlet to run in the environment.

Scenario 2 In scenario, the experimental setup creates the cloud environment with two data centers with one host that consists of the network topology with two cloudlets to run in the environment.

Scenario 3 In this scenario, the experimental setup creates two data centers with one host for each data center and run two cloudlets along with its corresponding network topology to be present.

Scenario 4 In this scenario, the experimental setup creates two data centers along with host for each data center and run two cloudlets on the cloud.

Scenario 5 Scenario 5 creates two data centers along with the host for each data center and run two cloudlets on the cloud with two users.

Scenario 6 Scenario 6 shows how to create scalable and affordable simulations for the cloud environment.

Scenario 7 This scenario shows how to create the simulation environment that can perform the pause/resume feature to the cloud environment and how to create the simulation entities for the corresponding cloud environment (the entity is data center in this example), dynamically.

Scenario 8 This scenario describes how to create the simulation entities in real-time environment using a global manager entity (Global Broker).

4 Result and Discussions

Table 1 shows the list of entities for various scenarios mentioned in Sect. 3.

From Fig. 2, it is evident that as the number of applications or the cloudlets in the cloud environment increases, the time taken for the particular application to run and deploy in the cloud environment increases. The execution time of a cloudlet depends on the number of cloudlets and the number of data centers present in the cloud environmental setup.

Table 1 Entities list for the various scenarios used in CloudSim

Scenarios	Entities list					Execution time
	Data center	Host	Cloudlet	Network topology	Pause/resume capabilities	
Scenario 1	1	1	1	Yes	No	0
Scenario 2	2	1	2	Yes	No	0
Scenario 3	2	2	2/1	Yes	No	0
Scenario 4	2	2	2	Yes	No	0
Scenario 5	2	2	2	Yes	No	0
Scenario 6	2	2	2	No	No	2
Scenario 7	2	2	2	No	Yes	6
Scenario 8	2	2	2	No	No	1

Fig. 2 Graphical representation of execution time for the corresponding scenario

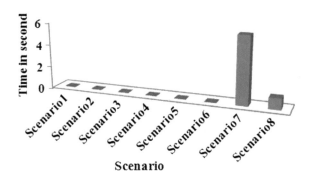

5 Conclusion

The CloudSim is a very effective simulator tool for the development and deployment of cloud computing environment. It helps to analyze the performance and utilization of various cloud computing resources and to obtain the execution time of the system under various cloud computing entities. This paper studied the cloud computing and cloud simulator tool known as CloudSim and presented an analysis of cloud environment using CloudSim.

References

1. http://www.cloudbus.org/cloudsim/
2. M. Satyanarayanan, P. Bahl, R. Cáceres, N. Davies, The case for VM-based cloudlets in mobile computing. IEEE Trans. Pervasive Comput. **8**(4), 14–23 (2009)
3. R. Malhotra, P. Jain, Study and comparison of CloudSim simulators in the cloud computing. SIJ Trans. Comput. Sci. Eng. Appl. (CSEA) **1**(4) (2013)

4. N. Khanghahi, R. Ravanmehr, Cloud computing performance evaluation: Issues and challenges. Int. J. Cloud Comput. Serv. Archit. (ijccsa) **3**(5) (2013)
5. R. Buyya, R. Ranjan, R. N. Calheiros, Modeling and simulation of scalable cloud computing environments and the CloudSim toolkit: challenges and opportunities. IEEE, Issue No: 978-1-4244-4907 (2009)
6. A. Gupta, P. Faraboschi, F. Gioachin, L.V. Kale, R. Kaufmann, B.-S. Lee, V. March, D. Milojicic, C.H. Suen, Evaluating and improving the performance and scheduling of HPC applications in cloud. IEEE Trans. Cloud Comput. 10.1109, 2339858 (2014)
7. S. Khurana, K. Marwah, Performance evaluation of virtual machine (VM) scheduling policies in cloud computing (Spaceshared & Timeshared). IEEE—31661, 4th ICCCNT (2013)
8. R.N. Calheiros, R. Ranjan, R. Buyya, Virtual machine provisioning based on analytical performance and QoS in cloud computing environments, in *International conference on Parallel Processing*, pp. 295–305 (2011)
9. L.S. Nishad, S. Kumar, S.K. Bola, S. Beniwal, A. Pareek, Round Robin selection for data center simulation technique CloudSim and CloudAnalyst architecture and making it efficient by using load balancing algorithm. IEEE 978-9-3805-4421 (2016)
10. M.A. Rodriguez, R. Buyya, Deadline based resource provisioning and scheduling algorithm for scientific workflows on clouds. IEEE Trans. Cloud Comput. 2314655 (2014)
11. X. Zhu, C. Chen, L.T. Yang, Y. Xiang, ANGEL: agent-based scheduling for real-time tasks in virtualized cloud. IEEE Trans. Comput. 10.1109, 2409864 (2015)
12. I.S. Moreno, P. Garraghan, P. Townend, J. Xu, Analysis, modeling and simulation of workload patterns in a large-scale utility cloud. IEEE Trans. Cloud Comput. **2**(2) (2014)
13. D. Bruneo, A. Lhoas, F. Longo, A. Puliafito, Modeling and evaluation of energy policies in green clouds. IEEE Trans. Parallel Distrib. Syst. 10.1109, 2364194 (2014)
14. N. Kumar, S. Saxena, Migration performance of cloud applications—a quantitative analysis, in *International Conference on Advanced Computing Technologies and Applications*. Proc. Comput. Sci. **45**,823–831 (2015) (Elsevier)
15. W. Long, L. Yuqing, X. Qingxin, Using CloudSim to model and simulate cloud computing environment, in *Ninth International Conference on Computational Intelligence and Security*, IEEE, Issue No: 978-1-4799-2548-3 (2013)

IoT Protocols to Manage and Secure Data for the Technological Revolution

B. N. Arathi and R. Pankaja

Abstract The organization of interrelated devices that perform calculations auto-matically, any object, automative, and electronic machines, animals or people embedded with electronics, software, sensors are provided with distinct identifiers and the capability to transport the data over a network without the need of either human-to-human or human-to-computer interaction is Internet of things. The IoT protocols used to manage and secure data are discussed in this paper.

Keywords Internet of things · IoT data link layer standards · Network and session layer standards · Security and privacy · IoT challenges

1 Introduction

The Internet of things (IoT) and device-to-device communication markets demand wireless networking standards that operate on long-range and low-power operation and manage in the sub-1 GHz spectrum. IoT affords effective connection of devices, facilities, and systems which crosses the boundary of device-to-device communi-cations and it includes different protocols, areas of specifications, and implementa-tions. The embedded devices which are interrelated is anticipated to advance the technology in almost every area, and applications like a smart health, smart agri-culture, and enlarging to areas such as smart cities. Internet of things is gain-ing appeal, and it promises to be a technological revolution that will invest every area: industrial, transportation, energy, everyday life, and so on. Thus, IoT is

B. N. Arathi (✉)
Department of CSE, Sri Venkateshwara College of Engineering,
Bangalore, Karnataka, India
e-mail: aradvg@yahoo.com

R. Pankaja
Department of ISE, Sri Venkateshwara College of Engineering,
Bangalore, Karnataka, India
e-mail: pankaja.ssu@gmail.com

© Springer Nature Singapore Pte Ltd. 2018
S. S. Dash et al. (eds.), *Artificial Intelligence and Evolutionary Computations
in Engineering Systems*, Advances in Intelligent Systems and Computing 668,
https://doi.org/10.1007/978-981-10-7868-2_33

Anything, Anywhere, Anytime, Anyway, Anyhow-5 A's. By 2020, IoT will have 50 billion devices including embeddable and wearable computing devices. In both industry and academics, IoT and its protocols are highly recommended topics [1]. The IoT technologies are emerged from micro-computing, internet-mobile, embedded, and machine-to-machine communication technologies [1]. Enormous amount is having been used on emerging IoT technologies and further much more is expected.

Internet of things technologies enable any devices and things to interact smoothly with one another and make individual decisions to communicate and collaborate and make single effective decision so that it is useful for certain applications [1]. To meet lots of challenges, IoT needs special standards and communicating protocols.

Highlights on IoT protocols that operate at medium access control (MAC), network, session layers and which can be more reliable for low-power devices and standards offered by IETF and IEEE which were suggested before half-decade to enable technologies that match the significant growth in IoT are discussed here.

1.1   Related Works

IoT Ecosystem

Seven-layer model of IoT ecosystem is shown in Fig. 1. The lowest layer specifies application computing, such as smart agriculture, smart city, or smart home. Sensors are in the second layer that makes the application operate, such as humidity sensors, temperature sensors [1]. The interconnection layer is the third layer which permits the generation of data by sensors to data centers, cloud, or computing facility. Then, data is grouped with other known sets of data. This collaborated data is then examined by using techniques such as data mining and machine learning [1]. The uppermost layer consists of facilities that enable the market level such as environmental, education, power utilizing, hospital, transportation managements. The management and security applications are considered for all the above layers.

Data link layer connects a set of sensors or sensor and gateway to the Internet. Many sensors are needed to interact and combine the data before getting to the network. For routing among the sensors, many specialized protocols have been designed which are a part routing layer. The protocols of the session layer help in messaging among various devices of the IoT communication system (Fig. 2).

2   Protocols of Data Link Layer of IoT

The data link layer protocol has physical (PHY) and MAC layer protocols, and in most of the standards, they are combined [2].

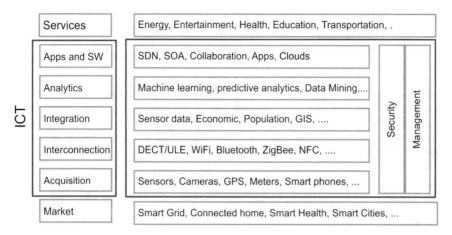

Fig. 1 IoT ecosystem

	Session		CoRE, DDS, AMQP, XMPP, MQTT, SMQTT, CoAP,...	Security	Management
Network	Encapsulation		6LowPAN, 6TiSCH, 6Lo, Thread, ...	TCG, Oath 2.0, SMACK, SASL, ISASecure, ace, DTLS, Dice, ...	IEEE 1905, IEEE 1451, ...
		Routing	RPL, CORPL, CARP, ...		
	Datalink		HomePlug GP, 802.11ah, 802.15.4e, G.9959, WirelessHART, DASH7, ANT+, LTE-A, WiFi, Bluetooth Low Energy, Z-Wave, ZigBee Smart, DECT/ULE, 3G/LTE, NFC, Weightless, LoRaWAN, ...		

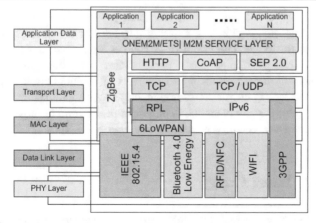

Fig. 2 Internet of things protocols

2.1 IEEE 802.11 AH

It is a revision of wireless networking standard IEEE 802.11-2007 and a wireless networking protocol. To extend the range of Wi-Fi networks which operates between 2 and 5 GHz bands, it uses Sub-1 GHz license-exemption bands. It covers the range to 1 km with minimum of 100 kbps throughput at 1, 2, 4, 8, and 16 MHz channels. For IEEE 802.11 ah, its maximum throughput may reach as high as 40 Mbps. The benefit of low power utilization permits the building of large groups of sites or sensors that interact with each other to share the signal, assisting the concept of the Internet of things [3]. Data packets occurring at intervals in short enable a short on time for distant or sensors that are powered by sensors and are deliberate in competition with Bluetooth with its low energy consumption and low throughput covering wider range.

2.2 Z-Wave

It is a transmission protocol which is wireless and used mainly for automation of residences. It covers up to 30 m communication. It is used for home control and automation market and planned to provide a easy and valid technique for lighting which is wirelessly controlled, wearable healthcare control, HVAC, security systems, spa, cinema at home, automatic operated windows, swimming pool, and garage [4].

2.3 Bluetooth Low Energy or Bluetooth

It is the category of Bluetooth with low energy which was constructed for the Internet of things (IoT). This task makes it absolute for the devices that work on power sources for longer interval of time, such as very small batteries or devices that harvest energy. Important characteristics of this protocol include industry-standard wireless protocol that operates in conjunction across platforms, extreme less peak, less power consumption, architecture with standard application development eases development and deployment of time and cost, permits some of the government-grade security with AES data encryption with 128-bit.

2.4 Zigbee Smart Energy

Zigbee smart energy, the major global accepted standard for products that interact with each other, detects, manages, and automates the transport and use of energy and water. It assists in creating greenery homes by providing the consumers facts and automation required to certainly reduce consumption of energy and saving of money. It pillars the different requirements of intercontinental ecosystem of useful features, government organizations, and industrial organizations that meet future energy and water requirements by operating in the 2.0–2.4 GHz frequency range with 250 kbits/s data rate. It covers up to the range of 10–100 m and includes maximum of 1024 number of nodes. ZigBee uses 128 bit AES encryption for security. It is extensively used for IoT applications such as smart homes, smart health [1].

2.5 DASH7 (DASH7 Alliance Protocol (D7A))

It is access allowable standard of extreme low-power middle range sensor and actuator wireless network protocol, which operates in the 433, 868, and 915 MHz ISM band/SRD band unlicensed. It is an "instant-on," which covers several kilometers range, increase in battery life for years, supports sensor and security, and directs device-to-device communications. The bit rate ranges from 9.6 to 167 kbit/s, with a latency under 15 ms to deliver a packet [5]. DASH7 devices operate effectively with each other and operate on a single global frequency. It is designed to support indoor location and outdoor use which is preferable for IoT requirements [1].

2.6 LTE-A (Long-Term Evolution Advanced)

It is a standard for communication in mobiles. It is designed for device-to-device communication and IoT applications The new functionalities introduced in LTE-Advanced support for relay nodes, aggregation of carriers (CA,), increased use of multi-antenna techniques. These are all designed to increase the strength, speed of LTE networks, bandwidth, and connections. Up to 100 MHz of spectrum, ultra-wide bandwidth is provided, assisting very high data rates, reliability, scalability, and low cost (Fig. 3).

2.7 LoRaWAN

Low power wide area network specification (LPWAN) purposive for wireless cell operated devices anywhere in any region, nation or global network. It focuses on essential needs of Internet of things like low cost, secured two-direction transmission, movability, and localized facilities. Without the need of complex local installations, LoRaWAN specification provides extreme interoperability among Smart Things [4].

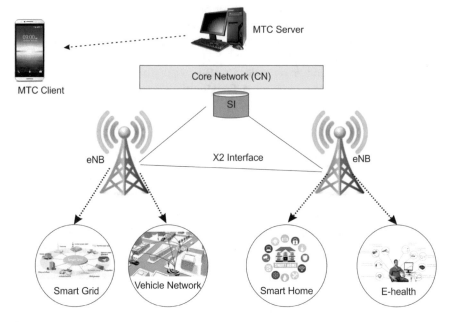

Fig. 3 LTE-A architecture

2.8 Weightless

Weightless is the name of Weightless Special Interest Group (SIG) and the method for interchanging data between a relay station which is at the center and hundreds to thousands of machines situated on every side. Weightless technology operates in both sub-1 GHz license exempt and licensed spectrum. It is helpful in building LPWANs for IoT applications.

2.9 DECT/ULE (Digital Enhanced Cordless Telecommunications Ultra-Low Energy)

It is a standard for wireless communication. It is used for constructing wireless sensor and actuator networks for smart home applications. DECT/ULE devices are used in home automation, home security, and climate control [1]. It uses a star network topology; i.e., one main device, i.e., base, is wirelessly connected to nodes, which are devices with committed functions, such as sensors, remote controls, actuators, smart meters, earthquake detectors, door locks, gas meters, electricity meters, traffic detectors [1].

3 Network Layer Routing Protocols

3.1 RPL

It is an IPv6 routing protocol for low power or lossy networks. Low lossy network routers usually operate with restriction on processing power, memory, and energy (battery power). It specifies the building of a destination-oriented directed acyclic graph (DODAG) by using a function that operates on by considering both, i.e., set of standards of measurement and constraints to calculate the best route.

3.2 CORPL (Cognitive RPL)

It is an extension of RPL for cognitive networks. DODAG is built same as in RPL. It forwards the packets by using forwarding technique by choosing many forwarders and interrelates between the nodes for choosing the ideal next jump to progress the packets [1]. Each node updates its neighbor with changes using DIO messages and maintains a forwarding set. Every node updates the priorities of its neighbor nodes dynamically by constructing new forwarder set.

3.3 CARP (Channel-Aware Routing Protocol)

It is a wireless networks for under water acoustic system, and it is a distributed routing protocol. Due to its lightweight packets, it can be used for IoT. The two scenarios with which it works are: network initialization and data forwarding.

4 Network Layer Encapsulation Protocols

4.1 6LoWPAN

6LoWPAN is an abbreviation of IPv6 over low-power wireless personal area networks. The 6LoWPAN concept emerged from the concept that the low-power devices with limited processing capabilities should be able to take part in the Internet of things, and the Internet Protocol should be applied to the smallest devices. 6LoWPAN is a mesh network which is wireless, low power network in which every node has its own IPv6 address, that allows it to connect directly to the Internet. 6LoWPAN helps in bridges more things to the cloud. It is a robust, scalable, and self-healing network.

4.2   *6TiSCH*

6TiSCH is an open standard-based architecture, highlights best practices, and standardizes according to industrial-grade performance in terms of jitter, latency, scalability, reliability, and low-power operation for IPv6 over IEEE802.15.4e. The 6TiSCH architecture will specify how packets that belong to a deterministic IPv6 flow are marked and routed or forwarded over the mesh within jitter and latency budgets. It will also cover security, link management for the IPv6 network layer, neighbor discovery, and routing. 6TiSCH group will reuse existing protocols such as IP6 neighbor discovery (ND), IPv6 low-power wireless personal area networks (6LoWPAN), and the routing protocol for low-power and lossy networks (RPL), with the minimum adjustment required to meet criteria for reliability and determinism within the mesh and scalability over the backbone [1].

5   Session Layer Protocols

5.1   *MQTT*

It is a lightweight messaging protocol helps in distribution of wireless transmission of data from remote sources by providing constrained resource network clients. The protocol, which uses a publish/subscribe communication pattern, is used for machine-to-machine (M2M) communication and plays an important role in the Internet of things (IoT). It is useful for mobile applications due to its small size, low energy, minimized data packets, and well-organized distribution of information to one or many receivers [6]. It helps minimize the resource requirements for IoT device and ensures reliability and assures the delivery of improved service (Fig. 4).

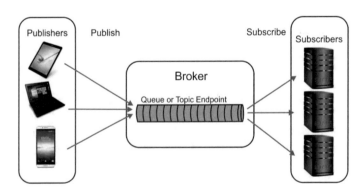

Fig. 4 Architecture of MQTT

5.2 XMPP (Extensible Messaging and Presence Protocol)

It is a set of allowable accessing XML technology for actual time communication, which potentials a vast span of applications such as instantly messaging, existence, lightweight, routing of XML data and collaboration. It is found to have new use in IoT technology such as dryers, refrigerators, washing machines [6].

5.3 AMQP (Advanced Message Queuing Protocol)

It is an open standard for applications or organizations for passing business messages. It bridges systems, supplies the business processes with the information they require, and transmits onwards the instructions that attain their goals. AMQP provides security, reliability, interoperability, standard, open features. It is mainly designed for economic industries [6] (Fig. 5).

5.4 CoAP

"The constrained application protocol (CoAP) designed by IETF is a specialized Web transfer protocol for use with constrained nodes and constrained networks in the Internet of things. The protocol is designed for machine-to-machine (M2M) applications such as smart energy and building automation." It is best for low-energy operable devices [7].

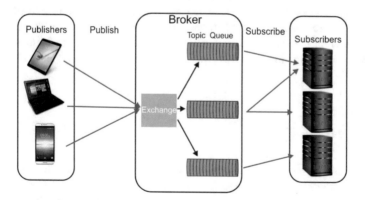

Fig. 5 Architecture of AMQP

5.5 DDS (The Data Distribution Service)

It is an Object Management Group (OMG) device-to-device standard for actual time systems that aims to provide scalability, real-time, dependability, durability, reliability, high performance, and interoperability of data exchanges using a publish–subscribe pattern. DDS is used in applications like financial process, air-traffic control, big data applications, transportation systems, vehicles, smartphone operating systems, and by healthcare providers. This protocol provides outstanding quality of service levels and reliability that is suitable for both IoT and M2M communication [8].

6 Security in IoT Protocols

Security is one of the most significant aspects of IoT applications and found in nearly all layers of the IoT protocols. Damage exists in data link, network, session, and application layers. In this part, security procedures that are built in the IoT protocols are discussed.

6.1 MAC 802.15.4

It implements several features which are used by the ZigBee protocol in the network and application layers. One of this feature is the security services. It uses the encryption algorithm when encrypting data to transmit, but it does not specify management of keys or kind of authentication policies implemented. All the above issues are taken care in the upper layers which are managed by technologies such as ZigBee [3]. Thus, it offers different security modes which include confidentiality, authentication, integrity, and time-synchronized and secured communications [1].

6.2 6LoWPAN

The devices operate either in secured mode or non-secured mode. Two security modes are defined in the specification in order to achieve different security objectives: access control list (ACL) and secure mode [1]. 6LoWPAN does not offer any mechanisms for security, but it includes analysis of threats in security, requirements, and proposal that should be considered in IoT network layer [4].

6.3 RPL

RPL protocol is the first and most important routing protocol suitable for the sensor network with IPv6 support. The information in the field indicates the level of cryptography algorithm used to encrypt the message and security, after the four-byte ICMPv6 message header. RPL takes care of selective forwarding, sink-hole, DoS, Sybil, wormhole attacks. RPL includes unsecured, preinstalled, and authenticated levels of security.

6.4 Application Layer

This layer security refers to techniques at the application layer for protecting Web applications from harmful attacks that may reveal confidential information. Safety concepts are applied to specifically protect against uncertified entries and attacks at the application layer. Encryption algorithms and point-to-point authentication handles diversed levels of security.

7 Challenges in IoT

It is not an easy task for facing multiple challenges such as mobility, management, availability, interoperability, security, reliability, scalability, and privacy for evolving a successful IoT application. The challenges are briefly described [1].

7.1 Security and Privacy

Security is an important pillar of the Internet and one of the most significant challenges for the IoT. Unrestricted access to data poses major security and privacy challenges, including insufficient authentication/authorization, lack of transport encryption, default credentials, lack of secure code practices privacy concerns in IoT.

7.2 Mobility

Devices in IoT system, based on their location need to change their IP address and networks as they move freely. Thus, all time the device goes away from or connects to the network; RPL has to reorganize the DODAG which creates lots of problem.

Also, mobility will enable in a change of service provider which adds one more complexity due to intervention in service and changing of gateways.

7.3 Reliability

All devices should absolutely work and furnish its job correctly. It is a very important feature in areas that demand immediate replies. The devices should be highly authenticated, quick in communicating, gathering data, and upgrading decisions in IoT system. Incorrect decisions sometimes can end with great damage.

7.4 Scalability

When billions and trillions of devices are associated and connected on the same network, then it becomes a challenge in IoT applications and management of distribution of data among them is not an easy job. Added to this, IoT applications should provide extensible services to the devices constantly joining the network and provide new services constantly joining the network.

7.5 Management

This is another challenge of IoT application. It has to manage and keep in course of lack of success and achievement of large number of devices in the system configurations, and account of their internetworked devices and should be managed by providers.

7.6 Availability

IoT refers to the availability of levels of software and hardware being provided always and everywhere for service members. Software provision means the facility provided to any authorized users. Hardware provision means the easy accessing of devices operating at current time and that is compatible with IoT functionality and protocols [1].

7.7 *Interoperability*

This is another challenge where different devices and protocols need to work and effectively operate internally with each other, as the IoT system has large number of different platforms.

References

1. IEEE 1905.1-2013, IEEE Standard for a Convergent Digital Home Network for Heterogeneous Technologies, 93 pp., April 12 2013
2. ITU-T, Short range narrow-band digital radio communication transceivers—PHY and MAC layer specifications (2012)
3. K. Malar, N. Kamaraj, Development of smart transducers with IEEE 1451.4 standard for industrial automation, in *2014 International Conference on Advanced Communication Control and Computing Technologies*
4. IEEE 802.15.4-2011, IEEE Standard for Local and metropolitan area networks–Part 15.4: Low-Rate Wireless Personal Area Networks (LR-WPANs), 314 pp., Sept 5 2011
5. O. Cetinkaya, O. Akan, A dash7-based power metering system, *in 12th Annual IEEE Consumer Communications and Networking Conference (CCNC)*, Jan 2015, pp. 406–411
6. OASIS, OASIS Advanced Message Queuing Protocol (AMQP) Version 1.0 (2012)
7. A. Aijaz, A. Aghvami, Cognitive machine-to-machine communications for internet-of-things: a protocol stack perspective. IEEE Internet Things J. **2**(2), 103–112 (2015)
8. Object Management Group, Data Distribution Service V1.4, April

Design of Smart Home Using Internet of Things

Shaik Naseera, Anurag Sachan and G. K. Rajini

Abstract Smart home is a practical technique to build the simplicity of life. It can be utilized to give assistance and fulfill the necessities of the elderly and the handicapped at houses. Home automation framework will enhance the ordinary living status at houses. The aim of this paper is to implement a central control framework which utilizes a remote Bluetooth gadget and gives wireless access to smart phones. This framework is intended to control electrical gadgets all through the house with ease of installing it, ease of use, and cost-effective design and implementation.

Keywords Cloud computing · Wi-fi · Smart home · Sensors · Arduino
IoT · Home automation

1 Introduction

Smart Home/Home automation is nothing but computerization of the home, housework, or family activity. Home automation may join a control unit for controlling of lighting, HVAC (warming, ventilation, circulating air and cooling), machines, and diverse structures, to give upgraded settlement, comfort, better energy saving, efficiency, and security. Home automation has been around for a long time, and things have been accessible for a significant number of years; however, no one's game plan has gotten through to the standard yet. Home

S. Naseera · A. Sachan (✉)
School of Computer Science and Engineering, VIT University,
Vellore, Tamil Nadu, India
e-mail: anurag.sachan2013@vit.ac.in

S. Naseera
e-mail: naseerakareem@gmail.com

G. K. Rajini
School of Electrical Engineering, VIT University, Vellore, Tamil Nadu, India
e-mail: rajini.gk@vit.ac.in

© Springer Nature Singapore Pte Ltd. 2018
S. S. Dash et al. (eds.), *Artificial Intelligence and Evolutionary Computations
in Engineering Systems*, Advances in Intelligent Systems and Computing 668,
https://doi.org/10.1007/978-981-10-7868-2_34

automation for the elderly and crippled can give extended individual fulfillment to people who may for the most part need parental figures or institutional consideration. It can similarly give a remote interface to home mechanical assemblies or the automation framework itself, through phone line, remote transmission, or the Web, to give control and observe and monitor by means of smart phones or a Web explorer program.

Home automation is a practical technique to build the simplicity of life. It can be utilized to give assistance and fulfill the necessities of the elderly and the handicapped at houses. Furthermore, home automation framework will enhance the ordinary living status at houses. The advancement of any technology is measured as to how much does it simplify human life. Home automation is an upcoming field where the comfort of the user is the foremost aim, followed by scalability and integration of all the processes that occur within any home. Home automation results in a smarter home and is used to provide a higher and healthier standard of living. The beauty of a home automation system is that it is highly scalable, flexible, and its capabilities are limited only by our imagination. With the IoT revolution just around the corner, it is high time to move toward widespread adoption of such systems.

2 Literature Survey

2.1 Survey of the Existing Models and Methods

Upon study of existing models of home automation which are used in the industry today, following are some of the points that have jumped up.

- All home automation systems in the world today are not customizable much after what the industry manufactures have allowed or designed [1].
- Very few models are cost-efficient and are not easy to install or maintain breakdowns [2].
- The Home automation framework utilizes Wi-fi innovation [3]. Framework comprises of three fundamental segments: Web server, which presents framework center that controls and screens clients' home furthermore, equipment interface module [Arduino PCB (instant), Wi-fi shield PCB, 3 input alerts PCB, and 3 yield actuators PCB], which gives fitting interface to sensors, and actuator of home automation framework. The system is better from the versatility and adaptability perspective than the industrially accessible home automation frameworks. The user may utilize a similar innovation to log in to the server electronic application. In the event that server is associated with the Web, so remote clients can get to server online application through the Web utilizing good Web program [4–6].

- Some applications have been created in light of the android framework [7]. An interface card has been created to guarantee correspondence between the remote client, server, raspberry pi card, and the home appliances. The application has been introduced on an android smart phone, a Web server, and a raspberry pi card to control the screen of windows. Android application on a cell phone issues charge to raspberry pi card. An interface card has been acknowledged to refresh signals between the actuator sensors and the raspberry pi card [8].

- Cloud-based home apparatus checking and controlling Framework outlines and actualize a home door to gather metadata from home apparatuses and send to the cloud-based information server to store on HDFS (Hadoop Distributed File Framework), prepare them utilizing MapReduce, and use to give a checking capacity to remote client.

- It has been actualized with raspberry pi through perusing the subject of E-mail and the calculation. Raspberry pi demonstrates to be an effective, monetary, and productive stage for actualizing the keen home automation. Raspberry pi-based home automation is superior to other home automation strategies in a few ways. For instance, in home automation through DTMF (double tone multi-recurrence), the call duty is a gigantic inconvenience, which is not the situation in their proposed strategy. Additionally, in Web server-based home automation, the outline of Web server and the memory space required is cata-pulted by this strategy, since it basically utilizes the as of now existing Web server benefit gave by G-mail. LEDs were used to demonstrate the exchanging activity. Framework is intelligent, effective, and adaptable [8].

- Shih-Pang Tseng et al. proposed Smart House Monitor and Administrator (SHMM), in light of the ZigBee, and all sensors and actuators are associated with a ZigBee remote system. They planned a straightforward shrewd attach-ment, which can remote control by means of ZigBee. PC host is utilized as an information authority and the movement detecting, and all detecting information is exchanged to the VM in the cloud. The client can utilize the PC or android smart phone to screen or then again control through the Internet to power-sparing of the house [9].

2.2 Drawbacks of the Existing Approaches

Many of the above proposed home automation systems are extremely expensive and not easy to implement in one's home. As opposed to that, the one that we create can be made within any specified budget and is easy to set up in the home and very easy to troubleshoot if anything goes wrong. The only reason why the industry is still hesitant to use home automation as a norm is due to the high costs associated with these, and that is the main problem that we try to address through our system [10].

3 Proposed Methodology

The objective of the proposed system is to implement a central control framework which utilizes a remote Bluetooth gadget and gives wireless access to smart phones. This, alongside the assistance of a microcontroller, computerizes almost all components of a home. This framework is intended to control electrical gadgets all through the house with ease of installing it, ease of use, and cost-effective design and implementation.

The objectives of the proposed system include the following:

- To make a home automation system within a budget
- To make an effective and easy to use home automation system
- To make a home automation system that is easily scalable.

3.1 Architecture for the Proposed System

The basic framework of the proposed system is as shown in Fig. 1. The main controlling point of the smart home will be the Arduino board, connected via the required power supply. To implement Bluetooth connectivity, it will be connected to the HC-05/06 Module, and to implement Wi-fi connectivity, it will either be connected to the Wi-fi shield or the Ethernet shield. All the components of the smart home such as the lights and fans can be controlled via the Arduino board with the help of simple relays. Others, such as LEDs, temperature sensors, and humidity sensors will be directly connected to the Arduino, through which all the information from the environment will be redirected to the user's smart phone or laptop.

The smart home system offers features such as environmental monitoring using the temperature, humidity, and gas and smoke sensors. It also offers switching functionalities to control lighting, fans/air conditioners, and other home appliances connected to the relay system. Another feature of this system is the intrusion detection which it offers using the motion sensor, and all these can be controlled from the android smart phone app or Web application.

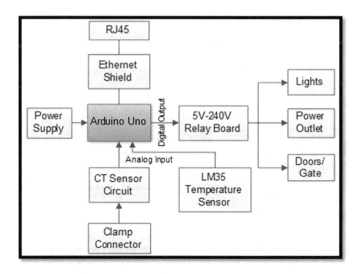

Fig. 1 Framework for smart home

3.2 Proposed System Model

The methodology used to implement this system follows a waterfall model for software processes. The waterfall model is a sequential (non-iterative) design process, used in software development processes, in which progress is seen as flowing steadily downward (like a waterfall) through the phases of conception, initiation, analysis, design, construction, testing, production/implementation, and maintenance. The life cycle of a smart home is depicted in Fig. 2.

4 Properties of the Proposed System

Home automation is a strategy for controlling home machines consequently for the comfort of clients. This innovation makes life less demanding for the client and spares vitality by using gadgets as per strict necessities. Controls can be as fundamental as darkening lights with a remote or as perplexing as setting up a system of things in the home that can be modified utilizing a primary controller or even by means of mobile phone from anyplace in the world.

A. Kind Interface

The most fundamental and pivotal necessity in a home automation framework, the interface is the essential correspondence convention and equipment blend utilized for sending and getting messages among gadgets and the client.

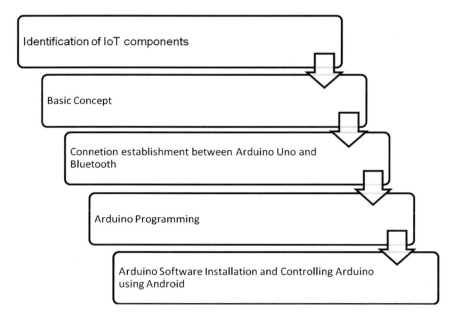

Fig. 2 Life cycle for smart home

B. Detecting Requirements

The planner needs to decide the detecting necessities of the client and choose the obliged sensor to play out the errand. He or she likewise needs to survey the sensor particulars required for various needs and convenience in various conditions. The scope of sensors that ought to be considered includes thermistors that can be utilized to control aeration and cooling systems, fridges, springs, warming framework, or in the event of flame.

C. Higher Security

Another significant necessity while planning the home automation framework is to make the whole framework secure so it cannot be effortlessly changed to give control of the house to unapproved clients. It ought to have the capacity to counteract most sorts of interruption. Regardless of the possibility that the framework is broken into, it ought to have the capacity to send signs to the client and the closest police headquarters.

D. Reliability

The system should be reliable in the sense that there should be no failure of components, and if that does happen, it should be immediately notified to the user. The application in the smart phone or Web application should not crash during usage leading to compromise of safety of the system. The application should be robust and easy to learn and use.

E. Usability

The scalability of the system should be such that the User Interface of the Web application and/or the android app should be easily understandable. No need of extra training for the users should be required. Complete control should be given to the users for all the appliances that have been configured into the smart home automation system.

5 System Design

The system consists of the following modules which will later be uploaded to the Arduino for implementation:

A. Capacitive Switch

Configure all the inputs and outputs in the setup() function. Create flipflop() function which uses flag to determine the current state of the output pin and accordingly changes the state. Create the loop() function to check the state from the flipflop() function and accordingly write to the output pin for the result.

B. Dimmer LED

Configure all the inputs and outputs in the setup() function. Create out() function that uses the switch case to determine amount of dimming in the connected LED with the help of dimmer module, connected via the Arduino. Call the out() function from the main loop for dimming accordingly. Use capacitive switch to choose the amount of dimming by the user.

C. Matrix Switch

Connect the matrix switch to the Arduino. Using the supplied data sheet, accordingly configure the matrix switch to respond to touch and this can then be connected to various other modules of inputs or outputs to control them via one single switch.

D. Relay

Connect the relay to the Arduino and set up all the Arduino pins in the setup() function. Create the main loop() function, and using DigitalWrite() inbuilt functions, run the relay according to desire. Other environment components such as fans and lights can be connected to the relay to control them.

E. Stepper Motor

Connect the stepper motor to the Arduino. Using the setup() function, set the motor speed of the stepper motor and any other input/output pins. Create the main loop(), and using inbuilt functions such as motor.step(), control the stepper motor according to needs.

F. Temperature and Humidity Sensor

Connect the temperature and humidity sensor to the Arduino. Configure all the inputs and outputs in the setup() function. Create the main loop() function, and using the inbuilt function for the humidity sensor model, such as DHT.read() functions, take the required readings and either display them in the serial monitor or any other display screen.

6 Results and Discussion

After the successful completion of the system, the home automation system is applied using Arduino concepts into a working model that is both user-friendly and cost-effective. It is user-friendly in the sense that setting up the smart home is not difficult for any layman and cost-effective in the sense that all the components and assembly are used and purchased keeping the optimal price point in mind.

7 Conclusion

Depending on the results, we conclude that it is much more efficient as it is easy to maintain and if it breaks down, no more charges will be required to call external parties for maintenance. The system is much more scalable due to the versatility of the microprocessor used. Thus, in the future, it will be easier to add on modules as per requirement of the users, as opposed to the present products available in the market. The modules of the system are easily integrated into current home systems without uprooting it as opposed to industry-grade home automation systems where complete overhaul of the home systems is required.

References

1. K.N. Vinay Sagar, S.M. Kusuma, Home automation using internet of things. Int. Res. J. Eng. Technol. **02**(03), 1965–1970 (2015)
2. Z.A. Jabbar, R.S. Kawitkar, Implementation of smart home control by using low cost Arduino & android design. Int. J. Adv. Res. Comput. Commun. Eng. **5**(2) (2014)
3. A. ElShafee, K.A. Hamed, Design and implementation of a WiFi based home automation system. World Acad. Sci. Eng. Technol. **68**, 2177–2180 (2012)
4. N. David, A. Chima, A. Ugochukwu, E. Obinna, Design of a home automation system using Arduino. Int. J. Sci. Eng. Res. **6**(6), 795–801 (2015)

5. S.R.M. Zeebaree, H.M. Yasin, Arduino based remote controlling for home: power saving, security and protection. Int. J. Sci. Eng. Res. **5**, 266–272 (2014)
6. H. Iyer, C. Wagaj, N. Newale, G. Mhatre, Arduino based home automation system: a survey. Int. J. Comput. Appl. **115**(8) (2015)
7. Kanchan, P. Agarwal, M. Vibhute, Home automation using android and bluetooth. Int. J. Sci. Res. (IJSR) ijsr.net
8. S. Paul, A. Antony, B. Aswathy, Android based home automation using raspberry pi. Int. J. Comput. Technol. **1**(1), 143–147 (2014)
9. S.P. Tseng, B.R. Li, J.L. Pan, C.J. Lin, An application of Internet of things with motion sensing on smart house, in *2014 IEEE International Conference on Orange Technologies (ICOT)* (IEEE, 2014), pp. 65–68
10. V.S. Gunge, P.S. Yalagi, Smart home automation: a literature review. Int. J. Comput. Appl. (0975-8887) National Seminar on Recent Trends in Data Mining (RTDM, 2016)

A Real-time Image Mosaicing Using Onboard Computer

K. Sai Venu Prathap, S. A. K. Jilani and P. Ramana Reddy

Abstract In some situations, the total scene of human vision cannot capture in a single shot of camera sensor, and image mosaic is a technique of forming a large image by combining more than one frame of a scene. This paper demonstrates a simple and novel approach for a real-time image mosaicing technique implemented with a standalone rapid prototype Raspberry Pi2 device with a camera sensor. Mosaicing many video frames of having similar scenes are a time-consuming process and lead to inconsistency between frames of the overlap region. Instead of mosaicing all capture video frames, only every nth frame is considered for mosaicing technique, and filtering video frames are processed for every nth frame of a video sequence. Image mosaicing process starts with corner point detection followed by feature point extraction, matching, geometric transformation, hamming distance and finally wind up with warping and blending.

Keywords Mosaicing · Corner points · Feature extraction · Geometric computation · Raspberry Pi

1 Introduction

Mosaicing is a process of merging more than one image to form a single image that represents every image information since the mosaic image has richer information than individual video frames with high resolution [1]. Mosaicing can obtain in two ways [2], they are a direct method and feature-based method. The Feature-based

K. S. V. Prathap (✉) · S. A. K. Jilani
Department of ECE, MITS, JNTUA, Ananthapuramu, Andhra Pradesh, India
e-mail: venu9459@gmail.com

S. A. K. Jilani
e-mail: drsakjilani@mits.ac.in

K. S. V. Prathap · P. R. Reddy
Department of ECE, JNTUA, Ananthapuramu, Andhra Pradesh, India
e-mail: prrjntu@gmail.com

© Springer Nature Singapore Pte Ltd. 2018
S. S. Dash et al. (eds.), *Artificial Intelligence and Evolutionary Computations in Engineering Systems*, Advances in Intelligent Systems and Computing 668,
https://doi.org/10.1007/978-981-10-7868-2_35

method is operated on features of an image like corners and edges. This paper focuses on proficiently sewing a constantly caught video frames for a real-time mosaicing with the general purpose rapid prototype hardware Raspberry Pi2 device. The blurred and discontinuous video frames are discarded by moving the camera slowly and smoothly, but it leads to a great number of frames with highly similar content. Mosaicing with many frames leads to a time-consuming process for feature extraction and corner matching; therefore, to accelerate the processing time, every nth frame of a video sequence is considered because consecutive frames have no much variation. Image mosaicing is executed in multiple phases. The features from accelerated segment test (FAST) has been employed to detect feature points in a frame [3]. FREAK binary descriptor extracts the information of each and every feature point in an image by considering an image patch around corner point. Feature matching is implemented with the hamming distance calculation between the feature points of distinct images. Traditional descriptors like scale-invariant feature transform (SIFT) [4] and speeded up robust features (SURF) employ Euclidean distance for feature matching. Hamming distance swapped Euclidean distance in binary descriptors. Geometric transformation maps the exact matching corner points of the distinct image. Image warping and blending correct the distortions of mosaic image and form a rich vicinity of the mosaic image [5]. The mosaic image has various applications like creating an astronomy, video summarization, 3-D scene reconstruction and robotic navigation, video compression, video matting, and inspection of natural disasters, to mosaic retinal [6] and tissue image [7]; maps are created with the underwater and aerial images [8], to create charming panoramic images. Teller and Sand matched frames of video by using detected features. Their aim was to find best matching frames in different videos [9].

2 Setup of Module

The total module setup of this project is shown in Fig. 1. The Webcam is an external USB module or Raspberry Pi camera module CSI interface of the Raspberry Pi2 device. Raspberry Pi camera features and experiment specifications are listed in Tables 1 and 2. Raspberry Pi2 is a small credit card-sized computer board, with a broad range of peripheral devices. Raspberry Pi2 is a viable, flexible, robust, and inexpensive device [10]. The total Simulink program blocks are a dump into Raspberry Pi2 to execute the process, and Raspberry Pi2 LCD touch screen is used to display the output.

Procedure for the mosaic technique is as follows:

1. The video frames are acquired by using an external Webcam of Raspberry pi.
2. Downsampling is used to select every nth frame from a video sequence.
3. Corners are detected in each frame.
4. Best corner matchings are found by using hamming distance.

Fig. 1 Total module setup

Table 1 Specifications of Raspberry Pi2 camera sensor

Parameters	Description
Still resolution	5 megapixels
Video modes	1080p30
Sensor	Omni vision OV56421
Sensor resolution	2592 × 1944 pixels
Dimensions	25 × 20 × 9 mm
Linux integration	V4L2 driver available
Horizontal FoV	53.50 ± 0.13°
Vertical FoV	41.41 ± 0.11°

Table 2 Specifications of an experiment

Specification	Description
Standalone hardware	Raspberry Pi2
Operating system	Raspbian
Frame size (RGB)	640 × 480
Memory	1 GB RAM
Camera sensor	5 Megapixel
Frame rate	30 frames per second
Downsampler rate	Select every 5th frame ($n = 5$)
Video composition time	35 s

5. RANSAC is applied to estimate geometric transformation matrix for mapping the feature points of one image with another image.
6. Overlay video frames one by one to form a mosaic image (Fig. 2).

The transformation matrix T_{n+1}^n is calculated from the correspondence of corner points denoted the transformation of the warped image from the frame $(n + 1)$ to n. The transformation matrix for mosaicing frame $n + 1$ to the mosaic image identified with the transformed matrix T_{n+1}^m.

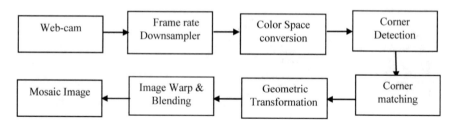

Fig. 2 Block diagram of mosaic system

$$T^m_{n+1} = T^m_n T^n_{n+1} = \prod_{j=1}^{n} T^i_{i+1} \tag{1}$$

The consecutive frames contain similar data of a scene since the stitching is processed with continuous frames. Downsampler is used to filter the frames by selecting every nth frame and its multiple frames of a video sequence. Downsampling has effective computation time just by considering few video frames rather than aggregate video frames that lead to efficient processing time. As sampling frame rate increases than the size, mosaic quality reduces and vice versa. Figure 3 depicts for $n = 5$ and its multiple frames. RGB has three individual primary colors each of 8-bits, the aggregate of 24 bits. Any function with RGB image ought to handle three colors independently, which devours additional time and numerous operations. RGB to intensity is converted over to identify corners effectively because the corners are same in both grayscale image and color image. A grayscale image can be represented with only 8 bits.

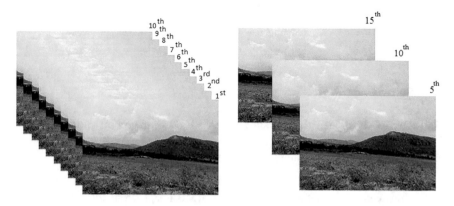

Fig. 3 Downsampling of a video sequence

3 Corner Detection

Corners are detected in each and every video frame. Salient corners are calculated with "corner strength" for all corners in each frame. The corners are represented by the local image descriptors proposed by Lowe and match them accordingly. Suppose that $(D_S^1 \ldots D_S^m)$ and $(D_t^1 \ldots D_t^m)$ are the descriptors for the detected features in two consequent video frames. The feature matching should satisfy the Eq. (2).

$$\left\| D_s^i - D_t^j \right\| < \varepsilon \quad \text{and} \quad \left| P(D_S^i) - P(D_t^j) \right| < K \tag{2}$$

where ε is a small threshold, $P(D)$ is the position of a pixel for feature D in the corresponding image and K symbolizes the window size. The video frames consist of distinct parts of the scene which are synthesized to generate a panoramic mosaic image [11] with high resolution. Corners are local maxima pixels and well-defined as minimum angle of image patch after local averaging [12]. Our basic strategy is to register rich feature points of two consecutive video frames. Figure 3 refers to a corner detection of one video frame from a video sequence, in which corner points are registered correctly. Many feature detectors and descriptors are proposed, first detect corners by using FAST algorithm [13].

This algorithm operates by a circle of 16 pixels around each and every pixel $P(x, y)$, there should be m contiguous pixels whose intensity should brighter than IP + t or darker than IP − t or same intensity as IP [14]. Here, IP represents intensity value of pixel P and t is the threshold value. Each location on circle Y belongs to (1, 2, ..., 16) pixels, then $P \rightarrow Y$ may have one of this conditions.

$$S_{P \rightarrow Y} = \begin{cases} I_{P \rightarrow Y} \leq I_P - t & \text{for Darker} \\ I_P - t \leq I_{P \rightarrow Y} \leq I_P + t & \text{for Similar} \\ I_P + t \leq I_{P \rightarrow Y} & \text{for Brighter} \end{cases} \tag{3}$$

In the above equation, $S_{P \rightarrow Y}$ operates in only one of the three conditions. The corner matrix is found by converting intensity video frames into binary video frames. The isolated point is determined by the length and slopes averaging calculation. A pair of chords having peaks with widths and a central gap is computed to define corners. Figure 4 illustrates a corner detection and segment test operated by 16 pixels around a corner pixel P. It utilizes local intensity comparison with a specified threshold level of intensity based upon Rosten and Drummond. It detects feature point as a corner if and only if it has a maximum of angle to be a corner is 157.5° and a minimum of greater than 10°. The strong corners are invariant to scale, orientation and affine transformations, and even for lighting difference. FREAK [15] is binary feature extractor, inspired by a human vision, especially from the retina. It extracts the information from the surrounding pixels of a corner point. This extracted data is analyzed for feature matching.

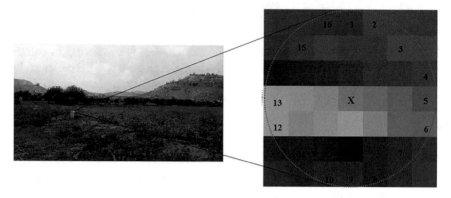

Fig. 4 Segment test operated by 16 pixels around a corner pixel P

4 Corner Matching

Traditional descriptors use nearest neighbor matching to detect best matching corners from two distinct frames. Each corner of one frame it finds the best matching corner in next immediate frame. Binary descriptors use hamming distance between two image patches descriptors P and Q with binary descriptor $B(P) \in \{0, 1\}n$ and $B(Q) \in \{0, 1\}n$ then hamming distance is

$$\text{Ham}(P, Q) = \sum_{i=1}^{n} B_i(P)\phi B_i(Q) \tag{4}$$

Matching measurements are used to find the best overlapping area. Matching [16] is utilized to find the best matching corners between the present video frame and a previous video frame of a video sequence.

5 Geometric Computation

Geometric transformation can be estimated with homography process. Random sample consensus (RANSAC) and least mean square methods perform better results in homography process [17]. If $Q(x, y, w)$ and $Q'(x', y', w')$ are the two homogeneous coordinate points of two distinct video frames, then the homography of two frames is given by $Qr = T Q$, here T is the homography matrix that relates the pixel coordinates in different frames. RANSAC was used to refine the homography matrix between images and also determine matching frames which share image overlaps. The homography matrix is estimated to localize current frame as a region of the mosaic image. The corresponding homography matrix is still estimated to localize this frame as the region of stitching interest on the mosaiced image and to weed out the wrong matching points. The corresponding homography matrix is still estimated to localize this frame as the region of stitching interest on the mosaicked image and to weed out the wrong matching points [18]. A point pair refers to a point in the previous frame and its related point in the current frame. To find the correct set of inliers a number trials T must be tried. Let P_c is the probability for given valid correspondence and P_{total} is the total successive probability after T trials. Then, the likelihood 'T' trials may fail is

$$1 - P_{total} = \left(1 - P_c^m\right)^T \tag{5}$$

and minimum number of trials required is

$$T = \frac{\log(1 - P_{total})}{\log(1 - P_c)} \tag{6}$$

where m is the random sample that is inliers.

RANSAC calculates the number of inliers and outliers. Inliers are common feature points in both previous frame and the current frame; outliers are not common in both images. Correspondence between feature points of two images having two different planes is called homography.

6 Mosaic Results

This project constructed a real-time video mosaicing from video frames of a video sequence with increased field of view (FoV), which has more information. The consequent video frames are correlated with each other since the previous frame is sufficiently utilized to accelerate mosaicing the current frame [19] (Fig. 5). Figure 6 is a Resultant mosaic image. The artifacts can be optimized with histogram equalization technique. Increasing the nth frame results in an efficient mosaic time and shown in Table 3.

Fig. 5 Frame 5th, 135th, 285th, 358th, 520th are processed from a video sequence

Fig. 6 Mosaic image with 150 video frames only

Table 3 Comparison of various parameters for distinct frame selection

nth frame selection	Frame rate	Total frames for mosaic	Execution time (s)
$n = 1$ (all frames)	30	1050 frames	86
$n = 2$	15	525	43
$n = 5$	6	210	18
$n = 10$	3	105	9

7 Conclusion

This paper proposes an efficient real-time wide Field of view (FoV) mosaic algorithm with standalone general purpose embedded Raspberry Pi2. This is an improved algorithm to obtain a wide mosaic image from video frames with FAST and FREAK algorithms. Since several consecutive frames capture similar scene so the total scene can be represented by only a few frames. The frame filtering algorithm accompanied by downsampling. The mosaic algorithm is implemented on the multicore processor to speed up computations of corner detection and corner matching. Homography is estimated between the current video frame and the mosaic frame. To reduce the computation time, the region of mosaic interest is identified by mosaicing every 5th frame of the video sequence. Moreover, for a better mosaic image quality, seams are reduced by image blending.

References

1. S.W. Chew, P. Lucey, S. Lucey, J. Saraqih, J.F. Cohn, I. Matthews, S. Sridharan, In the pursuit of effective affective computing: the relationship between features and registration. IEEE Trans. Syst. Man Cybern. Part B Cybern. **42**, 1006–1016 (2012)

2. M. Brown, D.G. Lowe, Automatic panoramic image stitching using invariant features. Int. J. Comput. Vis. **74**(1), 59–73 (2007)
3. B.S. Kim, S.H. Lee, N.I. Cho, Real-time panorama image synthesis by fast camera pose estimation, in *Proceedings of the Asia-Pacific Signal Information Processing Association Annual Summit and Conference (APSIPA ASC)*, Dec 2012, pp. 1–4
4. K.S.V. Prathap, S.A.K. Jilani, P.R. Reddy, A real-time image mosaicing using scale invariant feature transform. Indian J. Sci. Technol. **9**(12) (2016). https://doi.org/10.17485/ijst/2016/v9i12/88175
5. M. Sharma, Image mosaicing and producing a panoramic visibility. Int. J. Recent. Innov. Trends Comput. Commun. **2**(2), 198–201 (2014)
6. T.E. Choe, I. Cohen, M. Lee, G. Medioni, Optimal global mosaic generation from retinal images, in *The 18th International Conference on Pattern Recognition, ICPR 2006*, vol 03 (IEEE Computer Society, Washington, DC, USA, 2006), pp. 681–684
7. V. Tom, P. Aymeric, G. Malandain, P. Xavier, A. Nicholas, Robust mosaicing with correction of motion distortions and tissue deformation for in vivo fibered microscopy. Med. Image Anal. **10**(5), 673–692 (2006)
8. N.R. Gracias, J. Santos-Victor, Underwater video mosaics as visual navigation maps. Comput. Vis. Image Underst. **79**(1), 66–91 (2001)
9. P. Sand, S. Teller, Video matching. ACM Trans. Graph. **23**(3), 529–599 (2004)
10. http://www.raspberrypi.org/help/faqs/. Accessed 15 June 2014
11. S. Peleg, J. Herman, Panoramic mosaics by manifold projection, in *Proceedings of the Conference on Computer Vision and Pattern Recognition*, pp. 338–343 (1997)
12. M. Nixon, A. Aguado, *Feature Extraction and Image Processing*, 2nd edn. (Oxford-Butterworth, Heinemann, Newnes, 2008)
13. E. Rosten, R. Porter, T. Drummond, Faster and better: a machine learning approach to corner detection. IEEE Trans. Pattern Anal. Mach. Intell. **32**(1), 105–119 (2010)
14. T. Kekec, A. Yildirim, M. Unel, A new approach to real-time mosaicing of aerial images. Rob. Auton. Syst. **62**, 1755–1767 (2014)
15. R. Ortiz, FREAK: fast retina key point, in *CVPR 12 Proceedings of the 2012 IEEE Conference on Computer Vision and Pattern Recognition*, pp. 510–517, June 2012
16. Y. Yan, H. Xia, S. Huang, W. Xiao, An improved matching algorithm for feature points matching, in *ICSPCC-2014*, pp. 292–296 (2014)
17. K.S.V. Prathap, S.A.K. Jilani, P.R. Reddy, A critical review on image mosaicing, in *International Conference on Computer Communications (ICCCI16)*, Jan 2016. https://doi.org/10.1109/iccci.2016.7480028
18. L. Yu, Z. Yu, Y. Gong, An improved ORB algorithm of extracting and matching features. Int. J. Signal Process. Image Process. Pattern Recogn. **8**(5), 117–126 (2015)
19. M. El-Saban, M. Izz, A. Kaheel, Fast stitching of videos captured from freely moving devices by exploiting temporal redundancy, in *Proceedings of the IEEE International Conference on Image Processing* (Hong Kong, 2010), pp. 1193–1196

Rough Set and Multi-thresholds based Seeded Region Growing Algorithm for Image Segmentation

D. Anithadevi and K. Perumal

Abstract The segmentation of brain tumor from MRI (Magnetic Resonance Imaging) scan images is still demanding because it exhibits complex characteristics such as high diversity in tumor appearance and ambiguous tumor boundaries. In this paper, multi-thresholds and rough set-based region growing method for MRI brain image is been proposed as a fully automatic technique. The extracted Region of Interest (ROI) from the proposed method helps in improving the performance of the overall proposed system. The consequent features are been extracted from MRI images by applying the suitable feature extraction techniques to classify the tumor images from normal images. The results of various segmentation techniques are been compared and are proved experimentally. The performance is been evaluated based on the Jaccard distance and DICE coefficient, and it has been found that the proposed approach fairs better with high similarity and less computation time. The overall system achieved 98% accuracy.

Keywords Multi-thresholds · Region growing · Rough set · Feature extraction
Classification

1 Introduction

The medical images play a vital role for detecting the abnormalities; significantly, MRI scan images are a major requisite to detect abnormalities (i.e., cancer) in soft tissues. The requirement of techniques to detect the occurrence of a tumor is very much essential. Most of the image segmentation techniques play a key role in an image processing mainly used in object identification, which aims at partitioning

D. Anithadevi (✉) · K. Perumal
Department of Computer Applications, Madurai Kamaraj University,
Madurai, Tamil Nadu, India
e-mail: danithatce@gmail.com

K. Perumal
e-mail: perumalmkucs@gmail.com

© Springer Nature Singapore Pte Ltd. 2018   369
S. S. Dash et al. (eds.), *Artificial Intelligence and Evolutionary Computations
in Engineering Systems*, Advances in Intelligent Systems and Computing 668,
https://doi.org/10.1007/978-981-10-7868-2_36

the images into a set of meaningful regions and easier to analyze. The primary features of MRI images have translated by the scanner, like tissue-related information such as intensity and gray-level, respectively. Intensity alone cannot provide adequate information to extract the meaningful Region of Interest (ROI) if the ROI has anatomically complex shapes or it is contiguous to other regions that have similar intensities.

In fact, it is a quite common approach to derive and use other types of information like anatomical shapes and location for segmenting the significant ROIs. Due to different imaging types, conditions, and modalities, there is a significant difficulty in preparing a complete set of knowledge and algorithm that can correctly segment ROIs. Rough set representation has proved to be a key structure for the representation of ROIs under such an insufficient set of knowledge. In this paper, the idea is to state, that extent to which a given set of objects approximates the significant ROIs in MRI images automatically. The major advantage of approximation is an ability to represent an inconsistency between actual and expected shapes of ROI. Consequently, features for classifier obtained easily from the ROIs by the application of suitable feature algorithms.

2 Related Works

In this section, previous works of different authors and papers related to this proposed algorithm are discussed. In [1], the different types of fuzzy c-means (FCM) were analyzed. The rough set theory is applied for fuzzy c-means to meet the greatest segmentation results which will be evaluated based on image quality index. In medical images [2], the rough representation of region of interest was extracted based on the approximation. Prior knowledge is required to represent rough ROI from positive and boundary regions. Chavan et al. [3] discussed the extraction of textures' features based on gray-level co-occurrence matrix and classified with k-nearest neighbor classification algorithms. Rough set concepts were applied in fuzzy k-means algorithm [4]. Statistical features-based approach was adopted in [5]. Mean, standard deviation, area, and solidity of features can be extracted for classification. The combined method of gray-level co-occurrence matrix [4] and segmented-based fractal texture features of Hausdorff dimension [5, 6] are extracted based on MRI and CT images. Bayesian classification for brain tumor detection is explained in [7]. For this classification, segmentation-based fractal texture analysis (SFTA)-based texture features are used to classify as either tumor being present or absent.

Basic texture features were extracted for machine learning classification for tumor detection. Especially here, integrated density, aspect ratio was extracted for a decision tree classifier [8]. In [9], the fundamental concepts of sets, fuzzy sets, rough sets and its types how we applied rough sets for images and RST types and the quality of approximations were discussed. The segmentation technique based on region growing for MRI and CT images for the detection of tumors was discussed in [10]. In an image, a seed point can be selected based on the various seed point selection methods. According to the seed point, the appropriate regions are grown and merged. Those obtained regions are called as similar regions. Based on the survey, a novel multi-thresholds and rough set-based seeded region growing segmentation has been proposed.

3 Proposed Methodology

In general, tumor cells occur only because of excessive production of white matters. The tumor could be used to find the help of MRI scan images. In MRI, there are some common sequences, as T1, T2 weighted and Flair (Fluid Attenuated Inverse Recovery). Each sequence is determining the image contrast and brightness by its own properties. From three of them, T2 weighted and Flair sequences are having same characteristics; though, in T2 weighted sequence, normal brain images are misclassified as a tumor since cerebral spinal fluid (CSF) is bright. In fluid, abnormalities have remained bright but normal CSF is attenuated and made dark. However, many systems getting the failure to identify the actual tumor portion clearly in the fluid because of high-intensity variations. In order that the rough set-based region-growing technique experimentally shows that helps to extract actual tumor portion from the MRI image. The flow of the proposed method can be divided into three phases, namely pre-processing, segmentation, and classification. The overall flow of the process is diagrammatically illustrated in Fig. 1.

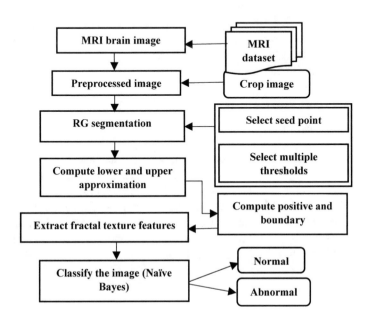

Fig. 1 Process flow representation of proposed system

3.1   Pre-processing

In this phase, suitable pre-processing techniques are been developed to remove the portions of the background pixels surrounding the object (i.e., brain image) [8, 11]. However, region growing segmentation works well with finding objects from the image, it's having a major drawback is time complexity. Since this segmentation method has to compare all the pixels in an image, this may increase the execution time. In order to improve the processing speed, there is a need to eliminate the background pixels [11]. The pre-processed image is used for further processing in terms of segmentation and classification.

3.2   Rough Set-based Seed Region Growing Segmentation

The segmentation is the process of separating the object from the background. The proposed approach for segmentation endeavors to roughly represent the shape of the ROI by approximating [1, 2] the regions based on the criteria derived from the prior knowledge of the image itself. For instance, some threshold values do not meet the exact ROI of images due to high variances of the intensity or gray-level. The rough set-based region growing algorithm achieved by the extraction of the precise ROI and the process of the proposed method has been derived as following steps. Each step is been explained in the following subsections.

Step 1: Fix seed point to growing region.
Step 2: Select multiple thresholds to compute heterogeneous regions.
Step 3: Compute lower and upper approximations.
Step 4: Define the positive and boundary regions.
Step 5: Extract the defined ROI.

3.2.1   Seed Selection

The seeded region growing approach is to segment an image into regions with respect to the seed point selection. In many ways, seed point can be selected automatically, such as center pixel selection, high-intensity pixel-based seed selection, histogram-based seed selection, and random seed selection. Based on the criteria, any one of the methods of seed selection can be followed to find the similar pixels to growing regions, such as Euclidean distance or threshold. Here, the high-intensity pixel has chosen as a seed point.

3.2.2 Threshold Selection

The threshold is selected by means of determining the homogeneous regions. Here, multiple thresholds are selected with the prior knowledge of MRI brain images [10, 12]. Based on the thresholds to build up the elemental criteria is derived according to the homogeneity of the low-level characteristics of the image itself. The knowledge about intensity defines an ideal shape of the ROI [1, 4] containing pixels that satisfy the thresholds criteria. Rough set-based region merging is discussed below.

Rough sets are similar to fuzzy sets [1]; additionally, it has a decision attribute [2, 9] based on the prior knowledge and this will help to find the appropriate ROI from an image. This process is known as supervised learning. Let U is a non-empty finite set called universe of discourse; X is a non-empty finite set and also a subset of U and R an equivalence relation on X. A decision system reveals all the knowledge about the model [4]. A decision system represented in the mathematical form,

$$Z = (U, X \cup \{d\}) \tag{1}$$

where $Z = (U, X), x : X \rightarrow V_x, \forall x \in X, d \notin X, d$ is a decision attribute, X is the whole image pixels, V_x is the values of set X. Based on the prior knowledge of the image and RST, the decision system can be formulated and is shown below.

From the above Table 1, X, V_x are image pixels with corresponding intensity values. $X = \{0, 50, 100, 200, 255\}$, $(V_x) = \{$gray value, distance$\}$ and $d = \{$yes, no$\}$ [2, 9]. From the Table 1, it observed the pixels X_3 and X_4 of two gray attributes having different decision attributes which make sure the values are not dependents on the gray values [1]. After separation of regions, each one is represented by R_i, where $i = 1, 2, \ldots, n$. Then, RST concepts apply to extract rough ROI. The set approximations of RST concepts are discussed below.

Table 1 A decision system for an image

Pixels (X)	Attributes (V_x)		Decision attribute (tumor region)
	Gray values	Distance	
X_1	0	High	No
X_2	50	High	No
X_3	100	Low	No
X_4	100	Low	Yes
X_5	200	Low	Yes
X_6	255	Low	Yes
X_7	200	High	No

3.2.3 Set Approximations

Approximation is induced to signify the roughness of the knowledge. In an equivalence relation R and a set of objects $X \in U$, the approximation of R-lower and R-upper of X is represented as

$$\underline{R}X = \cup \{Y \in U/R | Y \subseteq X\}, \overline{R}X = \cup \{Y \in U/R | Y \cap X \neq \Phi\} \tag{2}$$

The Eq. 2 represents lower and upper approximation, $\underline{R}X$ contains sets that are certainly included in X, and $\overline{R}X$ contains sets that are possibly included in X. According to the decision system, the positive, negative, and boundary regions [2, 9] are defined respectively,

$$POS_R(X) = \underline{R}X, NEG_R(X) = U - \overline{R}X, BN_R(X) = \overline{R}X - \underline{R}X \tag{3}$$

The positive-region ($POS_R(X)$) represents the pixels certainly included in the region. Negative region $NEG_R(X)$ refers the pixels certainly excluded in the region. Boundary region $BN_R(X)$ depends on the difference between upper and lower approximations, in order to obtain the pixels possibly included in the region [2]. The lower and upper approximations can be defined the exact shape of ROI, from a set of regions. After constructing the regions of lower and upper approximations, positive and boundary regions were defined. Based on the Eqs. 2 and 3, the exact ROI can be constructed [2]; it will be roughly represented by

$$ROI = POS_{ROI}(R_i) + BN_{ROI}(R_i) \tag{4}$$

The resultant image is rough ROI, which is inputted for a classification based on the extraction of segmented fractal texture features.

3.3 Classification

The classification is the process which labeled the class of unknown data on the basis of known data set. It should classify the given set of images and categorize those images based on the process of classifiers. In fact, the classification can be intended accurately to classify new images with tumor and non-tumor (normal) image [3]. The process of classification consists of two phases, training and testing. In the training phase, a maximum number of images in the dataset to be trained whereas in testing, new (non-trained) image will be fed to an input image based on the training set; it could find the actual result. The machine learning algorithms are used for classification of MRI brain image. Here, Naïve Bayes classifier is used; hence, it would find out whether the test image contains the tumor or not. Before proceeding to classification, features have to be extracted based on types of data.

3.3.1 *Feature Extraction*

Features are the properties which describe the whole image [5]. The aim is to reduce the size of the original dataset by measuring certain valuable features. Here it is worked, based on the features can extract by the segmentation-based fractal texture analysis (SFTA) [6, 8]. This algorithm is quite common to obtain statistical texture features. The following fractal texture features are calculated such as mean, standard deviation, Hausdorff dimension, and solidity.

3.3.2 *Naïve Bayes Classification*

The major aim of Naïve Bayes algorithm is to learn automatically and make intelligent decisions. It is a statistical method as well as supervised learning. This classifier has two phases, namely training and testing, which can be derived from the Bayesian classifier [7]. The suitable fractal texture features of brain image are obtained and trained with the Naïve Bayes (NB) by using the fit function. If the features cannot fit with this classifier, keep altering the features until it fit. This method may require only a small amount of training data to estimate the brain image either as a tumor or a normal. The time taken for training and testing for classification is less, compared with support vector machine (SVM).

4   Results and Discussion

The experimental results of real-time MRI brain images are been used to compare with some segmentation techniques, such as classical seeded region growing segmentation, classical threshold segmentation, hybridization of region growing and threshold segmentation, multiple thresholds-based seeded region growing segmentation and multi-threshold, and rough set-based region growing segmentation.

Most of the segmentation algorithms failed to detect but the rough set-based region growing method gave better results than others. Based on the multiple thresholds, the seeded region growing technique separates the brain image into the number regions. RST-based approximations were applied to extract tumor. According to the decision attribute, the similar groups of regions were approximated with the prior knowledge. Those approximated regions help to define the exact shape of the ROI. Finally, the exact shape of ROI is been defined based on positive and boundary regions. The expected shape of the ROI is shown in Table 2.

Table 2 Comparison of extracted tumor results of various segmentation techniques on brain MRI image

Images/ ROI Results	Actual tumor region marked as red	Result of proposed method	Result of region growing method	Result of hybrid method	Result of threshold method
Image 1					
Image 2					

4.1 Performance Measure for Segmentation

In medical volume segmentation, Jaccard and DICE coefficient metrics are most popular methods for validation. The outcome of the proposed method verified with the ground truth image. Especially, Jaccard distance is used as a statistical validation metric to evaluate the performance of both over and under segmentations. It mainly focuses on the dissimilarity of ground truth and rough ROI. The Jaccard distance value lies between 0 and 1, and it reaches the best value at 0 and worst value at 1. Jaccard and DICE similarity coefficients measure the overlap or similarity between those binary images. The values of these metrics should lies between 0 to 1.

The value 1 or nearest 1 denotes statistically best result whereas 0 means worst. The value of Jaccard distance should be low (i.e., nearer to 0) and the Jaccard and DICE coefficient should be high (i.e., nearer to 1). These two metrics are used to validate the rough ROI image for similarity. Table 3 clearly shows that the proposed method achieved high similarity. In Table 3, the values of the single image are mentioned and the best values are highlighted. To analyze the performance of the proposed method, the following quality metrics are involved such as Jaccard distance, Jaccard and DICE coefficient and processing time. Here, the Jaccard and DICE coefficients are used to measure the similarity between the extracted tumor region from the proposed method and the ground truth image (i.e., manually segmented tumor region). Jaccard distance is used to measure the dissimilarity between obtained results and ground truth images. The processing time is used to measure the average time taken by the overall system. Figure 2 clearly shows that the proposed algorithm would require minimum time to process that compared to the other algorithms.

Table 3 Comparison of different segmentation methods in terms of Jaccard Distance, Jaccard and DICE similarity coefficients and processing time

Segmentation methods	Jaccard distance	Jaccard coefficient	DICE coefficient	Processing time (in s)
Classical region growing	0.36	0.64	0.84	7.57
Threshold method	0.322	0.68	0.87	4.71
Hybridization of region growing and threshold method	0.29	0.71	0.92	4.53
Multiple thresholds-based region growing method	0.24	0.76	0.92	5.65
Rough set, multi-thresholds-based region growing method	**0.16**	**0.83**	**0.95**	**1.49**

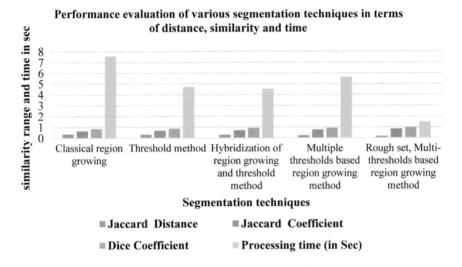

Fig. 2 Performance evaluation with various segmentation methods

The reasons behind the selective methods of proposed work are given respectively. The multi-thresholds facilitates to make a decision for grouping the homogeneous pixels into one region whereas rough set helps to define the lower and upper approximations, which are essential to make a valuable decision to extract ROI. Therefore, both the concepts are really helpful to make this algorithm more efficient to find tumor presence. The effectiveness of the overall process of the system has been evaluated using the following measures:

$$\text{Accuracy} = (\text{TP} + \text{TN})/(\text{TP} + \text{TN} + \text{FP} + \text{FN}) * 100 \qquad (5)$$

where TP—number of true positive cases (tumor images correctly classified), TN—number of true negative cases (non-tumor images correctly classified), FP—number

Table 4 Overall performance of the system

Classifiers	Processing time (in s)	Accuracy (%)
Naïve Bayes	**61.28**	**97.67**
SVM	102.56	96.01

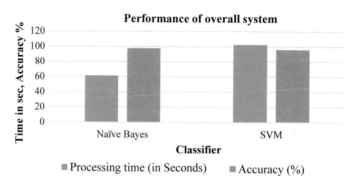

Fig. 3 Representation of overall performance of the system

of false positive cases (normal image classified as a tumor), FN—number of false negative cases (tumor images classified as normal), respectively. Accuracy is the proportion of correctly classified images from the total number of cases. Here, all the segmented methods those were discussed are compared with the SVM and Naïve Bayes classifiers. Totally, 452 images were trained for both classifiers. The overall performance of the accuracy rate is shown in Table 4. The high accuracy and the low processing time of the method are considered as the best method. Here, Naive Bayes classifier works as the best method for this system and attained 97.67 percent (%) in accuracy. This accuracy is high compared with the SVM classifier.

Figure 3 shows that Naïve Bayes classifier achieves greater accuracy that correctly predicted 446 images, from the overall images, six images are misclassified because of t2-weighted images. T2-weighted images having Cerebro Spinal Fluid (CSF) in white color whereas in flair images it is in darken. Hence, this classifier misclassified those CSF as tumor; otherwise, Naïve Bayes finds properly.

5 Conclusion

The rough sets are employed to seeded region growing segmentation. The lower and upper approximation sets inherent from the segmented regions are analyzed and the result of rough ROI is represented the approximation of two regions, positive and boundary regions. The proposed multi-threshold and rough-based seeded region growing algorithm has been found to be performing well compared to the existing segmentation techniques. The former corresponds to the region that can certainly be defined as the ROI with respect to both the given knowledge and the actual feature of the image. The Naïve Bayes classifier has been used to classify the images based on the SFTA features, and the performance is evaluated in terms of accuracy. It achieves a reduction in the overall processing time. The fractal texture features of ROI help to classify the images correctly with an accuracy of 98%. In the future, this proposed algorithm can be extended to find out the tumors with MRI t1- and t2-weighted images.

Acknowledgements The authors would like to place on records their heartfelt thanks to Rajaji Government Hospital, Madurai, and Government Hospital, High Ground, Tirunelveli, as an acknowledgment of the help rendered for this research by providing MR Images.

References

1. S. Patel, K.S. Patnaik, Analysis of cluster algorithms for MR image segmentation using IQI. Procedia Technol. **6**, 387–396 (2012)
2. S. Hirano, S. Tsumoto, Rough set representation of region of interest in medical images. Int. J. Approx. Reason. **40**, 23–34 (2005)
3. N.V. Chavan, B.D. Jadhav, P.M. Patil, Detection and classification of brain tumors. Int. J. Comput. Appl. **122**(8), 48–53 (2015)
4. E.V. Reddy, E.S. Reddy, Image segmentation using rough set based fuzzy K-means algorithm. Global J. Comput. Sci. Technol. **3**(6), 23–28 (2013)
5. B. Esmael, A. Arnaout, R.K. Fruhwirth, G. Thonhauser, A statistical feature-based approach for operations recognition in drilling time series. Int. J. Comput. Inf. Syst. Ind. Manag. Appl. **5**, 454–461 (2013)
6. R. Korchiyne, S.M. Farssi, A. Sbihi, R. Touahni, M.T. Alaoui, A combined method of fractal and GLCM features for MRI and CT scan image classification. Int. J. Signal Image Process. **5**(4), 85–97 (2014)
7. Q. Ain, I. Mehmood, S.M. Naqi, M.A. Jaffar, Bayesian Classification using DCT features for Brain Tumor Detection. KES **1**, 340–349 (2010)
8. J. Chen, G. Bai, S. Liang, Z. Li, Automatic Image Cropping: A Computational Complexity Study. Computer vision foundation, Proc: 2016 IEEE Conference on Computer Vision and Pattern Recognition (CVPR), EISBN: 978-1-4673-8851-1, Las Vegas, NV, USA, pp. 507–515 (2016)
9. T.P. Shewale, S.B. Patil, Detection of brain tumor based on segmentation using region growing method. Int. J. Eng. Innov. Res. **5**(2), 173–176 (2016)
10. N. Senthilkumaran, R. Rajesh, A study on Rough set theory for Medical Image Segmentation. Int. J. Recent Trends Eng. **2**(2), 236–239 (2009)
11. D. Anithadevi, K. Perumal, A hybrid approach based segmentation technique for brain tumor in MRI images. Signal Image Process. Int. J. (SIPIJ) **7**(1), 21–30 (2016)
12. P. Roy, S. Goswami, Image segmentation using rough set theory. Int J Rough Sets Data Anal **1**(2), 62–74 (2014)

Improvement of Security in CR Networks Using Relay Selection

Bhanu Prasad Eppe, Vamsi Krishna Inturi,
Yaswanth Chowdary Boddu, K. Rohith Kumar
and Ratna Dev Singh Didla

Abstract In this paper, we consider a cognitive radio network which consists of a secondary transmitter (ST), secondary destination (SD) and multiple secondary radios along with the existence of eavesdropper. Here, we proposed a relay selection schemes for the improvement of security. If the number of relays is increased then the security of cognitive radio (CR) transmission approaches to significantly improve and the chances of a data breach are low in the presence of an eavesdropper.

Keywords Single relay · Multiple relays · Secondary transmitter
Secondary destination · Eavesdropper

1 Introduction

This is actually about radio communication. Coming to the past of our topic, i.e. CR, this is a new long-term development. 'Actually, Cognitive Radio (CR) [1–3] is defined as the radio that is aware of its environment' and the internal state and the predefined objectives make and implement decision about its behaviour. But in some cases this CR system is used as a software-defined radio, to achieve optimal transmission technology for a given set of parameters. Here, the both CR technology and software-defined radio often lightly. The main purpose of this CR system is to improve the requirements for radio spectrum for improved communication and to protect from jamming [4]. CR networks face security threats during spectrum sensing [5, 6] as well as spectrum sharing [7], spectrum mobility [8] and spectrum management [9]. This CR system consists of a secondary transmitter (ST), secondary destination (SD), multiple secondary radios in the presence of eavesdropper the purpose of the secondary transmitter is to transmit to SD. Coming to the point of the eavesdropper is to intercept the transmission. In addition to an

B. P. Eppe (✉) · V. K. Inturi · Y. C. Boddu · K. R. Kumar · R. D. S. Didla
Department of ECE, KKR & KSR Institute of Technology and Sciences,
Vinjanampadu, Guntur 522017, Andhra Pradesh, India
e-mail: bhanuprasad87@hotmail.com

© Springer Nature Singapore Pte Ltd. 2018
S. S. Dash et al. (eds.), *Artificial Intelligence and Evolutionary Computations in Engineering Systems*, Advances in Intelligent Systems and Computing 668,
https://doi.org/10.1007/978-981-10-7868-2_37

eavesdropper, primary user emulation (PUE) [10], denial-of-service (DOS) attacks [11], brute force attack [12] are also security threats for CR networks. Additionally, physical layer security is also provided in CR networks [13]. In [14], the security of secondary transmission is achieved by quality-of-service (QoS) on primary transmission. So far protecting the SN's against eavesdropper [15] attacks we use a Single relay and multiple relay selection. Coming to the part of the Single relay as we all know that only one relay is taken from multiple sets of relays for secondary transmission from ST to SD. But multiple relay selection takes multiple relays it means SR's to transmit ST–SD. In this cognitive system, we use SRT which means security reliability trade-off for the analysis of SingleRS and Multiple RS schemes. From this, we can observe that the reliability of spectrum is increased. The above probability is reduced because of this the SDs of both Single RS and Multiple RS are improved. Here is the technique to protect the CR system. It is, relay system aided against eavesdropping CR.

Let us discuss how this technology protects the CR systems.

2 Relay Selection Aided Protection Against Eavesdroppers in CR Networks

Now the above system is introduced to protect the CR network. So in this part, we divided into four parts to protect the system.

1 System model.
2 Data direct transmission.
3 Single relay selection.
4 Multiple relay selection.

This transmission not only protects our system but also protect the SingleRS and MultipleRS schemes against the eavesdropping. Let us discuss each part.

2.1 System Model

The system model is shown in Fig. 1 in this, there are two networks which are internally linked each other. They are primary and secondary networks.

In the primary network, mainly consists of two parts. They are multiple primary user and primary base station. In this primary user communicate with the primary bus station.

Fig. 1: System model

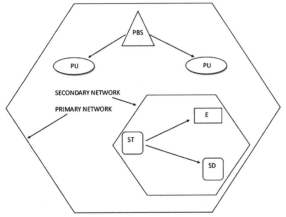

PBS: Primary Base Station PU: Primary User

ST: Secondary Transmitter SD: Secondary Destination

E: Eavesdropper

2.2 Direct Transmission

Up to now, we have discussed system model that means depending upon the status how the spectrum is going to give specified output.

In the existing system, there is only direct transmission between the users with minimum security by using cryptography techniques, i.e. more security is not there for the data that is being transmitted. Cracking of data may be done using brute force attack. So, it becomes easier for eavesdropping the network, i.e. listening to the transmission secretly without the knowledge of transmitter and receiver.

2.3 Single Relay

Here both the SD and eavesdropper are considered to be away from the coverage area of the ST. N secondary relays (SingleRS) are employed for assisting the cognitive ST–SD transmission. Decode-and-forward (DF) relaying using two different time slots is used. The ST first transmits its signal ks to the N SingleRS, which try to decode ks from the received signals and transmit them to the receiver.

$$y_i = hs_i\sqrt{Psks} + hpi\sqrt{\alpha\overline{P}pxp} + ni$$

2.4 Multi-relay System

This part uses a Multiple RS scheme, here multiple SingleRS are used for forwarding the source signal *ks* to SD at the same time. To be determined, ST first transmits *ks* to *N* SingleRS over a free spectrum channel. Let *D* represents the number of relays that are successful in decoding the signal *ks*. If *D* is empty that means the decoding is not successful and the data transmission to the destination is failed (Fig. 2).

Now let us discuss if there is a direct transmission then how spectrum is going to give a specified output. Here, let us consider two variables *tp* and *ts* which represent alternate symbols used by MultipleRS and ST is a particular time distance. This transmission part without losing in generally, let us assume

$$E[1 * pl^2] = E[1 * sl^2] = 1$$

Here E is an expected value.

Then the signal reduced at so it is given as

$$y_d = hsd\sqrt{p_s}x_s + hpd\sqrt{\alpha \cdot p_p \times p} + \text{nd} \tag{1}$$

Here, p_p represents the transmissions power of PRS and *hsd*, *hpd* are fading coefficiencies, nd represents naldon.

α is denoted as 0 for *H*0 and 1 for *H*1.

Now for protecting the ST and SD, we keep an observation on relay selection. Coming to the part of the relay as we all now that it acts as selector without this equipment nothing is going to operate in big projects. But relays are divided into two types and every relay has its own description and use. Nowadays, these relays are used for security purpose. Our main topic is about how to protect CR system using the relay. This relay not only useful for security purpose but also it has the ability to perform its required functions under given condition for a specified time.

Fig. 2: Multi-relay system

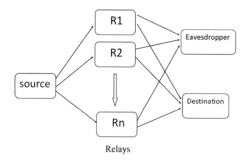

In this, there are three important parts for completing a technique of CR system. They are data transmission, Single relay selection and multiple relay selection. So, when we exit the data we keep a complete on how it is trying on Single relay selection and multiple relay selection.

So when we see the display, we can observe that the spectrum is unaccepted because the signal SR changes from rs to n Single RS which causes decode in rs from received signal. Then the given

N Single RS have 2^N possible subsets then sample space can be represented as

$$\eta = \{\varnothing, D1, D2, \ldots Dn\} \tag{2}$$

here $\varnothing$ is the empty set, D is nonempty set.

So, depending upon empty set and nonempty set the signal received at SRP is given as

$$y_i = hs_i \sqrt{p_s \times x_s} + hp_i \sqrt{\alpha x_p \times p_p} + n_i \tag{3}$$

Here, hs_i and hp_i are fading coefficients of ST and Single RS. So from (3) we can know the capacity of SR-Single RS channel as

$$c_{si} = 1/2 \left[\log 2 \left(1 + \frac{\|hs_i\| \sqrt{s}}{\alpha \|hp_i\| \gamma p + 1} \right) \right] \tag{4}$$

Coming to the new network, i.e. secondary network. This secondary network consists of one or many STs which can protect the spectrum in a time-saving mode. Now for a specified output, the ST should detect the spectrum which is occupied by PBS. Now, in this A represents the status of our spectrum. Now, depending upon the operation of A, we can get our spectrum output from the sensing spectrum. Now, let us assume two terms $H1$ & $H0$. These $H1$ and $H0$ defines whether the spectrum is occupied or not occupied.

Coming to the operation of $H1$ and $H0$ if $H1 = H0$ then the total spectrum is said to be unoccupied and if $H = H1$ then the total spectrum is occupied. So likewise the probability can be calculated in two ways, probability of correct direction (pd), and the probability of false alarm Pp. It can be stated in terms of the equation are $Pd = Px(H = H1/H1)$ and $Pf = Px(H = H1/H0)$, respectively.

3 Results

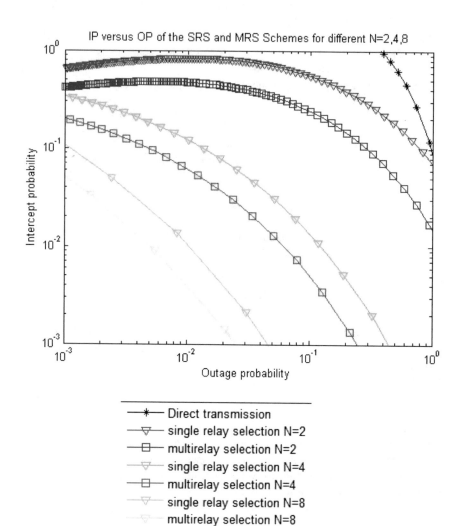

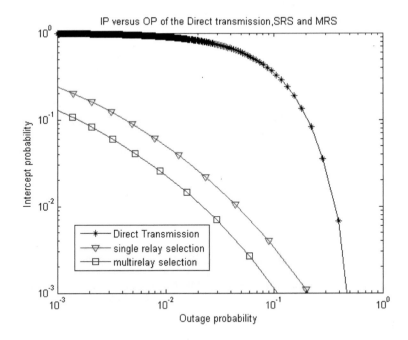

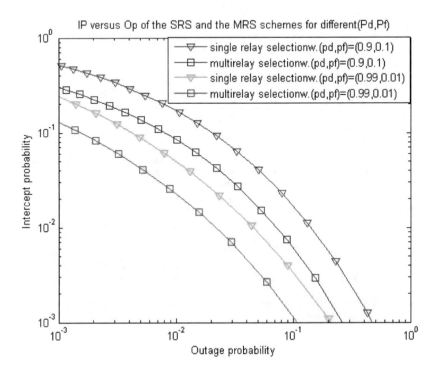

4 Conclusion

In this paper, we proposed relay selection schemes for a CR network consisting of a ST, a SD and multiple SingleRS communicating in the existence of an eavesdropper.

We examined the outage probability of the Single RS and Multiple RS assisted secondary transmissions. We also analyzed the outage probability of the existing direct transmission as a benchmark.

SingleRS and MultipleRS schemes improved the security over direct transmission and MultipleRS performed more reliably than SingleRS.

References

1. J. Mitola, G.Q. Maguire, Cognitive radio: making software radios more personal. IEEE Pers. Commun. **6**(4), 13–18 (1999)
2. IEEE 802.22 Working Group, IEEE P802.22/D1.0 draft standard for wireless regional area networks part 22: cognitive wireless RAN medium access control (MAC) and physical layer (PHY) specifications: policies and procedures for operation in the TV bands, Apr 2008
3. G. Baldini, T. Sturman, A.R. Biswas, R. Leschhorn, Security aspects in software defined radio and cognitive radio networks: a survey and a way ahead. IEEE Commun. Surveys Tuts. **14**(2), 355–379 (2012)
4. D. Cabric, S. M. Mishra, R.W. Brodersen, Implementation issues in spectrum sensing for cognitive radios, in *Proceedings 38th Asilomar Conference Signal, Systems Computers*, Pacific Grove, CA, USA (Nov 2004), pp. 772–776
5. H. Li, Cooperative spectrum sensing via belief propagation in spectrum heterogeneous cognitive radio systems, in *Proceedings IEEE WCNC*, Sydney, N.S.W., Australia (Apr. 2010), p. 1
6. J. Ma, G. Zhao, Y. Li, Soft combination and detection for cooperative spectrum sensing in cognitive radio networks. IEEE Trans. Wireless Commun. **7**(11), 4502–4507 (2008)
7. A. Ghasemi, E.S. Sousa, Fundamental limits of spectrum-sharing in fading environments. IEEE Trans. Wireless Commun. **6**(2), 649–658 (2007)
8. R. Southwell, J. Huang, X. Liu, Spectrum mobility games, in *Proceedings 31st INFOCOM*, Orlando, FL, USA (Mar 2012), pp. 37–45
9. I.F. Akyildiz, W.-Y. Lee, M.C. Vuran, S. Mohanty, A survey on spectrum management in cognitive radio networks. IEEE Commun. Mag. **46**(4), 40–48 (2008)
10. H. Li, Z. Han, Dogfight in spectrum: combating primary user emulation attacks in cognitive radio systems part I: known channel statistics. *IEEE Trans. Wireless Commun.* 9(11), 3566–3577
11. T. Brown, A. Sethi, Potential cognitive radio denial-of-service vulnerabilities and protection countermeasures: a multi-dimensional analysis and assessment, in *Proceedings of the 2nd International Conference CROWNCOM*, Orlando, FL, USA (Aug 2007), pp. 456–464
12. A. Mukherjee, S.A. Fakoorian, J. Huang, A.L. Swindlehurst, Principles of physical layer security in multiuser wireless networks: a survey. IEEE Commun. Surveys Tuts. **16**(3), 1550–1573 (2014)
13. Z. Shu, Y. Qian, S. Ci, On physical layer security for cognitive radio networks. IEEE Netw. Mag. **27**(3), 28–33 (2013)

14. Y. Pei, Y.-C. Liang, K.C. Teh, K. Li, Secure communication in multiantenna cognitive radio networks with imperfect channel state information. IEEE Trans. Signal Process. **59**(4), 1683–1693 (2011)
15. S. Lakshmanan, C. Tsao, R. Sivakumar, K. Sundaresan, Securing wireless data networks against eavesdropping using smart antennas, in *Proceedings of the 28th ICDCS*, Beijing, China (June 2008), pp. 19–27

Biorthogonal Wavelet-based Image Compression

P. M. K. Prasad and G. Umamadhuri

Abstract The image compression is required to reduce the size and transmission bandwidth. The wavelet transform-based image compression is more preferable than other techniques such as DCT. The biorthogonal wavelets are more preferable than orthogonal wavelets due to symmetry property and flexibility. This paper proposes image compression using biorthogonal wavelets. The various biorthogonal wavelets are applied to image compression. The bior1.3 wavelet has the highest PSNR and lowest computation time. The bior1.3 wavelet is superior wavelet out of all the biorthogonal wavelets for image compression.

Keywords Image compression · Biorthogonal wavelet · Wavelet decomposition PSNR

1 Introduction

Fourier transforms translate the time domain information into frequency domain. But, Fourier transform cannot provide the time of occurrence. The short-time Fourier transform (STFT) is used to provide time-frequency analysis [1]. The STFT has fixed window size, that is both low-frequency and high-frequency signals have the same window size. The wavelet transform is time-frequency analysis technique and is used instead of STFT, as the wavelet transform has flexible window size.

Please note that the AISC Editorial assumes that all authors have used the western naming convention, with given names preceding surnames. This determines the structure of the names in the running heads and the author index.

P. M. K. Prasad (✉)
Department of Electronics and Communication Engineering,
G.V.P. College of Engineering for Women, Visakhapatnam, India
e-mail: pmkp70@gmail.com

G. Umamadhuri
Department of Electronics and Communication Engineering, AIET, Viziangaram, India

© Springer Nature Singapore Pte Ltd. 2018
S. S. Dash et al. (eds.), *Artificial Intelligence and Evolutionary Computations in Engineering Systems*, Advances in Intelligent Systems and Computing 668,
https://doi.org/10.1007/978-981-10-7868-2_38

Generally, images require large amounts of data storage and high transmission bandwidth [2–4]. Therefore, it is required to compress the image by storing valid information. The redundant data must be reduced in order to compress the image [5, 6].

Wavelet transform decomposes the image into approximation and detail sub-images. The approximation image almost looks like original image. The detail images show horizontal, vertical and diagonal details of an image. The details of an image can be set to zero if they are very small. The threshold is selected based on the details of an image that is below the threshold, the details are considerably zero. The level of image compression depends on the number of zeros. If the number of zeros is higher, then the level of image compression is also high. The amount of retained information due to image compression is known as the 'retained energy'. If the image is reconstructed perfectly, then the retained energy is 100%, which results 'lossless image compression'. If the retained energy is lost, then it is lossy compression [7]. The wavelet transform performs superior when compared to discrete cosine transform for image compression. It gives better image quality due to its multiresolution. Wavelets are used for multiresolution analysis, where the signal is analysed at different resolution levels [8]. There are various wavelet transforms like orthogonal and biorthogonal wavelets. The orthogonal wavelet transforms like Haar, Daubechies, Coiflet and Symlets can be used for image compression. But they are unable to compress the image effectively and unable to work effectively under noisy conditions. So biorthogonal wavelets are used in place of orthogonal wavelets due to more flexibility. In biorthogonal wavelets, the orthogonality property is relaxed as the properties of orthogonality and symmetry conflict each other. So the design of biorthogonal wavelet is more flexible, and it is symmetric wavelet. The characteristics of biorthogonal wavelets can be varied by changing the various properties of the wavelet such as orthogonality, symmetry and vanishing moments. This paper proposes biorthogonal wavelet-based image compression using multiresolution analysis.

2 Biorthogonal Wavelet Theory

The orthogonal wavelet design is not flexible due to the orthogonality condition. The wavelet is characterized by wavelet function and scaling function [9]. In orthogonal wavelet, scaling and wavelet functions are orthogonal to each other. It results in complex design equations and prevents linear phase analysis. The biorthogonal wavelet design is more flexible as the orthogonality condition is relaxed to retain linear phase characteristics. The orthogonality and symmetry are conflicting each other, so relax orthogonality to get symmetry. Therefore, the biorthogonal wavelet is basically symmetric wavelet. The characteristics of wavelet can be improved by relaxing the orthogonality constraints [10]. The Haar wavelet is the only symmetric, finite length, orthogonal wavelet with the support interval is equal to one which is less [11, 12]. The Haar wavelet may not be able to detect the large changes in the input data due to less support interval. So it is required to design symmetric filters with support interval more than two [13, 14]. The biorthogonal wavelet has linear phase characteristics which are required for image reconstruction. It uses two wavelets, that is one wavelet for decomposition and another wavelet for reconstruction [15]. The decomposition filter is also known as analysis filter and the reconstruction filter is also known as synthesis filter. The biorthogonal wavelet order is specified by the order of these two filters. There are various biorthogonal wavelets such as bior1.1, bior1.3, bior1.5, bior2.2, bior2.4, bior2.6, bior2.8, bior3.1, bior3.3, bior3.5, bior3.7, bior3.9, bior4.4, bior5.5 and bio6.8. The order of synthesis filter is indicated by first number and analysis filter by second number.

3   Wavelet Decomposition

In wavelet decomposition, an image will be passed through an analysis filter bank followed by a decimation operation. The analysis filter bank consists of a low-pass filter and a high-pass filter at each decomposition stage. When a signal passes through these filters, it divides into two bands. The low-pass filter extracts the approximation information of the signal. The high-pass filter extracts the detail information of the signal. The output of the filtering operation is decimated by a factor of two. This multiresolution analysis allows decomposition of signal into various resolution levels. The wavelet transform decomposes the image into four sub-bands, that is LL, LH, HL and HH [16, 17]. The LL sub-band provides low-frequency information, which is often referred to as the average information. LH provides horizontal details of an image such as horizontal edges. HL provides vertical details of an image such as vertical edges. The HH sub-band provides diagonal details of an image such as diagonal edges [18]. This is called one-level wavelet decomposition. The LL sub-band is further decomposed into four sub-bands such as LL2, LH2, HL2 and HH2. This is called second-level decomposition. The process is repeated in accordance with the desired level. In this paper, second-level decomposition is used for compressing the images.

The discrete wavelet transforms are used in many applications like image denoising, image compression, edge detection and signal processing [19, 20]. In image compression, information may be lost. But, this information loss must be minimized to maintain the image quality. There is always compromization between the image quality and the compression size. There are two performance metrics for image compression, i.e. peak signal-to-noise ratio (PSNR) and compression ratio. A large value of PSNR indicates the high quality image. Equations 1 and 2 show the PSNR and MSE.

$$PSNR = 20 \; \log_{10} \frac{255}{\sqrt{MSE}} \tag{1}$$

where Mean Square Error

$$\text{MSE} = \frac{1}{mn} \sum_{y=1}^{m} \sum_{x=1}^{n} (p(x,y) - p'(x,y))^2 \tag{2}$$

$p(x,y)$ is the original image, $p'(x,y)$ is reconstructed image, and mn is the size of the image. Computation time is the time required to execute the program.

4 Image Compression Using Biorthogonal Wavelets

Figure 1 shows the block diagram of biorthogonal wavelet image compression. First, select the image and apply the various biorthogonal wavelets such as bior1.1, bior1.3, bior1.5, bior2.2, bior2.4, bior2.6, bior2.8, bior3.1, bior3.3, bior3.5, bior3.5, bior3.7, bior3.9, bior4.4, bior5.5 and bior6.8 to the image which generates the wavelet coefficients. Select the threshold to obtain the modified coefficients. Then apply inverse wavelet transform to reconstruct the image. The performance metrics like PSNR and computation time are computed and then compared for various biorthogonal wavelets. There are different biorthogonal wavelets such as bior1.1, bior1.3, bior1.5, bior2.2, bior2.4, bior2.6, bior2.8, bior3.1, bior3.3, bior3.5, bior3.7, bior3.9, bior4.4, bior5.5 and bio6.8.

Fig. 1 Biorthogonal wavelet-based image compression

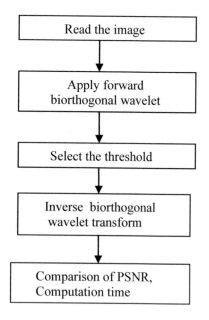

5 Results and Discussion

The results are simulated with the help of MATLAB. The test images like cell image (intensity image) with size 159 × 191 and Barbara image (indexed image) with size 256 × 256 are considered. In addition, it results in the influence of wavelet towards PSNR, with several types of biorthogonal wavelets. Tables 1 and 2 show the peak signal-to-noise ratio value and computation time for various types of biorthogonal wavelets' cell image and Barbara image.

From Tables 1 and 2, it was observed that the cell image has highest PSNR value of 52.5242 dB and lowest computation time of 0.9844 s for bior1.3 wavelet. The barbara image has highest PSNR value of 44.6206 dB and lowest computation time of 1.9344 s for bior1.3 wavelet. From Tables 1 and 2, it was observed that the cell image has lowest PSNR value of 48.9349 dB and highest computation time of 1.4123 s for bior3.1 wavelet. The barbara image has highest PSNR value of 40.8095 dB and lowest computation time of 2.5471 s for bior3.1 wavelet.

Figure 2 shows the bior3.1 wavelet image compression results for cell image, and Fig. 3 shows bior3.1 wavelet approximation and detail images for cell image.

Figure 4 shows the bior3.1 wavelet image compression results for Barbara image, and Fig. 5 shows bior3.1 wavelet approximation and detail images for Barbara image.

Table 1 PSNR values for various biorthogonal wavelets

Wavelets	PSNR (dB) cell image	PSNR (dB) Barbara image
bior1.1	52.5105	44.5183
bior1.3	52.5242	44.6206
bior1.5	52.4703	44.4889
bior2.2	50.9283	43.4227
bior2.4	50.9749	43.4750
bior2.6	50.9368	43.4597
bior2.8	50.8914	43.4261
bior3.1	48.9349	40.8095
bior3.3	49.7433	41.8626
bior3.5	49.9444	42.0891
bior3.7	50.0045	42.1651
bior3.9	50.0539	42.1622
bior4.4	51.7790	44.3776
bior5.5	52.0525	44.5349
bior6.8	51.7006	44.2869

Table 2 Computation time for various biorthogonal wavelets

Wavelets	Computation time (s) cell image (s)	Computation time (s) Barbara image (s)
bior1.1	1.929	2.0904
bior1.3	0.9844	1.9344
bior1.5	1.0625	2.1216
bior2.2	1.1094	2.0592
bior2.4	1.0469	2.0436
bior2.6	1.0781	2.1060
bior2.8	1.1094	2.1996
bior3.1	1.4123	2.5471
bior3.3	1.0120	2.0748
bior3.5	1.0313	2.0592
bior3.7	1.0625	2.1528
bior3.9	1.1875	2.2932
bior4.4	1.2124	2.1060
bior5.5	1.0781	2.1216
bior6.8	1.0781	2.3088

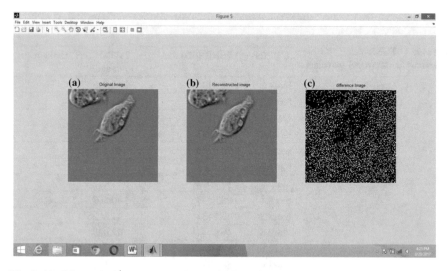

Fig. 2 bior3.1 wavelet image compression results for cell image: **a** original image, **b** reconstructed image and **c** difference image

Figure 6 shows wavelet versus PSNR values for various biorthogonal wavelets for cell image and Barbara image. Figure 7 shows wavelet versus computation time for various biorthogonal wavelets for cell image and Barbara image.

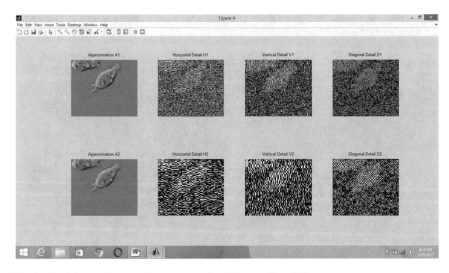

Fig. 3 bior3.1 wavelet approximation and detail images for cell image

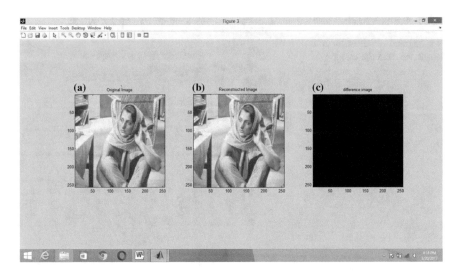

Fig. 4 bior3.1 wavelet image compression results for Barbara image: **a** original image, **b** reconstructed image and **c** difference image

The bior1.3 wavelet has high PSNR which indicates highest image quality among all the biorthogonal wavelet family. It means the reconstructed image is almost same as the original image. Figure 8 shows bior1.3 wavelet image compression results for cell image, and Fig. 9 shows bior1.3 wavelet approximation and

Fig. 5 bior3.1 wavelet approximation and detail images for Barbara image

Fig. 6 Wavelet versus PSNR values for various biorthogonal wavelets for cell image and Barbara image

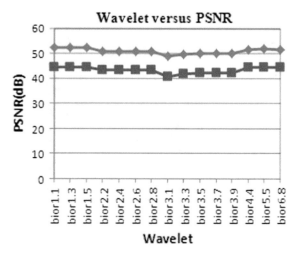

detail images for cell image. Figure 10 shows bior1.3 wavelet image compression results, and Fig. 11 shows bior1.3 wavelet approximation and detail images for Barbara image. The detail images include horizontal, vertical and diagonal details. Here, second-level decomposition is used.

Fig. 7 Wavelet versus computation time for various biorthogonal wavelets for cell image and Barbara image

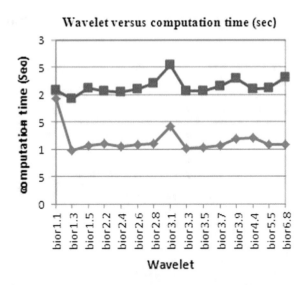

Fig. 7 Wavelet versus computation time for various biorthogonal wavelets for cell image and Barbara image

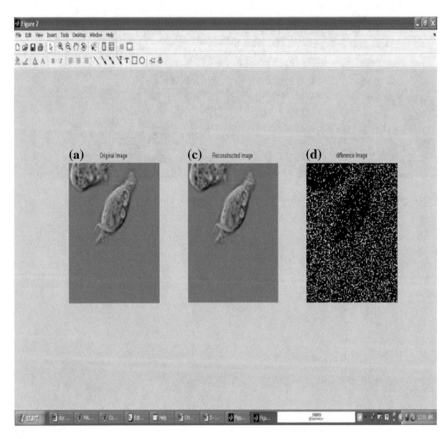

Fig. 8 bior1.3 wavelet image compression results for cell image: **a** original image, **b** reconstructed image and **c** difference image

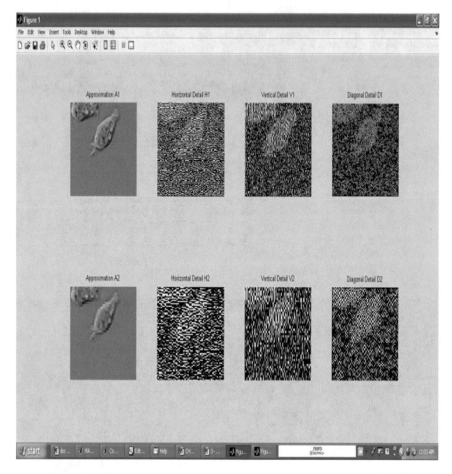

Fig. 9 bior1.3 wavelet approximation and detail images for cell image

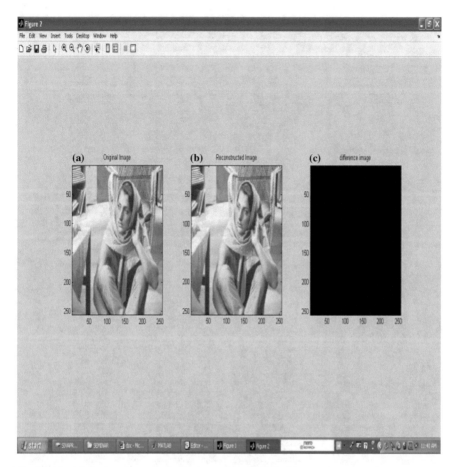

Fig. 10 bior1.3 wavelet image compression results: **a** original image, **b** reconstructed image and **c** difference image

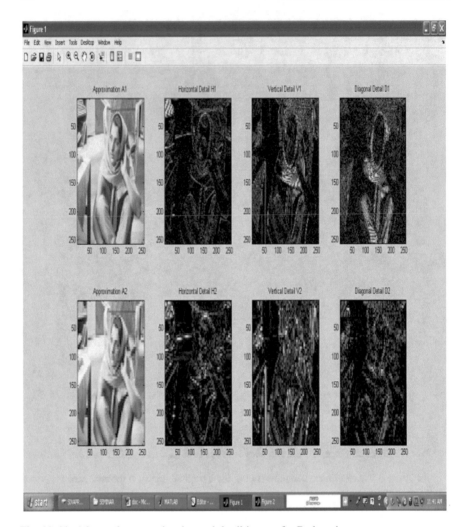

Fig. 11 bior1.3 wavelet approximation and detail images for Barbara image

6 Conclusion

In this paper, the various biorthogonal wavelets are applied to image compression. Wavelet transform is widely used for image compression with better image quality due to its multiresolution. The biorthogonal wavelet is more flexible to design when compared to orthogonal wavelet, and it can be applicable for image denoising, edge detection and image compression. The bior1.3 wavelet is superior for image compression as it has highest PSNR and lowest computation time among all the biorthogonal wavelets. The future extension of this research work includes selecting the best thresholding technique, exploring other wavelet families, compressing the image with higher levels of decomposition without losing the quality of information and implementing the image compression technique using neural networks.

References

1. S. Jayaraman Esakkirajan, T. Veerakumar, *Digital Image Processing* (Tata Mc Graw Hill Publication, 2009)
2. R.C. Gonzalez, R.E. Woods, *Digital Image processing,* 2nd edn. (Prentice Hall of India Ltd, 2004)
3. R.C. Gonzalez, R.E. Woods, S.L. Eddins, *Digital Image processing using MATLAB,* 2nd edn. (Prentice Hall of India 2003)
4. K.H. Talukder, K. dan Harada, *Haar Wavelet Based Approach for Image compression and Quality Assessment of Compressed Image,* IAENG Int. J. Appl. Mat. **36**(1), (2007)
5. A.K. Jain, *Fundamental of Digital Image Processing,* 4th edn. (Prentice Hall of India Private Ltd, 2000)
6. M. Sonka, V. Hlavac, R. Boyle, *Image Processing, Analysis, and Machine Vision*, 2nd edn. (Vikas Publishing House, 2001)
7. W.K. Pratt, *Digital Image Processing*, 3rd edn. (John Wiley & Sons Inc, 2001)
8. P.M.K. Prasad, D.Y.V. Prasad, G.S. Bhushana Rao, *Performance Analysis of orthogonal and Biorthogonal wavelets for Edge detection of X-ray Images,* Elsevier Procedia Computer science ISSN: 1877-0509, Vol. 87, 2016, pp. 116–121
9. K.P. Soman, K.I. Ramachandran, *Insight into Wavelets from Theory to Practice,* 2nd edn. (Prentice Hall of India, 2008)
10. L. Feng, C.Y. Suen, Y.Y. Tang, L.H. Tang, Edge extraction of images by reconstruction using wavelet decomposition details at different resolution levels. Int. J. Pattern Recognit. Artif. Intell. **14**(6), 779–793, (2000)
11. P.J. Vanfleet, *Discrete Wavelet Transformations an Elementary Approach with Applications.* (Wiley, 2011)
12. R.C. Gonzalez, R.E. Woods, *Digital Image Processing*, (Pearson Education, 2004)
13. F.Y. Cui, L.J. Zou, B. Song, Edge feature extraction based on digital image processing techniques in *Proceedings of the IEEE, International Conference on Automation and Logistics*, Qingdao, China September 2008, pp. 2320–2324
14. C.S. Burrus, R.A. Gopinath, Haitiao Guo, *Introduction to Wavelets and Wavelets Transforms*: *A Primer.* (China Machine Press, Beijing, 2005)
15. P. Singh, P. Singh, R.K. Sharma, JPEG image compression based on biorthogonal, coiflets and daubechies wavelet families. Int. J. Comp. Appl. (0975–8887). **13**(1), (2011)
16. D. Gnanadurai, V. Sadasivam, An efficient adaptive thresholding technique for wavelet based image denoising. Int. J. Electron. Commun. Engg. **2**(8), (2008)

17. S.G. Chang, B. Yu, M. Vetterli, Adaptive wavelet thresholding for image denoising and compression. IEEE Trans. Image Process **9**(9), (2000)
18. A. McAndrew, *Introduction to Digital Image Processing with MATLAB* (Cengage Learning India Private Limited, New Delhi, 2009)
19. S. Mallat, *A Wavelet Tour of Signal Processing* (Academic Press, USA, 1999)
20. M. Antonini, M. Barland, P. Mathien, I. Daubechies, Image coding using wavelet transform. IEEE Trans. Image Process. **1**, 205–220 (1992)

Image Encryption by Exponential Growth Equation

Purushotham Reddy M., Venkata Ramana Reddy B. and Shoba Bindu C.

Abstract This paper proposes the image encryption by using the exponential growth equation. This new algorithm encrypts the image through a two-stage process. In the first stage, a reference exponential growth function is generated for the foundation of the encrypted image. In the second stage, the random image matrix is used as a key for the encryption which is applied on the first-stage resultant matrix. The advantage of this method is encrypting the image efficiently. This method could be applied to provide more complexity against attackers, and it takes less time for encryption.

Keywords Exponential growth equation · Image encryption · Random image matrix · Attackers · Cryptosystem

1 Introduction

Nowadays, digital images play an important role in our daily lives and the transmission of data through communication network also important. The logarithmic function is used for encryption in the first stage and divides the resultant image into four sub-matrices. In the second stage, combine the four sub-matrices into a single matrix and exponential function used for image decryption [1]. In the diffusion stage, the pixel values are modified by using the XOR operation and Lagrange-least square interpolation used for image encryption [2]. Sudoku matrix is used for shuffling the pixel values and also for the encryption [3]. The 64-bit blowfish

Purushotham Reddy M. (✉) · Shoba Bindu C.
JNTUA University, Anantapuramu, India
e-mail: purushotham.mps@gmail.com

Shoba Bindu C.
e-mail: shobabindhu@gmail.com

Venkata Ramana Reddy B.
Narayana Engineering College, Nellore, India
e-mail: busireddy100@gmail.com

© Springer Nature Singapore Pte Ltd. 2018　　　　　　　　　　　　　　405
S. S. Dash et al. (eds.), *Artificial Intelligence and Evolutionary Computations
in Engineering Systems*, Advances in Intelligent Systems and Computing 668,
https://doi.org/10.1007/978-981-10-7868-2_39

function used as secret key and 16 iterations in a Feistel network are used to create cipher image [4]. XOR operation is used for changing the pixel values, and circular rotation is used for changing the places of pixel values in the image [5]. The image matrix multiplied by modular number as key and get cipher image [6]. The image encryption is done using singular value decomposition. This singular value decomposition used for scrambling the image using two different keys [7]. Random substitution using binary traversal with row and column operations is used for multimedia encryption [8]. Image encryption done using pseudorandom number permutation along with chaotic maps [9]. The confusion function is created using XOR keys and image encrption is done using chaotic logistic maps [10]. The image is read from the file. The logic chaotic map is used for generating keys and then encrypts the image [11]. Image encryption is based on skew tent chaotic map and permutations. In this study, P-box is chosen for shuffling the pixel values in the image. Skew tent chaotic is used for generating key stream and then encrypts the image [12]. Random number key enhances the security of the transmitted image. The quantum key distribution protocol is used for generating the key and then encrypts the image [13]. Linear feedback algorithm is used for shuffling pixel values, and other pseudorandom number is used for encrypting the image [14].

2   Exponential Growth Equation

The exponential growth equation is defined as

$$y = ab^x \tag{1}$$

where a is constant, $a \neq 0$, $b > 1$ and x is a variable.

Here, the exponential growth equation is used to modify the intensities of the pixel values in the image, and it scatters the elements differently.

The graph of the exponential growth equation is shown in Fig. 1.

3   Proposed Method

3.1   Encryption and Decryption Algorithm

In the presented encryption system, the exponential growth equation is used to modify intensities of the pixel values in the image block by block. The random image matrix is generated by using random number generation. This random image matrix is used as the key image. The XOR function is applied between resultant image matrix and key matrix. The resultant image is an encrypted image, and it is in unknown form, and also pixel values are approximately equal to the original image.

Fig. 1 Graph of exponential growth equation

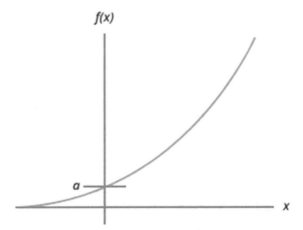

$f(x)$

a

x

In the decryption stage, the XOR function is applied to the encrypted image matrix and key matrix. After that inverse growth equation is applied on that then get the decrypted image which appears same like as the original image. The encryption algorithm consists of the following steps.

Step 1: Take the original image I is of size $M \times N$.

Step 2: To apply the exponential growth equation on the original image and to reduce the intensities of the pixel values in the image and it is represented as J.

Step 3: Convert the image J into the binary image.

Step 4: Generate the binary random image by using a random number generator, and it is represented as K. Here, K is treated as the key image.

Step 5: The exclusive or operation applied block by block between binary image J and K and the resultant matrix represented as L. Here, L is the encrypted image, and it contains pixel values same as the original image. L appears as an unknown image

$$L_i = P[J_i] \oplus P[K_i] \qquad (2)$$

Step 6: In the decryption stage, take L_i as input image of size $M \times N$.

Step 7: Take the random binary image K as key.

Step 8: The exclusive or operation applied block by block between binary image L and K and the resultant matrix represented as J.

$$C_i = P[L_i] \oplus P[K_i] \qquad (3)$$

Step 9: Convert the binary image C into the decimal image.

Step 10: To apply the inverse exponential growth equation on C and to get the decrypted image D. The image D is appeared like as the original image I.

Algorithm 1: Encryption

INPUT: I = M* N (Where M is the number of rows and N is the number of columns).

1. A = double(I)/255;
2. for i=1: M
3.   for j=1:N
4.   Bij = 0.4* 2^(A$_{ij}$);
5.   end for
6. end for
7. C = B*255;
8. D = de2bi(C);
9. K$_{ij}$ = randi([0 1], 8, 8);
10. blocksize = 8;
11.   for r = 1:blocksize: M
12.   for s = 1:blocksize: N
13.   E = D(row:row+blocksize-1,col:col+blocksize-1);
14.   F = E $\oplus$ K
15.   end for
16.   end for
17. Output F is the encrypted image.

Algorithm 2: Decryption

INPUT: F = M * N.

1. blocksize=8;
2. Key matrix K;
3. for r = 1:blocksize: M
4.   for s = 1:blocksize: N
5.   L = F(row:row+blocksize-1,col:col+blocksize-1);
6.   E = L $\oplus$ K
7.   end for
8. end for
9.   D = bi2de(E);
10.   C = double(D)/255;
11.   for i=1: M
12.   for j=1:N
13.   Bij = log$_2$(C$_{ij}$/0.4);
14.   end for
15.   end for
16. P = B*255;
17. P is the decrypted image.

4 Experimental Results and Analysis

4.1 Experimental Results

The experimental results of the proposed scheme are shown in the Fig. 2.

Figure 2a represents the original image, and Fig. 2b represents the modified pixel values of image by using the exponential growth equation. Figure 2c represents the encrypted image, and it contains the pixel values approximately equal to the original image. Figure 2d represents the decrypted image which appears like as

Fig. 2 **a** Original image, **b** low intensity image, **c** encrypted image and **d** decrypted image

the original image. This scheme provides high integrity for security and adding more complexity against attackers.

4.2   Histogram and Entropy Analysis

Figure 3 represents the original image and its histogram as well as encrypted image and its histogram, respectively.

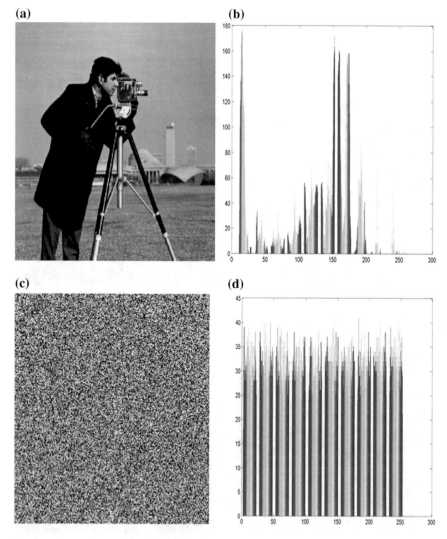

Fig. 3 **a** Original image, **b** original image histogram, **c** encrypted image and **d** encrypted image histogram

Table 1 Entropy of the original image and encrypted image

S.no.	Entropy analysis	Size	Original image	Cipher image
1	Cameraman	256 × 256	7.0097	7.9971
2	Peppers	384 × 512	6.9917	7.9991
3	Football	256 × 320	6.7134	7.9977

Table 1 shows the entropy of the original image and encrypted image, respectively.

The above histograms and entropy show the high integrity of security and adding more complexity against the hacker.

5 Conclusion

An effective image encryption algorithm with exponential growth equation is presented to enhance confusion function. Here, 8 * 8 random image is utilized as a key, and it enhances the diffusion function. To construct the resistance of the cryptosystem to differential attacks, the encoded pixel is evaluated by applying XOR between the image pixel and key image pixel. This demonstrates that it gives greater integrity to security and it includes greater multifaceted nature against programmers. This proposed method has fewer operations; thus, it requires less investment for execution.

References

1. A.AL. Rammahi, Calculus logarithmic function for image encryption. World Acad. Sci. Eng. Technol. Int. J. Math. and Comput. Sciences. **8**(3), 615–618 (2014)
2. M.A. Shareef, H.K. Hoomod, Image encryption using lagrange-least squares interpolation. Int. J. Adv. Comput. Sci. Inf. Technol. (IJACSIT) **2**(4), 111–131 (2013). ISSN 2296–1739
3. Y. Wu, Y. Zhou, J.P. Noonan, K. Panetta, S. Agaian, Image encryption using the sudoku matrix, in *Proceedings SPIE 7708, Mobile Multimedia/Image Processing, Security, and Applications 2010*, 77080P (28 April 2010). https://doi.org/10.1117/12.853197
4. A. Shah, A. Shah, T. Biradar, Image encryption and decryption using Blowfish algorithm in MATLAB. Int. J. Electron. Electr. Comput. Syst. IJEECS **4**(11), 13–18, ISSN 2348–117X (2015)
5. M.A.F. Al-Husainy, A novel encryption method for image security. Int. J. Secur. Appl. **6**(1), 1–8 (2012)
6. A. Al-Rammahi, Encryption image using small order linear systems and repeated modular numbers, in *Proceedings of the World Congress on Engineering 2014*, vol. II (WCE, London, U.K, 2014), 2–4 July 2014
7. N.K.E. Abbadi, A. Mohamad, M. Abdul-Hameed, Image encryption-based on singular value decomposition. J. Comput. Sci. **10**(7), 1222–1230 (2014). ISSN 1549–3636

8. P. Vidhya, Saraswathi and M. Venkatesulu, A block cipher algorithm for multimedia content protection with random substitution using binary tree traversal. J. Comput. Sci. 8(9), 1541–1546 (2012). ISSN 1549–3636

9. V. Patidar, N.K. Pareek, K.K. Sud, A new substitution–diffusion-based image cipher using chaotic standard and logistic maps. Commun. Nonlinear Sci. Numer. Simulat. 14, 3056–3307 (2009)

10. D. Rathore, A. Suryavanshi, A proficient Image Encryption using chaotic map approach. Int. J. Comput. Appl. 134(10), 1–5 (2016). (0975–8887)

11. R. Boriga, A.C. Dăscălescu, I. Priescu, A new hyperchaotic map and its application in an image encryption scheme. Signal Process: Image Commun. 29(8), 887–901 (2014)

12. Guoji Zhang, Qing Liu, A novel image encryption method based on total shuffling scheme. Optics Communications 284, 2775–2780 (2011)

13. O.K. Jasim Mohammad, S. Abbas, El-S.M. El-Horbaty, A-B.M. Salem, A new trend of pseudo random number generation using QKD. Int. J. Comput. Appl. 96(3), 13–17 (2014)

14. Vishal Kapur, Surya Teja. P, Navya. D, Two level image encryption using pseudo random number generators. Int. J. Comput. Appl. 115(12), 1–4 (2015)

IoT-Based Indoor Navigation Wearable System for Blind People

B. Vamsi Krishna and K. Aparna

Abstract This paper presents a wearable audio assistance indoor navigation system for blind people using IoT. In this model, visual markers are used for identification of the points of interest in the environment; in addition, we enhance this location status with information collected in real time by other sensors. The blind users wear glasses built with sensors like RGB camera, ultrasonic, magnetometer, gyroscope, and accelerometer to improve the quantity and quality of the available data. To improve the ultrasonic perception, we use two ultrasonic sensors. An audio assistance system provided for user uses an audio bank with simple known instructions. This makes the user to navigate freely in the prepared environment. The readings of the sensors are uploaded to the cloud, and it will send an alert message to the registered mobile whenever the readings are greater than the threshold value.

Keywords Audio assistance · Visual markers · Ultrasonic perception
IoT · Cloud

1 Introduction

People with visual impairment face many problems while moving in any environment. About 39 million people are blind and 249 billion people have moderate and severe visual impairment. Walking freely is a hectic task for blind people [1]. In previous models, they are using canes for navigation, that is, a very useful instrument for them. But they cannot navigate independently. There are many methods for navigation of blind people. In [2], a method was proposed by using

B. Vamsi Krishna (✉) · K. Aparna
Department of ECE, Malla Reddy Engineering College for Women,
Hyderabad, Telangana, India
e-mail: vamsi.kurnool@gmail.com

K. Aparna
e-mail: kunaaparna@gmail.com

© Springer Nature Singapore Pte Ltd. 2018   413
S. S. Dash et al. (eds.), *Artificial Intelligence and Evolutionary Computations
in Engineering Systems*, Advances in Intelligent Systems and Computing 668,
https://doi.org/10.1007/978-981-10-7868-2_40

ultrasonic sensors in cane to detect the obstacles on the ground. In [3] a dog robot was designed to help in the navigation of blind people. The robot gives indication through vibrations and jerks of the action to be performed. The robot also consists of a camera to capture the images of the obstacles, and they are recognized by using cascade classifiers. Tian et al. [4] and this team proposed a tool that uses the pre-estimated information collected from magnetic sensors, gyro sensors, and camera to teach the blind users. In this work, they used Kalman filters to process the information and for route calculation.

Based on the previous works and on a specific need of indoor navigation, we proposed a low-cost audio assistance indoor navigation system for blind people based on visual markers and ultrasonic obstacle perception using Internet of things. The obstacles are detected by using computer vision and ultrasonic perception. The paper is organized as follows: proposed system in the next section, experiments in Sect. 3 followed by results in Sect. 4, and Sect. 5 consists of conclusion and future scope.

2 Proposed System

In the proposed system, we are using raspberry pi 3 as a minicomputer to store the data collected from the sensors. The collected data is processed and uploaded to the cloud platform from the raspberry pi through Wi-Fi. The block diagram of the proposed system is shown in Fig. 1. In the proposed system, there are two software modules and one hardware module. Computer vision module and ultrasonic perception module are the two software modules. The software modules are used to collect the information from RGB camera and ultrasonic sensors. The data is processed, and it is given as audio instructions to the user through the headset [5]. The hardware module consists of glasses built with the pi camera at the center and two ultrasonic sensors at the two ends.

A low-cost mini PC is used to store the audio database and to run the Haar cascade classifiers and computer vision algorithms. A controller is used to control the sensors like magnetometer, accelerometer, gyro sensor, and ultrasonic sensors. The prototype of the hardware is shown in Fig. 2.

The software modules used in this proposed system are computer vision module and ultrasonic perception module. The proposed software module 1, that is, computer vision module, is used to recognize the markers and some obstacles in the working environment. The architecture of computer vision module is shown in Fig. 3. It consists of the following steps. The first step is improvement of visual search. In this step, the video stabilization is performed. This is performed to correct the motion which is caused by walking of the user.

The second step is image preprocessing. In this step, the brightness and luminous intensity changes can be done using radiometric calibration. Some noise reduction techniques like histogram equalization, morphological transformation, and Gaussian smoothing are used to increase the object perception.

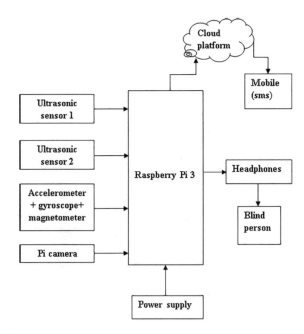

Fig. 1 Block diagram of proposed system

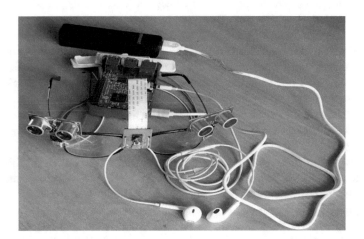

Fig. 2 Prototype of the wearable system

The next step is segmentation. In this step, we are using filters to separate or group the pixels into patterns. In this paper, we are using Sobel [6] filter to emphasize the edges of the objects. The canny [7] filter is used to rebuild the thin and thick edges of the objects.

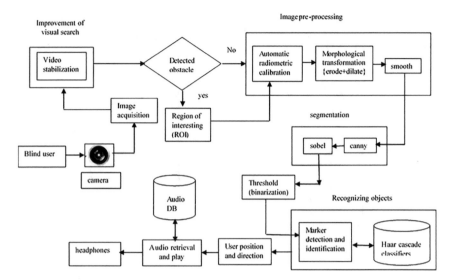

Fig. 3 Proposed computer vision module

The next step is recognizing objects. In this step, we are using Haar cascade classifiers. To create Haar-like classifiers, we require two types of image sets: one is positive image set, which contains the object one wants to map, and the other is negative image set, which contains other objects.

Open CV uses three algorithms to make the Haar-like cascade: object marker, create samples, and train cascade [8]. The object marker creates a text file containing the image name and markers of the selected area. The create samples tool is used to convert this text file into a vector. The create sample tool standardizes the brightness, lighting, and the image size for submission in the classification process. The train cascade learns the pattern submitted in the vector and builds a Haar-like tree. This proposed method uses 2000 high definition images and 18-level tree to reach a good classifier [8, 9].

The indoor environments may be workplaces or homes. The static objects are walls, doors, etc., in homes and workplaces. The dynamic objects are furniture. A map is created by placing the visual markers arranged at static and dynamic object places [10].

The attributes of the markers like IDs, audio information, and relationship with other markers are recorded in the database. We must give a unique ID for each marker. So, the markers can be easily identified by the system.

The ultrasonic perception module collects the information from the ultrasonic sensors and sends warning signals about the dynamic objects and static objects to the user and the computer vision. The ultrasonic perception module is shown in Fig. 4.

During navigation, when the system detects a marker, the information about your position, direction to other markers, arrival time, etc., are updated and

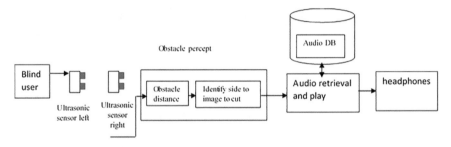

Fig. 4 Proposed ultrasonic perception module

improved. This method is called navigation by proximity based on the visual pattern. This method allows increasing the information about the states of the markers with the use. With an enough learning time, the system tends to stabilization and stops the modification of the data.

2.1 Guidance

To run the experiment, we select six markers (a, b, c, d, e, and f). The main aim is to place the markers at desired places by considering the height of users, ease of reading the markers in the room. Paste the markers at the selected locations on the floor. In this experiment, the markers were placed in the corridor as shown in Fig. 5.

The visual markers are considered as nodes in a bidirectionally connected graph. A weight value is given to each of the markers based on the direction and distance between them. Let us assume the weights of indirectly linked, directly linked, and route junction as 0, 1, and 2, respectively. The adjacent matrix is drawn based on the recorded values. The adjacent matrix is shown in Table 1.

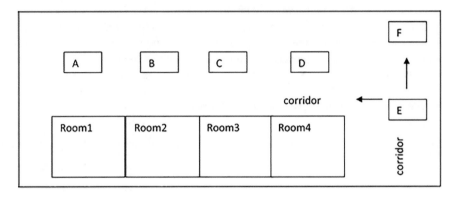

Fig. 5 Layout of free mode navigation

Table 1 Adjacent matrix from experimental results

	A	B	C	D	E	F
A	0	1	0	0	0	0
B	1	0	1	0	0	0
C	0	1	0	1	0	0
D	0	0	1	0	2	0
E	0	0	0	2	0	2
F	0	0	0	0	2	0

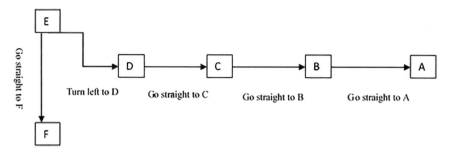

Fig. 6 Path between marker E and marker A with audio assistance

Multiple sound instructions are associated with each marker. Based on the user trajectory, the user receives accurate instruction. For example, consider the navigation from E (route junction) to the marker A, the system will build a trajectory and selects the audio instruction as shown in Fig. 6.

By using the adjacent matrix, the system can offer the best path and will make corrections if the user takes the wrong way. The system uses proximity algorithms to receive the information about the position of the markers.

The input readings from the ultrasonic sensors, accelerometer, gyro sensor, and magnetometer are collected and processed in raspberry pi 3, and these results are uploaded to the cloud platforms like ubidots platform [11]. The ubidots is a cloud platform which should be configured before we upload the readings. After pre-configuration, the readings are accepted to the cloud and showed in the dashboard. The readings are uploaded continuously whenever the user navigates in the prepared environment. This platform has a salient feature of configuring alerts for different events. Whenever the readings are greater than the specific value, we get the SMS alert from the registered mobile number.

3 Experiments

To evaluate the efficiency of the proposed system, ten blind users were tested. We have given the guidance about the system to the users. The profile of the users is as follows: average height of 1.8 m and average steps distance of 48 cm. The average speed walk is 1.76 steps in one second. The path chosen should be made compulsory for all the users with a mandatory return to the starting point. The starting point of the path is at the route junction marker E. It is followed by the corridor, and the destination is at the marker A. The user is guided to return to the initial point to complete the navigation.

In this experiment, we have done five navigations to each user. Hence, the total number of navigations for ten users is 50. We have used six markers for each navigation covering a total distance of 100 m. In each navigation, the time taken for the full path and the delivery time of the audio messages about the obstacles and the target were recorded. The path is also registered in the database.

4 Results

In all navigations, the users encountered 300 markers, out of which only 8 users were failed to detect due to light conditions. The success rate for detecting the markers is about 97.33%, and the success rate for the obstacle perception is about 98.33%. The average time to reach the destination is about 100 s.

To know the overall performance of the system, we ask feedback from the users. The users were asked six questions about the system. The first question is about the audio assistance of the system. The second question was about the localization or to what extent the system is helpful in providing the information about the current position of the markers. The third question was about how freely the user can navigate with the help of the system. The fourth question was about the system's response time. The fifth question was about the system's reliability. The sixth

Table 2 Feedback questions after experiments

S. no	Question	Performance evaluation level				
		Excellent (%)	Very good (%)	Good (%)	Satisfactory (%)	Poor (%)
1	Assistance quality	58	30	5	5	2
2	Localization	40	35	20	5	0
3	Independence	40	30	25	5	0
4	Response time	35	35	25	5	0
5	Reliability	50	35	10	5	0
6	Usability	25	50	15	20	0

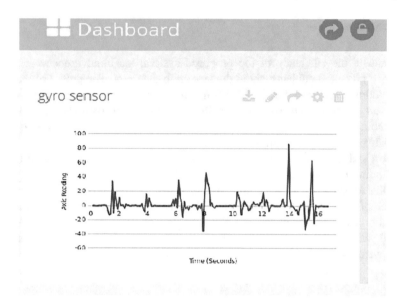

Fig. 7 Gyro sensor readings

question was about the usability of the system. All the above questions are tabulated as shown in Table 2.

Based on the two best results (excellent and very good), we can realize that 88% approved the assistance quality, 85% approved reliability, 75% approved the localization and usability, and 70% approved the independence and response time. These results show where the prototype must be improved to satisfy the users.

The readings of the sensors are uploaded in the cloud platform. The gyro sensor readings in the dashboards are as shown in Fig. 7.

When the axis reading is greater than 40, the event created in the cloud platform automatically sends a message to the registered mobile number.

5 Conclusion and Future Scope

In this paper, we suggested the better method for the blind people in a prepared environment by use of a wearable navigation system, which is inexpensive of cost. This system enables audio assistance when visual marker on the environment and gives alert to the user about obstacles. The cost of the system is low when compared to the conventional system. This system helps in alerting the neighbor of the user by sending a message to the registered mobile whenever the values are exceeding the threshold value. A pair of glasses with a camera and ultrasonic sensors is used as a wearable device for detection and identification of the points. In the future, we can get better results when we use an infrared camera instead of RGB camera.

6 Declaration

We the authors of the manuscript declare that there is no conflict of interest. The research work presented in this paper is tested in the laboratory without the involvement of blind participants, and there is no requirement for informed consent, with reference to this research, since the research work is related to assist devices.

References

1. S. Alghamdi, R. Van Schyndel, I. Khalil, Safe trajectory estimation at a pedestrian crossing to assist visually impaired people, in Engineering in Medicine and Biology Society (EMBC), 2012 Annual International Conference of the IEEE, pp. 5114–5117
2. S. Bharathi, S. Vivek, A. Ramesh, Effective navigation for visually impaired by wearable obstacle avoidance system, in International Conference on Computing, Electronics and Electrical Technologies (ICCEET), 2012
3. K. Xiangxin, L. Mincheol, W. Yuanlong, Vision based guide-dog robot system for visually impaired in urban system, in 13th International Conference on Control, Automation and Systems (ICCAS), 2013
4. Y. Tian, J. Tan, W.R. Hamel, Accurate human navigation using wearable monocular visual and inertial sensors. IEEE Trans. Instrum. Meas. 63(1), 203–213 (2014)
5. J. Coughlan, R. Manduchi, Functional assessment of a camera phone-based way finding system operated by blind and visually impaired users. Int. J. Artif. Intell. Tools 18(3), 379–397 (2009)
6. M.K. Vairalkar, S.U. Nimbhorkar, Edge detection of images using sobel operator. Int. J. Emerg. Technol. Adv. Eng. IJETAE 2(1), 1–3 (2012)
7. R.K., Sidhu. Improved canny detector in various color spaces. in IEEE 3dr International Conference on Reliability, Infocom Technologies and Optimization (ICRITO), India, 2014
8. R. Lienhart, A. Kuranov, V. Pisarevsky, in Empirical Analysis of Detection Cascades of Boosted Classifiers for Rapid Object Detection, 25th Pattern Recognition Symposium (Magdeburg, Germany, 2003)
9. L.G. YI, Hand gesture recognition using kinect. in IEEE 3rd International Conference Software Engineering and Service Science (ICSESS), 2012
10. H. Nishino, A split-marker tracking method based on topological region adjacency & geometrical information for interactive card games. in SIGGRAPH ASIA '09: ACM SIGGRAPH ASIA 2009 Posters, ACM. 2009, 2009
11. S. Saha, A. Majumdar, Temperature and Air Quality Monitoring in Hospitals Helping Quick Evacuation of Patients during fire- A solution with ESP8266 Based Wireless Sensor Network and Cloud Based Dashboard with Real Time Alert System. in 2nd International Conference on Engineering Technology, Science and Management Innovation (ICETSMI-2017)

Automated Raspberry Pi Controlled People Counting System for Pilgrim Crowd Management

P. Satyanarayana, K. Sai Priya, M. V. Sai Chandu and M. Sahithi

Abstract People counting system which can get the exact count of people either indoor or outdoor has a wide range of applications in view of permeating systems. These people counting systems are widely used for promotional evaluation, to get the demand ratio in a retail environment and in the crowd management surveillance. The crowd management, which is one of the emergency handling situations, is ought to be an arduous task at holy places during the religious festivals in India. This strenuous task can be transformed to unchallenging with the proposed system in this paper. This paper presents a model which displays the people count based on the head count of people either entering or leaving the area cropped by the camera of the system. Though there are many algorithms like popular face detection and face recognition algorithms, the developed system opts head detection method to confirm the count. The paper presents a detailed work on detection of persons based on their movement with respect to head using OpenCV implemented on Raspberry Pi which is apt for a set of specific applications.

Keywords Emergency handling situation · People count · Head
OpenCV · Raspberry Pi

P. Satyanarayana · K. Sai Priya · M. V. Sai Chandu (✉) · M. Sahithi
Department of Electronics and Communication Engineering,
K L University, Guntur 522502, India
e-mail: saichand1102516@gmail.com

P. Satyanarayana
e-mail: satece@kluniversity.in

K. Sai Priya
e-mail: kondapaneni.saipriya@gmail.com

M. Sahithi
e-mail: muppasahiti2013@gmail.com

© Springer Nature Singapore Pte Ltd. 2018　　　　　　　　　　　423
S. S. Dash et al. (eds.), *Artificial Intelligence and Evolutionary Computations
in Engineering Systems*, Advances in Intelligent Systems and Computing 668,
https://doi.org/10.1007/978-981-10-7868-2_41

1 Introduction

The reliable counting system which involves segmentation and tracking a moving object for the people flow through an area in video sequences attracts the number of customs for controlling the entry and accessing in the military, building security, visual surveillance, human–computer interaction, and commercial applications. There are certain approaches like rotary bar, light beams, and turnstiles unable to count the multiple people. This has been a formidable task in the research to rectify this troublesome problem, a number of image processing-based robust methods with various kinds of applications were proposed.

Moreover, there are many advantages of using these kinds of people counting systems. Barbara Winkler-Chimbor, Director of Global Education Market Development, Genetec said, "post-secondary education institutions are looking to develop people counting systems to assess the capacity at stadiums to plan evacuation procedures during emergency situations at the events such as fires and weather related emergencies". So, they are advantageous in real-world applications. Based on his thoughts, our idea is to implement this kind of surveillance systems in India where there is a probability of causing emergencies occasionally. Indian temples are such kind of places where crowd management for the temple department becomes arduous task during the occasions like festive eve (Hindu festival events, customs, and rituals). The knowledge of the density of people inside the temple is helpful in controlling the crowd and handling emergency situations.

There is a classification of this people counting systems into obstructive and non-obstructive. This turnstile is a good example of obstructive type systems which clogs the way of moving objects or people passing by its way. Cost-effective nature with low pliability snags these turnstiles from effective implementation. Also, the prime factor consumption of more time when there is a huge crowd is accountable for its effectivity. Non-obstructive systems like infrared beams or heat sensors do not obstruct the path but suffer the counting problem. Computer vision technology-based system is one of the alternatives with added edges like accuracy in counting, non-instructiveness, and inexpensive.

The paper is structured as follows: Sect. 2 gives the brief information about surveys and taxonomies or related work implemented previously, Sect. 3 briefs about the implementation of proposed system, and Sect. 4 gives an overview of results and accuracy of the system.

2 Surveys and Taxonomies

Over the few decades, the papers on counting the people confined to the computer vision environment have grown notably. Taxonomy is defined to get the overview of algorithms, approaches followed in the existing systems; it also helps to group them. Most of the current people counting systems focus on body motion detection, face detection that can produce good results but gets failed in case of crowded situations. These kinds of face detections become inaccurate in crowded areas and mislead the count.

The existing systems follow different methodologies for developing the counting module. The classifier is the common functional unit, which is used in the numerous existing systems. The work by Cai et al. [1] belongs to the map-based method. The classifier is trained offline with the provided snapshots. The positive samples of the heads from the top view and negative samples that reflect the background in an image are fed to the classifier with the SEMB-LBP feature. The input to this classifier is an input frame that is conceptually defined by the ROI, which is simply a prohibition or selected area for detection. The camera is focused based on the ROI. The fact of distinct sizes of heads of human beings allows the ROI search window to detect the targets within that area and inputs the classifier for judgment that sort outs each element in the frame separately as a district target. After the classification, a target tracker is activated by the 2nd input frame of which updates the 1st input frame and concentrates on the movement of the detected target from 1st frame to the 2nd. In this way, a people counting system in crowded scenes by video analyzing works.

Another avant-garde approach followed by Liu et al. [2], which counts the people with respect to their head and shoulder. The classifier in this practice is trained using linear SVM to count the pedestrians at aisles and stairs in the railway stations. The functionality used in the context based on contour which includes head and shoulder might lead to the false information, so as a solution, the approach is around the detection of head shoulder of the person passing by. This needs the camera fixing with some specific angular arrangements due to the involvement of shoulder detection. The approach has two stages: detection stage and tracking stage. In the detection stage, the linear SVM classifier is trained in two rounds for better judgment. Particle filter method, an expeditious algorithm, is adopted to track the detected objects from the classifier. These two stages are directed to work parallelly. The problem faced by authors during implementation is a repetitive counting of a same object, to resolve the raised issue, they computed the distance between the newly detected object with the existing tracked object so that the count gets incremented when the measured distance is greater than the threshold indicating the recognition of the person.

Pang and Hou [3] worked on challenging situation to detect the individuals in an enormous crowd. A neural network approach is opted to classify the individuals so that count increases according to the recognition. In this system, there are two phases: people counting and individual detection. There are two groups of people like stationary and moving. The foreground pixels give the information related to the total number of persons but are not capable of data about their exact location, which is so-called individual detection. The feature detection by Kanade–Lucas–Tomasi method which is a good corner detection method for tracking based on eigenvalues.

Tao Zhao, Ram Nevatia, and Bo Wu's work implying segmentation and challenging cases [4]. The foreground blobs are extracted and matched with the 3D human hypotheses, a model developed using ellipsoids that reflects the shape of the body. These 3D human models capture the gross shape of the human beings. Segmentation and tracking are implemented on the correlated blobs. They are opted Bayesian inference method for interpretation and MCMC Model to compute the MAP for tracking.

When the overhead cameras are used to detect the people, from [5] it is inferred that frame differencing-based background subtraction is the easiest practice for counting moving objects. The bounding box parameters are extracted and used for classifying these objects in motion. Gaussian mixture model [6, 7] is another foreground method for detecting moving objects as human beings. After extracting the foreground objects from the image and then normalizing it to find the edges using Gaussian process regression that yields to people count.

The haar features are extracted from the image by Philip and Georg [9] to count the people. From [8, 9], these extracted haar features are used to classify the human beings using cascade classifier. Human faces are counted by these haar features.

The Statistically Effective Multi-Scale Block Local Binary Pattern (SEMB-LBP) [10] classifier is used to detect the human heads and then classified for counting. The head features are extracted, and based on them, the number of head detections are marked which yields total count. Most recent techniques follow the histogram of oriented gradient-based approach where the histogram of gradient features of foreground objects is extracted and then classified using a classifier [11]. This methodology can be applied for both inclined view of camera and overhead cameras.

For tracking a target, there are more than one option such as Kalman filter [12], Mean shift [13], Camshift [14], optical flow [15], and so on. Kalman filter which is an optimal filter under the assumption of Gaussian noise and linear model but needs correction of the obtained state vector of the target. The mean shift algorithm is a kernel-based function with deterministic characteristics, and the camshaft is an adaptive mean shift algorithm.

3 Proposed System

3.1 Overview of the Proposed System

The proposed system is a non-obstructive computer vision technology overhead-based head detection system. Moving objects are to be only human beings, and they are confirmed based on their detected head. The use of classifier is prohibited in our system leading to instant detection and thereby incrementing the count after confirmation. Note that the suggested system is implemented using OpenCV over mini-computer Raspberry Pi mounted with Pi-camera module. The video surveillance is monitored from a display unit connected to the Raspberry Pi. As the system is connected at the entrance, the head by head gets detected and the

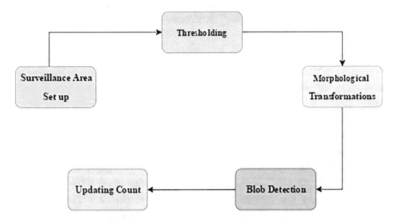

Fig. 1 Block diagram of the people counting system with the stages {*input frame at instance, Thresholding, Morphological Transformations, Blob detection and thereby count incrementing*}

blob is marked; based on the extracted features of the head, the head size parameter determines whether it is a human head or any other moving object. The following is the structure of the proposed system (Fig. 1).

3.2 Surveillance Area Setup

The area of detection should be initialized to start the counting so that whoever passes through the selected area gets detected overhead and the system count gets updated on identification. Two reference lines are set accordingly, so that camera focus is limited to the area. The heads passing by get identified.

3.3 Thresholding

The frame is background subtracted and converted to black and white using inbuilt methods of cv2 library. Then, based on the pixel intensity to normalize the image, a threshold value is set such that for the pixel which is lesser than the threshold is assigned with some value and that pixel with greater value than threshold is assigned with other value.

```
If (pixel value < threshold value)
assign black;
  #else if (pixel value > threshold value)
assign white;
```

3.4 Morphological Transformations

These morphological transformations are the operations performed on the binary images which are used to change the shape of the image to remove the noise with a specific structuring element called kernel. The kernel is user-defined structuring component so that based on the defined pattern, the original shape gets transformed. There are two morphological operations: erosion and dilation. The erosion is like the soil erosion that fades the edges which are broader, and dilation is the one which broadens the narrow parts of the image. In erosion, when the kernel slides through image, the pixel value is considered and marked to be one if and only if all the elements in the kernel are 1s otherwise that particular pixel with uneven binaries gets eroded to 0. At the end of erosion, all the pixels near the boundaries are discarded depending on the size of the kernel simply; the white region gets decreased in the image.

The dilation is quite reciprocal to erosion. In this process, the pixel value is rounded to 1 if any element in the kernel is 1. So, the foreground image size gets enlarged.

The erosion removes the noise and simultaneously shrinks the original image, so to retain the original image, the dilation is implemented soon after erosion.

3.5 Blob Detection

The detection of the blob is the major part of this system so that moving objects and human beings are differentiated based on identified parameters. *The default parameters like minimum threshold, minimum Area filter and maximum area filter, minimum circularity filter are set to identify the oval shaped structures* (Fig. 2).

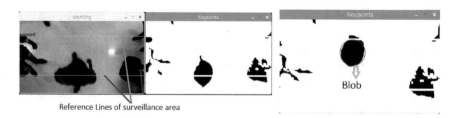

Fig. 2 Left-hand side of figure illustrates the surveillance area created with two imaginary lines and right-hand side of figure showing the detected blob

3.6 Key Points Identification for Tracking and Count Enhancement

The blob gets detected with filters applied, and its movement is tracked with three key points or coordinates of blob both x-coordinate and y-coordinate with respect to the center of the blob. The size of the blob decides the shape of the head, based on the minimum range of the size of the head, the count gets incremented when the blob moves in the direction of Y-axis the y-coordinate is monitored and change is predicted to decide the movement upwards (Inside) or downwards(outside) and displayed simultaneously.

4 Experimental Results and Strategy

The implementation of the proposed system provides a better way of evaluating the chosen algorithm. The camera used for implementation is Raspberry Pi camera module 8Mp V2 which is attached to the upper beam of the room or entrance of surveillance area. The frame size as stated it is fixed by means of reference lines.

The setup of the hardware system is illustrated in Fig. 3.

Figure 4 illustrates the detection of heads and count Updating.

4.1 System Performance and Accuracy

The performance of the system completely depends upon the processor of Raspberry Pi. The accuracy component plans the usage of the system. Since the

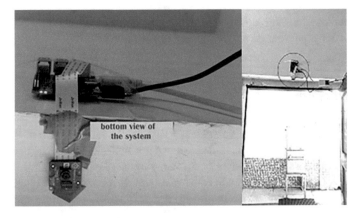

Fig. 3 Figure shows the Raspberry Pi setup affixed at the entrance of a room

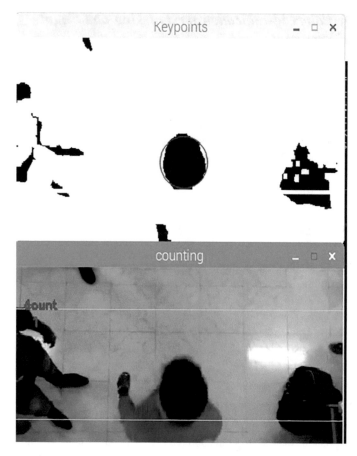

Fig. 4 Count increments simultaneously detecting the blob when the person enters surveillance area

system developed is irrespective of using any type of classifier's we achieved 94% optimum after a number of adulterations.

5 Conclusion

This paper presents the counting system based on the head detection. The minor errors faced during the implementation are that it is unable to count the people at the instance when more crowd enters at a time, similarly, persons with a dark color uniform. With the help of such a head-based counting method, the management can easily identify the count of people inside and take specific measure to direct the huge crowd.

As stated the theme of the project is to highlight the use of blob detection technique for overhead-based people count deduction. The traditional methods of using a specific classifier followed by the training it is a time-consuming process with memory allotment. The proposed system is opposite to the classifier-based mechanisms and achieved 94% accuracy that can be further modified.

Acknowledgements The authors would like to thank the management of K L University and the Honorable delegates of the university for providing the facilities that contributed to the smooth flow of the work within the stipulated time.

References

1. Z. Cai, Z.L. Yu, H. Liu, K. Zhang, in Counting people in crowded scenes by video analyzing, 2014 IEEE 9th Conference on Industrial Electronics and Applications (ICIEA)
2. J. Liu, J. Liu, M. Zhang, in A detection and tracking based method for real-time people counting, Chinese Automation Congress (CAC). IEEE (2013), pp. 470–473
3. G.K.H. Pang and Y.-L. Hou, People counting and human detection in a challenging situation. IEEE Trans. Syst. Man Cybern. 41(1) (2011)
4. T. Zhao, R. Nevatia, B. Wu, Segmentation and tracking of multiple humans in crowded environments. IEEE Trans. Pattern Anal. Mach. Intell. 30(7), 1198–1211 (2008)
5. K.-Y. Yam, W.-C. Siu, N.-F. Law, C.-K. Chan, in Effective bi-directional people flow counting for real time surveillance system, IEEE International Conference on Consumer Electronics (ICCE), vol. 11 (2011), pp. 863–864
6. B. Chan Antoni, V. Nuno, Counting people with low-level features and Bayesian regression. IEEE Trans. Image Process. 21, 2160–2177 (2012)
7. H. Xu, P. Lv, L. Meng, in A people counting system based on head-shoulder detection and tracking in surveillance video, IEEE International Conference on Computer Design and Applications (ICCDA), vol. 1 (2010), pp. V1–394
8. D.S. Chen, S. Z.K. Liu, in Generalized Haar-like features for fast face detection, International Conference on Machine Learning and Cybernetics, vol. 4 (2007), pp. 2131–2135
9. G. Philip, S. Georg, in A two-staged approach to vision-based pedestrian recognition using Haar and HOG features, IEEE Symposium on Intelligent Vehicles (2008), pp. 554–559 2008
10. C. Zeng, H. Ma, in Robust head-shoulder detection by PCA-based multilevel HOG-LBP detector for people counting, IEEE International Conference on Pattern Recognition (2010), pp. 2069–2072
11. B. Li, J. Zhang, Z. Zhang, Y. Xu, in A people counting method based on head detection and tracking, IEEE International Conference on Smart Computing (SMARTCOMP) (2014), pp. 136–141
12. M.S. Grewal, A.P. Andrews, *Kalman Filtering: Theory and Practice Using MATLAB* (Wiley, New York, 2001)
13. Y. Cheng, Mean shift, mode seeking, and clustering. IEEE Trans. Pattern Anal. Mach. Intell. 17(8), 790–799 (1995)
14. G.R. Bradski, Computer vision face tracking for use in a perceptual user interface. Intell. Technol. J. 2
15. B.D. Lucas, T. Kanade, et al., in An iterative image registration technique with an application to stereo vision, Proceedings of 7th International Joint Conference on Artificial Intelligence, vol. 81 (1981), pp. 121–130

A 2 × 2 Block Processing Architecture for a Two-Dimensional FIR Filter Using Scalable Recursive Convolution

Anitha Arumalla and Madhavi Latha Makkena

Abstract In this paper, a 2×2 block processing architecture for a two-dimensional FIR filter is proposed. A 2-parallel scalable recursive convolution algorithm is used to develop $p \times q$ block processing algorithm where p and q are multiples of 2. The algorithm is easily scalable to any order of 2×2 parallel block processing. Computational complexity in terms of multiplications is reduced by a factor of 9/16 at the cost of increased number of additions. However, the adder complexity reduces for window size greater than two times the block size.

Keywords Block processing · 2D FIR filter · Scalable recursive convolution

1 Introduction

Most of the portable communication systems are being developed along with enormous image and video processing applications. Image and video applications demand high computational speed, memory, and hardware resources. Multiplier and adder blocks are the major hardware resources in image filter. Block processing for 1D FIR filters was widely used to improve the processing speed of 1D signal while minimizing the hardware resources required for convolution. Fast FIR algorithm (FFA) is a block processing structure used for parallel computation of filter outputs.

A good effort was directed to the development of efficient filter architectures over the past decade. From the literature, it is evident that different synthesis of filter structure for the realization of same filter can result in different hardware requirements, throughput, accuracy, storage, and computational complexity. Distributed

A. Arumalla (✉)
Velagapudi Ramakrishna Siddhartha Engineering College, Vijayawada, India
e-mail: anithaarumalla83@gmail.com

M. L. Makkena
Jawaharlal Nehru Technological University Hyderabad, Hyderabad, India
e-mail: mlmakkena@yahoo.com

© Springer Nature Singapore Pte Ltd. 2018
S. S. Dash et al. (eds.), *Artificial Intelligence and Evolutionary Computations in Engineering Systems*, Advances in Intelligent Systems and Computing 668, https://doi.org/10.1007/978-981-10-7868-2_42

arithmetic (DA) [1–4] implementation of FIR has gained consideration for better resource utilization of FPGAs.

Systolic 1D and 2D FIR filter decomposition architectures using DA are proposed in [3] for power-/delay-/area-optimized implementation. But the memory requirement for these architectures increases exponentially with increase in order of the filter. In [4], partitioned DA FIR filter implementation is introduced. Partitioned ROM is used to store the combinational values of filter coefficients. This leads to low hardware cost while increasing the quantization error. Canonical signed digit (CSD) encoding is used to generate partial products [5] for programmable multiplierless filter. This reconfigurable architecture eliminated the need for custom design of filter for every new set of CSD filter coefficients. However, the design supports bit widths less than 16, and for higher bit widths, the architecture needs to be redesigned. Filter design using shift and two additions multiplier is presented in [6]. This method is limited to signal processing with bit widths up to 13 bits only.

A parallel computation of the convolution can result in reduced computational complexity [7–10]. Two variations of spectrum inspired fast FIR algorithm (FFA) architectures [7–9] are used for efficient hardware cost and power reduction in 1D FIR filters. These architectures have large logic depth and suitable for parallel computation of 1D filters. A block processing filter for motion estimation is presented in [10]. The algorithm is not regular for scaling the level of parallelism.

In this paper, a 2-parallel scalable recursive convolution algorithm (SRCA) is presented that can be recursively used to achieve high level of parallelism. A 2×2 block processing architecture is proposed using 2-parallel SRCA. The rest of the paper is organized as below. SRCA is detailed in Sect. 2, 2×2 block processing architecture for 2D Filter using SRCA is presented in Sect. 3, implementation details are presented in Sect. 4, and conclusions are drawn in Sect. 5.

2 Scalable Recursive Convolution Algorithm

A 2-parallel direct convolution, of a 2-tap filter, represented in (1) requires 4 multiplications and 2 additions. The direct convolution of (1) can be rewritten as reduced complexity convolution in (2) requiring 3 multiplications and 4 additions; that is, saving of one multiplier is achieved at the cost of 2 adders.

$$
\begin{bmatrix} y_n \\ y_{n+1} \end{bmatrix} = \begin{bmatrix} h_1 & h_0 & 0 \\ 0 & h_1 & h_0 \end{bmatrix} \begin{bmatrix} x_{n-1} \\ x_n \\ x_{n+1} \end{bmatrix}
\tag{1}
$$

$$
\begin{bmatrix} y_n \\ y_{n+1} \end{bmatrix} = \begin{bmatrix} 1 & 1 & 0 \\ 0 & 1 & 1 \end{bmatrix}_{2\times3} \operatorname{diag} \begin{bmatrix} h_1 \\ h_0 + h_1 \\ h_0 \end{bmatrix}_{3\times3} \begin{bmatrix} 1 & -1 & 0 \\ 0 & 1 & 0 \\ 0 & -1 & 1 \end{bmatrix}_{3\times3} \begin{bmatrix} x_{n-1} \\ x_n \\ x_{n+1} \end{bmatrix}_{3\times1}
\tag{2}
$$

Equation (2) can be further decomposed as

$$\begin{bmatrix} y_n \\ y_{n+1} \end{bmatrix} = \begin{bmatrix} 1 & 1 & 0 \\ 0 & 1 & 1 \end{bmatrix}_{2\times 3} \text{diag} \left\{ \begin{bmatrix} 0 & 1 \\ 1 & 1 \\ 1 & 0 \end{bmatrix}_{3\times 2} \begin{bmatrix} h_1 \\ h_0 \end{bmatrix}_{2\times 1} \right\}_{3\times 3} \begin{bmatrix} 1 & -1 & 0 \\ 0 & 1 & 0 \\ 0 & -1 & 1 \end{bmatrix}_{3\times 3} \begin{bmatrix} x_{n-1} \\ x_n \\ x_{n+1} \end{bmatrix}_{3\times 1}$$

(3)

$$Y_2^n = Q_2 \, \text{diag}\{G_2 H_2\} P_2 X_2^n \tag{4}$$

where Y_2^n is a 2×1 matrix of nth 2-parallel outputs, Q_2 is a post-processing matrix for outputs, G_2 is a coefficient pre-processing matrix, H_2 is a coefficient matrix, P_2 is a pre-processing matrix for inputs, and X_2^n is a 3×1 matrix of nth 2-parallel inputs including a sample from past input. G_2 is a transposed Q_2 matrix. To extend 2-parallel implementation of SRCA to multi-tap filters, each element of matrix H_2 will be a vector, that is, h_0, h_1 are decimated by 2 sub-filters. For example, the matrix elements of a 8-tap filter are shown in Eq. (5).

$$\begin{aligned} h_0 &= \{h_6', h_4', h_2', h_0'\} \\ h_1 &= \{h_7', h_5', h_3', h_1'\} \end{aligned} \tag{5}$$

The SRCA is closely inspired by 2-parallel FFA [7] and iterated short convolution algorithm (ISCA) [11]. Two new input samples along with a previous sample can generate two parallel filter output samples. Each sub-filter is a $t/2$ filter, where t is the number of taps in the implemented filter. ISCA is a short convolution algorithm to effectively decompose large convolutions. While ISCA in [11] is used to derive one-dimensional higher-order parallel filters, the transposed ISCA, that is, SRCA can be used to derive two-dimensional parallel filters. The SRCA decomposition is a regular structure that can be easily scaled for the decomposition of two-dimensional parallel filters unlike FFA and ISCA. Delay elements are evaded both in pre-processing and in post-processing matrix to maintain simple decomposition. The hardware implementation of 2-tap SRCA is shown in Fig. 1. The SRCA has smaller logic depth than FFA, thereby increasing the throughput of the FIR filter.

Fig. 1 Hardware implementation of 2-tap SRCA

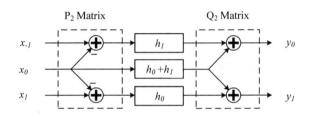

3 A 2 × 2 Block SRCA (BSRCA) Decomposition for 2D Filter

Most of the spatial and spectral image processing techniques use two-dimensional spatial filters. Due to high spatial neighborhood, large computational redundancy can be exploited from block/window processing. Conventional 2D filters fetch data of kernel size for the computation of every output pixel; that is, each input pixel is fetched several times depending upon the kernel size to compute output pixel. Block processing for 2D filter can greatly reduce multiple data fetch of similar data and increase the throughput of the filter. For a two-dimensional $p \times q$ block processing FIR filter, a $p \times q$ kernel is convolved on an $(2p - 1) \times (2q - 1)$ input data.

The 2D non-recursive filter is described using the Eq. (6)

$$y_{uv} = \sum_{i=0}^{p} \sum_{j=0}^{q} x_{u-pv-q} h_{pq} \tag{6}$$

where u, v represent the uth row, vth column respectively of two-dimensional vector, and h_{pq} is the kernel coefficient of pth row & qth column.

A direct 2×2 block processing 2D FIR filter is represented as in (7) that requires 16 multiplications and 12 additions. The input matrix is a 4×4 matrix with nine unique input samples.

$$
\begin{bmatrix} y_{uv} \\ y_{uv+1} \\ y_{u+1v} \\ y_{u+1v+1} \end{bmatrix} =
\begin{bmatrix}
x_{uv} & x_{uv-1} & x_{u-1v} & x_{u-1v-1} \\
x_{uv+1} & x_{uv} & x_{u-1v+1} & x_{u-1v} \\
x_{u+1v} & x_{u+1v-1} & x_{uv} & x_{uv-1} \\
x_{u+1v+1} & x_{u+1v} & x_{uv+1} & x_{uv}
\end{bmatrix}
\begin{bmatrix} h_{00} \\ h_{01} \\ h_{10} \\ h_{11} \end{bmatrix} \tag{7}
$$

Equation (7) can be effectively decomposed by exploiting common terms from four output pixel equations as in (8). Out of 16 product terms for four output equations, there are only nine unique product terms to be generated. These equations can be generalized with SRCA algorithm as recursive one-dimensional SRCA.

$$
\begin{aligned}
y_{uv} &= (x_{u-1v-1} - x_{u-1v} - x_{uv-1} + x_{uv})h_{11} + (x_{u-1v} - x_{uv})(h_{10} + h_{11}) \\
&\quad + \cdots (x_{uv-1} - x_{uv})(h_{01} + h_{11}) + x_{uv}(h_{00} + h_{01} + h_{10} + h_{11}) \\
y_{uv+1} &= (-x_{u-1v} + x_{u-1v+1} + x_{uv} - x_{uv+1})h_{10} + (x_{u-1v} - x_{uv})(h_{10} + h_{11}) \\
&\quad + \cdots (x_{uv+1} - x_{uv})(h_{00} + h_{10}) + x_{uv}(h_{00} + h_{01} + h_{10} + h_{11}) \\
y_{u+1v} &= (-x_{uv-1} + x_{uv} + x_{u+1v-1} - x_{u+1v})h_{01} + (x_{u+1v} - x_{uv})(h_{00} + h_{01}) \\
&\quad + \cdots (x_{uv-1} - x_{uv})(h_{01} + h_{11}) + x_{uv}(h_{00} + h_{01} + h_{10} + h_{11}) \\
y_{uv} &= (x_{uv} - x_{uv+1} - x_{u+1v} + x_{u+1v+1})h_{00} + (x_{u+1v} - x_{uv})(h_{00} + h_{01}) \\
&\quad + \cdots (x_{uv+1} - x_{uv})(h_{00} + h_{10}) + x_{uv}(h_{00} + h_{01} + h_{10} + h_{11})
\end{aligned} \tag{8}
$$

If two-dimensional BSRCA decomposition is applied in (7) and (8), the 2 × 2 direct block convolution (DBC) can be represented in the matrix form as in (9). The resultant decomposition is clearly a Kronecker product of each matrix with its replica excluding input and coefficient matrices.

$$
\begin{bmatrix} y_{uv} \\ y_{uv+1} \\ y_{u+1v} \\ y_{u+1v+1} \end{bmatrix} = \left\{ \begin{bmatrix} 1 & 1 & 0 \\ 0 & 1 & 1 \end{bmatrix} \otimes \begin{bmatrix} 1 & 1 & 0 \\ 0 & 1 & 1 \end{bmatrix} \right\}_{4\times9}
$$

$$
\mathrm{diag} \left\{ \left\{ \begin{bmatrix} 1 & 0 \\ 1 & 1 \\ 0 & 1 \end{bmatrix} \otimes \begin{bmatrix} 1 & 0 \\ 1 & 1 \\ 0 & 1 \end{bmatrix} \right\}_{9\times4} \begin{bmatrix} h_{00} \\ h_{01} \\ h_{10} \\ h_{11} \end{bmatrix}_{4\times1} \cdots \right\}_{9\times9}
$$

$$
\cdots \left\{ \begin{bmatrix} 1 & -1 & 0 \\ 0 & 1 & 0 \\ 0 & -1 & 1 \end{bmatrix} \otimes \begin{bmatrix} 1 & -1 & 0 \\ 0 & 1 & 0 \\ 0 & -1 & 1 \end{bmatrix} \right\}_{9\times9} \begin{bmatrix} x_{u-1v-1} \\ x_{u-1v} \\ x_{u-1v+1} \\ x_{uv-1} \\ x_{uv} \\ x_{uv+1} \\ x_{u+1v-1} \\ x_{u+1v} \\ x_{u+1v+1} \end{bmatrix}_{9\times1}
$$

(9)

$$
Y_4^n = \{Q_2 \otimes Q_2\} \, \mathrm{diag}\{\{G_2 \otimes G_2\}H_4\}\{P_2 \otimes P_2\}X_4^n \tag{10}
$$

where Kronecker product of two matrices X and Y is defined as $X \otimes Y = \begin{bmatrix} x_0 Y & x_1 Y \\ x_2 Y & x_3 Y \end{bmatrix}$ if $X = \begin{bmatrix} x_0 & x_1 \\ x_2 & x_3 \end{bmatrix}$.

The number of multiplications in the 2D BSRCA is given by the size of diagonal matrix, while the number of additions is contributed by post-processing and pre-processing matrices. Since the 2D BSRCA matrix is formed from 2-parallel SRCA, the hardware architecture is also a cascaded structure of 2-parallel SRCA. The post-processing and pre-processing matrix implementations are shown in Figs. 2 and 3, respectively. Similar scaling can be directly applied for higher order of 2 × 2 block decomposition also.

If block size is greater than window size, then each element of coefficient matrix is a 2 × 2 decimated window and each element is a sub-filter of decimated window coefficients.

If computational complexity is estimated in terms of hardware resources required, BSRCA proves to be more efficient if block size is greater than window size. For 2 × 2 DBC on a $p \times q$ kernel, the number of multipliers required is given

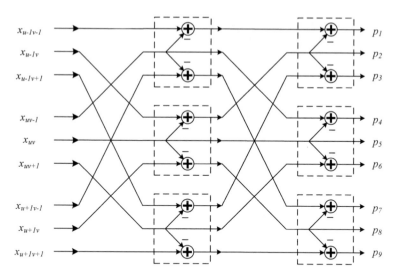

Fig. 2 P-Matrix implementation for 2 × 2 block processing

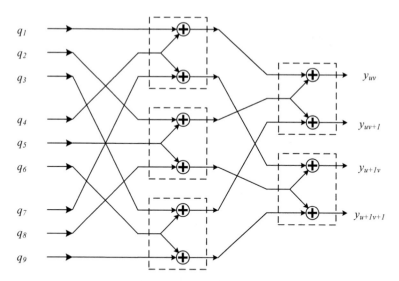

Fig. 3 Q-Matrix implementation for 2 × 2 block processing

by $4pq$ and the number of adders required is $4(pq - 1)$. For a 2×2 BSRCA, the number of multipliers required is given by $\frac{9pq}{4}$ and the number of adders required is $9\left(\frac{pq}{4} - 1\right)$. The reduction in multiplier complexity for BSRCA is twice the order of reduction in multiplier complexity in one-dimensional SRCA. The hardware resource estimate for 2×2 DBC and BSRCA is listed in Table 1 for different window sizes. It can be observed that hardware resource advantage is best achieved for large window sizes.

Table 1 Computational complexity of 2 × 2 DBC and BSRCA for different window sizes

Window size	2 × 2 DBC		2 × 2 BSRCA	
	No. of multipliers	No. of adders	No. of multipliers	No. of adders
2 × 2	16	12	9	22
4 × 4	64	60	36	49
8 × 8	256	252	144	157
12 × 12	576	572	324	337
16 × 16	1024	1020	576	589
20 × 20	1600	1596	900	913
24 × 24	2304	2300	1296	1309
28 × 28	3136	3132	1764	1777
32 × 32	4096	4092	2304	2317

4 Results

The proposed architecture is implemented in verilog HDL and synthesized for Virtex-6 FPGA. The design is validated on ML605 FPGA board. Table 2 shows the comparison results of 2 × 2 DBC and BSRCA for 8-bit and 16-bit data widths. It can be observed that 8-bit implementation of BSRCA occupied 3% less number of slices than 8-bit implementation of DBC, while 16-bit implementation of BSRCA occupied 7% less number of slices than 16-bit implementation of DBC.

Similarly, power results also indicate that 8-bit implementation of BSRCA dissipated 10.3% less power and 16-bit implementation of BSRCA dissipated 37.10% less power than the corresponding implementation of DBC. The reduction in the number of slices is due to computation sharing capability of the algorithm. The reduction in power is due to reduced resources. These architectures are implemented for 2 × 2 window size. From the complexity analysis, it can be observed that larger window size implementations perform better.

Table 2 Comparison results of 2 × 2 DBC and 2 × 2 BSRCA

Convolution algorithm		No. of slices	No. of slice LUTs	No. of slice reg	Power dissipation (W)
2 × 2 direct block convolution	8 bit	164	349	355	0.358
	16 bit	302	741	443	0.69
2 × 2 block SRCA	8 bit	158	487	321	0.321
	16 bit	280	810	377	0.434

5 Conclusion

A 2-parallel SRCA is presented in this paper, and design of two-dimensional FIR filter implementation is proposed using a 2×2 BSRCA. Higher-order filters are decomposed into lower-order sub-filter for reducing the computational complexity of the parallel filters. This algorithm is most suitable for parallel/block implementation of higher-order filters. No additional effort is required in designing the 2D BSRCA filters as it is a scaled version of 1D SRCA filter.

References

1. D. Llamocca, M. Pattichis, G.A. Vera, in A dynamically reconfigurable platform for fixed-point FIR filters, *Proceedings of the International Conference on ReConFigurable Computing and FPGAs (ReConFig'09)*. IEEE (2009), pp. 332–337
2. B.K. Mohanty, P.K. Meher, S.K. Singhal, M.N.S. Swamy, A high performance VLSI architecture for reconfigurable FIR using distributed arithmetic. Integr. VLSI J. **54**, 37–46 (2016)
3. P. Kumar Meher, S. Chandrasekaran, A. Amira, FPGA realization of FIR filters by efficient and flexible systolization using distributed arithmetic. IEEE Trans. Signal Process. **56**, 3009–3017 (2008)
4. C.F. Chen, Implementing FIR filters with distributed arithmetic. IEEE Trans. Acoust. **33**, 1318–1321 (1985)
5. R. Mahesh, A.P. Vinod, New reconfigurable architectures for implementing FIR filters with low complexity. IEEE Trans. Comput. Des. Integr. Circuits Syst. **29**, 275–288 (2010)
6. M. Mottaghi-Kashtiban, A. Jalali, FIR filters involving shifts and only two additions, efficient for short word-length signal processing. Microelectronics J. **49**, 57–63 (2016)
7. K. Parhi, *VLSI Digital Signal Processing Systems: Design and Implementation* (Wiley, New York, 2007)
8. D.A. Parker, K.K. Parhi, Low-area/power parallel FIR digital filter implementations. J. VLSI Signal Process. Syst. Signal Image. Video Technol. **17**, 75–92 (1997)
9. J.G. Chung, K.K. Parhi, Frequency spectrum based low-area low-power parallel FIR filter design. EURASIP J. Adv. Signal Process. **2002**, 944–953 (2002)
10. Y. Naito, T. Miyazaki, I. Kuroda, A fast full-search motion estimation method for programmable processors with a multiply-accumulator, in *Acoustics, Speech, and Signal Processing, ICASSP-96* (1996), pp. 3221–3224
11. C. Cheng, K.K. Parhi, Hardware efficient fast parallel FIR filter structures based on iterated short convolution. IEEE Trans. Circuits Syst. **51**, 1492–1500 (2004)

Implementation and Performance Analysis of SIFT and ASIFT Image Matching Algorithms

Rajasekhar D., Jayachandra Prasad T. and Soundararajan K.

Abstract Image registration helps in aligning corresponding points in images acquired under different conditions. The different contributory conditions can be sensor modality, time, and viewpoint. One critical aspect of image registration is matching corresponding positions in different images to be registered. Image matching signifies the difference between a successful registration or otherwise. With the increasing applications of image processing in solving real-world problem, there is a need to identify and implement effective image matching protocols. In this work, Scale-invariant Feature Transform (SIFT) and Affine—Scale-invariant Feature Transform (ASIFT) have been implemented and analyzed for performance. The performance analysis is done for different images with different attributes like change in tilt and illumination. Apart from calculating the number of matches, the accuracy of the correct matches has been calculated through manual visual inspection. The results demonstrate the efficiency of ASIFT over SIFT in delivering an enhanced performance.

Keywords Image registration · Image matching · SIFT · ASIFT
Tilt · Illumination

Rajasekhar D. (✉)
Department of Electronics and Communication Engineering,
Jawaharlal Nehru Technological University Anantapur,
Ananthapuramu 515002, Andhra Pradesh, India
e-mail: dkethanaraj@gmail.com

Jayachandra Prasad T.
Rajeev Gandhi Memorial College of Engineering and Technology,
Nandyal 518501, Andhra Pradesh, India

Soundararajan K.
Teegala Krishna Reddy Engineering College, Hyderabad
500097, Telangana, India

© Springer Nature Singapore Pte Ltd. 2018   441
S. S. Dash et al. (eds.), *Artificial Intelligence and Evolutionary Computations
in Engineering Systems*, Advances in Intelligent Systems and Computing 668,
https://doi.org/10.1007/978-981-10-7868-2_43

1 Introduction

The two primary tasks in photogrammetry are feature detection and image matching. They have multiple applications in different domains. Their applications extend from a simple photogrammetry task like feature recognition to a complex 3D modeling software. Image matching has remained an active area of research as can be observed from the amount of literature available in this specific area. Researchers are pushed toward the development of more techniques and methods in order to fulfill the demands. Even though the literature presents a wide variety of approaches for image matching, it is interesting to observe that there are no universally acceptable methods.

The approaches under image matching typically can be categorized under two categories of direct and feature-based approaches. Feature-based methods operate by extracting edges, corners, and other features. Local information like correlation is used very sparsely by these methods to match specific image patches [1, 2]. In the case of direct methods, images are iteratively aligned by using all the pixel values. A variant of these approaches called invariant approaches forms invariant descriptors for image matching and indexing. They arrive at these descriptors by employing large amounts of local image data around salient features.

Initial applications of image matching were developed by Schmid and Mohr [3] who demonstrated the application of Gaussian derivatives in developing a descriptor around a Harris corner which is rotationally invariant. Lowe [4] further improved the work by incorporating scale invariance and published his work as scale-invariant feature transform (SIFT) algorithm. He identified distinctive invariant features in an image and used them to identify robust and reliable matching. Reliable image matching and distinctive invariant features are two important attributes of image matching. A number of researchers have come up with a variety of approaches for identifying image descriptors that exhibit invariance under affine transformation [5]. These approaches include Harris corner detector, maximally stable region, and stable local phase structures [6]. Literature also presents different methods that have been evolved for evaluating these approaches. The evaluation has typically been carried in regard to repeatability of interest points [7] and performance of descriptors [8]. Approaches based on invariant features have also been applied successfully in wide range of problems like panoramic image stitching, object recognition, and identification of structure from motion [9, 10]. Lowe [11] was able to effectively detect features with the help of staged filters. Through his work, he demonstrated that it is possible to detect features that are partially invariant to illumination on the one hand and fully invariant to other aspects like rotation, translation, and scaling on the other. Image keys helped in representing local geometric deformation. Keys were generated with the help of multiple orientation planes at multiple scales. Through this process, blurred image gradients have also been represented. Image keys created in the process allowed for local geometric deformations. The allowance for local geometric gradients was accommodated by representing blurred image gradients. This was ensured by proper representation of these gradients in multiple orientation planes at different scales. This ensured the detection of points even in a very busy and noisy background. The keys that are created in course of the filtering step are used to identify the candidate matching objects using the nearest neighborhood indexing method. A low-resolution least square solution is eventually used to verify each match. Mikolajczyk and Schmid [12] proposed a new approach by combining Harris detector with Laplacian-based scales selection approach. This feature detector was able to select the points from a multi-scale transformation. Dorkó and Schmid [13] proposed a method that was capable of extracting specific objects from a set of images. Unlike other approaches which selected features in an entire image, this approach was based on constructing scale-invariant object parts.

Subsequently, many attempts were made to enhance the performance of SIFT and reduce the computational costs associated with it. Ke and Sukthankar [14] presented an "improved" version of SIFT descriptor by using principal component analysis (PCA). In their work, PCA was used to extract the local features unlike SIFT which employed smoothed weighted histograms. The PCA-SIFT enhanced speed of implementation of SIFT matching process by an order of magnitude. Even though it could speed up SIFT, it failed in delivering results that are as distinctive as that of the SIFT. Bay developed SURF [15], SURF stands for speeded-up robust features. This algorithm was intended to improve the capabilities of other feature detectors like SIFT and PCA-SIFT. SURF was able to improve the speed of detection process without compromising on the quality of the detected points. This is a relative improvement compared to PCA-SIFT. Smooth apparent deformations are characteristics of images obtained by camera of any physical object. These deformations are predominant especially when the object exhibits a smooth or piece-wise smooth boundary. Solid object recognition problems lead to computation of local features that are affine invariant. These invariant features can be obtained by normalization approaches. SIFT exhibits invariance to four parameters of an affine transform. ASFIT accounts for all the six parameters including the latitude and longitude angles which are not accounted by SIFT.

In this paper, we have presented a quantitative and qualitative comparison of SIFT and ASIFT image matching approaches for different types of images. These images have different degrees of tilt and illumination. The results demonstrate the suitability of SIFT and ASIFT, especially the improved performance of ASIFT over SIFT. This performance is visible in terms of both the number of matches and the accuracy of correct matches which has been evaluated manually.

2 Scale-invariant Feature Transform (SIFT)

SIFT is one of the pioneering methods used for extracting descriptor points in an image. SIFT operates by extracting points of interest in addition to extracting those features that are present around these points of interest. It is one of the most reliable approaches for matching different viewpoints of a scene or object in an image. SIFT approach is robust in that it is not only invariant to image orientation but also to image scale. Hence, it is capable of providing matching even in the presence of affine distortions. The performance of SIFT has also been found to be better in regard to 3D viewpoint, variation in illumination, and presence of noise. SIFT operates by extracting features from an image under study and storing them. Image matching is then performed by comparing feature of the image with those stored in the database. Euclidean distance is employed for finding candidate matches.

The important steps in the computation of SIFT are as follows:

1. Construction of scale space along with the construction of Gaussian space. This step also incorporates building difference of Gaussian space pyramids.
2. Key point localization—the key point candidates are identified from the extrema in the scale space. The key points are then selected on account of their stability.
3. Histogram of gradient direction computed using a 16×16 window is used to assign orientations to each key point.

SIFT compares two images by deducing one from each other. It employs operations like rotation, translation, and scale change to perform this deduction and eventual comparison. SIFT performs zoom in the scale space and achieves scale invariance, and it's also robust to viewpoint changes. Points of interest at the extrema are detected by SIFT in scale space representation. Smoothing parameter σ is introduced due to the scale space representation. In view of this, the images are smoothed at different scales to obtain;

$$w(\sigma,x,y) = G(\sigma,x,y)*I(x,y)$$

where $I(x, y)$ represents the input image and $G_\sigma(x,y) = G(\sigma,x,y) = \frac{1}{2\pi\sigma^2} e^{-(x^2+y^2)/2\sigma^2}$ represents the two-dimensional Gaussian function with a standard deviation σ. The notation * is to represent the convolution of 2D-space.

3 Affine Scale-invariant Feature Transform (ASIFT)

The main objective of the SIFT approach is combining the simulation and normalization approaches. SIFT detector operates by performing rotation and translation and simulates all the zooms out of search images. As mentioned earlier, SIFT is capable of accommodating only four of the six parameters attributed to affine invariance. ASIFT on the other hand can accommodate all the six properties and is capable of simulating with enough accuracy even the distortion caused by the camera optical axis direction.

ASIFT proceeds by the following steps.

(1) The first step is to transform the image to account for maximum number of affine distortions. These distortions could be introduced by the variations in optical axis rotation introduced by the camera position. Typically, longitude Φ

and the latitude θ influence these distortions to a great extent. Operation $(x, y) \rightarrow u(tx, y)$ can be used to transform an image with rotation followed by tilt 't' in the direction of 'x'. t-sub sampling is employed to simulate the tilt in these images. This is facilitated with the help of an anti-aliasing filter along 'x' direction. This is usually achieved through convolution by a Gaussian having a standard deviation of $c\sqrt{t^2 - 1}$. The value for c is 0.8 to produce small aliasing error.

(2) In the above step, a finite and small number of latitude and longitude angles define operations of tilt and rotation. The sampling steps thus chosen to see that the simulated images remain close to any other possible views that are generated by other values of Φ and θ

(3) The simulated images are then compared by using SIFT.

4 Implementation

In order to localize FPs, we have used multi-scale corner detector. There are two options:

(1) Harris corner detector

$$H_r = \begin{bmatrix} \frac{\partial^2 I}{(\partial x)^2} & \frac{\partial^2 I}{(\partial y)(\partial x)} \\ \frac{\partial^2 I}{(\partial y)(\partial x)} & \frac{\partial^2 I}{(\partial y)^2} \end{bmatrix}, \; H_r F = \det(H_r) - k \cdot (\text{trace}(H_r))^2 . \, (k \in [0.04, 0.06])$$

(2) Harmonic mean $(H_m = \det(H_r)/\text{trace}(H_r))$

Also, localization of FP is calculated with sub-pixel accuracy, with Taylor expansion/approximation. Then with normalized Laplace operator, we find 'characteristic scale' for each FP. If for FP there is no 'characteristic scale,' it's rejected. ('Characteristic scale' is a scale where Laplace operator receives local maximum, and also at this scale, FP is a 'corner.' Characteristic scale gives to descriptor invariance to scale/zoom.) After that characteristic scales, we found for all scales image is blurred with them. All FPs with specific characteristic scale are taken with their neighborhood of size according to their scale in the other image with this specific scale—which is standard deviation of blurring Gaussian. For each FP in those neighborhoods then calculated derivatives and with weighted histogram defined main orientation (MO). FP without MO is rejected. After MO is calculated, SIFT-like descriptor is calculated for each pair of FP and MO.

For matching FPs, the following options implemented; matching can be done by:

(1) Minimal distance
(2) Best match/second match
(3) Best match/second match and threshold distance of matched FP.

5 Results and Discussions

In order to validate the performance of the two different algorithms, different images are matched with different degrees of tilt, scale, etc. The results are described in this section.

Image pair 1 comprises of image having an absolute tilt of 20° compared with a frontal image with a tilt of 0°. It is observed from the results that ASIFT returns 592

Fig. 1 ASIFT matching for image pair 1

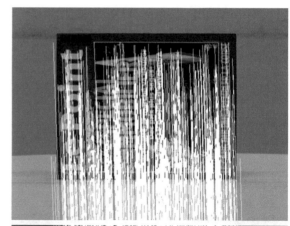

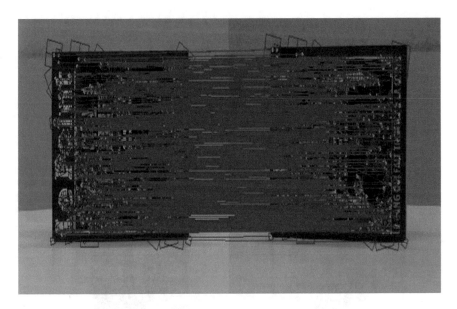

Fig. 2 SIFT matching for image pair 1

matches while SIFT returns 565 matches. The results are illustrated in the Fig. 1 for ASIFT and Fig. 2 for SIFT.

Image pair 2 represents images which are taken having a transitional tilt of $t = 2$ and having tilt angles of $0°$ and $40°$. It is observed from the results that ASIFT returns 748 matches while SIFT returns 5 matches. The results are illustrated in the Fig. 3 for ASIFT and Fig. 4 for SIFT.

Image pair 3 comprises of images with an absolute tilt of $90°$ in relation to the frontal image. It is observed from the results that ASIFT returns 20 matches while SIFT returns 0 matches. The results are illustrated in the Fig. 5 for ASIFT and Fig. 6 for SIFT.

Image pair 4 comprises of images having different degree of illumination; it can be inferred from the results that ASIFT returns 186 matches while SIFT returns 115 matches. The object in the image also has a degree of rotation (Figs. 7 and 8).

Image pair 5 comprises of images having different degree of illumination, one image being 50% brighter than the other image. It can be inferred from the results that ASIFT returns 228 matches while SIFT returns 82 matches (Figs. 9 and 10).

Table 1 presents the summary of the results; it can be observed from the table that ASIFT algorithm is capable of returning more number of image matches than SIFT approach.

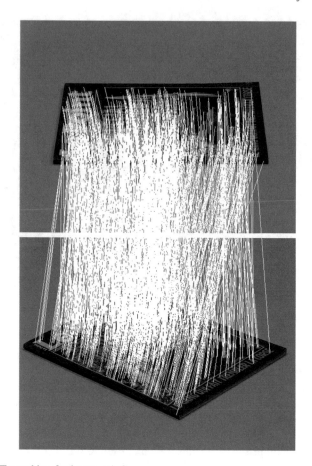

Fig. 3 ASIFT matching for image pair 2

Fig. 4 SIFT matching for image pair 2

Fig. 5 ASIFT matching for image pair 3

Fig. 6 SIFT matching for image pair 3

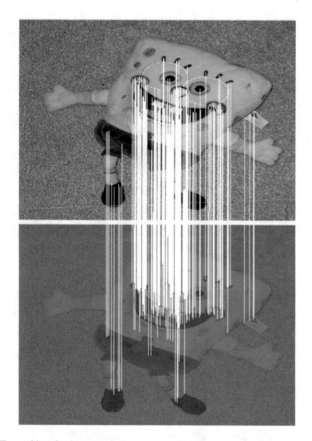

Fig. 7 ASIFT matching for image pair 4

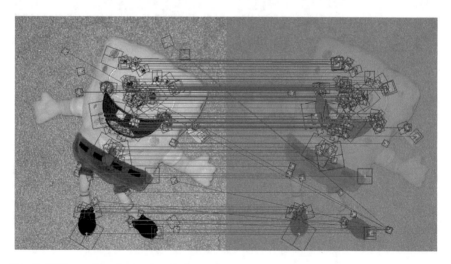

Fig. 8 SIFT matching for image pair 4

Fig. 9 ASIFT matching for image pair 5

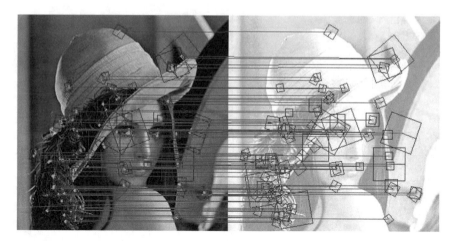

Fig. 10 SIFT matching for image pair 5

Table 1 Comparison
between SIFT and ASIFT

Test image	Feature	Number of matches	
		ASIFT	SIFT
Image pair 1	Tilt	592	565
Image pair 2	Tilt	748	5
Image pair 3	Tilt	20	0
Image pair 4	Illumination	186	115
Image pair 5	Illumination	228	82

6   Conclusions

The accuracy of the detection varies with the number of matches being detected. It can be observed that the ASIFT has an improved accuracy over SIFT when the number of matches delivered by the SIFT algorithm is appreciable. It can be concluded that the ASIFT delivers a better performance for different types of images under test. The performance is visible in terms of both number of matches and accuracy as observed through visual inspection. Even though the computational complexity of ASIFT is more than the SIFT considering the kind of processing power we have today, ASIFT presents a better face for image matching and image registration applications.

References

1. W. Förstner, A feature-based correspondence algorithm for image matching. Int. Arch. Photogramm. Remote Sens. **26**(3), 150–166 (1986)
2. C. Harris, Geometry from visual motion. *Active Vision* (MIT Press, USA, 1992), pp. 263–284
3. C. Schmid, R. Mohr, Local grayvalue invariants for image retrieval. IEEE Trans. Pattern Anal. Mach. Intell. **19**(5), 530–535 (1997)
4. D.G. Lowe, Distinctive image features from scale-invariant keypoints. Int. J. Comput. Vis. **60** (2), 91–110 (2004)
5. M. Brown, D. Lowe, Invariant features from interest point groups, in *Proceedings of the 13th British Machine Vision Conference (BMVC02)*, pp. 253–262 (2002)
6. G. Carneiro, A. Jepson, Multi-scale local phase-based features, in *Proceedings of the International Conference on Computer Vision and Pattern Recognition (CVPR03)* (2003)
7. C. Schmid, R. Mohr, C. Bauckhage, Evaluation of interest point detectors. Int. J. Comput. Vis. **37**(2), 151–172 (2000)
8. K. Mikolajczyk, C. Schmid, A performance evaluation of local descriptors, in *Proceedings of the International Conference on Computer Vision and Pattern Recognition (CVPR03)*, pp. 264–271 (2003)
9. F. Schaffalitzky, A. Zisserman, Multi-view matching for unordered image sets, or "How do I organise my holiday snaps?", in *Proceedings of the 7th European Conference on Computer Vision (ECCV02)*, pp. 414–431 (2002)
10. M. Brown, D. Lowe, Recognising panoramas, in *Proceedings of the 9th International Conference on Computer Vision (ICCV03)*, vol. 2, Nice, pp. 1218–1225, Oct 2003
11. D.G. Lowe, Object recognition from local scale-invariant features, in *The Proceeding of the Seventh IEEE International Conference on Computer Vision*, vol. 2, pp. 1150–1157 (1999)
12. K. Mikolajczyk, C. Schmid, Scale and affine invariant interest point detectors. Int. J. Comput. Vis. **60**(1), 63–86 (2004)
13. G. Dorkó, C. Schmid, Selection of scale-invariant parts for object class recognition, in *Proceedings on Ninth IEEE International Conference on Computer Vision*, Nice, France, pp. 13–16, Oct 2003
14. Y. Ke, R. Sukthankar, PCA-SIFT: A more distinctive representation for localimage descriptors, in *IEEE Computer Society Conference on Computer Vision and Pattern Recognition (CVPR'04)*, vol. 2, pp. 506–513 (2004)
15. H. Bay, A. Ess, T. Tuytelaars, L. Van Gool, *Speeded-up Robust Features (SURF)*. Computer Vision ECCV, Lecture Notes in Computer Science, vol. 3951, pp. 404–417 (2006)

Exudates Detection in Diabetic Retinopathy Images Using Possibilistic C-Means Clustering Algorithm with Induced Spatial Constraint

R. Ravindraiah and S. Chandra Mohan Reddy

Abstract Diabetic Retinopathy (DR) is a progressive ailment and is a prime cause of vision loss in diabetic patients. Presences of exudates in retinal fundus obstruct the vision. Exudate detection using image processing techniques are helpful for the ophthalmologists in prior screening and diagnosis. It helps to estimate the severity level of the condition, so that it drives the patient to control the pathos from progression. This paper exhibits a framework for the exudates detection in non-dilated fundus images using Possibilistic C means clustering algorithm with induced spatial constraint. The performance of the method is evaluated using statistical analysis.

Keywords Fuzzy C-means clustering (FCM) method · Possibilistic C-means clustering (PCM) method · Diabetic retinopathy (DR) · Exudates

1 Introduction

International Diabetes Federation has estimated that approximately 387 million people over the globe have Diabetes mellitus (DM) in 2014 and expected the rise to +205 million by 2035. Indian has a prevalence rate of 8.93% [1]. WHO predicted that Indian contribute 19% of the world's diabetic population and may increase to 80 million by 2030. It is estimated that 39/186 million new diabetic cases are expected from China and India [2]. Out of the many complications of DM, DR is the most common disease. Patients with DR are 25 times more prone to vision loss

R. Ravindraiah (✉)
Department of ECE, JNTUA, Anantapuramu, Andhra Pradesh, India
e-mail: ravindra.ranga@gmail.com

S. Chandra Mohan Reddy
Department of ECE, JNTUA College of Engineering, Pulivendula,
Kadapa District, Andhra Pradesh, India
e-mail: email2cmr@gmail.com

© Springer Nature Singapore Pte Ltd. 2018
S. S. Dash et al. (eds.), *Artificial Intelligence and Evolutionary Computations in Engineering Systems*, Advances in Intelligent Systems and Computing 668,
https://doi.org/10.1007/978-981-10-7868-2_44

than a nondiabetic individual. Proper maintenance with good screening by surgical interventions will delay sight threatening situation of DR.

DR in general characterized by microvascular variations in the retina. The capillary walls swell, fragile and then exudes blood and protein-based particles into retinal fundus, which causes retinal ischemia. The fine granules of ballooned out exudes are called microaneurysms leads to blurriness or floating spots in the eye. This condition is treated as Non-Proliferative DR (NPDR) [3]. Later, the blood vessels get blocked permanently and therefore tend to Proliferative DR (PDR) in which, new blood vessels proliferate so as to nourish the retina. But these newly grown blood vessels are weak and fragile and hence further complicate the situation [4].

It is a laborious process and requires a great deal of time for an ophthalmologist to analyze and diagnose the DR pathos. The manual analysis includes chemical dilation and even injection of organic compounds like fluorescein will result in unnecessary side effects on patients. The existing literature includes implementation of multiple image processing methods on DR images. The most notable works are as follows: Sophark et al. [5] employed FCM for coarse segmentation by allowing features like intensity, standard deviation on intensity, hue, and a number of edge pixels followed with morphological techniques for fine-tuning but their approach resulted in false alarms as some non-exudates are merged to lesion portion. Osareh et al. [6] applied local enhancement on I channel in HIS conversion and used FCM for coarse and fine segmentation. Later, they employed artificial neural networks for classification of exudates from non-exudates. But their method suffers when an evenly illuminated fundus images are utilized. Osareh et al. [7, 8] employed local contrast stretch and color normalization as preprocessing steps and done coarse segmentation using FCM algorithm. Further used MLP neural networks for classification of lesions from non-lesions. Kande et al. [9] utilized spatially weighted FCM method for exudates detection. Jaafar et al. [10] utilized morphological methods to refine the coarse segmentation results by an adaptive thresholding method. Their method suffers from the detection of non-exudate pixels as exudates as they have similar features as that of exudate pixels.

Welfer et al. [11] used the top-hat morphological method on L channel in Luv color. Walter et al. [12] utilized morphological methods to eliminate blood vessels and employed morphological reconstruction for exudate detection. Sopharak et al. [13, 14], used morphological closing to eliminate blood capillaries. Then used entropy feature to remove the optic disc, and extracted hard exudates using morphological dilation. Youssef et al. [15], used features like high intensity, yellow color, and high contrast to eliminate optic disc and attained an initial estimate of exudates. Then used morphological methods to get final estimate. Sanchez et al. [16, 17], used regional maxima algorithm and Hough transform to localize and segment the optic disc. Then segmented exudates using morphological opening and closing. Kumari et al. [18] and Ravishankar et al. [19] used morphological operation open and close to remove blood vessels and used open and close operation for exudate detection. Agurto et al. [20] proposed AM–FM texture feature extraction method for lesion classification. Sinthanayothin [21–23] proposed an automated detection of lesions in DR images using a window-based recursive region growing segmentation algorithm. Their system is good to detect hard exudates than faint exudates. Phillips et al. [24] proposed an automated DR detection using local and global thresholding algorithms for detection of hard and soft exudates.

2 Proposed Method

Clustering methods became more attractive due to their great capability to spot hypervolume clusters. They are predominantly used in wide applications (in particular pattern recognition and image segmentation) as they aim at making a crisp decision on the desired results [25]. Traditional FCM (proposed by Dunn [26] and modified by Bezdek [27]) and its derivatives suffer from the problem of outliers as the probabilistic constraint used in them assign equal probability forcefully to each pixel. Also, they are quite sensitive to noise. The problem improves if the dataset consists of more number of clusters. Krishnapuram et al. [25] proposed Possibilistic C-means clustering (PCM), used a modified objective function with relaxed membership function using a probabilistic constraint, η_i. Individual regions are connected set of pixels that are extremely interrelated with their neighbors. This characteristic is more helpful in the extrication of the noisy cluster from the desired clusters. This tends to modify the cost function and membership function used traditional clustering methods.

2.1 Spatial Possibilistic C-Means Clustering Algorithm (SPCM)

The proposed SPCM algorithm aims at minimizing the cost function with induced weighted membership function are given by:

$$J_{\text{SPCM}} = \sum_{i=1}^{c} \sum_{j=1}^{n} M_{ij}^{m} d_{ij}^{2} + \sum_{i=1}^{c} \eta_i \sum_{j=1}^{n} (1 - M_{ij})^{m} \qquad (1)$$

where d_{ij}^2 is the Euclidean distance metric between the data poits, $x_j = (x_1, x_2, x_3 \ldots x_n)$ and cluster centroids, $c_i = (c_1, c_2, c_3 \ldots c_c)$ are cluster centers. The green component of original input is used as it is associated more contrast.

The value η_i is probabilistic constraint which governs the 3 dB point of the membership value for a distance metric used. It is given by the following expression,

$$\eta_i = \frac{\sum_{j=1}^{n} M_{ij}^{m} d_{ij}^{2}}{\sum_{j=1}^{n} M_{ij}^{m}} \qquad (2)$$

where $j = 1, 2, 3 \dots n$.

The membership function used to converges the cost function is given by,

$$M_{ij} = \frac{1}{1 + \left(\frac{d_{ij}^2}{\eta_i}\right)^{\frac{-1}{m-1}}} \tag{3}$$

The membership function is weighted using the spatial function h_{ij} as

$$M_{ij}' = \frac{M_{ij}^m h_{ij}}{\sum_{k=1}^{c} M_{kj}^m h_{kj}} \tag{4}$$

where the spatial function h_{ij}, is the given by,

$$h_{ij} = \sum_{k \in NB_{x_j}} \sum_{l \in NB_{x_j}} \frac{1}{2\pi\sigma^2} e^{\frac{k^2 + l^2}{2\sigma^2}} M_{k,l} \tag{5}$$

h_{ij}, the spatial function which operates on x_j and its neighbors defined in a subimage subportion NB_{x_j} (of size 5×5 is chosen). This signifies probability of degree of belongingness of the pixel x_j to the ith cluster in of the image with x_j as it center.

2.2 Requirements

The algorithm is simulated in PC equipped with intel i5 5200 2.20 GHz CPU and Raspberry Pi 3 Model B with 1 GB RAM, quad core 1.2 Ghz processor kit. On PC the simulation of the proposed algorithm is done using Matlab R2015a and Octave 4.0 software. Octave 4.0 application with Image processing toolbox is installed in raspberry to run the proposed algorithm. The implementation of the proposed work on Raspberry Pi processor is shown in Fig. 1.

3 Results

The green component of the fundus images is applied to the proposed algorithm as they provide good contrast compared to red and blue components. The median filter is used to suppress the noise. The preprocessed image is then applied to the proposed algorithm. To avoid false alarms the optic disc was separated selectively. A total of 25 images taken from public databases *Diaretdb0* [28], *Diaretdb1* [29], *American Society of Retina Specialists* [30], and one private database provided by

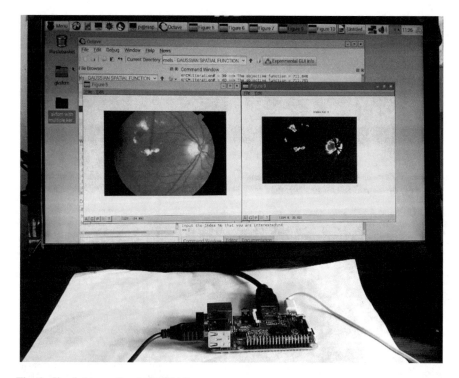

Fig. 1 Simulation on Raspberry Pi kit

Suthrama Eye Hospital, Madanapalle, A.P, India are used in this work. Sensitivity and Specificity are calculated using the clinician ground truth images.

Figure set 2 shows the typical DR images taken from different image databases used as input images in this work and Figure set 3 shows the respective output images of Fig. 2 obtained from the proposed method. The quantitative evaluation of the proposed method with the existing methods is shown in Table 1. The performance of the proposed method is statistically evaluated using clinician ground truth images. Sensitivity and specificity which are defined as (Fig. 3),

$$\text{Sensitivity} = \frac{TP}{TP + FN} \tag{6}$$

$$\text{Specificity} = \frac{TN}{TN + FP} \tag{7}$$

- TP (True Positive) : Correctly classified Exudate pixels
- TN (True Negative): Correctly classified non-exudate pixels

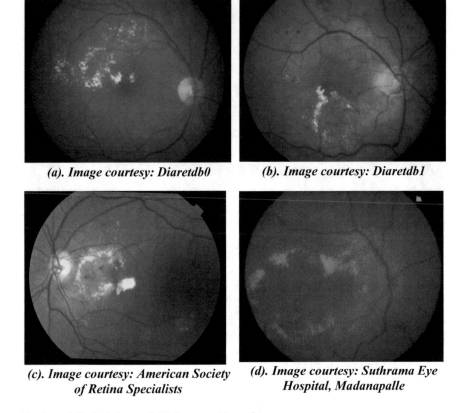

(a). Image courtesy: Diaretdb0   *(b). Image courtesy: Diaretdb1*

(c). Image courtesy: American Society   *(d). Image courtesy: Suthrama Eye*
of Retina Specialists   *Hospital, Madanapalle*

Fig. 2 **a–d** Typical abnormal DR images with exudates

Table 1 Quantitative evaluation

Author	Sensitivity	Specificity
Sopharpak et al.	87.28	99.24
Giribabu Kande et al.	86	98
Welfer et al.	70.48	98.84
Our proposed work	89.86	99.44

- FP (False Positive): Pixels correctly classified as Exudates but are non-exudates in ground truth image
- FN (False Negative): Pixels correctly classified as non-exudates but are Exudates in ground truth image.

The algorithm was simulated on PC equipped with intel i5 5200 CPU and Raspberri Pi 3 Model B processor. The elapsed time taken is given in Table 2.

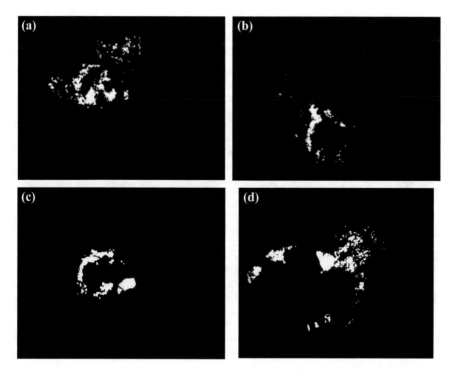

Fig. 3 a–d Respective output images for Fig. 2a–d

Table 2 Elapsed time in seconds for different processors

SPCM method		Figure 3a	Figure 3b	Figure 3c	Figure 3d
PC (Core Intel i5)	MATLAB R2015a	129.42	128.70	20.10	31.67
	OCTAVE 4.0	136.011	119.38	66.2146	39.60
Raspberry Pi 3 Model B	OCTAVE 4.0	720.8	6370.6	336.5	206.7

4 Conclusion

The presented results show that the proposed method good in extracting the lesions in fundus DR images and statistically good compared with the existing methods (Table 2). Further, this method can be made robust with the usage of kernel metrics.

Acknowledgements The authors thank **Dr. N. Praveen Kanth Reddy, M.B.B.S, M.S (Ophthalmology)**, Suthrama Eye hospital, Madanapalle, for his help and valuable suggestions in carrying out this work successful.

References

1. International Diabetes Federation. Diabetes atlas. 6th edn. http://www.idf.org/diabetesatlas
2. J.C. Chan et al., Diabetes in Asia: epidemiology, risk factors, and pathophysiology. JAMA **301**, 2129–2140 (2009)
3. A. Osareh et al., Automated identification of diabetic retinal exudates in digital colour images. Br. J. Ophthalmol. **87**(10), 1220–1223 (2003)
4. C. Sinthanayothin, et.al. Automated localization of the optic disc, fovea and retinal blood vessels from digital colour fundus images. Br. J. Ophthalmol. **83**, 231–238
5. A. Sopharak, B. Uyyanonvara, S. Barman, Automatic exudate detection from non-dilated diabetic retinopathy retinal images using fuzzy c-means clustering, 2148–216. doi:https://doi.org/10.3390/s90302148
6. A. Osareh, M. Mirmehdi, B. Thomas, R. Markham, Automated identification of diabetic retinal exudates in digital colour images. Br. J. Ophthalmology **87**(10), 1220–1223
7. A. Osareh, B. Shadgar, R. Markham, A computational–intelligence-based approach for detection of exudates in Diabetic Retinopathy Images. IEEE Trans. Inf. Technol. Biomed. **13**, 535–545 (2009)
8. A. Osareh, M. Mirmehdi, B. Thomas, R. Markham, *Automatic recognition of exudative maculopathy using fuzzy c-means clustering and neural networks* (Med. Image Understand. Anal., BMVA Press, UK, 2001), pp. 49–52
9. G.B. Kandel, P. Venkata Subbaiah, T. Satya Savithri, in segmentation of exudates and optic disc in retinal images, Sixth Indian Conference on Computer Vision, Graphics & Image Processing (2008)
10. H.F. Jaafar, A.K. Nandi, W. Al-Nuaimy, in Detection of exudates in retinal images using a pure splitting technique, 32nd annual international conference of the IEEE EMBS, Buenos Aires, Argentina, 2010
11. D. Welfer, J. Scharcanski, D. Ruschel Marinho, A coarse-to-fine strategy for automatically detecting exudates in color eye fundus images. Comput. Med. Imaging Graphics **34**, 228–235 (2010)
12. T. Walter, J.-C. Klein, P. Massin, A. Erginay, A contribution of image processing to the diagnosis of diabetic retinopathy—detection of exudates in color fundus images of the human retina. Trans. Med. Imaging **21**(10), 1236–1243 (2002)
13. A. Sopharak et al., Automatic detection of diabetic retinopathy exudates from nondilated retinal images using mathematical morphology methods. Comput. Med. Imaging Graph **32**, 720–727 (2008)
14. A. Sopharak et al., Fine exudate detection using morphological reconstruction enhancement. Appl. Biomed. Eng **1**, 45–50 (2010)
15. D. Youssef, et al., in New feature-based detection of blood vessels and exudates in color fundus images, *2010 2nd international conference on Image Processing theory tools and applications (IPTA)* (2010), pp. 294–299
16. C.I. Sanchez et al., Retinal image analysis based on mixture models to detect hard exudates. Med. Image Anal. **13**, 650–658 (2009)
17. C.I. Sanchez, et al., Improving hard exudate detection in retinal images through a combination of local and contextual information, 2010 IEEE International Symposium on Biomedical Imaging: from Nano to macro (2010), pp. 5–8
18. V. Kumari, N. Suriyanarayanan, Feature extraction for early detection of diabetic retinopathy, in International Conference on Recent Trends in Information, Telecommunication and Computing (2010), pp. 359–361
19. S. Ravishankar, et al. Automated feature extraction for early detection of diabetic retinopathy in fundus images. ISSN: 978-1-4244-3991-1/09/$25.00 ©2009 IEEE, pp. 210–217
20. C. Agurto et al., Multiscale AM-FM methods for diabetic retinopathy lesion detection. IEEE Trans. Med. Imaging **29**(2), 502–512 (2010). https://doi.org/10.1109/TMI.2009.2037146

21. C. Sinthanayothin, Image analysis for automatic diagnosis of diabetic Retinopathy. J. Med. Sci. **35**(5), 1491–1501 (2011)
22. C. Sinthanayothin, J.F. Boyce, T.H. Williamson, H.L. Cook, E. Mensah, S. Lal, Automated detection of diabetic retinopathy on digital fundus image. J. Diabetic Med. **19**, 105–112 (2002)
23. Boyce Sinthanayothin, Williamson and Cook, "Automated Detection of DiabeticRetinopathy on Digital Fundus Image". Diabet. Med. **19**, 105–112 (2002)
24. R.P. Phillips, J. Forrester, P. Sharp, Automated detection and quantification of retinal exudates. Graefe Arch. Clin. Exper. Ophthalmology **231**, 90–94 (1993)
25. R. Krishnapuram, J.M. Kellar, A possibilistic approach to clustering. IEEE Trans. Fuzzy Syst. **1**(2)
26. J.C. Dunn, A fuzzy relative of the ISODATA process and its use in detecting compact well-separated clusters. J. Cybern. **3**(3), 32–57
27. J.C. Bezdek, *Pattern recognition with fuzzy objective function algorithms* (Plenum, Newyork, 1981)
28. T. Kauppi, V. Kalesnykiene, J.K. Kamarainen, L. Lensu, I. Sorri, H. Uusitalo H. Kalviainen, J. Pietila, in Diaretdb0: evaluation database and methodology for diabetic retinopathy algorithms. Technical report Lappeenranta University of Technology Finland 2006
29. T. Kauppi, V. Kalesnykiene, J.-K. Kamarainen, L. Lensu, I. Sorri, A. Raninen, R. Voutilainen, H. Uusitalo, H. Kälviäinen, J. Pietilä, in DIARETDB1 diabetic retinopathy database and evaluation protocol, Proceedings of the 11th Conference on Medical Image Understanding and Analysis (Aberystwyth, Wales, 2007)
30. ASRS: Advocating for You and Your Patients in 2017, in *Retina Times*, , Vol. 35, No. 1, Issue 68, Spring 2017

Hand Gesture Recognition for Disaster Management Applications

R. Sumalatha, D. Rajasekhar and R. Vara Prasada Rao

Abstract This paper discusses about hand gesture recognition for disaster management like mining. In mining places, so many workers are always in danger situation. If the mine falls down suddenly then working people would be under the coal. If anyone is alive, but not able be to communicate with the rescue team in this situation working people may be dead. For saving their lives the proposed system provides facility to the working people send the information through hand gesture to rescue office. The proposed system is designed to recognize hand gestures using simple color feature extraction algorithm. After extracting the color feature of the test and training images we calculated the Euclidian distance to recognize the hand gesture. Finally, the recognized gesture image corresponding text can be converted into speech.

Keywords Disaster management · Hand gesture recognition · Mining
RGB to HSV · Text to speech · Euclidian distance

1 Introduction

In India's mining sector there has been a dead on every third day. Over past 5 years especially in coal mining, the accidents frequency has increased in the recent years. India has a major number of deaths when the roof and sides of underground mines suddenly fall. Therefore, in order to improve the security of miners in mine disaster,

---

Please note that the AISC Editorial assumes that all authors have used the western naming convention, with given names preceding surnames. This determines the structure of the names in the running heads and the author index.

R. Sumalatha (✉) · D. Rajasekhar
G. Pullaiah College of Engineering and Technology, Kurnool, India
e-mail: amrutha_suma@yahoo.com

R. Vara Prasada Rao
Rajiv Gandhi Memorial College of Engineering and Technology, Kurnool, India

S. S. Dash et al. (eds.), *Artificial Intelligence and Evolutionary Computations in Engineering Systems*, Advances in Intelligent Systems and Computing 668,
https://doi.org/10.1007/978-981-10-7868-2_45

this system provides communication between the alive underground miner and rescue team by sending the information through hand gesture only. In general, hand gestures are used when the person is unable to speak with other persons.

To provide communication between miner and rescue team persons so many monitoring systems are available in the market. But these systems are costly and persons must wear gloves. In gloves-based techniques sensors are used to measure the hand position and users press a button in one system. In general, injured miners are so weak that they are unable to press any button and most of them they cannot walk and speak. Hence, the main objective of the proposed system recognizes the hand gesture of the miner and corresponding predefined text can be converted into speech.

In general, hand gestures are used for communication between two persons. Physical movement of eyes and hands are used for gesture recognition. In this paper, hand gestures are used for mining people to communicate with the rescue team. Each hand gesture is having one information about the current position of a miner in mine. Based on the gesture, text and speech rescue team will take the decision about the current situation.

This paper is organized as follows Sect. 2 discusses about related work Sect. 3 describe about proposed system Sect. 4 gives the experimental results of the proposed system. Finally, Sect. 5 discuss about the conclusion and future work of the proposed system.

2 Related Work

Over the past decades, so many systems are implemented to save the miners life, but those systems are costly and complicated. Each and every miner must wear some gloves-or sensor-based systems. Rautary and Agarwal [1] proposed a low-cost interface device for interacting with objects in the virtual environment using hand gestures. Liao et al. [2] described vision-based dynamic hand gesture recognition system. Chen et al. [3] discussed about hand gesture recognition system using kinetic depth data. Chen et al. [4] presented a novel real-time work for hand gesture recognition. Ghosh and Ari [5] proposed an accelerometer-based smart ring and a similarity matching-based extensible hand gesture recognition algorithm. Xie et al. [6] discuss about a matching-based extensible hand gesture recognition.

3 Proposed System

The proposed system is as shown in Fig. 1 and it is implemented in offline mode. The proposed gesture recognition block diagram addressed in this paper involves following steps.

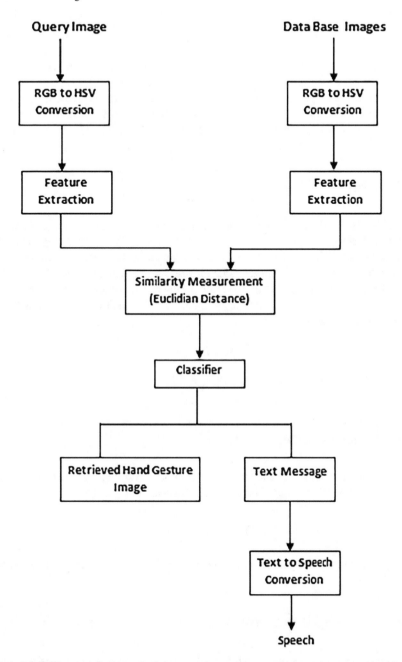

Fig. 1 Block diagram of proposed work

3.1   Preprocessing

In general, captured query image and database images are in RGB color map. In order to get the one-dimensional feature vector the RGB image can be converted into HSV color map. The RGB color map gives the information about red, green, and blue components. The HSV color map describes colors similarly how the human eye tends to perceive colors based on thecolor, vibrancy, and brightness.

3.2   Feature Extraction

In this paper, HSV colormap is used to extract the color feature vector. HSV stands for Hue, Saturation, and value. Hue is used to identify dominant color, Saturation gives the purity of color and value represents the intensity of color. Quantization of HSV colormap is used to extract color feature. Hue is quantized into 9 bins and is among [0, 8]. Saturation is quantized into 3 bins and is among [0, 2]. Value is also quantized into 3 bins and is among [0, 2]. Hence, the final one-dimensional color feature vector is coded as

$$C = 9 * H + 3 * S + V$$

where C is an integer value between 0 and 255.

3.3   Similarity Measurement

In this paper, Euclidian distance is used to find out the similarity between the query image and database images. The Euclidian distance is given below

$$ED = \sqrt{(X - Y)^2}$$

where X is feature vector of query image and Y is feature vector of database images.

3.4   Text to Speech Conversion

In the proposed method, hand gesture recognition can be classified by using simple classifier. After classifier as soon as the hand gesture is recognized the corresponding message can be converted into speech.

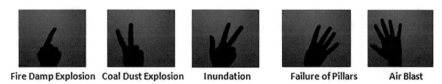

Fire Damp Explosion Coal Dust Explosion Inundation Failure of Pillars Air Blast

Fig. 2 Predefined gestures and corresponding text messages

Stepwise algorithm is given below

1. Test or Query image and training images can be converted into HSV model.
2. Extract the feature of test and training images based on the color.
3. For similarity measurement calculate the Euclidian distance between the test image and each training image.
4. Sort all the distances in ascending order and find the minimum distance of training image.
5. After classification, the recognized hand gesture will be displayed and corresponding predefined text can be converted into speech.
6. Plot the power spectral density for speech.

Figure 2 shows the predefined gestures and corresponding text messages.

4 Experimental Results

In the proposed system, five predefined hand gestures are used. Each hand gesture consists of 20 images with different angles totally 100 images are used for database, i.e., training images. Each test image feature vector is compared with 100 training images feature vector. Figure 3 shows the test images, power spectral density of predefined speech, and text.

5 Conclusion

The hand gesture recognition system is very easy and simple technique to communicate with the other people. Here, we presented a system for saving of miner's lives. So miners as early as possible send the current position in the mine to the rescue team. This system will be implementing in real time by using GPS system for tracking of a miner.

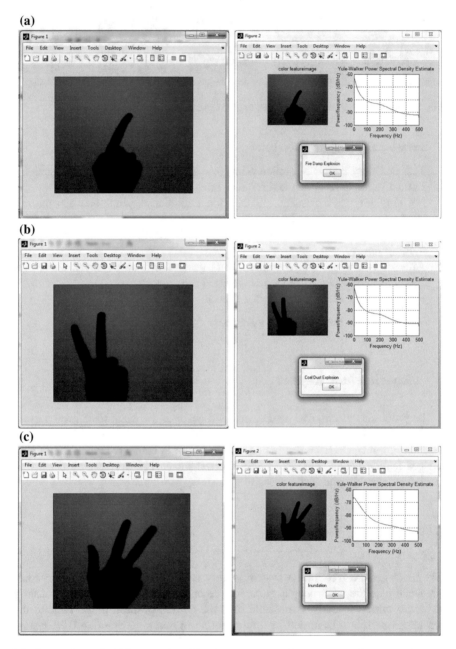

Fig. 3 **a–e** shows the different retrieved hand gesture recognitions corresponding text and power spectral density

(d)

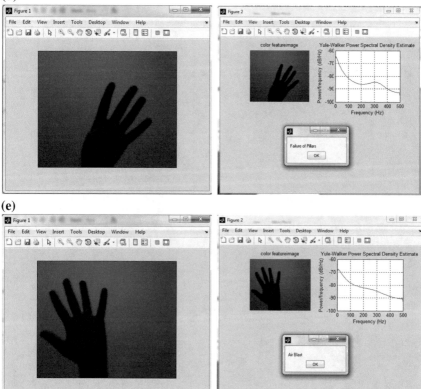

(e)

Fig. 3 (continued)

References

1. S.S. Rautary, A. Agarwal, Real time hand gesture recognition system for dynamic applications. Int. J. UbiComp (IJU) **3**(1) (2012)
2. C.-J. Liao, S.-F. Su, M.-C. Chen, in Vision-based hand gesture recognition system for a dynamic and complicated environment, IEEE International Conference on Systems, Man, and Cybernetics (2015), pp. 2891–2895
3. L. Chen, F. Wang, H. Deng, K. Ji, in A survey on hand gesture recognition, International Conference on Computer Sciences and Applications (2013), pp. 313–316
4. Z. Chen, J.-T. Kim, J. Liang, J. Zhang, *Real-time Hand Gesture Recognition Using Finger Segmentation* (Hindawi Publishing Corporation, Cairo, 2014)
5. D.K. Ghosh, S. Ari, in On an algorithm for Vision-based hand gesture recognition, *SIViP*, 30 June 2015
6. R. Xie, X. Sun, X. Cia, J.C. Cao, Similarity matching based extensible hand gesture recognition. J. Lteax Class Files **11**(4) (2012)

Edge Feature Extraction of X-Ray Images Based on Simplified Gabor Wavelet Transform

P. M. K. Prasad and Y. Raghavender Rao

Abstract Noadays, X-ray images play vital role in the field of medicine. The edge feature extraction technique which helps medical practitioners to detect the minute factures as it may not be possible by the necked eye. The edge feature extraction methods deals with detection of edges of an image. There are several wavelet-based methods to extract the features of an edge. The conventional Gabor Wavelet is a complex wavelet which is extensively used for edge feature extraction. This conventional Gabor wavelet may not be useful for real-time applications due to high-computational complexity. So simplified Gabor wavelet transform is used instead of conventional Gabor wavelet. It is a very efficient technique to detect location and orientation of edges. The proposed simplified Gabor wavelet transform performs well for the edge feature extraction in comparison with conventional Gabor wavelet, orthogonal and biorthogonal wavelets.

Keywords Conventional Gabor wavelets · Simplified Gabor wavelets
Edge feature extraction · X-ray image

1 Introduction

The digital images are extensively used by the medical practitioners for disease diagnosis. There are various medical imaging modalities, such as MRI, CT, ultrasound, and X-ray. The X-ray is the oldest and frequently used devices, as they are painless and economical [1]. The medical practitioners identify the fractures of X-ray images based on their past experience. But it may be difficult to detect the minute fractures of an image by the necked eye. So, the X-ray images need further processing to detect the minute fractures. Therefore, edge feature extraction

P. M. K. Prasad (✉)
Department of E.C.E, G.V.P. College of Engineering for Women, Visakhapatnam, India
e-mail: pmkp70@gmail.com

Y. Raghavender Rao
Department of E.C.E, JNTUH College of Engineering, Jagityal, Telengana, India

© Springer Nature Singapore Pte Ltd. 2018   473
S. S. Dash et al. (eds.), *Artificial Intelligence and Evolutionary Computations
in Engineering Systems*, Advances in Intelligent Systems and Computing 668,
https://doi.org/10.1007/978-981-10-7868-2_46

techniques are used to detect the minute fractures. These techniques provide better treatment to the patients. The edges of an image are detected by the edge feature extraction and it is also known as edge detection. Edge feature extraction deals with the extraction of the edges of an image by detecting the discontinuities in the gray level [2]. It is the basic step for image analysis and processing [3]. The noise corrupts the original image during the acquisition of image. This effects the blurring of the image and therefore, it is difficult to identify the features of an image [4]. The classical derivative operators, such as Roberts, Prewitt, Sobel, and Laplacian of Gaussian can be used for edge feature extraction [5]. These operators are not reliable as it detects false information about the edges of an image. Both the edges as well as noise of an image contain high-frequency components. So edge detection is not a simple task [6]. Therefore, wavelet transform technique is used to detect the edges of an image efficiently due to its multiresolution property. The quality of edge detection depends upon the wavelet properties. The Gabor wavelet is complex wavelet which is very efficient when compared to other wavelets, such as Haar, daubechies, coiflet, symlets, and biorthogonal wavelet transforms for the edge detection of X-ray bone images. The conventional Gabor wavelet can be used for edge detection but it has high-computational complexity. So simplified Gabor wavelet is used instead of conventional Gabor wavelet. This paper mainly discusses about the application of simplified Gabor wavelet (SGW) to detect edges of an image.

2   Wavelet Transform Theory

The wavelet transform is used to localize in both the time and frequency domain. It is characterized in terms of scaling and wavelet function. The scaling and wavelet functions are like twin functions. There are two operations performed by wavelet transform like dilation and translation. It is used in many applications, such as edge detection, image denoising, and compression. The wavelet transform decomposes the image into four subband, such as approximation, horizontal, vertical, and diagonal bands. There are various orthogonal wavelets, such as Haar, daubechies, coiflet, symlets can be used for edge detection. In orthogonal wavelets, scaling function, and wavelet function are orthogonal to each other. These orthogonal wavelets are not reliable for edge detection and also sensitive to noise. So biorthogonal wavelets are used instead of orthogonal wavelets as they are flexible to design [7]. The complex wavelet transforms such as Gabor wavelet is more preferable compared to orthogonal wavelets and biorthogonal wavelet for edge detection due to more visual quality.

2.1 Complex Wavelet Transform

The complex-valued extension of discrete wavelet transforms results complex wavelet transform. There are two advantages of complex wavelet transform when compared with discrete wavelet transform. They are high degree of shift invariance and directional selectivity [8–10]. It also has multiresolution property. Complex Wavelets Transforms decomposes the real or complex signals into real and imaginary parts in transform domain. It computes amplitude and phase information. The amplitude of these coefficients describes the strength of the singularity while the phase indicates the location of singularity. One of the very important complex transform is Gabor wavelet.

3 Gabor Wavelet Transform

The Gabor wavelet is a linear filter used for edge feature extraction. The Gabor wavelet is complex wavelet. The arbitrary signal can be represented as a sum of translated and modulated Gaussian functions. Gabor wavelet transform has both multiresolution and multi-orientation properties. Generally, Gabor filter bank is created with various scales and rotations. The signal is convolved with these filters and it is known as Gabor space [11]. It has so many applications in image processing, such as edge feature extraction, iris recognition, and finger print recognition [12, 13].

The one-dimensional Gabor wavelet is obtained by Gaussian kernel function modulated by a sinusoidal plane wave [14]. The one-dimensional Gabor wavelet is represented by the Eq. 1.

$$S(x) = \frac{1}{2\pi\sigma^2} \exp\left[-\frac{x^2}{2\sigma^2}\right] \sin(\omega x), \tag{1}$$

where σ is the standard deviation and ω is the spatial frequency. The imaginary part of one-dimensional Gabor wavelet are quantized to a different level since there is continuous values of the imaginary part. Gabor wavelet is the only wavelet with orientation selectivity [15, 16]. It chooses higher frequency information hence the edge is maximized. There are four orientations are selected to get the edges for a rectangular domain that is along 0°, 90°,270°, and 360°.

3.1 Conventional Gabor Wavelet

The two-dimensional Gabor wavelet is also known as conventional Gabor wavelet. The two-dimensional Gabor wavelet is represented by the Eq. 2. This Gabor filter is basically a Gaussian with variances along x-and y-axis [17]. The Gabor wavelet combines both time and frequency information [18, 19].

$$G(x, y) = \left[-\frac{x^2 + y^2}{2\sigma^2} \right] \exp\left[j\omega(x \cos \theta + y \sin \theta) \right] \qquad (2)$$

The different Gabor kernels will be obtained by choosing various center frequencies and orientations which in turn extracts the image features. The edge features are extracted by convolving the input image $P(x, y)$ and $G(x, y)$ which is given by the Eq. 3

$$P'(x, y) = G(x, y) \otimes P(x, y) \qquad (3)$$

The two-dimensional convolution is represented by $\otimes$. The Gabor wavelet strongly detects the edges if the direction of the edge is perpendicular to the vector $(\omega \cos \theta, \omega \sin \theta)$ [20].

4 Proposed Simplified Gabor Wavelet Transform

The conventional Gabor wavelet uses both real and imaginary parts of a function. Hence, it takes more computation time and therefore it is not used in real-time applications. The Simplified Gabor Wavelet transform is an efficient technique used for edge detection [21]. It is used in real-time applications as it has less computational complexity as it considers only imaginary part [8]. The imaginary part of the Simplified Gabor Wavelet is represented by the Eq. 4

$$S(x, y) = \exp\left[-\frac{x^2 + y^2}{2\sigma^2} \right] \sin[\omega(x \cos \theta + y \sin \theta)] \qquad (4)$$

4.1 Quantization

The quantization level of Simplified Gabor Wavelet transform is determined in such a way that one of the quantization levels is set to zero. The number of quantization levels for both positive and negative values are same and it is represented by n_1 as the imaginary part of Gabor function is antisymmetrical. Suppose 'A' is the largest magnitude of the SGW is "A", then quantization levels are given by the Eqs. 5 and 6.

$$q_+(k) = \frac{A}{2n_1 + 1} \cdot 2k \qquad (5)$$

$$q_-(k) = -\frac{A}{2n_1 + 1} \cdot 2k \tag{6}$$

where $k = 1, 2, \dots n_1$

5 Performance Factors of Edge Feature Extraction

5.1 Peak Signal-to-Noise Ratio

Peak Signal-to-Noise Ratio (PSNR) is used for comparing the two images quantitatively. It is defined in terms of the mean squared error (MSE), where $P(x, y)$ is the original image and $p'(x, y)$ is the edge detected image. The size of the image is "mn". The MSE is given by the Eq. 7.

$$\text{MSE} = \frac{1}{mn} \sum_{y=1}^{m} \sum_{x=1}^{n} (p(x, y) - p'(x, y))^2, \tag{7}$$

The PSNR is given by the Eq. 8

$$\text{PSNR} = 20 \, \log_{10} \frac{255}{\sqrt{\text{MSE}}} \tag{8}$$

When the pixels are represented using eight bits per sample, then the maximum possible pixel value of the image is equal 2^8 which is equal to 255.

5.2 Ratio of Edge Pixels to Size of Image (REPS)

A pixel can be considered as an edge pixel, if its value is more than the threshold. The 0.09–1 threshold is considered for edge feature extraction as it gives more number of edge pixels. The REPS gives edge information available in the image.

$$\text{REPS } (\%) = \frac{\text{No. of Edge Pixels}}{\text{Size of an image}} \times 100 \tag{9}$$

5.3 Computation Time

The computation time is the time taken to execute the program.

6 Procedure for Simplified Gabor Wavelet-Based Edge Feature Extraction

The procedural steps for SGW-based edge feature extraction are as follows. Figure 1 shows the block diagram for SWG wavelet-based edge feature extraction. Figure 2 shows the results for various wavelet-based edge feature extraction methods.

1. Read the X-ray bone image. In this paper, left-hand X-ray bone image with a minute fracture is considered. The X-ray bone image collected from the hospital with the size 447 × 274 (JPEG) is Considered.
2. Consider different orientations and central frequencies.

$$\theta = \frac{\pi}{4}, \frac{3\pi}{4} \text{ degrees} \quad \text{and} \quad \omega = 0.3, 0.625 \text{ rad/s}$$

3. Substitute these values in SGW kernel.
4. Apply three level quantization.
5. Convolve SGW $G(x, y)$ with the above orientations and central frequencies with the image $P(x, y)$ to obtain the edge features of an image.
6. Apply five level quantization.
7. Repeat the step 4 to obtain the edge features of an image.
8. Compute PSNR, REPS, and computation time.

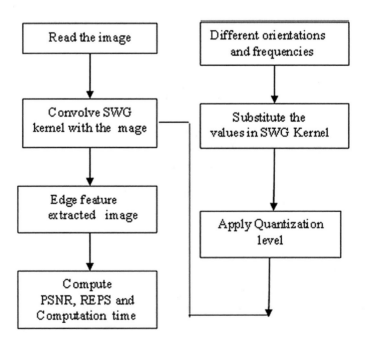

Fig. 1 Block diagram for simplified Gabor wavelet-based edge feature extraction

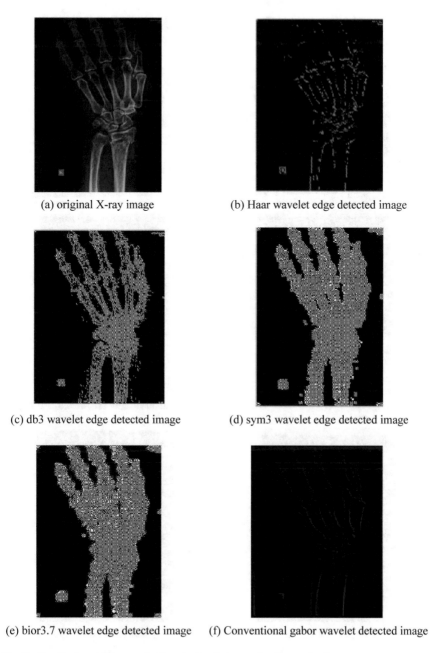

(a) original X-ray image   (b) Haar wavelet edge detected image

(c) db3 wavelet edge detected image   (d) sym3 wavelet edge detected image

(e) bior3.7 wavelet edge detected image   (f) Conventional gabor wavelet detected image

Fig. 2 Results for various wavelet-based edge feature extraction methods

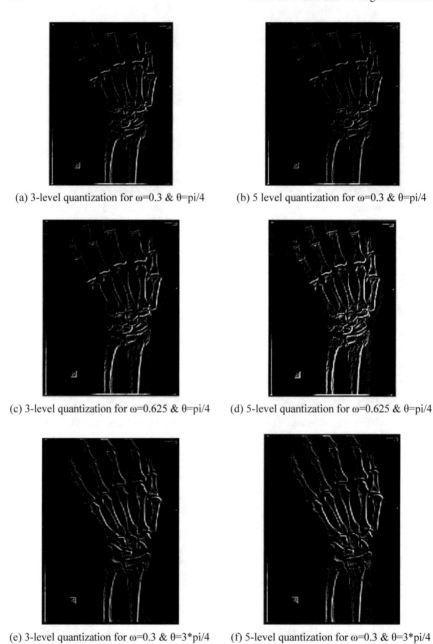

(a) 3-level quantization for ω=0.3 & θ=pi/4   (b) 5 level quantization for ω=0.3 & θ=pi/4

(c) 3-level quantization for ω=0.625 & θ=pi/4   (d) 5-level quantization for ω=0.625 & θ=pi/4

(e) 3-level quantization for ω=0.3 & θ=3*pi/4   (f) 5-level quantization for ω=0.3 & θ=3*pi/4

Fig. 3 Results for SGW-based edge feature extraction methods

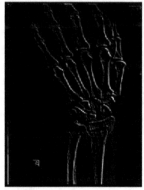 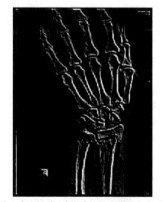

(g) 3-level quantization for ω=0.625 & θ=3*pi/4   (h) 5-level quantization for ω=0.625 & θ=3*pi/4

Fig. 3 (continued)

7 Results and Discussion

The results are simulated with the help of MATLAB 7.9. The various wavelets, such as Haar, db3, sym3, bior3.7, conventional Gabor wavelet, and SGW are applied to the X-ray image to extract the edge features of an image. Figure 3 shows the results for Simplified Gabor wavelet based edge feature extraction methods.

Table 1 shows the PSNR values for different orientations and various quantization levels of SGW-based edge feature extraction. It was observed that PSNR value is 26.7178 dB which is high for five level quantization with the orientation $3\pi/4$ degrees for the central frequency 0.625 rad/s. Table 2 shows the PSNR values and computation time of various wavelet edge detection methods. The SGW has highest PSNR value compared to other wavelets. It was observed that computation time is 23.3377 s which is lowest compared to another wavelet-based edge feature extraction methods. Table 3 shows REPS values for various wavelet-based edge feature extraction methods. The REPS value of SGW-based edge detection is 7.75

Table 1 PSNR values for different orientation with various quantization levels for SGW-based edge feature extraction

S. No	Level of quantization	Orientation (°)	Central frequency (rad/s)	PSNR (dB)
1	3 level	$\pi/4$	0.3	26.7050
2	3 level	$\pi/4$	0.625	26.7056
3	3 level	$3\pi/4$	0.3	26.7069
4	3 level	$3\pi/4$	0.625	26.7077
5	5 level	$\pi/4$	0.3	26.7054
6	5 level	$\pi/4$	0.625	26.7059
7	5 level	$3\pi/4$	0.3	26.7100
8	5 level	$3\pi/4$	0.625	26.7178

Table 2 PSNR values and computation time of various wavelet-based feature extraction methods

S. No	Wavelet edge feature extraction method	PSNR (dB)	Computation time (s)
1	Haar wavelet	21.11	29.8348
2	Daubechis (db3)	21.11	29.6359
3	Symlets (sym3)	19.73	29.4345
4	Biorthogonal (bior 3.7)	21.52	28.6513
5	Conventional Gabor wavelet	23.66	35.3377
6	Simplified Gabor wavelet	26.71	23.3377

Table 3 REPS values of various wavelet-based edge feature extraction methods

S. No	Wavelet edge feature extraction method	No. of edge pixels	REPS (%)
1	Haar	8524	6.95
2	Daubechies (db3)	8942	7.3
3	Symlet (Sym3)	8942	7.3
4	Biorthogonal (bior 3.7)	9283	7.57
5	Conventional Gabor wavelet	9389	7.66
6	Simplified Gabor wavelet	9493	7.75

and it is high when compared to all the other wavelets. It has more edge pixels and so it can detect the small edges. The visual quality of SGW-based edge feature extraction is superior, high PSNR, less-computation time, and high REPS when compared to other various wavelet-based edge feature extraction methods. The simplified Gabor wavelet is superior method for edge feature extraction of X-ray images.

8 Conclusion

In this paper, SGW transform is proposed for the efficient edge feature extraction of X-ray images. The SGW is the modified version of conventional Gabor wavelet. It gives information about both edge orientation and edge location. The various quantization levels and orientations are applied to improve the quality of the edge feature extraction. The visual quality of SGW-based edge feature extraction is high when compared with the conventional Gabor wavelet and other wavelet-based edge feature extraction methods. The SGW has high PSNR, high REPS, and fast compared with other wavelet methods. The results show that SGW is superior and efficient edge feature extraction method to extract the edges of X-ray images, so that it can be used to detect the minute fractures effectively.

References

1. S.K. Mahendran, S. Santhosh Baboo, in *Enhanced Automatic X-Ray Bone Image Segmentation using wavelets and Morphological Operators*. 2011 International Conference on Information and Electronics Engineering IPCSIT, vol. 6 (IACSIT Press, Singapore, 2011)
2. P.S. Addison, in *The Illustrated Wavelet Transform Handbook*. IOP publishing Ltd 2002
3. Lei Lizhen, Discussion of digital image edge detection method. Mapping Aviso **3**, 40–42 (2006)
4. F.y. Cui, L.J. Zou, B. Song, in *Edge Feature Extraction Based on Digital Image Processing Techniques*. Proceedings of the IEEE, International Conference on Automation and Logistics, (Qingdao, China Sept 2008), pp. 2320–2324
5. R.C. Gonzalez, R.E. Woods, in *Digital Image Processing*, 2nd edn. (Prentice Hall Publications, 1992)
6. R.C. Gonzalez, R.E. Woods, in *Digital Image Processing Using MATLAB*, 2nd edn. (Paretice Hall India Limited, 1992)
7. P.M.K. Prasad, G. Sasi Bhushana Rao, M.N.V.S.S. Kumar, in *Analysis of X-Ray images using MRA based Biorthogonal wavelets for detection of minute fractures*. International Conference on Science, Engineering and Management Research (ICSEMR 2014) IEEE
8. W. Jiang, K.M. Lam, T.Z. Shen, Efficient edge detection using simplified gabor wavelets. IEEE Trans. Syst. Man Cybern. Part: Cybern. **39**(4), Aug 2009
9. L. Debnath, *Wavelet Transforms & Their Application*. (Birkhauser Boston, 2002)
10. K.P. Soman, K. I. Ramachandran, *Insight Into wavelets from theory to Practice*, 2nd edn. (Prentice Hall of India, 2008)
11. H. Cheng, N. Zheng, C. Sun, in *Boosted Gabor Features Applied to Vehicle Detection*. 18th International (A.3) Conference on Pattern Recognition, pp. 662–666, 2006
12. R. Mehrotra, K.R. Namuduri, N. Ranganthan, Gabor filter-based edge detection. Pattern Recogn. **25**, 1479–1494 (1992)
13. W.P. Choi, S.H. Tse, K.W. Wong, K.M. Lam, Simplified gabor wavelets for human face recognition. Pattern Recogn. **41**, 1186–1199 (2008)
14. S.M. Salve, Mammographic image classification using gabor wavelet. Int. Res. J. Eng. Technol. **03**(03), pp. 202–207, 03 Mar 2016, ISSN: 2395-0072
15. C. Liu, H. Wechsler, Gabor feature based classification using the enhanced fisher linear discriminant model for face recognition. IEEE Trans. Image Process. **11**, 467–476 (2002)
16. H. Cheng, N. Zheng, C. Sun, *Boosted Gabor Features Applied to Vehicle Detection*. Proceedings 18th International Conference Pattern Recognition, vol. 1, pp. 662–666, 2006
17. W. Jiang, T.Z. Shen, J. Zhang, Y.Hu, X.Y. Wang, in *Gabor Wavelets for Image Processing*. IEEE International Colloquium on Computing, Communication, Control, and Management, vol. 1, pp.110–114, 2008
18. X. Xie, K.M. Lam, Gabor-based kernel PCA with doubly nonlinear mapping for face recognition with a single face image. IEEE Trans. Image Process. **15**(9), 2481–2492 (2006)
19. R. Mehrotra, K.R. Namuduri, N. Ranganthan, Gabor filter-based edge detection. Pattern Recognit. **25**(12), 1479–1494 (1992)
20. Y.P. Guan, Automatic extraction of lips based on multi-scale wavelet edge detection. Comput. Vis. **2**(1), 23–33, (2008)
21. W. Jiang, K.M. Lam, T.Z. Shen, in *Edge Detection Using Simplified Gabor Wavelets*. IEEE International Conference Neural Networks & Signal Processing Zhenjiang, China, June 8, 2008

An Enhanced Viola–Jones Face Detection Method with Skin Mapping & Segmentation

P. Satyanarayana, N. Jaya Devi, S. K. Sri Hasitha and M. Sesha Sai

Abstract People count is one of the tedious tasks if it involves human intervene. As an alternative, automation of the process is done using image processing as it provides high accuracy. The face detector is based on a state-of-the-art cascade of boosted integral feature. This algorithm overcomes the drawbacks of Viola–Jones algorithm and relatively faster than Hog detection. The proposed system makes use of Haar-cascade classifiers to detect the face in real time, and this alone is error prone. These errors are eliminated by eliminating the non-face regions detected by mapping the skin colour of image with detected regions. Two different threshold methods are used to reduce the search space. Algorithm is evaluated on many databases. Experiments on these databases reveal good performance of the proposed algorithm.

Keywords Haar-cascade classifier · Hog · Viola–Jones · Non-face region Segmentation · Skin mapping

1 Introduction

As people count has vast applications, this has turned into an area of research and it still remains as a challenging task. People in image are present at different positions having different skin tones. So as to get accurate people count, every person in the

P. Satyanarayana · N. Jaya Devi · S. K. Sri Hasitha (✉) · M. Sesha Sai
Department of Electronics & Communication Engineering, K L University,
Guntur, India
e-mail: hasithasrimathkandada@gmail.com

P. Satyanarayana
e-mail: satece@kluniversity.in

N. Jaya Devi
e-mail: n.jaya0007@gmail.com

M. Sesha Sai
e-mail: seshasamajeti.1996@gmail.com

© Springer Nature Singapore Pte Ltd. 2018　　　　　　　　　　　　　485
S. S. Dash et al. (eds.), *Artificial Intelligence and Evolutionary Computations*
in Engineering Systems, Advances in Intelligent Systems and Computing 668,
https://doi.org/10.1007/978-981-10-7868-2_47

image should be identified. In most of the cases, detection of a person is generally done by frontal face detection. Proposed system also highlights the face detection mechanism with some advancement. Detection also differs by image parameters like contrast, pixel density, more amount of light (aperture). Firstly, human detection algorithm should segregate humans as region of interest for perfect detection of human. Face detection is one among the object detection classifications. This proposed face detection technique considers skin detection so as to target it as human. There are several face detection algorithms and Viola–Jones face detection algorithm falls under the category of appearance-based face detection algorithm.

2 Related Work

Brief review of existed algorithms gives the overview of proposed algorithm. The first face detection technique, Viola–Jones [1], is the first face detection framework capable of achieving real-time performance. In the Viola–Jones face detection framework, a window slides over the entire image from top to bottom and then from left to right. In every window, Haar features are calculated using the already trained cascade classifier and decision is taken whether to reject or pass the window it to next stage of the cascade. Scaling the size of the scanning window is done to detect faces of different size. No doubt, it detects at a faster rate but it is more prone to errors in case of huge crowd in image. Proposed algorithm even works for the above-mentioned criteria. Viola–Jones detects some non-face regions as face region. This error will be rectified in proposed algorithm and is explained in methodology part.

Histograms of Oriented Gradients (HOG) feature extraction algorithm [2] gives good results as it includes overlapping grid mechanism. It is efficient when compared to many other face detection techniques. This paper explains various parameters which resulted in making best scale gradient, orientation matrix, and coarse space for overlapping.

Automated face detection using colour-based mechanism [3] algorithm gave its best performance in usual environmental conditions. Due to undesirable environmental conditions which leads to illumination variations in image, this effects the identification of skin tone. In spite of these imperfections, it detects faces wisely based on skin tone. This consideration of skin tone mapping is also included in proposed algorithm.

Face region detection based on skin region properties [4], this detection mechanism works with the help of face detection and skin segmentation techniques. This segmentation is done on normalized space binning by using probability concept. This combined technique will overcome problems aroused due to facial imperfections. In order to solve the above-mentioned problems, we proposed an efficient approach that combines skin colour mapping feature with concealed face detection. The main contributions of this work can be stated as follows: (1) A face detection algorithm for the real-world environment is proposed, (2) The skin colour feature and segmentation are merged.

3 Methodology

3.1 Pre-processing

The image which is captured from the camera device is of higher dimensions and higher resolution and this increases the computational time of the process so the image is reduced to required size and required resolution so that a perfect detection can be implemented with lesser computations.

3.2 Face Detection

After the Pre-processing stage, the control flows to face detection module. This module identifies the faces from image and a bounding box is drawn on the face. It also returns the left corner position of the bounding box. The Viola and Jones algorithm is used for the purpose of face detection. It is also referred as the Ada-Boost algorithm for face detection which is created by Viola P. and M. 1. Jones.

In this section, the methodology is explained, and Fig. 1 summarizes the process.

3.3 Thresholds

Application of Viola–Jones algorithm results in face regions, and in some cases non-face regions also detected as face regions, this can be omitted by maintaining cut-off limits for the dimensions of bounding boxes. As the dimensions of face will lie in a particular range, this range can be estimated based on the position of the camera and the dimensions of the room.

3.4 Skin Detection

Skin detection is applied to the pre-processed image, and this involves colour change to white in skin identified regions and remaining portions with black. Morphological operations are applied to eliminate noise in the image.

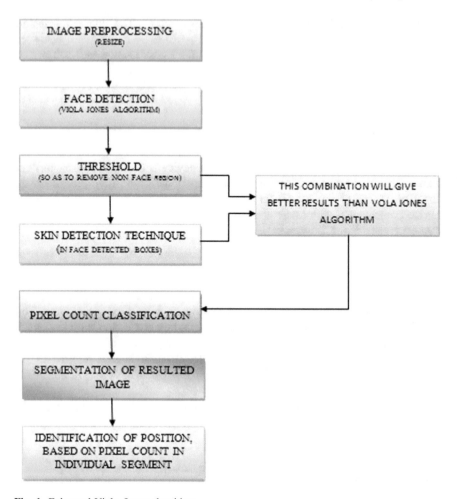

Fig. 1 Enhanced Viola–Jones algorithm

3.5 Pixel Count in Bounding Box

In each Bounding Box, pixel density is counted and based on the pixel density value, we can estimate whether the detected box consists of a face or non-face region. This helps in identifying the actual count of faces in an image.

3.6 Segmentation and Position identification

The Pre-Processed image is segmented into regions, and positions are assigned to each individual region. Based on the pixel density in individual regions, the face is identified, the respective positions are identified, and finally, the position of faces is declared.

3.7 Algorithm

Input: Image of classroom captured by camera.

Output: Attendance is allotted to students if face is identified in the prescribed segment.

Step 1: Start

Step 2: Seating plan of classroom is taken well in advance.

Step 3: Camera to be installed in a class room.

Step 4: The input image is taken from camera.

Step 5: Pre-processing of image

    i. Resizing of image.

    ii. Reducing resolution.

    iii. Eliminating noise.

Step 6: Face Detection

    i. Viola–Jones algorithm.

    ii. Select thresholds.

Step 7: Mapping of skin detection with face detection

    i. Skin Detection

    ii. Fusion of skin detection with face detection.

Step 8: Segmentation

    i. Partition of image based on seating plan.

    ii. Face is detected based on pixel count.

    iii. Position is displayed on image.

Step 9: End

4 Implementation

The proposed algorithm is applied on student attendance system. This can be implemented for huge database. The proposed algorithm is tested on KLU database. A classroom of interest is chosen and algorithm is applied on it.

First and foremost, input image is pre-processed to reduce the number of computations and pre-processed image is passed through Viola–Jones algorithm. After application of Viola–Jones algorithm, the results are not up to the mark, as it has identified a non-face region as face region and two faces are not identified due to occlusion. The pre-processed image is applied to skin detector which helps in identifying the skin and non-skin region. The pixel density is counted in each bounding box and if it is beyond the threshold, it is confirmed that it is a face. If any non-face region boxes' dimensions are equal to that of face regions' dimension, then skin mapping should be applied to confirm it as face (Figs. 2, 3, 4, 5 and 6).

This resulted image after skin mapping has eliminated errors generated by Viola–Jones algorithm. Enhanced Viola–Jones algorithm purely detected the face regions. After detecting the face regions completely, location of face region in desired segment treated as presence. And if desired segment does not have any face region, it is treated as absent. Accordingly, roll number assigned to a student is displayed.

5 Result

The proposed algorithm is implemented on different databases and results are tabulated.

From the above statistics, two persons were not identified due to the occlusion in image, and this can be overcome by placing the camera at proper position so that occlusion does not occur (Table 1).

Fig. 2 Originally captured image (with consent)

Fig. 3 Viola–Jones

Fig. 4 Skin detection

Fig. 5 Skin detection mapped with face detection

Fig. 6 Position identification

6 Conclusion

Present algorithm is advancement of Viola–Jones algorithm as it eliminates non-face regions by using thresholds and skin mapping techniques. Skin mapping in detected regions resulted in satisfactory results. Another advantage of the proposed algorithm is it works at a faster rate when compared to Hog algorithm. Many

Table 1 Comparsion between face detection algorithms

Algorithm	Table column head		
	Viola–Jones	HOG	Proposed algorithm
Faces/non-faces	19/3	21/0	21/0
Actual count	23	23	23
Execution time	Low	High	Low

algorithms need huge database for their storage. This efficient technique occupies less space. The efficiency of this proposed detection algorithm is found to be 84.68%.

Acknowledgement The presented work was supported by the principal of KL University and also the support of management means a lot. Thanks to the students who gave consent for using their image in paper.

References

1. Y.Q. Wang, An analysis of the viola-jones face detection algorithm. (2014)
2. N. Dalal, B. Triggs, Histograms of oriented gradients for human detection. INRIA Rhone-Alps, 655 avenue de lEurope, Montbonnot 38334
3. Y. Tayal, R. Lamba, S. Padhee, Automatic Face Detection Using Color Based Segmentation. Department of Electrical and Instrumentation Engineering Thapar University, Patiala-147004, Punjab
4. K.A. Bozed, A. Mansour, O. Adjei, Face region detection using skin region properties, Institute for Research in Applicable Computing, Department of Computer Science and Technology, University of Bedfordshire, Park Square, Luton, LU1 3JU, UK
5. P. Viola, M. Jones, Rapid object detection using a boosted cascade of simple features. Mitsubishi Electric Research Labs Compaq CRL 201 Broadway, 8th FL One Cambridge Center Cambridge, MA 02139 Cambridge, MA 02142
6. T.Y. Chen, C.H. Chen, D.J. Wang,Y.L. Kuo, A people counting system based on face-detection. IEEE Trans. Image Process.
7. O. Barnich, M. Van Droogenbroeck, ViBe: a universal, background subtraction algorithm. IEEE Trans. Image Process. **20**(6), 1709–1724, (June 2011)
8. P. Viola, M. Jones, Rapid object detection using a boosted cascade of simple features. Comput. Vis. Pattern Recogn. **1**, 511–518, (2001)
9. Q. Zhu, S. Avidan, M.C. Yeh, K.T. Cheng, Fast human detection using a cascade of histograms of oriented gradients. Comput. Vis. Pattern Recogn. **2**, 1491–1498, (2006)
10. S. Kawato, J. Ohya, Automatic skin color distribution extraction for face detection and tracking. IEEE Trans. Image Process. (Aug 2002)

Hardware Implementation of Variable Digital Filter Using Constant Coefficient Multiplier for SDR Applications

P. Srikanth Reddy, P. Satyanarayana, G. Sai Krishna and K. Divya

Abstract Software Defined Radio (SDR) is widely used in wireless communication where Variable Digital Filter (VDF) plays a major role in extracting signals. The VDF allows the low pass, high pass, band pass and band stop frequency responses as per the given filter coefficients. In this paper, we implemented a variable digital filter (VDF) based on decimation using Constant Coefficient Multiplier (CCM). The proposed method is synthesized using Verilog in Xilinx Vivado software and implemented on Zynq-7020(XC7Z020-1CLG484) development board. The performance parameters are compared with previous implemented architectures in terms of area, power and latency. By the analysis it is clear that the proposed method is less complex and utilizes less area. The maximum operated frequency for the proposed VDF is 132.22 MHz.

Keywords Variable digital filter (VDF) · Decimation · Zynq board
Constant coefficient multiplier · FIR

1 Introduction

Fusion is a method of retrieving relevant data from multiple images into a solitary image. In pixel level fusion [1] it is important to combine set of pixels which are necessary for future level representation [1]. In decision level fusion [1] from the merging information with higher level of abstraction final fusion image is yielded.

P. Srikanth Reddy (✉) · P. Satyanarayana · G. Sai Krishna · K. Divya
Department of ECE, K L University, Guntur, Andhra Pradesh, India
e-mail: palagani.srikanthreddy@gmail.com

P. Satyanarayana
e-mail: satece@kluniversity.in

G. Sai Krishna
e-mail: sai369g@gmail.com

K. Divya
e-mail: kolladivya8@gmail.com

© Springer Nature Singapore Pte Ltd. 2018
S. S. Dash et al. (eds.), *Artificial Intelligence and Evolutionary Computations
in Engineering Systems*, Advances in Intelligent Systems and Computing 668,
https://doi.org/10.1007/978-981-10-7868-2_48

Variable Digital Filters (VDF's) are applied in Software Defined Radio (SDR) [1] for extracting the radio channels. FIR digital filter plays a vital role while dealing with the signal processing applications. In order to deal with the advanced technology of SDR [2, 3] we mainly concentrate on designing variable FIR filter [4] which produces low complexity and flexibility is more. The designed filter is capable of passing all the coefficients based on the specifications like low pass, high pass and band pass that's why we entitled it as variable digital FIR filter. In Variable Digital Filter the term digital filter arises because the filter operates on discrete-time signals. The digital filter has its usage for signal analysis, computation, selection of band [5] etc. By finding the coefficients and filter order based on the required specifications a FIR filter [6] can be designed. For a particular frequency response there are several methods to design the FIR filter. Generally the FIR filters are mostly used for removing the unwanted parts of the input signal like random noise or components of a given frequency content. These are usually used in noise reduction and also channel equalization.

The main aim is to design the variable digital filter with low complexity in [7, 8] that should be applicable for the SDR'S [4] which is current trending technology that should be implemented for high speed and efficient data. In Software define radio instead of implementing in hardware we can directly implement it in the software. SDR can be applicable in military, mobile services where we use different radio protocols in real time. The concept of SDR is more flexible such that it can be used to avoid the limited spectrum. The advantage of SDRs usage is it utilizes less hardware resource such that the size and cost can be reduced. For the hardware implementation we use zed board (zynq 7020). This is the Field Programmable Gate Array (FPGA) implementation [9] which is used for SDR realization. The main advantages of the FPGA implementation are reusability, efficient energy usage, more speed and long time availability.

We discuss the implementation of FIR filter using constant coefficient multiplication method and realized using decimation method. The performance of the FIR filter before and after the decimation is compared. And also the implemented method is compared with the filter using arithmetic multiplier. The comparison is in terms of complexity, size and frequency.

The decimation method in [10] can be used to reduce the sampling rate of a signal. The different types of decimation methods like Coefficient Decimation Method (CDM), Modified Coefficient Decimation Method (MCDM) and Improved Coefficient Decimation Method (ICDM) are applied to the proposed Variable FIR digital filter. These methods are differentiated by the selection of the filter coefficients such that variable frequency responses can be obtained.

The rest of the paper is organized as follows, the theoretical concepts of FIR filter, constant coefficient multiplier, and decimation method are discussed in Sect. 2. Section 3 presents the implementation part and the comparison results of Constant Coefficient Multiplier. Section 4 represents the hardware implementation Sect. 5 concludes the paper.

2 Literature Review

2.1 FIR Filter

The FIR filter is abbreviated as the Finite Impulse Response which is used in digital signal processing. The FIR filter design implementation is very stable and easy. The FIR filter output is expressed as the following equation:

$$Y[n] = \sum_{i=0}^{N} b_i X[n-i]$$

Here 'Y' is the filter output, 'X' is the input signal, 'b' is the filter coefficients. 'N' is the order of the filter. As we increase the value of 'N' the filter design will be more complex. Let us consider the input sequence 'Xin' is of length 'm' and the coefficient sequence 'H' is of length 'n'. Now the final response of the FIR filter 'Y' will be of length '$m+n-1$'.The simple FIR filter can be shown as following (Figs. 1, 2).

The above example is a 4-tap filter which consists of 4 coefficients. The size of the input sequence chosen is of 24 bits and the coefficients are of 12 bits as per the requirement. If we want to handle the inputs of higher range then we can simply increase the size of inputs and the intermediate variables.

Initially we need the filter coefficients based on the required specifications. The filter coefficients are generated from 'sptool' which exists in Matlab software. Here we select the cut-off frequency and the sampling rate based on the specifications. Here we also consider the rectangular window for the design of the filter. Later the inputs are generated for a particular frequency using sine wave generation. In order to take the inputs and the coefficients of the filter in decimal number representation we consider the fixed point notation. Fixed point notation is very much used in representation of fractional values, usually of base 10 or 2. Here the fractional

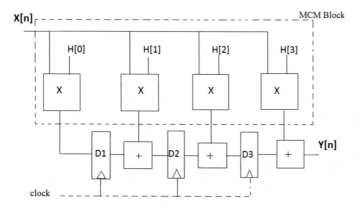

Fig. 1 FIR filter architecture

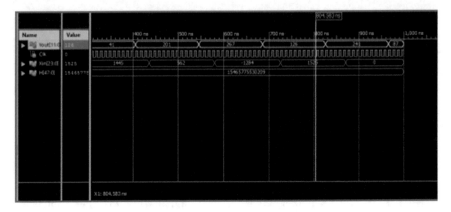

Fig. 2 Simulated waveform for constant coefficient multiplication

inputs are firstly multiplied by power of 2 i.e. inputs are left shifted such that there must be difference between the inputs. Later the final output is divided by the same power of 2 i.e. output is right shifted. In order to get the efficient filter [6] arithmetic multiplier is replaced with constant coefficient multiplier.

2.2 Constant Coefficient Multiplier

The Constant coefficient multiplier [10] is used to reduce the hardware complexity rather than considering the normal arithmetic multiplier. Because it utilizes less no: of LUT's, flip flops and slices i.e. less area, less complex and less time delay. The operation of the constant coefficient multiplier can be expressed as the following:

$$R = P * Q = \sum_{i=0}^{N-1} (P * q_i) \ll i$$

Here 'P' and 'Q' are the inputs to the multiplier, Where 'R' is the output of the multiplier. Here 'P' is multiplied with ones present in 'Q' and then left shifted by the position where the one's are present.

The design of filter generally involves the arithmetic multiplier which takes much time, more area, more no: of flip-flops. But here we design filter using constant coefficient filter which consists of only shifters and registers. Here we consider the coefficients to be constant and based upon the one's position of the coefficients we shift and add the inputs. Therefore this constant coefficient multiplier is more efficient than the normal multiplier.

In order to implement this multiplier in verilog we have used the state machine concept. There are several methods for designing multiplier like using for loop or switch case or if-else conditions etc. But here we considered the Moore state machine method for designing the multiplier such that the complexity, power and area are reduced.

2.3   Decimation

In order to reduce the hardware complexity [11] decimation method is applied. The sampling rate of the input signal can be reduced by applying decimation. The decimation methods involved are CDM, MCDM and ICDM. CDM is the Coefficient Decimation Method which is classified into 2 types as CDM-1 and CDM-2. In CDM-1 the Mth coefficient will be retained and the remaining values are noted as zeros. In CDM-2 the Mth coefficient will be retained and the remaining values are discarded, where 'M' is the decimation factor. MCDM is the Modified Coefficient Decimation Method which is classified into 2 types as MCDM-1 and MCDM-2. In MCDM-1 the Mth coefficient is retained, the alternate coefficient is sign reversed to generate the high pass filter and the remaining values are noted as zeros. In MCDM-2 the Mth coefficient is retained, the alternate coefficient is sign reversed and the remaining values are discarded. ICDM is the Improved Coefficient Decimation Method which is classified into 2 types as ICDM-1 and ICDM-2. ICDM-1 is the combination of both CDM-1 and MCDM-1. ICDM-2 is the combination of CDM-2 and MCDM-2.These methods only differs in the selection of the filter coefficients. In proposed VDF the ICDM is used such that variable frequency responses are obtained.

In the common design of FIR filter, all the coefficients are considered based on the required specifications like order of the filter. But in decimation based upon the method and the decimation factor conceived we consider only some of the coefficients that are required. Here the decimation factor mainly depends upon the sampling rate of the signal and order of the filter.

3   Implementation, Results and Comparisons

The filter coefficients and the input sequence which are generated from the Matlab software are given as the inputs through the test bench. Because as we are designing the variable digital filter, whenever the specifications got changed the coefficients will be changed. Later the constant coefficient multiplier is used for the multiplication operation such that the code is optimized.

Table 1 represents the time complexities, area, flip-flops and frequencies of the implemented Constant Coefficient and arithmetic multiplier. From the analysis, it is clear that number of LUT's which represents the area occupied, frequency and power consumed are effectively reduced in the implementation of constant Coefficient multiplier compared to arithmetic multiplier. Later the decimation method is applied to the filter with constant coefficient multiplier such that the area

Table 1 Performance analysis of implemented architecture

S. No	Performance parameters	Decimated filter with CCM	Filter with arithmetic multiplier
1	No. of input bits	24	24
2	No. of 4 input LUTs	765	1692
3	No. of flip flops	390	421
4	No. of occupied slices	510	992
5	Frequency (MHz)	118.91	90.473
6	No. of IOB's	96	145

Fig. 3 Hardware implementation for proposed architecture

occupied is much reduced than the before implementation. The complexity and area occupied are reduced by implementing the decimation method effectively. The performance analysis and the simulated result can be shown as:

Figure 3 depict the software output of the proposed method.

4 Hardware Implementation

The Variable Digital Filter is implemented on zed board (zynq 7020). It consists of mainly two parts programmable system (PS) and programmable logic (PL). Programmable system (PS) consists of two ARM cortex A9 processors, 512 MB

DDR3 memories and programmable logic (PL) consists of Xilinx 7 series FPGA. Inside PS, four high performance ports and two general purpose ports to transfer the data between PS and PL. Initially the data is stored in flash and then transferred to FPGA (PS or PL). As the part of SDR variable filters are very important. If filters are operated purely in processors, because of their sequential nature operation will be slow. As part of our project we are off-loading some of the arithmetic computations through FPGA in which there will be maximum throughput. Based on booting sequence jumper settings will be changed. The output of the hardware implementation is shown as.

5 Conclusion

In this paper the Constant Coefficient Multiplier architecture for FIR filter is implemented effectively using various techniques like decimation such that the area, complexity [8] and power can be reduced. The results are compared and analyzed with arithmetic multiplier and the Constant Coefficient Multiplier architecture which occupies less area compared to arithmetic multiplier. It is also observed that the area occupied is less after applying decimation technique compared to the implementation without decimation.

Acknowledgements The authors would like to the management of KL University and Honorable delegates of university for providing the facilities that contributed to the smooth flow the work within the stipulated time.

References

1. A. Ambede, S. Shreejith, in *Design and Realization of Variable Digital Filters for Software-Defined Radio Channelizers Using an Improved Coefficient Decimation Method*. IEEE Transactions on Circuits and Systems—II: Express Briefs, vol. 63, no. 1, Jan 2016
2. J. Mitola, The software radio architecture. IEEE Commun. Mag. **33**(5), 26–38, (May 1995)
3. E. Buracchini, The software radio concept. IEEE Commun. Mag. **38**(9), 138–143, (Sept 2000)
4. A.P. Vinod, E. Lai, Low power and high-speed implementation of FIR filters for software defined radio receivers. IEEE Trans. Wireless Commun. **5**(7), 1669–1675 (2006)
5. K.K. Parhi, *VLSI Digital Signal Processing Systems: Design and Implementation* (John Wiley & Sons Inc, New York, 1999)
6. A.G. Dempster, M.D. Macleod, Use of minimum-adder multiplier blocks in FIR digital filters. IEEE Trans. Circuits Syst.– II. **42**(9), 569–577, (Sept 1995)
7. K.-H. Chen, T.-D. Chiueh, A low-power digit-based reconfigurable FIR filter. IEEE Trans. Circuits Syst. II, Exp. Briefs **53**(8), 617–621 (2006)
8. R. Mahesh, A.P. Vinod, in *New Reconfigurable ARCHITECTURES for Implementing FIR Filters With Low Complexity*. IEEE Transactions on Computer-Aided Design of Integrated Circuits Systems, vol. 29, no. 2, pp. 275–288, Feb 2010

9. V. Bhatnagar, G.S. Ouedraogo, M. Gautier, A. Carer, O. Sentieys, in *An FPGA Software Defined Radio Platform with a High-Level Synthesis Design Flow*. Proceedings of 77th IEEE VTC–Spring, (Dresden, Germany, June 2013), pp. 1–5
10. O. Gustafsson, L. Wanhammar, in *A Novel Approach to Multiple Constant Multiplication Using Minimum Spanning Trees*. Proceedings of IEEE Midwest Symposium Circuits Systems, vol. 3 (Tulsa, OK, Aug 4–7 2002), pp. 652–655
11. K.G.S. Ambede, A.P. Vinod, in *A Modified Coefficient Decimation Method to Realize Low Complexity Fir Filters With Enhanced Frequency Response Flexibility and Passband Resolution*, Proceedings of 35th International Conference TSP, Prague, Czech Republic, pp. 658–661, July 2012
12. A. Johnson, B. Manohar, Modified MAC based FIR filter using carry select adders. Int. J. Eng. Sci. Innovative Technol. (IJESIT) **4**(3), (May 2015)
13. T. Solla, O. Vainio, in *Comparison of Programmable FIR Filter Architectures for Low Power*. Proceedings of 28th European Solid State Circuits Conference, pp. 759–762, Sept 24–26 2002

Enhanced Joint Estimation-Based Hyperspectral Image Super Resolution

R. Sudheer Babu and K. E. Sreenivasa Murthy

Abstract A new hyperspectral image super-resolution method from a Low-resolution(LR) image and an HR reference image of the same scene is proposed. The estimation of the HR Hyperspectral image is formulated as a joint estimation of the hyperspectral dictionary and the sparse course based on the prior knowledge of the spatial–spectral scarcity of the hyperspectral image. The hyperspectral dictionary representing prototype reflectance spectra vectors of the scene in first learned from the input LR image. Specifically, an efficient nonnegative dictionary learning algorithm using the block-coordinate descent optimization technique is proposed. Then, the sparse codes of the desired HR hyperspectral image with respect to learned hyperspectral basis are estimated from the pair of LR and HR reference images. To improve the accuracy of nonnegative sparse coding, a clustering-based structured sparse coding method is proposed to exploit the spatial correlation among the learned sparse codes. The experimental results on public datasets suggest that the proposed method substantially outperforms several existing HR hyperspectral image recovery techniques in the literature in terms of both objective quality metrics and computational efficiency.

Keywords Hyperspectral images · High-resolution reconstruction
Nonnegative dictionary learning · Clustering-based sparse representation

R. Sudheer Babu (✉)
Department of ECE, Rayalaseema University, Kurnool 518007, India
e-mail: sudheergprec@gmail.com

K. E. Sreenivasa Murthy
Department of ECE, G. Pullaiah College of Engineering and Technology,
Kurnool 518452, India
e-mail: kesmurthy@gmail.com

© Springer Nature Singapore Pte Ltd. 2018
S. S. Dash et al. (eds.), *Artificial Intelligence and Evolutionary Computations
in Engineering Systems*, Advances in Intelligent Systems and Computing 668,
https://doi.org/10.1007/978-981-10-7868-2_49

1 Introduction

1.1 Hyperspectral Imaging

Hyperspectral imaging is an emerging modality that can simultaneously acquire images of the same scene across a number of different wavelengths. Obtaining dense hyperspectral bands is important to remote sensing [1] and computer vision applications including object segmentation, tracking, and recognitions [2–4]. While Hyperspectral imaging can achieve high-spectral resolution, it has severe limitations in spatial resolution when compared against regular RGB cameras in the visible spectrum. This is due to the fact that hyperspectral imaging systems need a large number of exposures to simultaneously acquire many bands within a narrow spectral window. To ensure sufficient signal-to-noise ratio, long exposures are often necessary, resulting in the sacrifice of spatial resolution. While high-resolution (HR) hyperspectral images are desirable in real-world applications, it is often challenging to enhance the spatial resolution of those images due to various hardware limitations. Simply increasing the spatial resolution of image sensors would not be effective for hyperspectral imaging because the average amount of photons reaching the sensor would be further reduced leading to even lower signal-to-noise ratio. Consequently, signal processing-based approaches have been proposed for obtaining a HR hyperspectral image by combining a low-resolution (LR) hyperspectral image with a HR panchromatic image [5, 6]. In [7], the multispectral image is first transformed from the RGB color space to the intensity, hue, saturation (IHS) domain, and then the intensity channel is replaced by the HR panchromatic image. After resampling the hue and saturation channels, one can obtain the reconstructed HR multispectral image by inverse IHS transformation. While the technique does improve the spatial resolution to some extent, it often introduces spectral distortions in the reconstructed multispectral images. To further improve the reconstruction quality, other fusion methods such as [8] improved linear transformations. The HR hyperspectral image was then reconstructed using the learned basis and sparse coefficient computed from the HR RGB image [9]. In this paper, we propose a nonnegative structured sparse representation (NSSR) approach to recover a HR hyperspectral image from a LR hyperspectral image and a HR RGB image. The estimation of HR hyperspectral image is formulated as a joint estimation of spectral basis and sparse coefficients with the prior knowledge about spatio–spectral sparsity of the hyperspectral image. The contributions of this paper are twofold.

First, an efficient nonnegative dictionary learning algorithm using the block-coordinate descent optimization algorithm is proposed, which updates an atom per iteration via a closed-form solution. Second, to improve the accuracy of nonnegative sparse coding over the learned basis, a structural sparsity constraint is proposed to exploit the clustering-based sparsity of hyperspectral images—namely reconstructed spectral pixels should be similar to those learned centroids. The performance of our proposed method is verified using simulated low-resolution hyperspectral images from public datasets.

2 Problem Formulation

We aim at recovering a HR hyperspectral image $Z \in \mathbb{R}^{L \times N}$ from a LR hyperspectral image $X \in \mathbb{R}^{L \times n}$ and a HR RGB image $Y \in \mathbb{R}^{3 \times N}$ of the same scene where $N = W \times H$ and $n = w \times h$ denote the number of pixels in the HR hyperspectral image Z and LR hyperspectral image X respectively; and L is the number of spectral bands of $Z(L \gg 3)$ Both X and Y can be expressed as linear combinations of the desired hyperspectral image Z.

$$X = ZH, \quad Y = PZ, \tag{1}$$

where $H \in \mathbb{R}^{N \times n}$ denotes the degradation operator of blurring and downsampling associated with LR image X; P a transformation matrix mapping the HR hyperspectral image Z to its RGB representation Y.

$$Z_i = D\alpha_i + e_i, \tag{2}$$

where $D \in \mathbb{R}_+^{L \times K}$ $(K \geq L)$ is the spectral dictionary, $\alpha_i \in \mathbb{R}_+^K$ is the fractional abundance vector assumed to be sparse (i.e., $\| \alpha \|_0 < T$), and e_i is the approximation error. As the observed hyperspectral image $X = ZH$, each pixel $x_i \in \mathbb{R}^L$ of X can be written as

$$x_i = \sum_{j \in W_i} h_j Z_j = D \sum_{j \in W_i} h_j \alpha_i + v_i = D\beta_i + v_i, \tag{3}$$

where h_j denotes the weighting coefficients of a window W_i centered at the location i. Here, we assume that at each pixel location x_i of the LR hyperspectral image X, only a few distinct materials can be presented. Therefore, $\beta_i \in \mathbb{R}_+^k$ is a sparse vector. For each pixel y_i of the HR RGB image Y, we have

$$y_i = Pz_i = PD\alpha_i \tag{4}$$

From which we can see that for a fixed spectral dictionary D the sparse fractional abundance vectors α_i of the HR hyperspectral image Z can be estimated from the HR RGB image Y, as we will elaborate in the next section.

3   Proposed Method

Using the linear mixing model of Eq. (2), we can rewrite the desired HR hyperspectral image Z as

$$Z = DA + E \qquad (5)$$

where $A = [\alpha_1, \ldots, \alpha_N] \in \mathbb{R}_+^{K \times N}$ is the coefficient matrix. However, both spectral dictionary D and coefficients matrix A in Eq. (5) are unknown.

3.1   Spectral Dictionary Learning

As each pixel in the LR hyperspectral image X can be written as the linear combination of a small number of spectral signatures, we can estimate spectral dictionary v from X. In the matrix form, Eq. (3) can be rewritten as

$$X = DB + V, \qquad (6)$$

where $B = [\beta_1, \ldots \beta_N] \in \mathbb{R}_+^{K \times n}$ is the coefficient matrix and V denotes the approximation error matrix, which is assumed to be additive Gaussian. Both D and B are unknown in Eq. (6). Therefore, spectral dictionary D can be estimated by solving the following sparse nonnegative matrix decomposition problem

$$(D, B) = \arg\min_{D,B} \frac{1}{2} \| X - DB \|_F^2 + \lambda \| B \|_1, \quad \text{s.t. } \beta_i \geq 0, d_k \geq 0 \qquad (7)$$

In this paper, we propose a computationally efficient nonnegative dictionary learning (DL) algorithm, which updates each atom per iteration via a closed-form solution. For a fixed D, the subproblem with respect to B becomes,

$$B = \arg\min_B \frac{1}{2} \| X - DB \|_F^2 + \lambda \| B \|_1, \quad \text{s.t.} \beta_i \geq 0 \qquad (8)$$

For fast convergence rate, we use ADMM technique to solve Eq. (8). To apply ADMM, we reformulate Eq. (8) into

$$B = \arg\min_B \| X - DS \|_F^2 + \lambda \| B \|_1, \quad \text{s.t.} B = S, \beta_i \geq 0 \qquad (9)$$

The augmented Lagrangian function of Eq. (9),

$$L_\mu(B, S, U_1) = \frac{1}{2} \parallel X - DS \parallel + \lambda \parallel B \parallel_1 + \mu \parallel S - B + \frac{U_1}{2\mu} \parallel_F^2, \quad (10)$$

where U_1 is the Lagrangian multiplier ($\mu > 0$). Then, solving Eq. (9) consists of the following alternative iterations:

$$S^{(j+1)} = \arg\min_S L_\mu(B^{(j)}, S, U_1^{(J)}),$$

$$B^{(j+1)} = \arg\min_B L_\mu(B, S^{(j+1)}, U_1^{(j+1)}), s.t.\beta_i \geq 0, \quad (11)$$

where j is the iteration number and Lagrangian multiplier is updated by,

$$U_1^{(j+1)} = U_1^{(j)} + \mu(S^{(j+1)} - B^{(j+1)}) \quad (12)$$

Both subproblems in Eq. (11) admit closed-form solutions, namely

$$S^{(j+1)} = (D^T D + 2\mu I)^{-1} \left(D^T X + 2\mu \left(B^j - \frac{U_1^{(j)}}{2\mu} \right) \right),$$

$$B^{(j+1)} = \left[\text{Soft}\left(S^{(j+1)} + \frac{U_1^{(j)}}{2\mu}, \frac{\lambda}{2\mu} \right) \right]_+ \quad (13)$$

where Soft($\cdot$) denotes a soft-shrinkage operator and $[X]_+ = \max\{x, 0\}$. For a fixed B D is updated by solving,

$$D = \arg\min_D \parallel X - DB \parallel_F^2, \text{ s.t.} d_t \geq 0 \quad (14)$$

In this paper, we propose to solve Eq. (14) by using block-coordinate descent, i.e., during each iteration we update one column of D while keeping the others fixed under the nonnegative constraint. Let $D^{(t)}$ denote the dictionary obtained after the tth iteration and let $d_k^{(t+1)} = d_k^{(t)} + \triangle d_k$. Then, $\triangle d_k$ can be obtained by solving

$$\triangle d_k = \arg\min_{\triangle d_k} \parallel X - D^{(t)}B - \triangle d_k \beta_k \parallel_F^2 \text{ s.t.}(\triangle d_k + d_k^{(t)}) \geq 0, \quad (15)$$

The overall algorithm for nonnegative dictionary learning is summarized below in Algorithm-I.

Algorithm-I Non-Negative Spectral Dictionary Learning	Algorithm-II NSSR-Based HR Hyperspectral Image Super-Resolution

1. Initialization:
  (a) Input X, λ and μ ;
  (b) Initialize D based on X with Normalization
2. Outer loop: for t=1,2,.....,T_1 **do**
  (a) **Inner loop** : for j=1,2,.....,T_1 **do**
      [1] Update $B^{(j+1)}$ and $S^{(j+1)}$
      [2] Update $U_1^{(j+1)}$
      [3] Update $\mu := \rho\mu(\rho > 1)$
      [4] Output $B = B^{(j+1)}$ if j=J.
     End for
  (b) **Inner loop** : for k=1,2,.....,T_1 **do**
      [1]Update $d_k^{(t+1)}$
      [2] Output $D^{(t+1)}$ if k=K.
     End for
  (c) Output $D^{(t+1)}$ if t= T_1.
End for
	1. Initialization:
	  (a) Learn the spectral dictionary D from X using Algorithm 1.
	  (b) Obtain the sample set S_q using the HR color image Y.
	  (c) Set parameters η_1 *and* η_2 and initialize U=0;
	2. Solving Eq. (22) via ADMM: for t=0,1,2,.....,T_2 **do**
	  [1] Compute $A^{(t+1)}, Z^{(t+1)}$ and $S^{(t+1)}$ via Eq. (26)
	  [2] Update the Lagrangian multipliers $V_1^{(t+1)}$ and $V_2^{(t+1)}$ via Eq. (25)
	  [3] Update $\mu := \rho\mu(\rho > 1)$
	  [4] Update U using $A^{(t+1)}$ and Eq. (20)
	  [5] Output $Z^{(t+1)}$ if t= T_2-1.
	     End for

3.2 Sparse Codes Estimation via Nonnegative Structured Sparse Coding

Once the spectral dictionary D is estimated, sparse codes α_i for each pixel Z_i of the desired HR hyperspectral image Z can be estimated and then $\hat{Z}_i$ can be reconstructed as $Z_i = D\alpha_i$. Since both observed X and Y can be expressed as linear combination of the desired Z, X and Y can then be expressed as

$$Y = PDA + W = \tilde{D}A + W_1, X = DAH + W_2, \qquad (16)$$

where $\tilde{D} = PD$ denotes the transformed spectral dictionary, and W_1 and W_2 denote the approximation error matrix. From the above equation, we see that sparse codes α_i can be estimated from the HR RGB image Y and the low-resolution hyperspectral image X. With the sparsity constraint, sparse coefficient matrix A can be estimated by solving the following nonnegative sparse coding problem

$$A = \arg\min_{A} \parallel Y - \widetilde{D}A \parallel + \parallel X - DAH \parallel_F^2 + \eta \parallel A \parallel_1, \quad s.t. \alpha_i \geq 0 \qquad (17)$$

In Eq. (17), sparse codes of each pixel are estimated independently. We propose the following clustering-based nonnegative structured sparse representation (NSSR) model

$$A = \arg\min_{A} \parallel Y - \widetilde{D}A \parallel_F^2 + \parallel X - DAH \parallel_F^2$$
$$+ \eta_1 \sum_{q=1}^{Q} \sum_{i \in S_q} \parallel D\alpha_i - \mu_q \parallel_2^2 + \eta_2 \parallel A \parallel_1, \ s.t. \alpha_i \geq 0, \qquad (18)$$

where μ_q denotes the centroid of the qth cluster of the reconstructed spectral pixel Z_i. In addition to the l_1 sparsity regularization, the proposed NSSR model also exploits a structural prior that the reconstructed spectral pixels should be similar to those learned centroids. First, the nonnegative sparse decomposition of a hyperspectral image is exploited—note that material coefficients corresponding to the surface albedo in the physical world can only take nonnegative values. Second, instead of using a set of orthogonal PCA basis, a non-orthogonal dictionary is learned; such relaxation is beneficial to improve the accuracy of sparse reconstruction. Third, instead of using the l_1—norm, we adopt the l_2—norm to exploit the above-mentioned structural prior. The centroid vector of the Algorithm 2 NSSR-Based HR Hyperspectral Image Super-Resolution Cluster in Eq. (18) is then computed as

$$\mu_q = \sum_{i \in S_q}^{n} \omega_i(D\alpha_i), \qquad (19)$$

where $\omega_i = \frac{1}{c}\exp(- \parallel \widetilde{y}_i - \widetilde{y}_q) \parallel_2^2 /h$, $\alpha_i = \hat{\alpha}_i + e_i$, wherein e_i denotes estimation errors observing a Gaussian distribution with zero-mean. Then, Eq. (21) can be rewritten into

$$\mu_q = \sum_{i \in S_q} \omega_i(D\hat{\alpha}_i) + \sum_{i \in S_q} \omega_i(De_i) = \widehat{\mu_q} + n_q, \qquad (20)$$

where n_q denotes the estimation error of μ_q. As e_i is assumed to be zero-mean and Gaussian, n_q would be small. Therefore, μ_q can be readily estimated from the current estimate of α_i. In our implementation, we recursively compute μ_q using the previous estimate of α_i after each iteration. In our implementation, we use the k-Nearest Neighbor (k-NN) clustering method to group similar spectral pixels for each spectral pixel. Due to the structural similarity between Z and Y, we perform

k-NN clustering on the HR RGB image patches to search for similar neighbors of α_q, i.e.,

$$S_q = \{i/ \parallel \tilde{y}_q - \tilde{y}_i \parallel < T\} \tag{21}$$

After estimating μ_q, Eq. (20) can be

$$A = \underset{A}{\arg\min} \parallel Y - \tilde{D}A \parallel_F^2 + \parallel X - DAH \parallel_F^2 +$$
$$\eta_1 \parallel DA - U \parallel_2^2 + \eta_2 \parallel A \parallel_1 \quad \text{Where } U = [\widehat{\mu_1}, \ldots \widehat{\mu_N}] \quad , \text{s.t.} \alpha_i \geq 0 \tag{22}$$

For fast convergence, we use ADMM technique to solve Eq. (22) instead. More specifically, we obtain the following augmented Lagrangian function:

$$L_\mu(A, Z, S, V_1, V_2) = \parallel Y - \tilde{D}S \parallel_F^2 + \parallel X - ZH \parallel_F^2 + \eta_1 \parallel DS - U \parallel_2^2$$
$$+ \eta_2 \parallel A \parallel_1 + \mu \parallel DS - Z + \frac{V_1}{2\mu} \parallel_F^2 + \mu \parallel S - A + \frac{V_2}{2\mu} \parallel_F^2 , \text{ s.t.} \alpha_i \geq 0 \tag{23}$$

In Eq. (23) V_1 and V_2 are Lagrangian Multipliers.
Minimizing the augmented Lagrangian function leads to the following iterations:

$$A^{(t+1)} = \underset{A}{\arg\min} L_\mu\left(A, Z^{(t)}, S^{(t)}, V_1^{(t)}, V_2^{(t)}\right),$$
$$Z^{(t+1)} = \underset{Z}{\arg\min} L_\mu(A^{(t+1)}, Z, S^{(t)}, V_1^{(t)}, V_2^{(t)}), \text{ and} \tag{24}$$
$$S^{(t+1)} = \underset{S}{\arg\min} L_\mu(A^{(t+1)}, Z^{(t+1)}, S, V_1^{(t)}, V_2^{(t)}),$$

where the Lagrangian multipliers are

$$V_1^{(t+1)} = V_1^{(t)} + \mu(DS^{(t+1)} - Z^{(t+1)}) \text{ and } V_2^{(t+1)} = V_2^{(t)} + \mu(S^{(t+1)} - A^{(t+1)}) \tag{25}$$

All subproblems in Eq. (24) can be solved analytically, i.e.,

$$A = \left[\left(Soft(s^{(t)} + \frac{V_2^{(t)}}{2\mu}, \frac{\eta_2}{2\mu})\right)\right]_+ \quad Z = \left[XH^T + \mu\left(DS^{(t)} + \frac{V_1}{2\mu}\right)\right](HH^T + \mu I^{-1})$$
$$S = \left(\hat{D}^T\hat{D} + (\eta_1 + \mu)D^T D + \mu I^{-1}\right)\left[\hat{D}^T Y + \eta_1 D^T U + \mu D^T\left(Z^{(t)} + \frac{V_1^{(t)}}{2\mu}\right) + \mu\left(A^{(t)} + \frac{V_2^{(t)}}{2\mu}\right)\right], \tag{26}$$

As the matrix to be inverted in the equation of updating Z are large, we use conjugate gradient algorithm to compute the matrix inverse. The overall algorithm for estimating the HR hyperspectral image is summarized below in Algorithm-II.

4   Experimental Results

To verify the performance of our proposed method, we have conducted extensive experiments on both simulated LR hyperspectral images and real-world LR hyperspectral images. The basic parameters of the proposed NSSR methods are set as follows: the number of atoms in dictionary D is $K = 80$; the maximal iteration numbers in Algorithm 1 and 2 are $T = 10$, $J = 70$ and $T_2 = 25$, $\eta_1 = 0.015$ and $\eta_2 = 0.1 \times 10^{-3}$. We have compared the proposed method with several leading hyperspectral image super-resolution methods, including Matrix Factorization method (MF) method [9], coupled nonnegative matrix factorization (CNMF) method [10], Sparse Nonnegative Matrix Factorization (SNNMF) method [11], Generalization of Simultaneous Orthogonal Matching Pursuit (G-SOMP+) method, and Bayesian sparse representation (BSR) method [12].

Experiments on Simulated LR Hyperspectral Images

Two different public datasets of hyperspectral images, i.e., the CAVE and the Harvard datasets are used to assess the performance of our proposed method. The CAVE dataset consists of 512×512 hyperspectral images of everyday objects, which are captured using 31 spectral bands ranging from 400 to 700 nm at an interval of 10 nm. The Harvard dataset contains hyperspectral images of real-world indoor and outdoor scenes, which are acquired using 31 spectral bands ranging from 420 to 720 nm with an incremental of 10 nm. Some test images used in this paper from the two datasets are shown in Fig. 1.

Fig. 1 The HR RGB images from the CAVE

Fig. 2 The dictionary trained on the corrupted image

The hyperspectral images from the two datasets served as ground-truth images are used to generate simulated LR hyperspectral images and HR RGB images. The original HR hyperspectral images Z are downsampled by averaging over disjoint $S \times S$ blocks to simulate the LR hyperspectral images X, where s is the scaling factor (e.g., $s = 8, 16, 32$). HR RGB images Y are generated by downsampling the hyperspectral images Z along the spectral dimension using the spectral transform matrix F derived from the response of a camera. To evaluate the quality of reconstructed hyperspectral images, four objective quality metrics—namely peak signal-to-noise ratio(PSNR), root-mean-square error (RMSE), and spectral angle mapper (SAM) are used in our study (Fig. 2).

The average RMSE and PSNR results of the recovered HR hyperspectral images of the Harvard dataset are shown in Table 1. It can be observed that the proposed **NSSR** method also outperforms other competing methods. It can be seen that the PSNR gains of the proposed method over other methods increase for the spectral bands corresponding to longer wavelengths. We can be seen that all the competing methods can well recover the HR spatial structures of the scene, but the proposed method achieves the smallest reconstruction errors.

In the above experiments, the uniform blur kernel of size $s \times s$ is applied to Z before downsampling. In practice, the optics blur may be generated, which can be modeled by a Gaussian function. Table 2 shows the quality metric values of the test methods. We can see that proposed **NSSR** method still outperforms other competing methods. It can be seen that the proposed **NSSR** method outperforms the CNMF [10] and BSR [12] methods in recovering fine image details. Parts of the reconstructed HR HSI images are shown in Fig. 3.

Table 1 The average and standard deviation of PSNR, RMSE, SAM, and ERGAS results of the test methods for different scaling factors on the Harvard dataset

Method	GSOMP	MF [9]	SNMF [11]	CNMF [10]	BSR [12]	Proposed NSSR
S = 8						
PSNR	38.89 ± 5.94	43.74 ± 3.79	43.86 ± 3.67	44.35 ± 3.81	44.51 ± 4.07	45.03 ± 3.57
RMSE	3.79 ± 3.39	1.83 ± 0.90	1.79 ± 0.78	1.70 ± 0.76	1.71 ± 0.96	1.56 ± 0.67
SAM	4.00 ± 2.14	2.66 ± 0.92	2.63 ± 0.87	2.51 ± 0.79	2.48 ± 0.85	2.37 ± 0.71
ERGAS	1.65 ± 1.15	0.87 ± 0.45	0.85 ± 0.44	0.78 ± 0.32	0.84 ± 0.47	0.76 ± 0.31
S = 16						
PSNR	38.56 ± 5.72	43.30 ± 3.99	43.31 ± 3.87	43.56 ± 4.26	43.80 ± 4.37	44.51 ± 3.59
RMSE	3.83 ± 3.10	1.94 ± 0.95	1.93 ± 0.90	1.93 ± 1.18	1.91 ± 1.33	1.65 ± 0.72
SAM	4.16 ± 2.23	2.85 ± 1.02	2.82 ± 1.01	2.74 ± 1.05	2.69 ± 1.09	2.48 ± 0.78
ERGAS	0.77 ± 0.43	0.47 ± 0.29	0.45 ± 0.24	0.42 ± 0.18	0.45 ± 0.290	0.41 ± 0.20
S = 32						
PSNR	38.02 ± 5.71	43.19 ± 3.87	42.03 ± 3.61	43.00 ± 4.44	43.11 ± 4.59	44.00 ± 3.63
RMSE	4.08 ± 3.55	1.96 ± 0.97	2.20 ± 0.94	2.08 ± 1.34	2.10 ± 1.60	1.76 ± 0.79
SAM	4.79 ± 2.99	2.93 ± 1.06	3.17 ± 1.07	2.91 ± 1.18	2.93 ± 1.33	2.64 ± 0.86
ERGAS	0.41 ± 0.24	0.23 ± 0.14	0.26 ± 0.27	0.23 ± 0.11	0.24 ± 0.15	0.21 ± 0.12

Table 2 The average and standard deviation of PSNR, RMSE, SAM and ERGAS Results of the test methods on the Cave and Harvard Datasets (Gaussian Blur Kernel, Scaling Factor 8)

Method	GSOMP	MF [9]	SNMF [11]	CNMF [10]	BSR [12]	Proposed NSSR
CAVE dataset						
PSNR	33.69 ± 3.07	36.35 ± 3.73	36.13 ± 3.10	40.39 ± 3.17	43.96 ± 3.76	45.41 ± 3.55
RMSE	5.62 ± 2.11	4.31 ± 2.37	4.24 ± 1.55	2.61 ± 1.06	1.79 ± 0.88	1.50 ± 0.68
SAM	11.99 ± 5.53	11.04 ± 3.54	14.01 ± 5.01	8.01 ± 2.83	3.33 ± 1.04	2.95 ± 0.84
ERGAS	2.96 ± 1.36	2.19 ± 0.95	2.21 ± 0.92	1.37 ± 0.68	0.99 ± 0.51	0.82 ± 0.41
Harvard dataset						
PSNR	38.77 ± 5.41	40.40 ± 4.98	40.48 ± 4.85	43.42 ± 4.72	44.55 ± 4.00	44.95 ± 3.55
RMSE	3.60 ± 2.53	2.91 ± 1.92	2.86 ± 2.00	2.05 ± 1.61	1.69 ± 0.90	1.57 ± 0.67
SAM	3.92 ± 1.98	4.03 ± 1.99	4.14 ± 2.24	2.83 ± 1.22	2.47 ± 0.83	2.38 ± 0.73
ERGAS	1.56 ± 0.93	1.23 ± 0.78	1.17 ± 0.54	0.84 ± 0.39	0.83 ± 0.47	0.76 ± 0.32

Corrupted image, 6.6901dB   Original image   Restored image, 45.3119dB

Fig. 3 Parts of the reconstructed HR HSI images

5 Conclusion

HR hyperspectral imaging is challenging due to various hardware limitations. In this paper, we propose an effective sparsity-based hyperspectral image super-resolution method to reconstruct a HR hyperspectral image from a LR hyperspectral image and a HR RGB image of the same scene. The hyperspectral dictionary representing the typical reflectance spectra signatures of the scene is first learned from the LR hyperspectral image. Specifically, an efficient nonnegative dictionary learning algorithm is proposed using a block-coordinate descent algorithm. The sparse codes of the HR hyperspectral image with respect to the learned dictionary are then estimated from the corresponding HR RGB image. To improve the accuracy of estimating sparse codes, a new clustering-based nonnegative structured sparse representation framework is proposed to exploit both the spatial and spectral correlations. The estimated sparse codes are then used with the spectral dictionary to reconstruct the HR hyperspectral images. Experimental results on both public datasets and real-world LR hyperspectral images show that the proposed method can achieve smaller reconstruction errors and better visual quality on most test images than existing HR hyperspectral recovery methods in the literature.

References

1. J.M. Bioucas-Dias, A. Plaza, G. Camps-Valls, P. Scheunders, N.M. Nasrabadi, J. Chanussot, Hyperspectral remote sensing data analysis and future challenges. IEEE Geosci. Remote Sens. Mag. **1**(2), 6–36 (2013)
2. Y. Tarabalka, J. Chanussot, J.A. Benediktsson, Segmentation and classification of hyperspectral images using minimum spanning forest grown from automatically selected markers. IEEE Trans. Syst. Man Cybern. B Cybern. **40**(5), 1267–1279 (2010)

3. H.V. Nguyen, A. Banerjee, R. Chellappa, in *Tracking Via Object Reflectance Using a Hyperspectral Video Camera*. Proceedings IEEE Conference Computer Vision Pattern Recognition Workshops, June 2010, pp. 44–51
4. M. Uzair, A. Mahmood, A. Mian, in *Hyperspectral Face Recognition Using 3D-DCT and Partial Least Squares*. Proceedings of British Machine Vision Conference (BMVC), 2013, pp. 1–10
5. L. Alparone, L. Wald, J. Chanussot, C. Thomas, P. Gamba, L.M. Bruce, Comparison of pansharpening algorithms: outcome of the 2006 GRS-S data-fusion contest. IEEE Trans. Geosci. Remote Sens. **45**(10), 3012–3021 (2007)
6. Z. Wang, D. Ziou, C. Armenakis, D. Li, Q. Li, A comparative analysis of image fusion methods. IEEE Trans. Geosci. Remote Sens. **43**(6), 1391–1402 (2005)
7. W.J. Carper, T.M. Lillesand, R.W. Kiefer, The use of intensityhue-saturation transformations for merging SPOT panchromatic and multispectral image data. Photogram. Eng. Remote Sens. **56**(4), 459–467 (1990)
8. B. Huang, H. Song, H. Cui, J. Peng, Z. Xu, Spatial and spectral image fusion using sparse matrix factorization. IEEE Trans. Geosci. Remote Sens. **52**(3), 1693–1704 (2014)
9. R. Kawakami, J. Wright, Y. W. Tai, Y. Matsushita, M. Ben-Ezra and K. Ikeuchi, in *High-Resolution Hyperspectral Imaging via Matrix Factorization*. Proceedings IEEE Conference Computer Vision Pattern Recognition, Jun. 2011, pp. 2329–2336
10. N. Yokoya, T. Yairi, A. Iwasaki, Coupled nonnegative matrix factorization unmixing for hyperspectral and multispectral data fusion. IEEE Trans. Geosci. Remote Sens. **50**(2), 528–537 (2012)
11. E. Wycoff, T. H. Chan, K. Jia, W. K. Ma, Y. Ma, in *A Non-Negative Sparse Promoting Algorithm for High Resolution Hyperspectral Imaging*. Proceedings of IEEE International Conference Acoustics, Speech Signal Process (ICASSP), May 2013, pp. 1409–1413
12. Q. Wei, J. Bioucas-Dias, N. Dobigeon, J.Y. Tourneret, Hyperspectral and multispectral image fusion based on a sparse representation. IEEE Trans. Geosci. Remote Sens. **53**(7), 3658–3668 (2015)

Restoring EEG Signals by Artifact Suppression with Different Independent Component Analysis Techniques

M. Sreenath Reddy and P. Ramana Reddy

Abstract Signals from eye developments and squints can be requests of extent bigger than mind produced electrical possibilities and are one of the fundamental wellsprings of curios in electroencephalographic (EEG) information. The best in class in electroencephalographic sign extraction is assessed and constraining components and testing issues are cleared up. Since the historical backdrop of the issue is old, there are numerous strategies in the writing. Moreover, some of the current techniques have utilized a mix of systems to conquer the multifaceted nature of the issue. As opposed to already investigated ICA-based techniques for artifact rarity expulsion, this strategy is robotized. Also, look at changed calculations portrayed in this can separate corresponded electro-ocular segments with a high level of exactness. In this paper, Comparison of the raw information, Common ICA artifact concealment strategy and wICA artifact dismissal strategies. Despite the fact that the attention is on disposing of visual artifacts in EEG information, the methodology can be reached out to different wellsprings of EEG tainting, for example, cardiovascular signs, natural commotion, and anode float, and adjusted for use with magneto-encephalographic (MEG) information, an attractive correspond of EEG.

Keywords Independent component analysis (ICA) · Common ICA
wICA · Artifacts

M. Sreenath Reddy (✉)
Department of ECE, JNTUA, Anantapuramu, Andhra Pradesh, India
e-mail: sreenathamie@gmail.com

P. Ramana Reddy
Department of ECE, JNTUACEA, Anantapuram, Andhra Pradesh, India
e-mail: prrjntu@gmail.com

S. S. Dash et al. (eds.), *Artificial Intelligence and Evolutionary Computations in Engineering Systems*, Advances in Intelligent Systems and Computing 668,
https://doi.org/10.1007/978-981-10-7868-2_50

517

1 Introduction

To enhance the seizure identification and go to a usable and solid calculation at the bedside, these artifacts must be checked. In this article, we concentrate on the correlation calculations for expulsion of artifacts to bring down the quantity of false-positive recognitions of the beforehand created neonatal seizure finder without bringing down the affectability. We will not manage a wide range of artifacts as there are extremely numerous and a great deal of them are novel to the particular environment in which the EEG is recorded (e.g., artifacts of close-by component). We will address the three sorts of organic artifacts that are, in our experience, the most aggravating ones for computerized seizure discovery and that are moderately autonomous of the particular checking environment: ECG spikes and vein throb and breath antiquities. These antiquities frequently prompt false-positive identifications because of their comparative morphology to seizures and dreary nature (see Fig. 1). In the established visual appraisal of the neonatal EEG, extra polygraphy flags, for example, ECG, breath, development sensors, are recorded to bolster the neurophysiologist in perceiving the relics. Truth be told, these consolidated enlistments are the brilliant standard while recording neonatal EEG. The human onlooker can perceive the antiquities by contrasting the EEG with at the same time recorded polygraphy signals. In outlining our curio dismissal calculations for the robotized handling of the neonatal EEG, we intended to copy this conduct. On account of volume conduction through cerebrospinal liquid, skull, and scalp, EEG

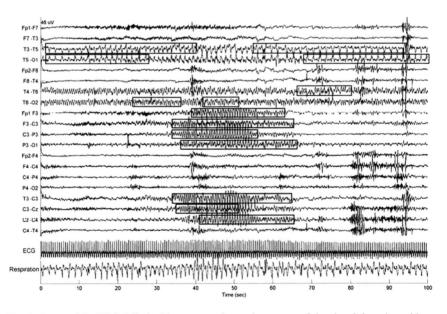

Fig. 1 Areas of the EEG delimited by a rectangle are those parts of the signal detection without artifact removal

signals gathered from the scalp are a blend of neuronal and artifactual exercises from different mind or additional cerebrum sources. Every one of these sources has covering scalp projections, time courses, and spectra. In this manner, their unmistakable components cannot be isolated by straightforward averaging or ghastly separating, and more propelled systems should be utilized.

2   Various Independent Component Analysis Techniques

Here, we are well known about noise, i.e., artifacts according to knowledge on artifacts we can develop the restoration techniques (Fig. 2). It is a subjective descriptor so that depends upon the goodness of the restoration technique restored signal more closer to the reference signal.

2.1   Independent Component Analysis (ICA)

The theory that the sources are commonly measurably autonomous is regularly made. In view of volume conduction through cerebrospinal (CSF) fluid, skull, and scalp, EEG signals gathered from the scalp are a blend of neuronal and artifactual exercises from numerous mind or additional cerebrum sources (Fig. 3). Every one of these sources is a more grounded necessity than the uncorrelatedness between

Fig. 2 Block diagram of proposed method

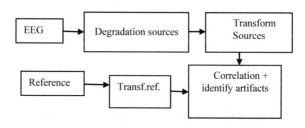

Fig. 3 Outline of the use of blind source separation (BSS) to EEG

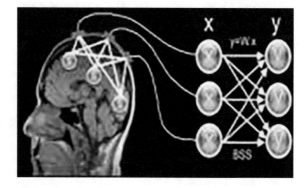

sources. Covering scalp projections, time courses and spectra. Along these lines their unmistakable components cannot be isolated by basic averaging or ghastly sifting and more propelled procedures like ICA should be utilized [1].

A few suppositions should be fulfilled before ICA can be connected to EEG

- The summation of various source signals at the sensors is straight.
- The engendering delays in the blending medium are irrelevant.
- The sources are measurably free.
- The number of autonomous sources is the same or less as the quantity of sensors.
- The blending network is stationary.

The principal suspicion is legitimate as the EEG blending procedure is direct, despite the fact that the procedure producing it might be exceedingly nonlinear. The second supposition is guaranteed by the biophysics of volume conduction at EEG frequencies. The third supposition of autonomy is perfect with physiological models for the era of EEG. The suspicion EEG progression can be demonstrated by feebly connected and in this way factual autonomous mind forms. Late discoveries about the nearness of stage synchrony as a clarification of how mind territories interface may debilitate this third presumption. Be that as it may, for neonates, stage synchrony is not an issue a third suspicion is legitimate. Suspicion four is flawed, since we do not have the foggiest idea about the quantity of measurably autonomous sources.

Two fundamental ways to deal with ICA have been proposed to date. The free parts can be removed mutually or at the same time and then again in a steady progression. It is for the most part recognized that because of blunder amassing all through progressive emptying stages that joint calculations beat deflationary calculations, however at a higher computational burden.

2.2 FastICA

FastICA is depicted as a computationally proficient strategy for playing out the estimation of ICA and is prevalent in neuroscience. It utilizes an alleged settled point cycle plot that has been ended up being 10–100 times quicker than customary inclination plummet strategies. The sources are extricated one by one in light of their kurtosis to make the dispersions of the sources as non-Gaussian as could be allowed. RobustICA is the same as FastICA, however with a little enhancement [2].

2.3 WICA

Wavelet system to improve the execution of the ICA artifacts concealment technique. At the point when managing genuine EEGs, ICA assessed autonomous parts catching counterfeit sources, other than of unequivocally present ancient rarities, much of the time contain a lot of cerebral movement. Dismissal of such segments assumes lost a part of the cerebral action and, thusly, contortion of the relic-free EEG as indicated by the ICA suppositions this part ca excludes different antiques free of the visual. The part can be part into a high adequacy artifacts $a(t)$ and a low plentifulness leftover neural sign $n(t)$

$$S(t) = a(t) + n(t) \qquad (1)$$

Evaluating the continuing neural sign $n(t)$ we can encourage subtract it from the segment and along these lines amend the ICA remaking of the antique-free EEG recording. From the earlier, as it happens in ICA, the decay of the free segment into artifactual and neural action is obscure. Be that as it may, utilizing properties of the signs $a(t)$ and $n(t)$ we can assess them. For sure, the antique, $a(t)$, has high greatness (control) and is restricted in the time and/or in recurrence areas, while $n(t)$ is of low sufficiency and has an expansive band range [3].

3 Simulation Results

Read raw data that is collected from the EEG laboratories. First by using conventional filters, remove the noise and plot the raw data as shown in Fig. 4.

Second step find the independent components and investigating the negative activations as shown in Fig. 5.

By using common ICA artifact suppression method, remove the artifact components and rebuild signal. ICA-cleaned EEG signal is shown in Fig. 6.

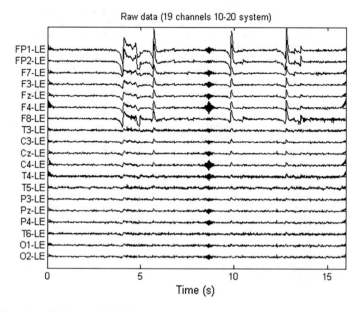

Fig. 4 Raw data (19 channels 10–20 system)

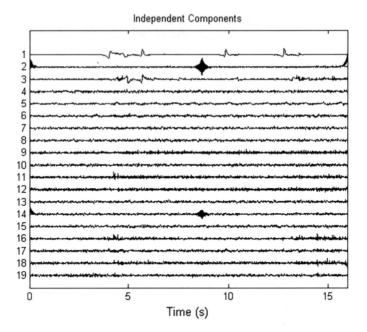

Fig. 5 Independent components

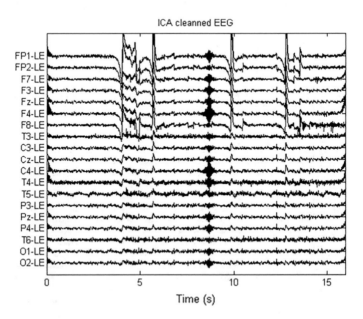

Fig. 6 ICA-cleaned EEG

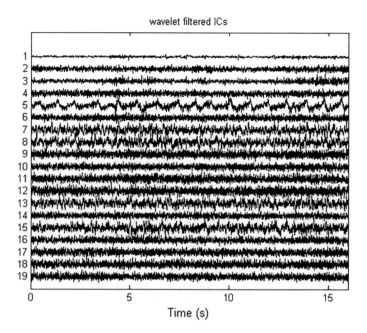

Fig. 7 Wavelet-filtered independent components

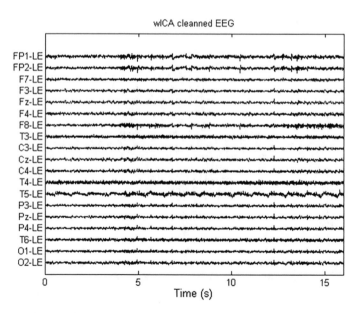

Fig. 8 wICA-cleaned EEG plot

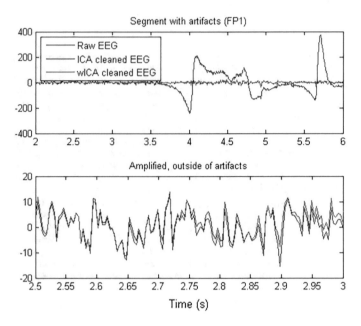

Fig. 9 Segment with artifacts

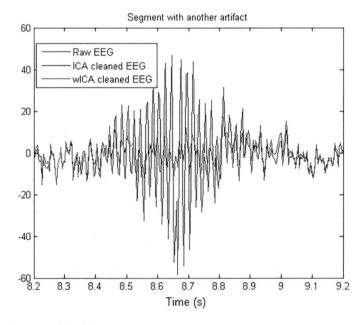

Fig. 10 Segment with artifacts

Next by using wICA artifact rejection method, remove the strong artifacts and plot the wavelet-filtered independent components and wICA-cleaned EEG signal as shown in Figs. 7 and 8.

Then make a comparative plot we plot a segment with (ocular) artifacts and an artifact-free segment. Observe in the artifact-free segment a good artifact suppression method should not perturb the original EEG signal. Comparative plots are shown in Figs. 9 and 10.

4 Conclusion

To begin with, we have demonstrated that ICA-adjusted EEG may halfway lose the cerebral action. In reality, ICA disintegrates EEG into segments of simulated and neural starting points and rejects the previous. Such a division is legitimate for free, directly blended sources when their aggregate number does not surpass the quantity of recording terminals by and by these suspicions can be disregarded prompting a "hole" of the cerebral action into segments esteemed simulated. Complete dismissal of such a segment assumes a halfway loss of the neural sign.

wICA depends on ICA signal disintegration and incorporates as a halfway stride the wavelet thresholding of the autonomous parts. This progression recuperates the low plentifulness, wide band neural action determined in the segments distinguished as in charge of relics. In this way, the consequent erasure of just the artifactual part of the segments does not bend the basic neural movement in the wICA-amended EEG. Note that wICA not just recuperates the cerebral action outside of the antique scenes, and additionally, it permits significant recouping of the neural sign under curios. At long last look at different independent part investigation calculations and further research papers we think about more calculations and comparisons.

References

1. A. Hyvärinen, J. Karhunen E. Oja, *Independent Component Analysis* (Wiley, New York, 2001)
2. M. Ungureanu, C. Bigan, R. Strungaru, V. Lazarescu, Independent component analysis applied in biomedical signal processing, Meas. Sci. Rev. **4**(2), 1–8 (2004)
3. A. Kanagasabapathy, A. Vasuki, Image fusion based on wavelet transform. Inter. J. Biomed. Signal Process. **2**(1), 15–19 (2015)

An Approach of Automatic Data Collection and Categorization

Gajendra Sharma, Manish Kumar and Shekhar Verma

Abstract A novel automatic data collection and categorizing (ADCC) technique is proposed, and it captures data based on the content of the acoustic signal. An important problem in automatic acoustic event detection system is to find the start and finish point indexing of acoustic event signal in the recorded signal. Firstly, a pre-profiling-based acoustic signal capturing technique is proposed. In the non-stationary environments (low signal-to-noise ratio), conventional approaches for start and finish indexing detection often fail. It is observed that the accuracy of the automatic event detection technique also degrades. To overcome this problem, secondly, feature extraction is done using Mel Filter Cepstral Coefficient (MFCC). A combined approach for classification is employed using K-means with 1-Nearest Neighbor (1NN). It is expected that ADCC approach integrated with classifier is computationally economical.

Keywords SNR · MFCC · K-Mean · 1-NN and ADCC

1 Introduction

The acoustic event detection techniques are useful when they report in real time. In continuous acoustic signal, it is difficult to locate the starting and the finishing point of the event signal. To deal with this problem, a threshold value is fixed for the starting point at the time of capturing the signal. After getting the threshold value, the signal is recorded for predefined time duration. In this work, an assumption is

G. Sharma (✉) · M. Kumar · S. Verma
Department of IT, Indian Institute of Information Technology Allahabad,
Allahabad, Uttar Pradesh, India
e-mail: rs118@iiita.ac.in

M. Kumar
e-mail: manish@iiita.ac.in

S. Verma
e-mail: sverma@iiita.ac.in

© Springer Nature Singapore Pte Ltd. 2018   527
S. S. Dash et al. (eds.), *Artificial Intelligence and Evolutionary Computations
in Engineering Systems*, Advances in Intelligent Systems and Computing 668,
https://doi.org/10.1007/978-981-10-7868-2_51

made that an event by-product sound falls under predefined time duration. The approach often fails for acoustic signal have unequal time duration. These signals may be considered as quasiperiodic when it is limited to a short time interval. Also, they are not stationary in nature, and their statistical characteristics vary from time to time such as tree cutting (the change of axe impact on tree trunk). The ambiance noise mapping in recorded acoustic signal is performed if signal-to-noise ratio is high. The algorithm proposed in this work is fairly simple for extracting event signal, and it can be used for recording the acoustic event signal in forest environment. There are two considerations in the algorithm which are measured for noise removal in the algorithm. These are based on signal-to-noise ratio and average RMS energy in the short-term frame. The decision criteria for the setting up the relevant threshold for background noise are estimated through experiment.

The acoustic events consist of constant frequency tones of same duration for, e.g., striking of an axe to the tree trunk. In the forest area, the wind noise and the variation of the distance from source to sensing element also effect the intensity and frequency of acoustic signal. Sometimes, the acoustic event signals merge with audible noise and reverberation. For example, bird's singing, animal's roaring, and insect's song are recorded with the event signal. In previous approaches, the background noise in the segmented audio signal is recognized on the basis of less RMS energy, zero-crossing rates, and the combination of both. However, these methods are having their own limitations. Bi-threshold values are taken for the recognition of the audio event signal in the proposed algorithm. The background noise is classified into two subclasses such as ambience noise and clutter noise. The problem of automatic event detection is based on the assumption that one can locate the region of event in the signal. Although it is difficult to extract the information about the start and finish point of the event in the presence of noise, it is important for various purposes in the area of acoustic event detection and classification. This algorithm must offer less computation time and high reliability.

2 Literature Review

The work done in the previous decades related to this problem is summarized in Table 1.

An algorithm based on logarithm of kurtosis of LPC residual of speech signal was proposed in [8]. The proposed scheme adopts two modes:

- A simple mode using SNR
- Enhanced mode using higher-order statistics such as low-band to high-band energy ratio and SNR.

They proposed maximum a posteriori condition for the voice activity and Likelihood ratio test was applied to the observed spectrum and voice activity decision of previous frame. For making decision of activity, inter-frame correlation

Table 1 History

Sr. No	References	Summary of work
1	Lamel et al. [1]	This work used energy and zero-crossing rate of a shorter period (frame) of speech signal to detect the end point
2	Qi and Hunt [2]	A multilayer feed-forward network compared with maximum likelihood classifier
3	Taboada et al. [3]	An explicit detector proposed to resolve the problems created by quasi-stationary background noise and the noise generated by the speaker
4	Sohn et al. [4]	In this work, the Hidden Markov Model was implemented, and speech detection and false-alarm probabilities are analyzed
5	Wang et al. [5]	They used RMS energy, volume standard deviation, and volume dynamic range for detection of silent ZCR
6	Davis and Nordholm [6]	A statistical method for voice activity detection method is proposed. A threshold value was calculated by low variance spectrum for making voice activity decision
7	Prasad et al. [7]	They compared time domain and frequency domain VAD algorithm. In time domain, linear energy-based detector, adaptive linear energy-based detector, and weak fricatives detector were used

is measured. Instead of conventional MAP criterion, highest probability condition and the speech activity decision are considered as hypothesis in previous frames [9]. The sound files (400) are used in [10] as database to verify the proposed approach. Root mean square (RMS) and four spectral features of sound are considered as feature set. Statistics (mean, variance, and autocorrelation at small lags) of each feature trajectory is stored for classification of all sounds. A voting algorithm proposed for detecting the existence of speech in audio signal. Spectral peak pattern, peak-valley difference, and spectral flatness measure features are calculated for speech, and silence decision TIMIT data are used for experiment in [11].

They used entropy feature to differentiate speech and silence in the speech signal. An adaptive threshold was used to minimize misclassification. The performance of proposed approach was compared with built-in voice activity detection of AMR codec in [12]. The signal-to-noise ratio feature for biometric sample extraction from the audio signal is used in [13]. Audio signal consists of TIMIT speech data with different noises such as car, office, construction, and stadium noise that were recorded for three minutes at eight kHz sixteen bit per samples and used as raw audio signal. They proposed an energy-based sound detection in which spectral centroid, spectral flatness, spectral roll-off, zero-crossing rates, and the first 24 MFCC parameters are used for feature extraction. Support vector machine and Gaussian Mixture model are used for classification of the audio signal. Different types of audio event include claps, door slam, etc. RWCP sound scene database and real-time recording of different audio signal are used in [14]. Long-term spectral flatness method proposed in [15] was suitable for 10 dB and lower signal-to-noise ratio. They tested twelve types of noise and compared with ESTI, AMR, LTSV voice activity approaches.

3 Proposed Methodology of ADCC

This section consists of description about the methodology of proposed algorithm. However, the forest environment poses lower signal-to-noise ratio as compared to the laboratory environment. Therefore, interpretation of the event becomes more complex at low signal-to-noise ratio as compared to high signal-to-noise ratio. In the proposed method, a predefined threshold is applied to locate the starting point of the pattern. Similarly, the end point is selected based on two scenarios as follows:

- The local SNR drops below a threshold.
- The duration of event pattern (Fig. 1).

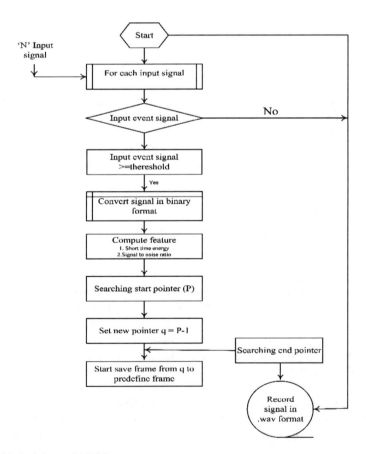

Fig. 1 Methodology of ADCC

The average energy of background noise per frame is calculated based on consecutive five frames of background noise. Signal-to-noise ratio of input frame was monitored for deciding the threshold for audio events. If the SNR per frame is higher than threshold, the input frame (i) will be dissimilar to background noise. The measured SNR is sufficient to determine the category of current frame. In case of low signal-to-noise ratio, start and finish indexing is difficult due to the loss of the initial and final parts of audio event signal. To solve this complication and extraction of start and finish indexing, the SNR pattern matching is proposed. SNR pattern is defined by the pre-profiling of the desired audio signal. To reduce the complexity, two distinctive features are compared in this work. The best suited feature sets used in the proposed algorithm are as follows:

- Short-time RMS energy
- Signal-to-noise ratio.

Two profiles are proposed in this algorithm

- Short-time acoustic event
- Longtime acoustic event.

Short-time acoustic event is related to sudden sharp sounds like axe impact audio signal. Longtime acoustic event is related to linearly increase in sounds like vehicle passing audio signal.

3.1 Preprocessing

A microphone along with preamplifier circuit is connected with sound card of the computer. The audio signal is amplified by the internal amplifier of the sound card. This signal is read by the binary read function, and the stream of data is stored in temporary memory. The data stream stored in temporary memory is the entire input signal and divided into frames of length M. Each of these frames is multiplied by a window function. Figure 2 shows the concept of the signal frame, where each frame represents parts of the previous frame. Hamming window function with shorter frame length of 1024 samples and overlapping choice is 50%. RMS energy and signal-to-noise ratio are selected as features. The threshold energy is selected on the basis of consecutive noise frame. In AAD block, start and finish pattern selection is performed. If the extracted signal is very tiny, then consider it as a noise, otherwise save it as an audio signal. After performing the necessary preprocessing on the input signal, the next step is feature extraction. The level of noise and signal describes the quality of signal. The SNR measured in dB gives better resolution which forecast the difference between noise level and signal level.

Fig. 2 ADCC block diagram

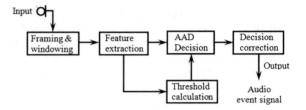

3.2 Feature Extraction

Features' set provides information about the event present in acoustic signal; therefore, properties of features set can be used to differentiate an event signal and background noise in audio signal. The interpretation of feature set requires an approach to reduce the computational complexity. It appears as non-stationary noises such as a bird chirping, monkey chatter, and insect sound (Figure 3).

The feature selection for searching mechanism of auditory event signal is restricted for reliability, speed (real-time processing), adaptive to the environment change. The signal-to-noise ratio compares the level of a desired signal to the level of background noise.

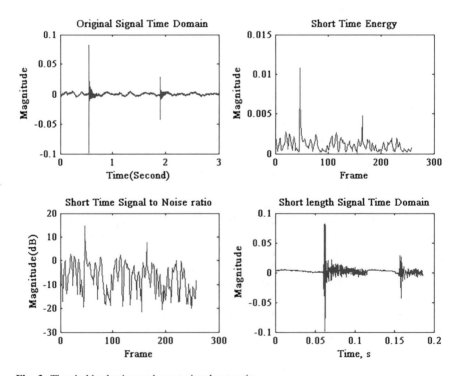

Fig. 3 Threshold selection and event signal extraction

3.3 Feature Extraction: Mel Frequency Cepstral Coefficient

Algorithm: **MFCC Feature Extraction**

Input: *Signal, $S_t = []_{88200}$; t=88200, Frame_Size=1024; Over_lapp=256; Step_Size=Frame_Size-Over_lapp; Number_of_frames= ((t-Frame_Size)/Step_Size) +1=114; FFT_size=Frame_size/2; $F_{s=}44100$*

Number of MFCC Coefficients, m=12; $f_l=100$, $f_{h=}F_s/2$;
Input frame, X(n) =Frame signal*Hamming (Frame_Size);
Output: *MFCC feature vector= $[]_{m*Number of frames}$*

For a=1: Number_of_frames   //Applied Fourier transform

$$X_a[FFT\_Size] = \sum_{n=1}^{Frame\_Size} x[n]e^{\frac{-2\pi i}{Frame\_Size}FFT\_Size*n}, 0 \le FFT\_Size < Frame\_size$$

For b=1: m   //Now making combination of filters

$$H_m = \begin{cases} 0 --- FFT\_Size < f[m-1] \\ \dfrac{(FFT\_Size - f[m-1])}{(f[m]-f[m-1])} ---- f[m-1] \le FFT\_Size < f[m] \\ \dfrac{(f[m+1] - FFT\_Size)}{(f[m+1]-f[m])} ---- f[m] \le FFT\_Size \le f[m+1] \\ 0 --- FFT\_Size > f[m+1] \end{cases}$$

End For

For b=1: m

$$f[m] = (\frac{Frame\_size}{F_s})B^{-1}(B(f_l) + m\frac{B(f_h)-B(f_l)}{M+1}) \text{ //Frequency of f[m] received}$$

$$M^{-1}(m) = 700(\exp(m/1125)-1) \quad \text{// Conversion of value of frequency to Mel scale accordingly}$$

$$S[m] = \ln(\sum_{FFT\_Size=1}^{FFT\_Size} |X_a[FFT\_Size]|^2 H_m[FFT\_Size]), ---- 0 \le m < M$$

End for //We calculate energy for each Fame_size

End for

Return *MFCC feature vector= $[]_{m*Number of frames}$*

3.4   Classification

Algorithm: K-Mean Clustering

Input: $D_{nr} = [d_1............d_n]$ //here D presents master data set extracted from acoustic signal using MFCC Algorithm
  $n=24, r=12$
Output: $D_c = [d_1........d_k]$ {L (d) |d=1......n}// set of cluster level

---

  Initialize randomly K prototype = $(R_1............R_K)$
  Repeat:
    For i=1 ...n   do
    Find k such that $||d_{ir} - R_k|| \leq || d_{ir}-R_j||$
    For j=1 ...K
      Set $P_k = P_k U \{d_i\}$
  End For
  For j=1 ...K   do
   Set R_j =the mean point in P_j
  End For
  Until it converges

Algorithm: 1-Neareset Neighbors

Input: $D_c = [d_1........d_k]$
        $T_c = [d_j]$
Output: A_m =Adjacency matching matrix

  For 1: length of D_c // each sample point d_i
  For each $d_i \in$ 1-NN (d_j) // If d_j is the neighbor of d_i
  $A_m (i, j)=1$
  End

---

4   Results and Discussion

The experimental results of this work are based on the signals collected in forest environment. This experiment consists of different classes of audio signals, such as felling of trees by axe and vehicles passing. The signals are sampled at a sampling rate of 44.1 kHz 16-bit resolution. The decision criteria for the proposed method are fixed by experimental analysis as given in Table 2. Start and finish point of different classes of audio event signals is obtained through listening and visual inspection test, and time space in which event occurs is obtained by finish point detection algorithm. In spite of failure detection of acoustic event segment, experiment is performed within allowed tolerance. Allowed tolerance is the time between start and finish point length. To analyze the performance of the proposed algorithm, a series of experiments are conducted; signal-to-noise ratio feature based in time domain and MFCC based in frequency domain. In raw audio signals, an event signal can be detected by the characteristics of signal-to-noise ratio.

In the open environment, when the signal is captured through device located at various places, the signal-to-noise ratio varies accordingly. During the experiment, some time sensor is covered by leaf of tree, which reduces the amplitude of input signal from actual amplitude. This condition reduces the performance and capturing noise as a false signal. Thus, a Mel frequency filter-based unsupervised

Table 2 Decision criteria

Sr. No	Event name	SNR threshold	ADCC decision	Duration
1	Axe impact	−10 dB	Frame(i)&&Frame (i + 1) > −10 dB	250 ms
2	Saw scratching	−25 dB	Frame(i)||Frame(i + 1) > −25 dB	5 s
3	Vehicle movement	−15 dB	Frame(i)||Frame(i + 1) > −10 dB	20 s

classification approach is proposed. This unsupervised approach uses K-mean and 1-NN (nearest neighbor) classification approach. This technique works like a human thinker. First, it detects the class to which signal belongs and then identifies the similar signals of the same class. The results are having variation in accuracy for the tree cutting and vehicle movement event. The average accuracy of the proposed algorithm is 60%.

5 Conclusion

In this work, an approach based on start and finish point detection technique for acoustic event is presented. This method uses signal-to-noise ratio between ambiance noise and generated acoustic signal during the anthrophonic event. It assumes the noise effect is less than event signals. The threshold is uniquely specified through precise analysis for selected class of event signal. The performance of the proposed method is better than the conventional methods for acoustic signal collection in noisy forest environment. Experimental results show that the proposed Mel Filter Cepstral Coefficient with K-mean and 1 NN method is quite efficient in case of considerable signal-to-noise variation.

References

1. L. Lamel, L. Rabiner, A. Rosenberg, J. Wilpon, An improved endpoint detector for isolated word recognition. IEEE Trans. Acoust. Speech, Signal Process. **29**(4), 777–785 (1981). T.F., Smith, M.S., Waterman, Identification of common molecular subsequences. J. Mol. Biol. **147**, 195–197 (1981)
2. Y. Qi, B.R. Hunt, Voiced-unvoiced-silence classifications of speech using hybrid features and a network classifier. IEEE Trans. Speech Audio Process **1**(2), 250–255 (1993)
3. J. Taboada, S. Feijoo, R. Balsa, C. Hernandez, Explicit estimation of speech boundaries. IEEE Proc.-Sci. Meas. Technol. **141**(3), 153–159 (1994)
4. J. Sohn, N.S. Kim, W. Sung, A statistical model-based voice activity detection. IEEE Signal Process. Lett. **6**(1), 1–3 (1999)
5. Y. Wang, Z. Liu, J.-C. Huang, Multimedia content analysis-using both audio and visual clues. IEEE Signal Process. Mag. **17**(6), 12–36 (2000)

6. A. Davis, S. Nordholm, A low complexity statistical voice activity detector with performance comparisons to ITU-T/ETSI voice activity detectors, in *Proceedings of the 2003 Joint Conference of the Fourth International Conference on Information, Communications and Signal Processing, 2003 and Fourth Pacific Rim Conference on Multimedia*, vol 1 (IEEE, 2003), pp. 119–123

7. R.V. Prasad, A. Sangwan, H.S. Jamadagni, M.C. Chiranth, R. Sah, V. Gaurav, Comparison of voice activity detection algorithms for VoIP, in *Seventh International Symposium on Computers and Communications, 2002 Proceedings ISCC 2002*, pp. 530–535

8. L. Ke, M.N.S. Swamy, M.O. Ahmad, An improved voice activity detection using higher order statistics. IEEE Trans. Speech Audio Process. **13**(5), 965–974 (2005)

9. J.W. Shin, H.J. Kwon, S.H. Jin, N.S. Kim, Voice activity detection based on conditional MAP criterion. IEEE Signal Process. Lett. **15**, 257–260 (2008)

10. G. Wichern, J. Xue, H. Thornburg, B. Mechtley, A. Spanias, Segmentation, indexing, and retrieval for environmental and natural sounds. IEEE Trans. Audio Speech Lang. Process. **18**(3), 688–707 (2010)

11. M.H. Moattar, M.M. Homayounpour, N.K. Kalantari, A new approach for robust realtime voice activity detection using spectral pattern. In *2010 IEEE International Conference on Acoustics, Speech and Signal Processing* (IEEE, 2010), pp. 4478–4481

12. R. Muralishankar, R.V. Prasad, S. Vijay, H. N. Shankar, Order statistics for voice activity detection in VoIP, in *2010 IEEE International Conference on Communications (ICC)* (IEEE, 2010), pp. 1–6

13. F. Beritelli, S. Casale, R. Grasso, A. Spadaccini, Performance evaluation of SNR estimation methods in forensic speaker recognition, in *2010 Fourth International Conference on Emerging Security Information, Systems and Technologies* (IEEE, 2010), pp. 88–92

14. N. Cho, E.-K. Kim, Enhanced voice activity detection using acoustic event detection and classification. IEEE Trans. Consum. Electron. **57**(1), 196–202 (2011)

15. Y. Ma, A. Nishihara, Efficient voice activity detection algorithm using long-term spectral flatness measure. EURASIP J. Audio Speech Music Process. **2013**(1), 1–18 (2013)

Neural Network Approach for Inter-turn Short-Circuit Detection in Induction Motor Stator Winding

Gayatridevi Rajamany and Sekar Srinivasan

Abstract This work deals with neural network approach for automatic detection of stator winding fault of the induction motor. The problem is faced through modeling of induction motor with short circuit in stator winding. Instantaneous phase voltages, peak values of phase currents, and parameters derived from these data are used to train artificial neural network. The output of the neural network classifies the condition of the stator winding. The proposed architecture performs with the selection of a significant feature set, and accurate fault detection results have been obtained.

Keywords Induction motor · Artificial neural network · Modeling
Inter-turn short circuit · Fault detection

1 Introduction

Early detection of faults in induction motors has been an important work for researchers and manufacturers to ensure safe and economic operation of industrial processes [1, 2]. Induction motors are frequently exposed to non-ideal and even detrimental operating environments. These circumstances include overload, recurrent start/stop, insufficient cooling, inadequate lubrication, and other undesirable stresses. During operation, a variety of faults occur within the three-phase induction motor. Studies reveal that 35–40% of induction motor breakdowns are because of stator winding faults [3–5]. Stator winding faults produce hot spots, making the

G. Rajamany (✉)
Department of EEE, Asan Memorial College of Engineering
and Technology, Chennai, India
e-mail: gayatridevir@gmail.com

S. Srinivasan
Department of EEE, Hindustan Institute of Technology
and Science, Chennai, India
e-mail: ssekar.pt@hindustanuniv.ac.in

© Springer Nature Singapore Pte Ltd. 2018
S. S. Dash et al. (eds.), *Artificial Intelligence and Evolutionary Computations
in Engineering Systems*, Advances in Intelligent Systems and Computing 668,
https://doi.org/10.1007/978-981-10-7868-2_52

failure spread rapidly in the winding [6]. Constant online monitoring of this fault is an important tool to reduce costs and save the machines, as the stator winding inter-turn short circuit takes just a few minutes to evolve. Early detection of minor faults in stator winding would ensure repair actions and save production costs.

The stator winding inter-turn short circuit (SWITSC) takes a short time to develop gradually and to damage motor completely. Fenger and Thomson [7] test's result using a low-voltage three-phase asynchronous motor reveals that there exists a time of few minutes to develop winding inter-turn short circuit. In such case, early detection increases the feasibility of repairing the machine by rewinding it or, in large motors, removing short-circuited coils. In the worst case, an early stop avoids electrical arcs and explosion risks.

The conventionally adopted fault detection techniques are thermal monitoring, noise monitoring, partial discharge, and vibration monitoring. Presently, signal conditioning techniques like motor current signature analysis, fast Fourier transform, wavelet analysis, and artificial intelligence techniques are considered for online monitoring [8, 9].

Major benefit of artificial neural network is its accuracy in approximating a nonlinear function as comparing with conventional methods. Artificial neural networks are easily adaptable to new data elements or parameters. Artificial intelligence (AI) is recognized as knowledge-based methodology for both research work and development. Mainly for fault detection and identification in the motor, artificial neural network has been recommended by several researchers [10].

Motor current signature analysis (MCSA) is one of the most important condition monitoring techniques using a stator current sensor. Faults like bearing problem, broken rotor bars, air gap eccentricity abnormalities, and short circuiting in windings generate additional components or variation in the amplitude of the stator current [11].

The authors in [12] focused on broken rotor bar fault detection of induction motor based on time domain analysis of one-phase steady-state and transient current signal and artificial neural network for classification.

Another similar method is motor square current signature analysis(MSCSA) in which spectral analysis of the instantaneous square stator current is carried out. In [13], Eernao Pires and Daniel Foito applied MSCSA to stator winding short-circuit fault and presented a comparison with instantaneous power spectrum.

The work of [14] deals with induction motor bearing fault detection by combining time domain signals (current and voltage) and structures of artificial neural network. Two neural structures composed with inputs {I,V} and {I,V,I2,V2,IxV} are compared in this work.

The objective of this paper is to present a strategy to detect stator inter-turn fault in induction motor based on artificial neural network platform with input from time domain, only considering the stator phase current of the motor.

2 Modeling of Stator Winding Fault

A mathematical model of the three-phase induction motor in abc reference frame is analyzed here for the study of operation of the motor under stator winding fault. The shorted phase winding is divided into two parts—the healthy coil part and the short-circuited part. The turns of shorted phase winding are reduced to y% by stator turn fault.

Flux linkages in stator and rotor are given, respectively, as

$$\lambda s = [\lambda as \quad \lambda bs \quad \lambda cs]T, \quad \lambda r = [\lambda ar \quad \lambda br \quad \lambda cr]T \tag{1}$$

The motor dynamics is represented by

$$P\lambda s = vs - Rs \ is, \quad P\lambda r = vr - Rr \ ir \tag{2}$$

$P\lambda s$ first derivative of stator flux linkage
$P\lambda r$ first derivative of rotor flux linkage.

The voltage and current equations are as follows:

$$is = [ias \quad ibs \quad ics]T, \ ir = [iar \quad ibr \quad icr]T \tag{3}$$

$$vs = [vas \quad vbs \quad vcs]T, \quad vr = [var \quad vbr \quad vcr]T \tag{4}$$

The steady-state equation for flux linkages in terms of the winding inductances and current is represented by Eq. (5)

$$\begin{bmatrix} \lambda_s \\ \lambda_r \end{bmatrix} = \begin{bmatrix} L_s(\theta) & L_{sr}(\theta) \\ L_{sr}(\theta) & L_r(\theta) \end{bmatrix} \begin{bmatrix} i_s \\ i_r \end{bmatrix} \tag{5}$$

where

L_s stator inductance
L_r rotor inductance
L_{sr} mutual inductance between stator and rotor
θr rotor position
Lms magnetizing inductance of stator
Lmr magnetizing inductance of rotor
Lls leakage inductance of stator
Llr leakage inductance of rotor.

Resistance of the shorted winding, a-phase winding resistance is expressed as

$$Ras\_fault = (1 - y\%)Ras\_nominal \tag{6}$$

The self-inductance of a-phase winding could be given by

$$(Lls + Lms)\text{fault} = (1 - y\%)2(Lls + Lms)\text{nominal} \tag{7}$$

Mutual inductance of a-phase winding with other two windings is given by

$$Lms\_new = (1 - y\%)2Lms\_nominal \tag{8}$$

The torque developed in the motor can be expressed using motor parameters and state variables as in Eq. 9

$$T_e = i_s^T \frac{\partial}{\partial \theta} L_{sr} i_r \tag{9}$$

The equation of motion is given by

$$T_e = J\omega' + B\omega + T_L \tag{10}$$

J moment of inertia
ω' time derivative of the rotor speed
$B\omega$ damping coefficient of the motor
T_L load torque.

3   Simulation of Stator Winding Fault Using Distributed Parameter Model

Induction motor modeling under fault conditions is an important factor in predicting its complete behavior. This led to the transition from steady-state operating models to refined models of transient operation. A universal model of a three-phase induction motor is represented in Fig. 1 Each coil of the stator winding is represented by a distributed π model.

The simulation of stator inter-turn fault is built using distributed parameter model as mentioned above. The simulated motor is a three-phase, 3 HP, 4 pole, 415 V, 50 Hz squirrel cage induction motor with 36 slots and 6 coils per phase. Various percentages of turn level short circuits in different phases have been simulated in MATLAB/Simulink environment. Figure 2 shows the Simulink model and submodels for stator inter-turn fault in the R-phase winding. Various percentages of shorting's can be modeled by shorting the resistance of each coil. At first, healthy condition of motor is studied without shorting any coil. During healthy condition, all phase currents were same and balanced.

A short circuit between the turns in a stator phase causes an unbalance in stator phase currents. Various percentages of shorting's like 16.67, 33.33, 50, 66, 83%

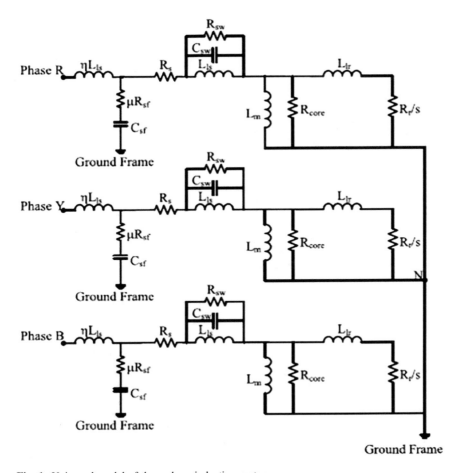

Fig. 1 Universal model of three-phase induction motor

and normal conditions in different phases have been simulated. Figure 3 shows variation in phase currents with five coils shorted (83%).

Figure 4 shows phase currents with three coils short-circuited (50%). Figure 5 shows phase currents with one coil shorted (16.67%).

Simulation shows clear variation in phase currents, Fig. 3. Peak value of R-phase current is 43A which is very higher than the normal value 20A. There is a clear difference between the peak values of phase currents.

From Fig. 4, stator phase currents for a short in three coils or 50% of shorting are found. It makes difference between the peak values of phase currents. One of the phase currents shows a peak value of 32A, while other phase currents are normal. Phase currents with one coil shorted or 16.67% of shorting are shown in Fig. 5 R-phase shows a peak value of 24A, while other phase currents are normal.

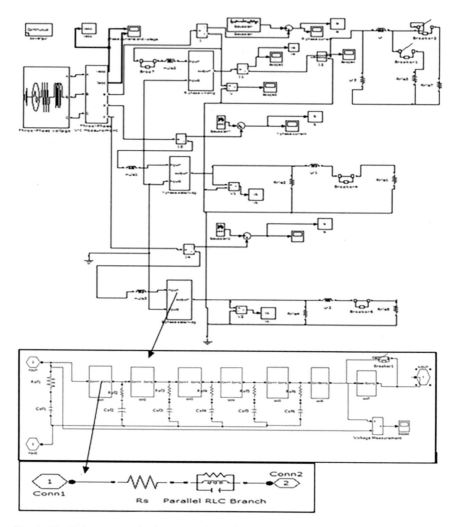

Fig. 2 Simulink system and subsystems of stator inter-turn fault

It became clearly verified that there is a clear increase in the peak value of shorted phase current while increasing the percentage of shorting from 0 to 16.67%, 33.33, 50, 66, and 83%.

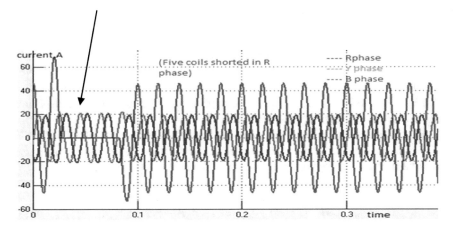

Fig. 3 Stator phase currents with 83% of shorting

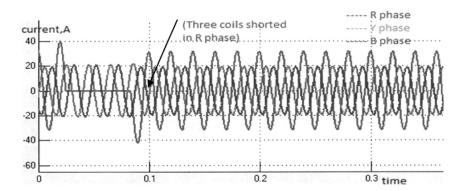

Fig. 4 Stator phase currents with 50% of shorting

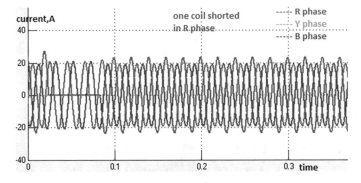

Fig. 5 Phase currents with one turn short-circuited

4 Neural Network Development

Classification using ANN has proved for the solution of a series of problems involving power systems as well as electrical machines [15]. In this work, ANNs are applied to detect stator winding short circuiting in IM. Multilayer perceptron (MLP) ANN with one hidden layer are applied where there is an unknown non-linear relation between input and output.

Back propagation is the most commonly used algorithm to train the MLP ANN. The learning algorithm has two phases of computation, a forward pass and a backward pass. The forward pass output is given by

$$y_k^{(o)}(t) = \varphi_k \left[\sum_{i=0}^{Q} m_{ki} \varphi_i \left(\sum_{j=0}^{P} w_{ij} x_j(t) \right) \right], \tag{11}$$

P is the size of the input vector, Q is the number of neurons in the hidden layer, M is the number of neurons in the output layer, w_{ij} is the weight connecting the input j to the hidden neuron I, $\varphi(.)$ are the activation functions, m_{ki} is the weight connecting the hidden neuron I to the output neuron k ($k = 1,2,\ldots, M$).

In the backward pass, error value generated $e_k^{(o)}(t)$ is

$$e_k^{(o)}(t) = d_k(t) - y_k^{(o)}(t), \ k = 1, \ldots, M \tag{12}$$

$$\phi_k' \left[u_k^{(o}(t) \right] = \partial \phi_k / \partial u_k^{(o)} \tag{13}$$

Local gradient of the output neuron k

$$\delta_k^{(o)}(t) = \phi_k' \left[u_k^{(o)}(t) \right] e_k^{(o)}(t) \tag{14}$$

Local gradient of the hidden neuron i is given as

$$\begin{aligned} \delta_i^{(k)}(t) &= \phi_i' \left[u_i^{(h)}(t) \right] \sum_{k=1}^{M} m_{ki}(t) \delta_k^{(o)}(t) \\ &= \phi_i' \left[u_i^{(h)}(t) \right] e_i^{(h)}(t), \quad i = 0, \ldots Q \end{aligned} \tag{15}$$

Updating the synaptic weights of the neurons,

$$\begin{aligned} m_{ki}(t+1) &= m_{ki}(t) + \eta \delta_k^{(o)}(t) y_i^{(h)}(t), \ i = 0, \ldots Q, \\ w_{ij}(t+1) &= m_{ij}(t) + \eta \delta_i^{(h)}(t) x_j(t), \ j = 0, \ldots P, \end{aligned} \tag{16}$$

where η is the learning rate.

The back propagation algorithm is based on the least mean square (LMS) method and applies a correction in the synaptic weights to the synaptic weights.

5 Classification of Failure Severity Levels During Inter-Turn Short Circuit Using ANN

Here, we use maximum values of phase currents and phase voltages in the time domain. The short circuits are applied in any one of the three phases during the data collection. Maximum value of phase currents (Irm, Iym, Ibm) is stored in column vectors. Differences between these peak phase currents are calculated, and absolute values of these differences are also saved in column vectors. Last parameter stored is the sum of the absolute values. Extracted parameters are tabulated in Table 1 for four samples.

Data collection is made from simulation output as well as from experiments using two motors of rating 5 and 1 HP three-phase induction motor. In this work, detection of stator inter-turn fault is done by data acquisition, selection of data, and treatment of data. Peak values of phase voltages and peak values of phase currents are collected. Differences between the peak values of phase currents are calculated for each sample. Absolute value of this differences and sum of these absolute values are given as the input vector to neural network. Figure 6 gives the block diagram of processes involved in this work of fault detection.

Samples are collected randomly from different conditions such as short circuit in different number of coils of each phase as well as without short circuit in coils. Among these 125 samples, 100 were used for training and 25 for validation. The desired output values are fixed and stored in target vector as shown in Table 2.

Table 1 Input parameters

Parameter extracted	Input 1	Input 2	Input 3	Input 4
Vr	282.4498	296.5794	268.3337	282.4565
Vy	282.8372	296.9788	268.6951	282.8370
Vb	282.4489	296.5713	268.3257	282.4481
Irm	19.96920	20.98110	21.15730	22.66110
Iym	20.61030	19.91630	19.99390	18.95510
Ibm	21.13820	21.20860	20.06790	21.13820
$\lvert$Irm $-$ Iym$\rvert$	0.641100	1.064800	1.163400	3.706000
$\lvert$Iym $-$ Ibm$\rvert$	0.527900	1.292300	0.074000	2.183100
$\lvert$Ibm $-$ Irm$\rvert$	1.169000	0.227500	1.089400	1.522900
$\lvert$Irm $-$ Iym$\rvert$ + $\lvert$Iym $-$ Ibm$\rvert$ + $\lvert$Ibm $-$ Irm$\rvert$	2.338000	2.584600	2.326800	7.412000

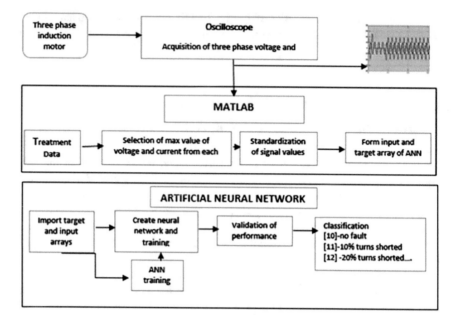

Fig. 6 Block diagram of processes

Table 2 Desired output from trained network

S. No	Severity level	Target value
1	Healthy winding	10
2	16.67% shorting	11
3	33.33% shorting	12
4	50% shorting	13
5	66% shorting	14
6	S3% shorting	15

Table 3 Neural network training parameters

Network functions	
Training	Levenberg–Marquardt
Learning	Gradient descent
Performance	MSE-mean squared error
Transfer	Hyperbolic tangent sigmoid

Data were imported to three feed-forward back propagation network architectures. Network parameters used for training are tabulated in Table 3. Input and output simulation characteristics of each network architecture are given in Table 4.

Table 4 Simulation parameters

Network	Nw1	Nw2	Nw3
Number of layers	1	1	2
Number of neurons in layer 1	8	16	16
	–	–	1
Number of neurons in layer 2	5.9418e-006 at epoch 8	0.00016132 at epoch 1	0.0017803 at epoch 54
Best validation performance			

6   Results and Validation

After training the performance and regression plots, all three NNs are obtained. Among these trained neural networks, network 3 shows least mean squared error while comparing the performance plot of three neural networks. The performance plot and regression plot of nw3 are given in Figs. 7 and 8. Net 3 reached stopping with least mean squared error of 2×10^{-4} at epoch 54. Here, epoch-100, show-100, and max-fail-100 are fixed for nw1, nw2, and nw3.

Validation test results of some samples are tabulated in Table 5.

Validation test is done for all three trained networks. The developed ANN-based stator short-circuit fault detection method is tested with several data with the healthy motor as well as that having stator winding fault. Network 3 obtained 100% accuracy and so confirmed the capability of generalization. The network can classify the different levels of severity of short circuiting in stator windings.

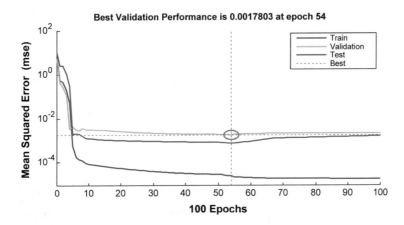

Fig. 7 Performance plot of trained network 3

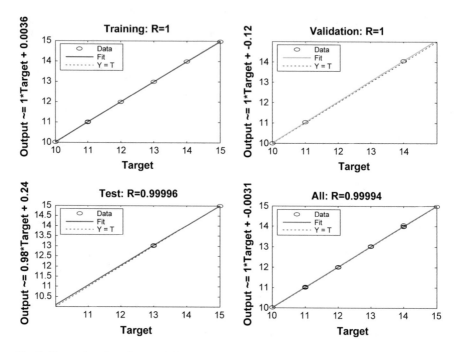

Fig. 8 Regression plot of network 3

S. No	Inputs	Targets	outputs
Table 5 Performance validation Of NW3			
1	2.33800	10	10.0064
2	2.58460	10	10.0086
3	2.32680	10	10.0063
4	7.41200	11	10.9799
5	7.61980	11	11.0437
6	7.49400	11	11.0738
7	11.9096	12	12.0917
8	11.1292	12	11.9946
9	11.2692	12	12.0142
10	45.8806	13	12.9999
11	46.3744	13	13
12	45.3866	13	13
13	71.0920	14	13.9999
14	74.6466	14	14
15	73.5376	14	14
16	100.756	15	14.9988
17	99.7940	15	14.9985
18	101.718	15	14.999

7 Conclusion

This paper introduces a diagnosis methodology applied to stator inter-turn faults using data from time domain analysis. From the acquired data, feature calculation process is performed. Application of selection process results in a selected data set of the most remarkable features which maximize the discrimination between levels of severity of shorting in stator winding. The performance of the proposed system has been tested on a 3 HP, 4 pole, three-phase 50 Hz asynchronous motor. The input parameters extracted and the output are found to be matching with the trained NN desired values. Present system can be extended to the feature addition to continuous monitoring, adding detection of other type of faults in asynchronous motor. Early identification of stator fault increases the possibility of repairing the machine. More sophisticated neural network structures may also be designed in the future for extending the fault detection scheme.

References

1. Z. Wang, C.S. Chang, Y. Zhang, A feature based frequency domain analysis algorithm for fault detection of induction motors, in *Proceedings 6th IEEE Conference on Industrial Electronics and Applications (ICIEA)* , pp. 27–32 (2011)
2. Y.H. Kim, J.H. Sun, D.H. Hwang, D.S. Kang, Y.W. Youn, High-resolution parameter estimation method to identify broken rotor bar faults in induction motors. IEEE Trans. Ind. Electron. **60**(9), 4103–4117 (2013)
3. G.C. Stone, A perspective on online partial discharge monitoring for assessment of the condition of rotating machine stator winding insulation. IEEE Electr. Insul. Mag. **28**(5), 8–13 (2012)
4. A. Shaeboub, S. Abusaad, F. Gu, A.D. Ball, Detection and diagnosis of motor stator faults using electric signals from variable speed drives, in *Proceedings of the 21st International Conference on Automation & Computing*, University of Strathclyde, Glasgow, UK, 11–12 Sept 2015
5. P. O'Donnell, Report of Large Motor Reliability Survey of Industrial and Commercial installations – Part 1, IEEE Transactions on Industry Applications, vol. 21, No. 4, pp. 853–864 July 1985
6. R.M. Tallam et al., A survey of methods for detection of stator related faults in induction machines. In *Proceedings of the IEEE International Symposium on Diagnostics for Electric Machines, Power Electronics and Drives (SDEMPED'03)*, [S.l.: s.n.], pp. 35–46 (2003)
7. W.T. Thomson, M. Fenger, Current signature analysis to detect induction motor faults. Ind. Appl. Mag. IEEE **7**(4), 26–34 (2001)
8. E.H. El Bouchikhi, V. Choqueuse, M. Benbouzid, Induction machine faults detection using stator current parametric spectral estimation. Mech. Syst. Signal Pocess. **52–53**, 447–464 (2015)
9. T. Yang, H. Pen, Z. Wang C.S. Chang, Feature knowledge based fault detection of Induction motors through the analysis of stator current data. IEEE Trans. Instrum. Measur. **65**(3), 549–558 (2016)

10. B. Wangngon N. Sittisrijan S. Ruangsinchai Wanich, Fault detection technique for identifying broken rotor bars by artificial neural network method, in *International Conference on Electrical Machines and Systems(ICEMS), Hangzhou, China*, 22–25 Oct 2014, pp. 3436–3439

11. C. Pezzani, P. Donolo, G. Bossio, M. Donolo, Detecting broken rotor bars with zero setting protection, in *48th Industrial and Commercial Power Systems Technical Conference IEEE* (2012)

12. R.A. Patel, B.R. Bhalja, Induction motor rotor fault detection using artificial neural network, in *International Conference on Energy Systems and Applications, Pune, India*, Oct 2015

13. V. Fernao Pires, D. Foito, J.F. Martins, A.J. Pires, Detection of stator winding fault in induction motors using a motor square current signature analysis (MSCSA), in *IEEE 5th International Conference on Power Engineering, Energy and Electrical Drives (POWERENG)* (2015)

14. W.F. Godoy, I.N. da Silva, A. Goedtel, R.H.C. Palcios, Neural approach for bearing fault classification in induction motor by using motor current and voltage, in *International Joint Conference on Neural Network (IJCNN)*, Beijing, China, July 2014

15. M. Seera, C.P. Lim, D. Ishak, H. Singh, Fault detection and diagnosis of induction motors using motor current signature analysis and a hybrid fmm-cart model. IEEE Trans. Neural Netw. Learn. Syst. **23**(1), 97–108 (2012)

Genetic Algorithm-based Multi-objective Design Optimization of Radial Flux PMBLDC Motor

Amit N. Patel and Bhavik N. Suthar

Abstract Genetic algorithm (GA)-based multi-objective optimal design procedure of radial flux permanent magnet brushless DC (PMBLDC) motor is presented in this paper. Three objective functions are considered, i.e., efficiency, weight, and combination of both. The first two fitness functions are single-objective, and the third one is multi-objective. Multi-objective function is combinational function which incorporates both efficiency and weight of the motor into single fitness function. Design of motor is optimized using these three functions separately. Average flux density (B_g), torque to rotor volume ratio (K_{trv}), air gap length (l_g), motor aspect ratio (A_r), and motor split ratio (S_r) are design variables to optimize. To validate optimized design obtained from the algorithm, finite element analysis is carried out.

Keywords Genetic algorithm · Finite element analysis (FEA) · Multi-objective optimization · Brush-less motor · CAD

1 Introduction

Permanent magnet motors have become widely accepted because of rapid development in the field of permanent magnet materials and semiconductor devices. Application of permanent magnet materials enhances performance of motors significantly. High-energy rare-earth magnets are popular type of permanent magnet materials. Increasing need to develop compact and energy efficient motors is witnessed since long. PMBLDC motors are used in many industrials as well as domestic applications where high efficiency and less weight are desired perfor-

A. N. Patel (✉)
Institute of Technology, Nirma University, Ahmedabad, India
e-mail: amit.patel@nirmauni.ac.in

B. N. Suthar
Government Engineering College, Bhuj, India
e-mail: bhavikiitd@gmail.com

© Springer Nature Singapore Pte Ltd. 2018
S. S. Dash et al. (eds.), *Artificial Intelligence and Evolutionary Computations in Engineering Systems*, Advances in Intelligent Systems and Computing 668,
https://doi.org/10.1007/978-981-10-7868-2_53

mance parameters. Desired performance may not be imparted by permanent magnet (PM) motors designed using conventional design procedures. Use of optimization technique in motor design can enhance performance. This paper presents GA-based design optimization of 2.2 kW, 1450 rpm radial flux PMBLDC motor. Initially, CAD program is developed for motor design. After initial design, GA-based single-objective optimization and multi-objective optimization are carried out. Results obtained from optimization are validated using FE modeling and analysis.

2 Design of Radial Flux PMBLDC Motor

Permanent magnet motor design comprises four basic steps, i.e., main dimensions calculation, design of stator, design of rotor, and estimation of performance [1]. According to sizing equations, CAD algorithm is prepared for designing the motor. All the design stages stated above are comprised, and additional two loops for correction in assumed air gap flux density and efficiency are also included. Average flux density, torque to rotor volume, slot loading, current density, aspect ratio, split ratio, air gap length, stator teeth flux density, stator yoke flux density, rotor yoke flux density, and factor of leakage are various assumed variables in CAD program [2].

3 Genetic Algorithm

Powerful and reliable optimization technique to solve various engineering design problems is GA-based optimization technique. This technique is based on the fundamental of Darwin's principle, survival of the fittest, to optimize particular fitness function. Based on the parametric analysis, five influential design variables are identified and shown in Table 1 with range. Fitness value is calculated for each string with one value of each variable. Several strings are combined together to form a population.

Table 1 Range of variables

Sr. No	Variable	Range	
		Minimum	Maximum
1.	Average flux density (B_g)	0.4 T	0.9 T
2.	Torque to rotor volume ratio (K_{trv})	12 kN/m^2	42 kN/m^2
3.	Motor aspect ratio (A_r)	0.5	1.40
4.	Motor split ratio (S_r)	0.4	0.7
5.	Air gap length (l_g)	0.3 mm	1.0 mm

Following are the main four operators of GA-based optimization technique. Figure 1 illustrates flowchart of optimization process.

- Generate population: Population from available ranges of design variables is randomly generated by this operator. Out of entire population, only one of the strings taken from the initially developed CAD program.
- Selection: It retains multiple copies of string with highest objective function, and it discards string with the lowest objective function.
- Crossover: Two different strings from particular population are interchanged randomly to generate new different strings.
- Mutation: Sudden changes in selected string from population will be done by this operator. Mutation brings positive change leading to generation of better string.

Fig. 1 Flowchart of GA-based optimization technique

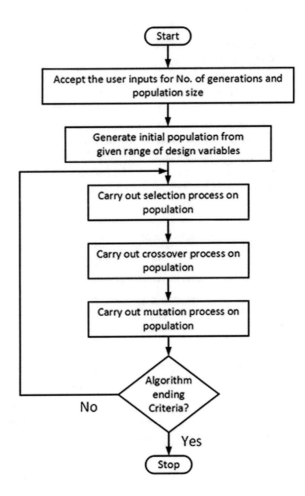

4 Efficiency Optimization

Objective function considered for optimization is efficiency. Initially, B_g and K_{trv} are two design variables considered and optimization is done. Same process of optimization is repeated considering increased number of design variables.

From the result of Table 2, it is observed that as a number of design variables increase, optimum efficiency also increases. Optimum value of efficiency of 97.32% converged after 52 iterations. It is observed that efficiency of GA-based constraint design is higher than CAD-based design; however, efficiency of GA-based unconstraint design is marginally higher than GA-based constraint design as shown in Table 3. In constraint optimization of efficiency, axial length is taken as constraint. Per unit relative comparison of both GA and CAD results considering CAD results as a base for efficiency optimization is shown in Fig. 2. It is observed that resistance, copper loss, and inductance are less in GA-based constraint optimized design as compared to CAD design. Weight of iron and weight of PM are more in optimized design than initial design.

FEA is carried out to validate optimized design. FE model is prepared based on the dimensions and parameters obtained from CAD- and GA-based optimized designs, and analysis is done. Comparison is done between results from FEA and GA techniques, and results are shown in Table 4.

Comparison of flux densities in various sections and average motor torque for initial and optimized designs are shown in Table 4. Figure 3a shows flux density distribution of CAD-based designed PMBLDC motor, and Fig. 3b shows flux density distribution of optimized PMBLDC motor. FEA results validate assumed flux densities in various sections of motor.

Table 2 Efficiency and weight optimization

Design variables	Optimum efficiency (%)	Optimum weight (kg)
B_g, K_{trv}	97.17	6.75
B_g, K_{trv}, l_g	97.20	6.67
B_g, K_{trv}, l_g, A_r	97.27	6.62
B_g, K_{trv}, l_g, A_r, S_r	97.32	6.0

Table 3 Performance comparison of designed motor for efficiency optimization

Parameters	CAD-based design	GA-based optimization	
		Constraint	Unconstraint
Efficiency (%)	95.8	96.34	97.32
Diameter (outer) (mm)	168	249.2	262.6
Length (axial) (mm)	102	100	63.7
No. of turns/slot	29	18	11
Motor weight (kg)	10.4	12.4	13.46

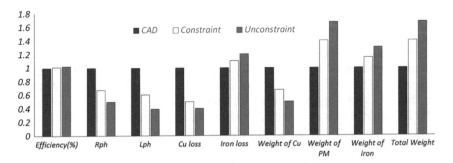

Fig. 2 Comparison of CAD- and GA-based designs for efficiency optimization

Table 4 Comparison of CAD- and GA-based designs for efficiency optimization

Parameters	CAD-based design		GA-based design (unconstraint)	
	CAD	FEM	GA	FEM
Avg. torque (N-m)	14.48	14.4	14.48	14.5
Flux density—air gap (T)	0.7	0.71	0.69	0.67
Flux density—stator core (T)	1.5	1.48	1.5	1.49
Flux density—teeth (T)	1.6	1.56	1.6	1.61
Flux density—rotor core (T)	1.5	1.44	1.5	1.50
Phase inductance (mH)	11.7	10.1	4.58	4.49

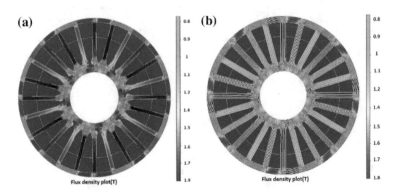

Fig. 3 **a** Field plot of CAD-based design, and **b** field plot of efficiency-optimized (unconstraint) design

5 Weight Optimization

Objective function considered for optimization is weight. To obtain minimum optimized weight of motor is objective here. Table 2 shows optimized motor weight for different number of design variables. As a number of design variables increase, optimized weight reduces. Optimum motor weight of 6.0 kg converged after 29 iterations. It is observed that weight of GA-based constraint design is lesser than CAD-based design; however, weight of motor for GA-based unconstraint is marginally lesser than GA-based constraint design as shown in Table 5. Per unit relative comparison of both GA and CAD results considering CAD results as a base for weight optimization is shown in Fig. 4. In constraint optimization of weight, outer diameter is taken as constraint.

FEA is carried out to validate design optimized for weight. FE model is prepared based on dimensions and parameters obtained from CAD- and GA-based optimized designs for weight, and analysis is done. Comparison is done between results from FEA and GA techniques, and results are shown in Table 6. Flux density distribution in various sections of optimized PMBLDC motor is shown in Fig. 5a. FEA results validate assumed flux densities in various sections of motor.

Table 5 Performance comparison of designed motor for weight optimization

Parameters	CAD-based design	GA-based optimization	
		Constraint	Unconstraint
Efficiency (%)	95.8	89.62	86
Diameter (outer) (mm)	168	130	86.4
Length (axial) (mm)	102	53.2	120.6
No. of turns/slot	29	32	34
Motor weight (kg)	10.4	6.8	6

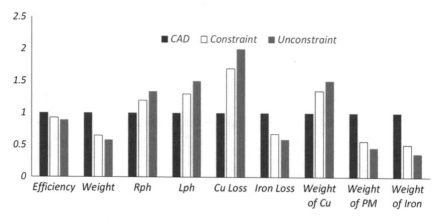

Fig. 4 Comparison of CAD- and GA-based designs for weight optimization

Table 6 Comparison of CAD- and GA-based designs for weight optimization

Parameters	CAD-based design		GA-based design (unconstraint)	
	CAD	FEM	GA	FEM
Avg. torque (N-m)	14.48	14.4	14.48	14.5
Flux density—air gap (T)	0.7	0.71	0.71	0.7
Flux density—stator core (T)	1.5	1.48	1.5	1.52
Flux density—teeth (T)	1.6	1.56	1.6	1.59
Flux density—rotor core (T)	1.5	1.44	1.5	1.5
Phase inductance (mH)	11.7	10.1	13.1	12.1

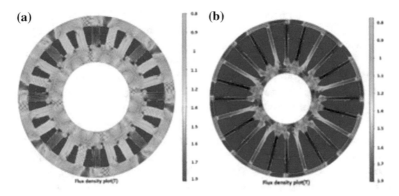

Fig. 5 **a** Flux density plot of weight optimized unconstraint PMBLDC motor, and **b** flux density plot of unconstraint multi-objective optimized PMBLDC motor

6 Multi-objective Optimization

Multi-objective optimization of motor is presented in this section. Optimization of both efficiency and weight is multi-objective fitness function considered here. Reduction in motor weight causes decrement in dimensions of motor; hence, efficiency and weight are two opposite performance parameters of motor. On account of reduction in weight, area available for copper conductors reduces resulting into increased resistance of winding and increased copper losses subsequently [3]. Likewise, reduction in weight of iron increases maximum flux density (B_{max}) and iron losses subsequently.

$$f_{\min(x)} = \left(\frac{WT}{WT_{\min}} \right) - \left(\frac{\eta}{\eta_{\max}} \right) \tag{1}$$

Multi-objective function used for GA is given in Eq. 1. Single-objective optimization gives optimized weight of WT_{min} and optimized efficiency of η_{max}.

Table 7 Performance comparison of designed motor for multi-objective optimization

Parameters	CAD-based design	GA-based optimization	
		Constraint	Unconstraint
Efficiency (%)	95.8	96	96.2
Diameter (outer) (mm)	168	210	127.7
Length (axial) (mm)	102	92	157.1
No. of turns/slot	29	26	25
Motor weight (kg)	10.4	8.5	8.04

Minimization of this multi-objective function gives optimum efficiency and weight simultaneously [4]. Optimized values design variables obtained are $B_g = 0.684\ T$, $K_{trv} = 41.69\ kN/m^3$, $L_g = 0.4\ mm$, $A_r = 1.25$, and $S_r = 0.416$. After 55 iterations converged, optimum efficiency is 96.2% and optimum weight is 8.04 kg. FEA is carried out to validate multi-objective optimized design. FE model is prepared based on the dimensions and parameters obtained from GA-based optimized design, and analysis is done. Multi-objective optimization of outer diameter is taken as constraint. Figure 5b shows flux density distribution of optimized PMBLDC motor. Comparison of flux densities in various sections and average motor torque for initial design and optimized design are shown in Tables 7 and 8. FEA results validate assumed flux densities in various sections of motor. Overall FEA results are fairly matching with results obtained from CAD and GA. Per unit relative comparison of both GA and CAD results considering CAD results as a base for multi-objective optimization is shown in Fig. 6. Table 7 illustrates performance comparison of designed motor for multi-objective optimization. It is observed that efficiency of GA-based constraint optimized design is higher than CAD-based design, and efficiency of GA-based unconstraint optimized design is marginally higher than constraint design, while weight of constraint design is lesser than CAD design, and weight of unconstraint is marginally lesser than constraint design.

Table 8 Comparison of CAD- and GA-based designs

Parameters	CAD-based design		GA-based design (unconstraint)	
	CAD	FEM	GA	FEM
Avg. torque (N-m)	14.48	14.4	14.48	14.5
Flux density—air gap (T)	0.7	0.71	0.684	0.68
Flux density—stator core (T)	1.5	1.48	1.5	1.47
Flux density—teeth (T)	1.6	1.56	1.6	1.61
Flux density—rotor core (T)	1.5	1.44	1.5	1.52
Phase inductance (mH)	11.7	10.1	9.1	8.5

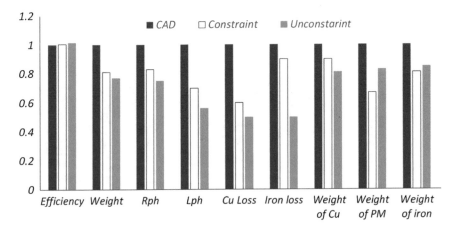

Fig. 6 Comparison of CAD- and GA-based designs for multi-objective optimization

7 Conclusion

Genetic algorithm-based design procedure to obtain optimum design of radial flux PMBLDC motor is discussed. Design optimization is carried out considering three fitness functions: efficiency, weight, and multi-objective. The objective functions improve as a number of design variables increase in all cases. Results and subsequent analysis prove that multi-objective optimization gives overall better design solution. Efficiency is enhanced from 95.8 to 96.2%, and motor weight is decreased from 10.42 to 8.04 kg in unconstraint multi-objective optimization. To confirm methodical optimized design, FEA is carried out. Results of FEA fairly agree with results obtained from optimization.

References

1. D.C. Hanselman, *Brushless Permanent Magnet Motor Design* (McGraw-Hill, New York, 1994)
2. P.R. Upadhyay, K.R. Rajagopal, FE analysis and CAD of radial flux surface mounted permanent magnet brushless DC motors. IEEE Trans. Magn. **41**(10) (2005), pp. 3952–3954
3. J.L. Hippolyte, C. Espanet, D. Chamagne, C. Bloch, P. Chatonnay, Permanent magnet motor multiobjective optimization using multiple runs of an evolutionary algorithm, in IEEE Vehicle Power and Propulsion Conference, Harbin, China November 2008, pp. 1–5
4. R. Ilka, A.R. Tialki, H. Asgharpour-Alamdari, R. Baghipour b jymmx, Design optimization of permanent magnet-brushless DC motor using elitist genetic algorithm with minimum loss and maximum power density. Inter. J. Mechatron. Elect. Comput. Technol. **4**(10) (2014), pp. 1169–1185

Enhancement of ATC Using PSO by Incorporating Generalized Unified Power Flow Controller

Srinivasa Rao Balusu and Lakshmi Narayana Janaswamy

Abstract In the power industry, the word deregulation means transforming the present form of the power market into another form so that better efficiency can be obtained by drawing private sector investments. The deregulated market is mainly concerned about the transmission pricing, congestion management, and available transfer capability (ATC). Among these interesting challenges, ATC should be the one to be taken care for providing a solid open access transmission service. ATC can be improved either by using new transmission facilities or by using FACTS devices. This paper concerns about the FACTS device application, namely generalized unified power flow controller (GUPFC), to maximize the power transfer for a proposed transaction during normal circumstances. Particle swarm optimization (PSO) is used as an optimization tool for obtaining optimal control settings of GUPFC, so that the audacious job of establishing new transmission system can be prevented for enhancement of ATC. Studies on IEEE 6 bus and IEEE 30 bus test systems are done to illustrate the results obtained in all the cases by the use of the proposed method which is the best.

Keywords ATC enhancement · Deregulation · Particle swarm optimization (PSO)
Optimal power flow · PTDF · GUPFC modeling

1 Introduction

Deregulated market operation is being adopted by many electric power utilities under the process of deregulation. It is the responsibility of the Independent System Operator (ISO) to render an indiscriminate open access transmission network to all the producers and distributors of electric power. For optimal utilization of the

S. R. Balusu (✉) · L. N. Janaswamy
V. R. Siddhartha Engineering College, Vijayawada, Andhra Pradesh, India
e-mail: balususrinu@gmail.com

L. N. Janaswamy
e-mail: jnarayanaln@gmail.com

© Springer Nature Singapore Pte Ltd. 2018
S. S. Dash et al. (eds.), *Artificial Intelligence and Evolutionary Computations
in Engineering Systems*, Advances in Intelligent Systems and Computing 668,
https://doi.org/10.1007/978-981-10-7868-2_54

existing transmission lines, the generating and distributing companies should provide accurate information to ISO so that it can post in advance the transfer capability of the network. Hence, ATC comes into use as a measure to utilize the transmission network at its full capacity. Among many methods proposed, (i) linear approximation methods, (ii) optimal power flow (OPF)-based methods, and (iii) continuation power flow (CPFLOW)-based methods are prominent in the calculation of ATC. Linear approximation methods [1] involve solving DC power flow equations and linear constraints. Compared to other two methods, the linear approximation methods are faster with less accuracy. The nonlinear power flow equations as well as nonlinear constraints such as the thermal limit along with voltage violation constraints can be solved using OPF-based methods [2]. However, the voltage stability and transient stability constraints cannot be directly dealt by the linear approximation methods and OPF-based approach. Here, the CPFLOW-based methods come into picture which can address the thermal limits and voltage violation constraints along with voltage collapse for each contingency study.

The inter-area tie lines are basically designed to maintain the system security, reliability, and system restoration in an integrated power market. But in the deregulated environment, the integration of various systems is becoming a market need. Thus, bulk power transfers are dependent on the inter-area tie lines on a regular basis from sources of low-priced generation to loads. It can also be said that, because of deregulation, the pattern of grid integration has shifted from regional self-sufficiency to best utilization of resources across vast geographical areas. Thus, the calculation of ATC of the network becomes a significant part of ISO, and the same is allotted efficiently to various market participants.

In general, depending upon the information of ATC, decisions can be taken for the next set of transactions. Therefore, calculation of ATC has become an important aspect under deregulated market structures. In the USA, the values of ATC estimated for the next hour as well as for each hour into the future would be displayed by an ISO on a Web site known as the open access same-time information system (OASIS) [3]. If one who wants to transact power on ISO's transmission system can access the OASIS Web pages, the data available on ATC is used to ascertain whether the transmission system could put up the transaction or not by maintaining necessary reserve for transmission service.

To calculate the amount of power that can be transacted across the transmission network in a particular direction, within the security constraints, ATC is used as a measure. The conventional methods to increase the transfer capability of the transmission lines are to install phase shifting transformers and SVC. FACTS devices have much better voltage control capabilities than those of the SVC. One of the key concepts in deregulated power industry is the fast and accurate calculation of ATC of the transmission system. Linear ATC calculation using DC power transfer distribution factor (DCPTDF) is given in [4] and is used to distribute real power in the transmission lines. In [5], the method for enhancement of ATC incorporating STATCOM using CSO optimization algorithm is given. An approach for enhancement of ATC using different FACTS devices like TCSC, SVC, and UPFC whose settings are obtained by PSO is given in [6].

In this paper, the multitype FACTS device, generalized unified power flow controller (GUPFC) for the enhancement of ATC has been presented using linear static method. The GUPFC is modeled as power injection model, and the control parameters are set in such a manner that the reactive power is compensated, thus enhancing the ATC. The power injection model of GUPFC is presented in Sect. 2. Introduction of PSO has been described in Sect. 3. The PSO algorithm for enhancement of ATC is presented in Sect. 4. Section 5 deals with results and discussions of two different test systems. Finally, some valid conclusions are drawn at the end.

2 Power Injection Model of GUPFC

FACTS controllers have the ability to change the network parameters fast and efficiently so that better performance can be achieved. GUPFC is one such device. GUPFC is also known as multiline UPFC. GUPFC has the ability to control power flow in more than one line. The basic structure of GUPFC is shown in Fig. 1.

The basic structure of GUPFC consists of totally three converters, one is connected in shunt and the other two are connected in series with the transmission line. So the GUPFC is able to control a total of five power system quantities like bus voltage, active and reactive power flow across the lines [7].

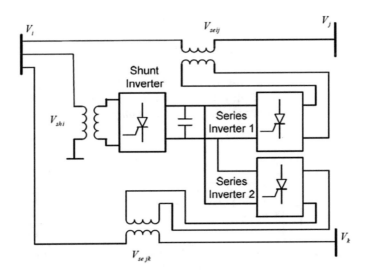

Fig. 1 Generalized unified power flow controller

2.1 Equivalent Circuit and Power Flow Constraints of GUPFC

The power injection model focuses on steady-state mathematical representation of GUPFC into Newton Raphson power flow equations. The main objective of the GUPFC under steady-state operating condition is to have control over the voltage and power flow. The equivalent circuit of GUPFC consists of one shunt controllable voltage source injection at bus and two series controllable voltage sources injected in lines as shown in Fig. 2. In GUPFC, exchange of real power between the AC terminal of shunt and series converters via the common DC link can be achieved. If the converter losses are neglected, sum of the real power exchange must be always equal to zero. By using GUPFC, it is always possible that more degree of control freedom of series converters results in more control over the objectives.

The injected powers [8] at bus i, j, and k can be expressed as

$$P_{i,g} = 0.03V_i^2 \left[\sum_{q=j,k} rB_{se,iq} \sin \gamma \right] - 1.03rV_iV_jB_{se,ij} \sin(\delta_i - \delta_j + \gamma)$$
$$- 1.03rV_iV_kB_{se,ik} \sin(\delta_i - \delta_k + \gamma). \tag{1}$$

$$Q_{i,g} = -V_i^2 \left[\sum_{q=j,k} rB_{se,iq} \cos \gamma \right] - Q_{sh}. \tag{2}$$

$$P_{p,g} = rV_iV_pB_{se,ip} \sin(\delta_i - \delta_p + \gamma)\forall p = j,k. \tag{3}$$

$$Q_{p,g} = rV_iV_pB_{se,ip} \cos(\delta_i - \delta_p + \gamma)\forall p = j,k. \tag{4}$$

where $B_{se} = 1/X_{se}$, r and γ are respective per unit magnitude and phase angles of the series voltage sources operating within the limits as $0 \le r \le 0.1$, $0 \le \gamma \le 2\pi$. The coefficient 1.03 represents the converter switching loss factor.

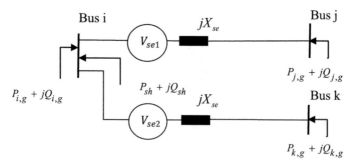

Fig. 2 Equivalent circuit of GUPFC

Power injection model has been implemented into load flow equations. A GUPFC model in power system is also same as that of the UPFC in steady state. The basic GUPFC can be represented by three voltage sources, which can control both magnitude and phase angle of the voltages injected and the impedances which are leakage reactance of three coupling transformers. Let us consider that GUPFC is connected between the buses i, j, and k as shown in Fig. 2. The following assumptions are taken into consideration for analysis purpose:

1. Voltage at bus i is taken as, $V_i = V_i \angle \delta_i$
2. The two controllable series voltage sources are identical, i.e., $V_{se1} = V_{se2} = V_{se} = rV_i e^{j\gamma}$ where 'r' and 'γ' are respective per unit magnitude and phase angles of series voltages and which are operating in the specified limits as $0 \le r \le r_{max}$ and $0 \le \gamma \le \gamma_{max}$.
3. The leakage reactance of the two series coupling transformers is equal.

Moreover, the GUPFC power injection model can be easily incorporated into the steady-state power flow analysis. Power injection equations for GUPFC are given in Eqs. 1–4. The power mismatches in Newton Raphson load flow method are modified using the Eqs. 5 and 6

$$\Delta P_{i,new} = \Delta P_{i,old} + P_{i,GUPFC}. \tag{5}$$

$$\Delta Q_{i,new} = \Delta Q_{i,old} + Q_{i,GUPFC}. \tag{6}$$

3 Particle Swarm Optimization

Particle swarm optimization (PSO) is a simple and modern evolutionary programming algorithm. James Kennedy who was a psychologist along with Russ Eberhart an electrical and computer science engineer developed a new algorithm called PSO with a combination of social science and computer science [9]. In PSO algorithm, random population is initialized with random solutions which search for optima by updating the current population. The current population may be nearer to local best or global best but finally all the population has to reach the global best. The random population updated with the past experiences will be generated until the global best solution is reached. Since the PSO is a population-based EP, a minimal or maxima search will be done using the randomly generated population. For each randomly generated population, a different weight or inertia will be given by

$$\omega = (maxit - it)/maxit. \tag{7}$$

Velocity of the particles is given by

$$
\begin{aligned}
\mathrm{velx}_{it}(pp, ig) = {} & \omega * \mathrm{velx}_{it-1}(pp, ig) \\
& + (c_1 * rx_1(pp, ig) * (\mathrm{localxg}(pp, ig) - \mathrm{DA}(pp, ig))) \\
& + (c_2 * rx_2(pp, ig) * (\mathrm{globalxg}(pp, ig) - \mathrm{DA}(pp, ig))).
\end{aligned}
\tag{8}
$$

where pp is population; ig is the particle; rx_1 and rx_2 are the random numbers of 1st and 2nd values; localxg, globalxg are local and global best values of the previous iteration; $DA(pp, ig)$ is the randomly generated particle of ppth population and igth particle. The position of the particle is updated using the velocity formula, as

$$
DA_t = DA_{t-1} + \mathrm{vel}_t.
\tag{9}
$$

Each updated particle is substituted in the objective function. The fitness is calculated by taking inverse of objective function, the value where the fitness is more will be chosen as the best fitness value.

4 Available Transfer Capability

4.1 Linear Static Method

Some of the methods used for calculation of ATC are continuous power flow, linear static methods, and optimal power flow methods. In continuous power flow method for the calculation of ATC from one node to another node, DC load flow is performed by increasing the amount of transaction until the limit of any of the corridor is reached. However, this method is computationally inefficient and time-taking process, therefore power transfer distribution factors (PTDF) are used for calculating ATC. Linear static methods include ACPTDF and DCPTDF for the calculation of ATC. ACPTDF method is accurate but computationally slow and becomes complex as the number of buses increases. DPTDF method is approximate method; however, it is a fast method because of the reason that it uses the network impedance matrix which does not change with time.

Assume that power is injected at bus i and is drawn at bus j, $\mathrm{PTDF}_{lm,ij}$ gives the fraction of that power transferred that ends up flowing in line k connected between bus l and bus m given by

$$
\mathrm{PTDF}_{lm,ij} = \frac{\Delta P_{lm}}{\Delta P_{ij}}.
\tag{10}
$$

Using the DC model [10], PTDF can be approximated as

$$
\mathrm{PTDF}_{lm,ij} = \frac{X_{li} - X_{mi} - X_{lj} + X_{mj}}{x_{lm}}.
\tag{11}
$$

ATC is calculated by observing the new flow on the line connecting bus l and bus m, because of the transaction from bus i to bus j.

$$P_{ij,lm}^{max} = \frac{P_{lm}^{max} - P_{lm}^0}{PTDF_{lm,ij}}. \tag{12}$$

where P_{lm}^0 is the base case real power flow on the line and P_{ij} is the magnitude of power for proposed transaction. If the maximum power that can be transferred without overloading line lm connected between bus l and bus m is $P_{ij,lm}^{max}$, then,

$$ATC_{ij} = min\left(P_{ij,lm}^{max}\right)\forall lm. \tag{13}$$

For the enhancement of ATC, the limiting line is taken into consideration and the base case flow in that line is optimized using PSO. The objective function taken is minimization of base case power flow in limiting line, i.e., line lm.

$$f = Min\left(P_{lm}^0\right). \tag{14}$$

The following steps describe the algorithm for proposed method.

Step 1: Read Data.

Step 2: Generate minimum and maximum velocity limits of the random particles to be generated.

Step 3: Generate random particles and velocities.

Step 4: Perform load flow analysis for all the randomly generated particles and calculate the objective function value for each particle.

Step 5: Local fitness value for each particle and global fitenss value among all the particles are stored.

Step 6: Update the particle velocities and perform the load flow for all the updated particles, and at the end of load flow, calculate the objective function value for each particle, i.e., the ATC. The limiting line for ATC is considered, and the base case flow is reduced.

Step 7: The global fitness and local fitness values are updated according to the objective function.

Step 8: From step 6 is repeated for a predefined number of iterations.

5 Results and Discussions

The proposed method has been implemented using MATLAB program with GUPFC and tested on two different test systems. GUPFC is incorporated in IEEE 6 bus system between buses 6, 4, and 5. The IEEE 6 bus test system is shown Fig. 3 for evaluation of ATC in PowerWorld Simulator. The transaction is assumed

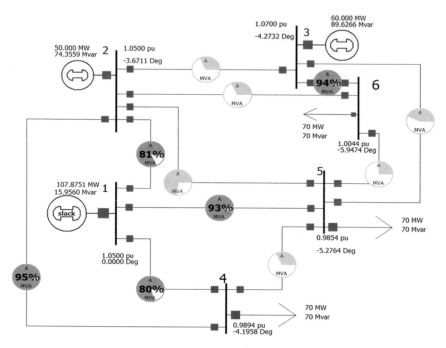

Fig. 3 PowerWorld simulation diagram IEEE 6 bus system

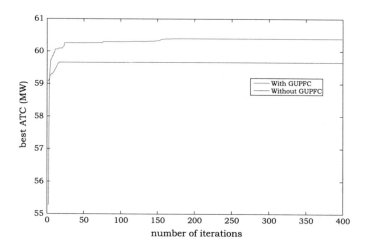

Fig. 4 Comparison of ATC for IEEE 6 bus system with and without GUPFC

between bus 3 and bus 6. ATC before optimization is 50.8247 MW and is optimized to 59.6581 MW using PSO and is further enhanced to 60.3914 MW by the placement of GUPFC. The control parameters settings of the GUPFC are obtained from PSO in accordance with the ATC optimization. Optimal settings of 0.1 per unit for series voltage magnitude and 1.6245 radians of series voltage angle are obtained after the use of PSO algorithm. The convergence plot for ATC enhancement using PSO for IEEE 6 bus system is shown in Fig. 4. It can be observed that the maximum value of ATC is obtained within 150 iterations, and by the use of PSO, there has been a significant enhancement of ATC when compared with the base case value.

The PTDF results have been verified with [11] as well as with the PowerWorld Simulator. The results of ATC for some sample transactions in IEEE 6 bus system and their comparison with the PowerWorld Simulator are shown in Table 1. The control variables and corresponding cost and power losses without and with GUPFC for IEEE 6 bus system are given in Table 2.

In IEEE 30 bus system, GUPFC has been placed between buses 12, 15, and 16. The control parameters settings of the GUPFC are set accordingly with respect to optimization of ATC using PSO. The transaction is proposed between bus 12 and bus 15. The convergence plot for ATC enhancement using PSO for IEEE 30 bus system

Table 1 Comparison of ATC results with PowerWorld Simulator for IEEE 6 bus system

Transaction between buses (i–j)	ATC values in MW	
	With PowerWorld	With MATLAB
1–6	14.784	14.7838
2–6	52.954	52.9525
3–6	50.8245	50.8247
4–6	37.654	37.6534
1–4	24.3413	24.3407
2–5	22.8311	22.8306
3–4	45.2736	45.2738

Table 2 ATC results for transaction between bus 3 and bus 6 with and without GUPFC for IEEE 6 bus system using PSO

Control variables	Without GUPFC	With GUPFC
P_{g1} (MW)	56.5268	76.7207
P_{g2} (MW)	117.8405	95.3911
P_{g3} (MW)	45.00	45.00
V_{g1} (p.u.)	1.05	1.1
V_{g2} (p.u.)	1.05	1.1
V_{g3} (p.u.)	0.95	1.038
r (p.u.)	–	0.1
γ (rad)	–	1.6245
Qsh	–	0
ATC (MW) (3–6)	59.6581	60.3913
Cost ($/MW)	418.0882	379.7271
Power loss (MW)	9.3669	7.1118

is shown in Fig. 5. The value of ATC before optimization is 20.5557 MW. The value of ATC after optimization using PSO is 31.2869 MW. The results of ATC for some sample transactions in IEEE 30 bus system and their comparison with the PowerWorld Simulator are shown in Table 3.

The optimal settings of GUPFC obtained corresponding to the ATC are series voltage source magnitude of 0.1 p.u. and angle of 2.9743 radians. After incorporating GUPFC, the value of ATC has been enhanced to 32.7082 MW. The ATC increased by 1.4213 MW after incorporation of GUPFC as shown in Fig. 5. The IEEE 30 bus test system is modeled in PowerWorld Simulator as shown in Fig. 6. The results of ATC for some sample transactions and their comparison with the PowerWorld Simulator for IEEE 30 bus system are shown in Table 4. The control variables and the corresponding cost and losses without and with GUPFC for IEEE 30 bus system are given in Table 4.

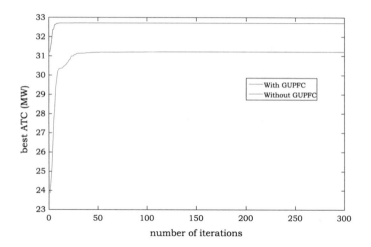

Fig. 5 Comparison of ATC for IEEE 30 bus system with and without GUPFC

Table 3 Comparison of ATC results with PowerWorld Simulator for IEEE 30 bus system

Transaction between buses (i–j)	PowerWorld Simulator (MW)	MATLAB (MW)
5–29	13.666	13.666
6–14	36.592	36.597
1–26	12.455	12.4547
2–15	27.898	27.9021
4–23	19.158	19.1594
12–15	20.553	20.5557

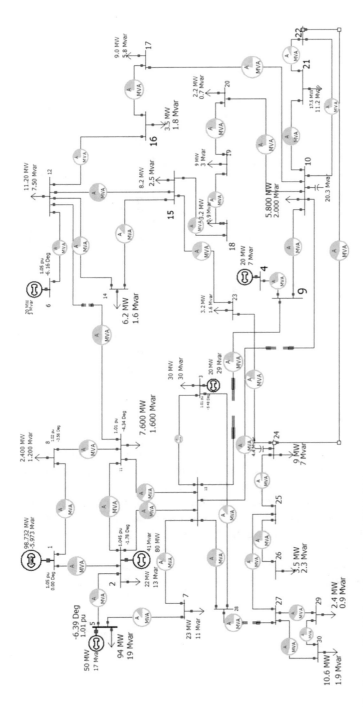

Fig. 6 PowerWorld simulation diagram IEEE 30 bus system

Table 4 ATC results for transaction between bus 12 and bus 15 with and without GUPFC for IEEE 30 bus system using PSO

Control variables	Without GUPFC	With GUPFC
P_{g1} (MW)	76.4001	102.1873
P_{g2} (MW)	80	80
P_{g3} (MW)	34.9998	28.2819
P_{g4} (MW)	30	28.3597
P_{g5} (MW)	50	40.2188
P_{g6} (MW)	12	12
V_{g1} (p.u.)	1.1	0.9925
V_{g2} (p.u.)	1.0981	1.1
V_{g3} (p.u.)	1.0578	1.1
V_{g4} (p.u.)	1.0393	0.9511
V_{g5} (p.u.)	1.0569	1.0722
V_{g6} (p.u.)	0.9484	1.0145
Tr1	0.9501	0.9129
Tr2	0.9	0.9016
Tr3	1.1	1.0772
Tr4	0.9	0.9382
r (p.u.)	–	0.1
γ (rad)	–	2.9743
Qsh	–	0.0452
ATC (MW) (12–15)	31.2869	32.7082
Cost ($/MW)	852.6209	880.2223
Loss (MW)	5.8598	7.5865

6 Conclusions

The ATC value aids as an important tool that points the system performance. In this paper, effect on ATC using GUPFC is studied on IEEE 6 bus and IEEE 30 bus test systems. The results are compared with the PowerWorld Simulator software. From the results, it can be observed that location and the control parameters of the GUPFC affect the ATC. The optimal control settings of GUPFC parameters are obtained by using PSO algorithm. It is shown that the ATC is enhanced significantly by using PSO algorithm with incorporation of GUPFC. The control of FACTS device offers an effective way to boost the system power transfer capability in turn improving the transmission services in deregulated market.

References

1. NERC Rep, Available transfer capability Definitions and determinations, *North American Electric Reliability Council (NERC)*, June 1996
2. J.C.O. Mello, A.C.G. Melo, S. Granville, Simultaneous transfer capability assessment by combining interior point methods and Monte Carlo simulation. IEEE Trans. Power Syst. **12** (2), 736–742 (1997)
3. J.H. Chow, F.F. Wu, J.A. Momoh, Applied mathematics for restructured electric power systems, in *Applied Mathematics for Restructured Electric Power Systems* (Springer, US, 2005), pp. 1–9
4. R.D. Christie, B.F. Wollenberg, I. Wangensteen, Transmission management in the deregulated environment. Proc. IEEE **88**(2), 170–195 (2000)
5. T. Nireekshana, G. Kesava Rao, S. Siva Naga Raju, Enhancement of ATC with STATCOM using Cat Swarm Optimization, in *ELSVIER Transactions on Proceedings of International Conference on Control, Communication and Power Engineering*, vol. 2, pp. 422–428 (2013)
6. B.V. Manikandan, S. Charles Raja, P. Venkatesh, Available transfer capability enhancement with FACTS devices in the deregulated electricity market. J. Electr. Eng. Technol. **6.1**, 14–24 (2011)
7. B.S. Rao, K. Vaisakh, Application of ACSA to solve single/multi objective OPF problem with multi type FACTS devices, in *Power and Energy Engineering Conference (APPEEC), 2013 IEEE PES Asia-Pacific*. IEEE (2013)
8. C.V. Suresh, S. Sivanagaraju, Increasing the loadability of power system through optimal placement of GUPFC using UDTPSO. J. Electr. Syst. **11.1** (2015)
9. R. Eberhart, J. Kennedy, A new optimizer using particle swarm theory, in *Micro Machine and Human Science, 1995. MHS'95, Proceedings of the Sixth International Symposium on*. IEEE (1995)
10. D.P. Kothari, J.S. Dhillon, *Power System Optimization* (Prentice Hall of India, 2004)
11. A.J. Wood, B.F. Wollenberg, *Power Generation, Operation, and Control* (Wiley, New York, 2012)

Electric Field Computation of Epoxy- Nano and Micro Composite Conical Spacer in a Gas Insulated Busduct

Y. Swamy Naidu and G. V. Nagesh Kumar

Abstract In gas insulated systems, the breakdown strength of SF_6 gas is badly affected by locally enhanced electric fields. Polymer nano-composites are the recent advancements in alternatives for the existing insulating materials. Polymer nano- and micro-composite mixture exhibits excellent electrical, thermal, and mechanical properties. In GIS, the reliability can be enhanced up to a great extent with epoxy resin along with polymer nano- and micro-filler mixture as dielectric coating material. The addition of nano- and micro-filler can further enhance insulation properties of epoxy resin. In this paper, the electric field distribution and calculation of relative permittivity of a single-phase common enclosure for an optimized design of GIS with nano-composites are carried out. Inorganic nano-fillers like alumina (Al_2O_3) and titanium (TiO_2) with 100 μm dielectric coating thickness are added to epoxy, and the resultant permittivity is calculated. The electric field distribution with AC as applied voltage is calculated at the surface of the cone type spacer and the distribution. Finite element method (FEM), one of the proven numerical methods, is used for computing the electric fields at various points under consideration and is plotted. The results are presented and analyzed for various filler concentrations.

Keywords Gas insulated substation (GIS) · Nano- and micro-fillers
Epoxy nano- and micro-composites

1 Introduction

Sulfur hexafluoride gas insulated frameworks (GIS) have been in operation for AC frameworks for more than 40 years. GIS has been created with the perspective of using the high dependability, compaction, and economization accomplished by gas

Y. Swamy Naidu
Vizag Institute of Technology, Visakhapatnam, Andhra Pradesh, India

G. V. Nagesh Kumar (✉)
Vignan's Institute of Information Technology, Visakhapatnam, Andhra Pradesh, India
e-mail: drgvnk14@gmail.com

© Springer Nature Singapore Pte Ltd. 2018
S. S. Dash et al. (eds.), *Artificial Intelligence and Evolutionary Computations in Engineering Systems*, Advances in Intelligent Systems and Computing 668,
https://doi.org/10.1007/978-981-10-7868-2_55

protection [1]. It was demonstrated that even with clean protection conditions, the DC flashover voltage of the bolster epoxy spacers can be out of the blue lessened [2–4]. For an AC GIS, the issues important to spacers can be unraveled via cautious plan of spacers regarding electric field appropriation. Regularly, unadulterated SF_6 at high weights is utilized to protect the framework. The nearness of spacers results in complex-dielectric field circulation. It frequently heightens the electric field especially on the spacer's surface. The protection capacity of SF_6 is exceedingly delicate to the most extreme electric field, and besides, the protection quality along a spacer's surface is generally lower than that in the gas space. They ought to be definitely intended to acknowledge pretty much uniform field conveyance along their surfaces. Also to decrease its value as low as possible keeping in mind the optimum leakage path. Spacer's profile is viewed as the fundamental variable, which controls the field dissemination, and henceforth, field consistency can be accomplished by receiving the proper profile.

Epoxy gum is a standout among the most usually utilized thermosetting materials in high-voltage contraption as protection because of its superb mechanical, electrical properties, and concoction dependability. Epoxy saps are broadly utilized as a part of fiber wound composites and are appropriate for trim prepress. They are sensibly steady to concoction assaults and are brilliant followers having moderate shrinkage amid curing and no discharge of unstable gasses. These preferences, be that as it may, make the utilization of epoxies rather costly. Additionally, they cannot be normal past a temperature of 140 °C. Their utilization in high-innovation regions where benefit temperatures are higher, subsequently, is precluded. The usage of polymer nano-composites in electrical insulation was inspired by Lewis in 1994 [5]. Generally, any polymer composite contains three main constituents. They are (1) polymer matrix, (2) fillers, and (3) interaction zone (which play a major role in enhancing the polymer composite properties). The epoxy nano-composites are in the range of nanometers in size less than 100 nm and small in quantity less than 10 wt% of the total material. The epoxy micro-sized filler particles were of an average size of 16 μm which is 50 wt% of the total material [6]. The nano- and micro-epoxy resin composites have high thermal conductivity, high heat resistance, and dielectric breakdown strength [7]. Generally, one-dimensional fillers, clays or layered silicates, or three-dimensional fillers, silica (SiO_2), alumina (Al_2O_3), and titanium (TiO_2), are used for insulation purpose. Fillers exhibit excellent electrical, thermal, and mechanical properties.

In this paper, the finite element method (FEM) is utilized to compute the electric field on the spacer's surface. FEM worries about minimization of the vitality of the entire field area of intrigue, regardless of whether the field is electric or attractive, by isolating the locale into triangular components for two-dimensional issues or tetrahedrons for three-dimensional issues. For enhancing insulation properties, the nano- and micro-fillers, namely inorganic fillers like alumina (Al_2O_3) and titanium (TiO_2), are added to epoxy resin and the resultant permittivity is calculated. The electric field is determined for different compositions of epoxy nano- and micro-composites, and the presence of nano-composites improved the electrical characteristics of the gas insulated busduct.

2 Electric Field Along Spacer Surface

Electrostatic field improvement of the profile of the Spacer-SF_6 gas interface was examined as a method for enhancing the dielectric execution of epoxy spacers. A composite-formed spacer shown in Fig. 1 was demonstrated which joins the benefit of the long spillage separation of a cone molded profile with that of the semi-uniform field conveyance of a plate molded profile. The streamlining system depends on the control of the field appropriation (reliant on geometry parameters just) at the spacer surface by forming the spacer profile. Significance with respect to handle improvement is given to the inward side of spacer as the intersection framed by dielectric-SF_6-electrode, generally called as triple intersection, at cathode end is more powerless against flashovers. Dielectric execution of the spacer is enhanced misleadingly by decreasing the nearby field power at the metal-epoxy-SF_6 intersections, therefore limiting their capacity to start flashovers. For geometry, ro (outer-fenced in area range)–ri (Inner conductor span) is taken to be 100 mm.

Fig. 1 Profile of composite conical spacer

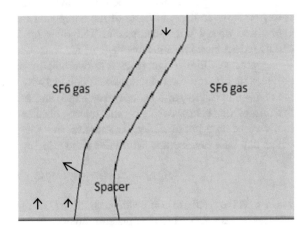

Fig. 2 Electrical field distribution along spacer surface

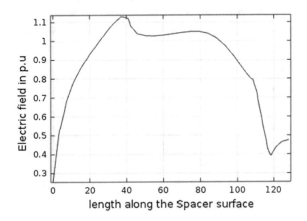

A 1 V is connected to anode while the cathode is grounded. The relative permittivities of spacer material and SF_6 are taken to be 4.5 and 1.005 individually. The
electric field dissemination for the planned spacer is shown in Fig. 2.

3 Calculation of Relative Permittivity

If a material with a high dielectric constant is placed in an electric field, the
magnitude of that field will be measurably reduced within the volume of dielectric.
In coaxial cable, polyethylene can be placed between the center conductor and
outside shield. Epoxy/epoxy-based nano- and micro-composites are preferred
insulating materials for electrical applications for bushings, GIS spacers, etc. In
epoxy nano- and micro-composites, nano- and micro-composites play a vital part in
the enhancement of the properties of epoxy because the permittivities of fillers are
high.

Due to the higher individual permittivities of the fillers and on combining with
epoxy resin, overall permittivity of the composite increases when compared to net
epoxy and epoxy micro-composite. The filler loading can be considered up to
certain extent based on the advantage of the interaction zone. If filler concentration
is increased to a high value, it leads to overlapping of the interaction zone between
polymer matrix and filler due to which conductivity increases. The overlapping of
the nano- and micro-particles in epoxy nano- and micro-composites depends upon
the rate of dispersion of nano- and micro-particles in the epoxy resin. The permittivity of two-phase dielectric satisfies the Lichtenecker–Rother mixing rule
which can be extended and written as shown in Eq. (1)

$$\text{Log } \varepsilon_c = x \text{ Log } \varepsilon_1 + y \text{ Log } \varepsilon_2 + z \text{ Log } \varepsilon_3 \tag{1}$$

where ε_c is the resultant composite permittivity, ε_1, ε_2, ε_3 are the permittivities of the
filler and epoxy x, y, z are the concentrations of filler and polymer. The permittivity
of the epoxy is 3.60, and the permittivity of alumina (Al_2O_3) is 9.2, and then, the
effective permittivity at filler loading 5wt% nano and 65wt% micro is 16.759; the
permittivity of epoxy nano- and micro-composites is high compared to filler-free

Table 1 Relative permittivity values with filler concentration

Filler concentration (wt%)-nano	Filler concentration (wt%)-micro	Alumina (Al_2O_3)	Titanium (TiO_2)
0	0	3.6	3.6
5	0	4.018	4.532
5	60	15.016	71.829
2.5	62.5	15.016	71.829
0	65	15.016	71.829
5	65	16.759	90.427

epoxy matrix. From Table 1, as the overall wt% of the nano- and micro-fillers increases, only the relative permittivity of the epoxy resin nano- and micro-composites increases.

4 Results and Discussions

The values of the relative permittivity with various filler concentrations are calculated and plotted in Table 1. From Table 1, it is observed that as the filler concentration increases, the relative permittivity increases, and it is more for titanium than alumina. The maximum electric field for different nano- and micro-filler concentrations of alumina are determined and presented in Table 2. The electric field distributions for different permittivities obtained for different filler concentrations of alumina are plotted and presented in Figs. 3, 4, 5, and 6. It is observed that with increase in filler concentration of alumina, there is a gradual increase in the electric field distribution. The electric field value is observed to be the least for zero concentrations of alumina for a relative permittivity of 3.6 as shown in Fig. 3 and its value is 1.15 p.u. A noticeable change in electric field of 1.3% is observed for a filler concentration of 5% (nano) and 65% (micro) of alumina shown in Fig. 6.

Table 2 Maximum electric field values for alumina filler concentration

Filler concentration (wt%)-nano	Filler concentration (wt%)-micro	Alumina (Al_2O_3)	Maximum electric field in p.u
0	0	3.6	1.15
5	0	4.018	1.197
5	60	15.016	1.273
2.5	62.5	15.016	1.273
0	65	15.016	1.273
5	65	16.759	1.28

Fig. 3 Variation of electric field strength having relative permittivity $\varepsilon_r = 3.6$ for alumina

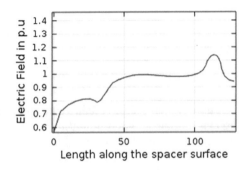

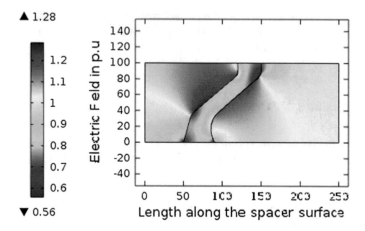

Fig. 4 Electric field distribution along the surface having relative permittivity $\varepsilon_r = 3.6$ for alumina

Fig. 5 Variation of electric field strength having relative permittivity $\varepsilon_r = 15.016$ for alumina

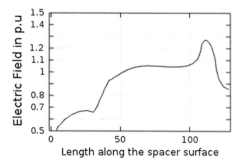

Fig. 6 Variation of electric field strength having relative permittivity $\varepsilon_r = 16.759$ for alumina

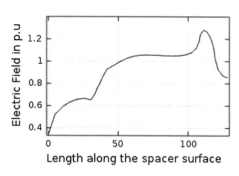

The maximum electric field values for different nano- and micro-filler concentrations of titanium are determined and presented in Table 3. It is observed that the maximum electric field occurs for a filler concentration of 5% (nano) and 65% (micro) of titanium. The maximum electric field value is obtained to be 1.34 p.u as shown in Fig. 9. There is an increase of 1.9% in the maximum electric field value

Table 3 Maximum electric field values for titanium filler concentration

Filler concentration (wt%)-nano	Filler concentration (wt%)-micro	Titanium (TiO$_2$)	Maximum electric field in p.u
0	0	3.6	1.15
5	0	4.532	1.21
5	60	71.829	1.335
2.5	62.5	71.829	1.335
0	65	71.829	1.335
5	65	90.427	1.34

Fig. 7 Variation of electric field strength having relative permittivity $\varepsilon_r = 3.6$ for titanium

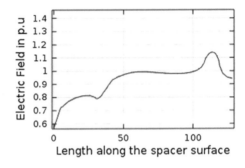

Fig. 8 Variation of electric field strength having relative permittivity $\varepsilon_r = 71.829$ for titanium

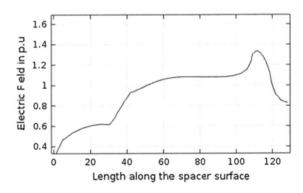

from zero concentration to 65% (micro) concentration of titanium. The electric field distribution along the spacer surface for different concentrations of Titanium is plotted in Figs. 7, 8, 9, and 10. However, for similar concentrations of nano- and micro-composites, alumina is subjected to less electric field stress than titanium. This is clear from the values tabulated in Tables 2 and 3.

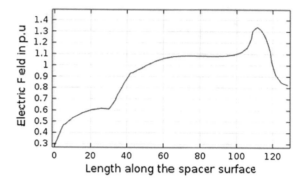

Fig. 9 Variation of electric field strength having relative permittivity $\varepsilon_r = 90.427$ for titanium

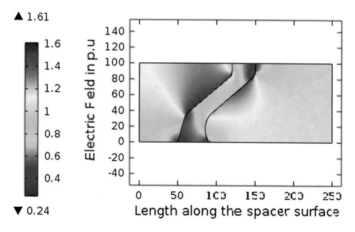

Fig. 10 Electric field distribution along the surface having relative permittivity $\varepsilon_r = 90.427$ for titanium

5   Conclusion

A key angle in the plan and enhancement procedure of high-voltage mechanical assembly is the exact recreation and geometric streamlining of the electric field dispersion on terminals and dielectrics. Exact learning of the electric field appropriation empowers the electrical specialist to avoid flashovers in critical areas of the device being designed. The geometrical state of the anodes affects the subsequent electrostatic field. When planning and enhancing high-voltage segments, it is imperative to know the electric field appropriation on the terminals and dielectrics. Initially, a composite cone sort spacer is upgraded to get uniform field along the curved side of the spacer. The model has been developed for a single-phase common enclosure. From the obtained results, it is observed that nano- and micro-composites exhibit high thermal conductivity and high electrical insulation properties. Due to the addition of these fillers such as alumina and titanium to base epoxy resin, relative permittivity and thermal conductivity of the composite are high compared to base epoxy resin and epoxy micro-composite. Further the maximum electric field values' variation with different concentrations of nano- and micro-composites of alumina and titanium was a little. However, the increase in electric field is more prominent in the case of titanium than alumina. For an optimal profile spacer, the electric field distribution with different concentrations of nano- and micro-compositions of alumina and titanium, alumina is subjected to a lesser stress than titanium for similar compositions.

References

1. K. Polivanov, *Theoretical Foundations of Electrical Engineering, Part 3* (Energiya, Moscow, 1969)
2. K. Nakanishi, A. Yoshioka, Y. Arahata, Y. Shibuya, Surface charging on epoxy spacer at DC stress in compressed SF_6 gas. IEEE Trans. Power Appar. Syst. **PAS-102**, 3919–3927 (1983)
3. B. Weedy, DC conductivity of voltalit epoxy spacers in SF_6. IEE Proc. **132**, pt. A, 450–454 (1985)
4. V. Varivodav, E. Volpov, Study of SF_6/epoxy insulation properties at high direct stress, in *6th International Symposium an High Voltage Engineering, (ISH)*, New Orleans, USA, paper 32.35 (1989)
5. H. Fujinami, T. Takuma, M. Yashima, Mechanism and effect of DC charge accumulation on SF_6 gas insulated spacers. IEEE Trans. Power Deliv. **4**, 1765–1772 (1989)
6. G.V. Nagesh Kumar, J. Amarnath, B.P. Singh, K.D. Srivastava, Particle initiated discharges in gas insulated substations by random movement of particles in electromagnetic fields, Int. J. Appl. Electromagn. Mech. **29**(2), 117–129 (2009)
7. G. V. Nagesh Kumar, J. Amarnath, B.P. Singh, Behavior of metallic particles in a single phase gas insulated electrode systems with dielectric coated electrodes, in *IEEE International Conference on "Condition Monitoring and Diagnosis"* CMD-2008, 377–380, Beijing, P. R. China, 21–24 Apr 2008

Comparative Analysis of Fault Diagnosis in Distribution System with the Aid of DWT-FFNN and DWT-RBFNN

T. C. Srinivasa Rao, S. S. Tulasi Ram and J. B. V. Subrahmanyam

Abstract In this paper, the fault location and types of them are recognized and analyzed in the distributed system by using wavelet and feed-forward neural network and radial basis function neural network. When fault occurring in the system, the system behaviors are examined and signals are measured which can be seen as distorted waveforms. These distorted waveforms are self-possessed of different frequency components and are needed to be represented in time–frequency domain for fault analysis. For this representation of signal, discrete wavelet transform (DWT) is offered. It extracts the features and the datasets are formed which are forwarded to FFNN and RBFNN for classification of fault occurred in the distributed power system. The proposed method is implemented in MATLAB/ simulink platform. This work is validated using statistical parameters, such as accuracy, sensitivity, and specificity and is compared with each other.

Keywords Discrete wavelet transform · FFNN · RBFNN · Fault location and fault type

T. C. S. Rao (✉)
J.N.T.U. College of Engineering, Hyderabad, India
e-mail: vas_cnu@yahoo.com

T. C. S. Rao
Department of EEE, Vardhaman College of Engineering,
Shamshabad Mdl., R.R. District, Hyderabad, India

S. S. T. Ram
Department of EEE, JNTU, Hyderabad, India

J. B. V. Subrahmanyam
TKREC, Hyderabad, India

© Springer Nature Singapore Pte Ltd. 2018
S. S. Dash et al. (eds.), *Artificial Intelligence and Evolutionary Computations in Engineering Systems*, Advances in Intelligent Systems and Computing 668,
https://doi.org/10.1007/978-981-10-7868-2_56

1 Introduction

As a result of different kinds of faults, the electric power distribution system's reliability and energy quality are exaggerated [1, 2]. The protection plans are important to uphold the system control and to diminish the consumer and network indemnity with the economical losses [3].

Gradient descent scheme that is at the heart of BP uses the only first derivative. Newton's method enhances the performance by making use of the second derivative which is computationally intensive. Gauss Newton's method approximates the second derivative with the help of first derivatives and hence, provides a simpler way of optimization.

In addition, the fault location methods play a vital position in the fast and loyal power system restoration process. Nowadays, one of the most important challenges is the fault location in electric power distribution systems because of its exact topological and operational features [4]. Moreover, the fast learning rates and generalization capabilities of radial basis function neural networks (RBFNN) have showed excellent accuracy in microcalcification detection task [5, 6]. The advantages of RBFNN are a simple structure, good performance with approaching nonlinear function, and fast convergence velocity. Thus, it has been widely used in pattern recognition and system modeling [7]. Moreover, the irrelevant components in the inputs will decrease the generalization performance of RBFNN [8].

Incidentally, the interior condition or start of each oscillator enhances during an interval of time till it arrives at a specific threshold [9]. Fault detection of systems have surfaced as a vital topic of intelligent systems applications during the course of the past 20 years and achieved incredible progress in the application of artificial intelligence to power systems [10]. Fault detection, in essence, is a binary decision procedure ascertaining the incidence or otherwise of a flaw in a particular mechanism [11]. A radial basis function NN is employed for fault detection along with the signal conditioner and an adaptive filter as a pre-classifier with an eye on mining the signal traits [12].

2 Related Work

Numerous related works are already existed in the literature which based on the fault diagnosis in the electrical distribution system. Some of them reviewed here.

De Oliveira et al. [13] have resourcefully dealt with the faulty branch identification significant for power distribution systems function and restitution. The technique employs traveling waves and autocorrelation theory.

El-Zonkoly [14] has amazingly evolved an innovative distributed generation (DG) in distribution power system which has a significant impact on the conservative fault current intensity and traits. As a result, the time-honored safety arrangements designed in distribution utilities were extremely hard to coordinate.

Bretas et al. [15] have amazingly offered further configuration data on an executed hybrid fault detection technique method for unbalanced underground distribution system (UDS). The innovative method was a hybrid-based approach, by duly blending both wavelet transforms (WTs) and artificial neural networks (ANNs) for the fault detection task.

3 Problem Formulation

In modern times, disparate kinds of fault location techniques for power transmission and distribution systems have been applied. However, those techniques do not completely reflect on the features of power distribution systems such as the presence of intermediate loads and laterals, unbalanced operation, and time changeable load profile. Hence, optimal fault location algorithm is required for finding out the distribution system fault. To make out the fault locations and types, a DWT and FFNN/RBFNN methods are applied in this document. The fault location variables are optimized by employing the suggested method.

3.1 Fault Identification and Localization in Distribution System with the Aid of DWT and FFNN/RBFNN

In these only four categories of faults like phase A-G, B-G, C-G, and ABC-G faults are taken into account. Also, the locations of faults are identified by using the proposed method. Here, the sending end voltages are specified for describing the fault Single line to ground fault conditions,

$$V_{sp}^f = V_p^F + \alpha \sum_i Z_{pi}^f I_{si}^f \tag{1}$$

where, $i = a, b, c$ and Z_{pi}^f is denoted as the fault impedance between the line to ground fault in the faulted phase, and I_{si}^f is the phase fault current. The fault current can be evaluated by using the following equation,

$$I_F = I_s^f - I_L \tag{2}$$

From the above equation, I_s^f and I_L are denoted as the three-phase sending end current vector and the current vector of load, respectively. For analyzing the faults, two processes are carried, such as fault location identification and fault classification. Figure 1 represents the block diagram of the proposed controller technique. Here, the voltage and current signals are measured at the terminal of the line before and during the fault with the use of measuring (or) monitoring device. Then these signals are given to the input of the DWT. According to their input signals, the datasets are formed and the data information contains the signals is in the form of

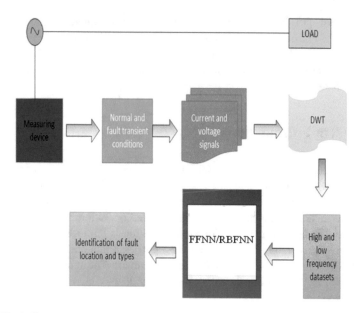

Fig. 1 Block diagram representation of proposed controller

wavelet detail and approximate coefficients. With the use of datasets, the FFNN/RBFNN network is trained and identified the fault locations and their types of the distribution system.

3.1.1 Exploitation of DWT Technique to Analyze the Normal and Faulty Signals

The discrete wavelet transform (DWT), in essence, represents a linear alteration which functions on a data vector. It is effectively calculated with a cascade of filtering ensued by a factor 2 subsampling like high-and low-pass filters correspondingly. Here, the original and fault transient current and voltage signals are decomposed into two components namely, high-and low-frequency components. The high-frequency component of the signal is represented as low-scale decomposition which is denoted as detailed coefficients. And the low-frequency component of the signals is precised as high-scaled decomposition which is called as approximate coefficients. Then the signals should be normalized to diminish the difficulty involved due to line magnitudes.

$$X_{V_p}^{\text{dwt}}\left[X_{V_p}(t)\right] = \int X_{V_p}(t)\lambda(m,n)\mathrm{d}t \tag{3}$$

$$X_{C_p}^{\text{dwt}}\left[X_{C_p}(t)\right] = \int X_{C_p}(t)\lambda(m,n)\mathrm{d}t \tag{4}$$

From the above Eqs. (3) and (4) represents the voltage and current signals of DWT, respectively. Where, $X_{V_p}(t)$ and $X_{C_p}(t)$ is denoted as normalized voltage and current signal at three phases then, $\lambda(m, n)$ is specified as the basic analyzing function. The mother wavelet is defined as follows:

$$\lambda(m, n) = \left(\frac{1}{\sqrt{a_0^m}} \lambda \left(\frac{t - na_0^m b_0}{a_0^m} \right) \right) \tag{5}$$

where, m and n are integers and $a = a_0^m$ and $b = na_0^m b_0$ are denoted as the real numbers which represent the scaling (dilation) and translation (shift) parameters. In DWT, after sampling, the normalized line voltage and current signals pass through two complementary filters, namely high-pass and low-pass filters and emerge as two signals, called detail coefficients, $X_p^{d,DWT}$. And coefficients of approximation, $X_p^{a,DWT}$.

With the use of DWT, two types of data are collected in normal and faulty case, such as low-and high-frequency signal data. These outputs are used for analyzing the fault types and locations. The output of the DWT algorithm described as the following equations, which contains the low-and high-frequency datasets of current and voltage.

$$DS_{dwt} = \begin{bmatrix} LF^1 \\ LF^2 \\ \cdot \\ \cdot \\ \cdot \\ LF^n \end{bmatrix}^{c,v} \begin{bmatrix} HF^1 \\ HF^2 \\ \cdot \\ \cdot \\ \cdot \\ HF^n \end{bmatrix}^{c,v} \tag{6}$$

where, $LF_n^{c,v}$ and $HF_n^{c,v}$ are specified as the low-and high-frequency voltage and current signals. These datasets are given to the input of the FFNN/RBFNN.

3.1.2 Feedforward Neural Network

The feedforward neural network was the first and simplest type of artificial neural network devised. In this network, the information transfers in only one direction, forward, from the input nodes, to the output nodes through the hidden nodes. Networks without feedback loops (cycles) are called a feedforward networks (or perceptron). FFNN networks may have more hidden layers. Typically in FFNN hidden and output neurons share a common neuron model. Hidden and output layers of FFNN are usually nonlinear. Each hidden neuron in a FFNN has an activation function whose argument computes the inner product of input vector and the synaptic weight vector of that neuron. FFNN construct global approximations to nonlinear I/O mapping.

FFNN has been used successfully to various applications, such as control, signal processing, and pattern classification. Neurons in the hidden layer receive weighted inputs from a previous layer and transfer output to the neurons in the next layer in FFNN, and these computations can be described as

$$y_{net} = \sum_{i=1}^{n} x_i \omega_i + \omega_0 \tag{7}$$

$$y_{out} = f(y_{net}) = \frac{1}{1 + e^{-y_{net}}} \tag{8}$$

$$E = \frac{1}{2} \sum_{i=1}^{k} (y_{obs} - y_{out}) \tag{9}$$

where ω_0 be the bias, w_i be the each input neuron's weight, x_i be input neuron, y_{net} be composed of the summation of weighted inputs, y_{out} is the output of system, $f(y_{net})$ denotes the nonlinear activation function, y_{obs} is the observed output value of neural network, and E is the error between output value and network result.

3.1.3 Radial Basis Function Neural Network

Generally, a Radial basis function (RBF) network is constructed with three layers of an input layer, a hidden layer with a nonlinear RBF activation function and a linear output layer. A function is said to be radial basis (RBF), if its output depends on the distance between the input and a given stored vector. In a RBF network, if one hidden layer uses neurons with RBF activation functions recitating local receptors, then one output node is used to unite the outputs of the hidden neurons linearly.

Learning Algorithm for RBFNN

For learning the RBF network parameters different learning algorithms may be used. There are three possible methods for learning centers, spreads, and weights.

- From the training set, Opt the centers randomly.
- Using the normalization method, Work out the spread for the RBF function.
- using the pseudoinverse method, get the weights.

The property of the radial function for the hidden layer of RBFNN is determined by its data center. Thus, for the construction of RBFNN, the main problem is to determine the position, number, and width of the radial function. The data center can be found by supervised and unsupervised methods. According to supervised method, it needs us to specify the number of the hidden layers has to be specified which requires prior knowledge. But in most of the cases, though there are large amounts of data but their class is not clearly known. So there is a need to determine the centers by using the unsupervised method.

RBFNN is a three-layer feedforward network, with only one hidden layer whose activation function is the radial basis function, for example, the Gauss function. Its output layer is a simple linear function.

Let $x_i \in R^N$, $i = 1, 2, \ldots, n$ is the ith learning sample, each sample is N-dimensional, and then RBFNN's output can be expressed by the following formulae (10) and (11):

$$y_i = f_{i(x)} = \sum_{k=1}^{M} \omega_{ik} \varnothing_k(x, C_k) = \sum_{k=1}^{M} \omega_{ik} \varnothing_k(\|x - C_k\|) \quad i = 1, 2, 3, \ldots, n \quad (10)$$

$$\varnothing_k(\|x - C_k\|) = \exp\left[\frac{(x - C_k)(x - C_k)}{\sigma_k^2}\right] \quad (11)$$

where y_i be the network actual output according to the ith learning sample; w_{ik} be the weight of the ith hidden layer node to the kth output node. φ_k be the Gauss Radial Function; C_k be the data center in the hidden layer; $\|\cdot\|$ represents the Euclidean distance; σk be the kth width of the radial basis function.

4 Analysis of Results

The proposed technique is implemented in MATLAB/Simulink platform with the model network as shown in Fig. 2 and their performances are evaluated. Here, the proposed technique is based on the DWT and FFNN/RBFNN for classifying the faults and identify the faults location. The performances of the proposed method is evaluated and compared with one another and the statistical measures, such as accuracy, sensitivity, and specificity values are analyzed.

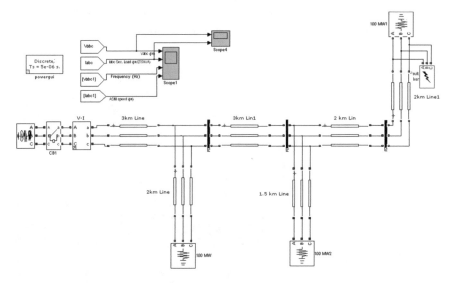

Fig. 2 Model simulink network

4.1 Performance Analysis of the Fault Signals in Different Locations

The performance of the proposed technique on the test system is evaluated by programming different types of faults. Here, the faults implemented on test system are, Phase A to Ground (AG) fault, BG fault, CG fault, Phase ABC to Ground (ABCG) faults. The performance of the normal signal is shown in Fig. 3, in which, *x*-axis represent the time in seconds and *y*-axis represents the currents in amperes for three phases. Here, totally five sets of results are analyzed and compared to the existing methods.

After the implementation of the faults on a test system, DWT is used to extract the best features of the line signals. It decomposes the line signals into low-frequency and high-frequency components. These are formed as a dataset for classifying the types of faults. The frequency components of the faulted line signals for above-mentioned faults (Fig. 4) are shown in Fig. 5.

For analyzing the performance of proposed technique, the low-and high-frequency datasets are applied to the FFNN or RBFNN network. From the output of FFNN or RBFNN network, the fault location, and their types are correctly identified. Then, the TP, TN, FP, and FN are evaluated from the testing output of FFNN or RBFNN network. Then the accuracy, sensitivity, and specificity of the proposed technique is analyzed from the True positive (TP), False positive (FP), True negative (TN), and False negative (FN) values. Then the performance of proposed technique (DWT-FFNN AND DWT-RBFNN) is determined and compared with one another. The evaluated output of proposed technique is tabulated in Tables 1 and 2 and also plotted as in Figs. 6 and 7.

The above comparison reveals the performance analysis of proposed technique. The proposed method has high accuracy and specificity with DWT-RBFNN when compared with DWT-FFNN as described in Fig. 8. The accuracy, sensitivity, and

Fig. 3 Normal current signal

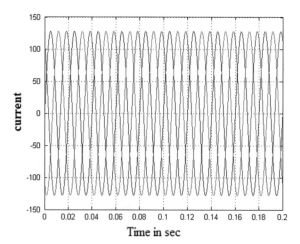

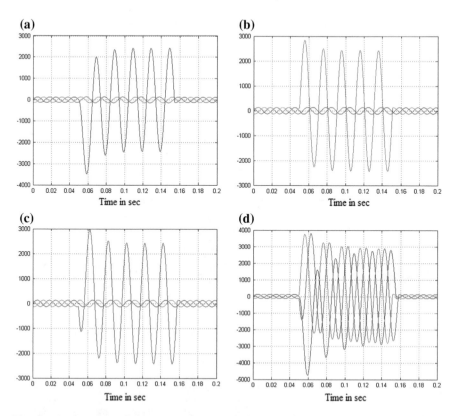

Fig. 4 a Performance of phase A to ground fault. **b** Performance of phase B to ground fault. **c** Performance of phase C to ground fault. **d** Performance analysis of three phase to ground fault

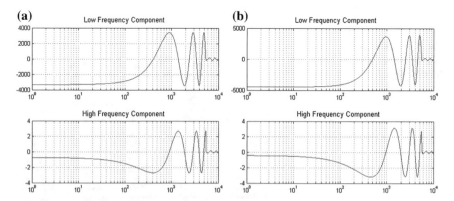

Fig. 5 a After applying DWT, the performance analysis of AG fault. **b** After applying DWT, the performance analysis of ABCG fault

Table 1 Accuracy, sensitivity, and specificity for DWT-RFFNN method

Types of fault	TP	FP	FN	TN	Accuracy	Sensitivity	Specificity
AG fault	3	1	2	4	0.7	0.6	0.8
BG fault	5	2	0	3	0.8	1	0.6
CG fault	4	1	1	4	0.8	0.8	0.8
ABCG fault	4	2	1	3	0.7	0.8	0.6
Normal signal	4	1	1	4	0.8	0.8	0.8

Table 2 Accuracy, sensitivity, and specificity for DWT-FFNN method

Types of fault	TP	FP	FN	TN	Accuracy	Sensitivity	Specificity
AG fault	4	1	1	4	0.8	0.8	0.8
BG fault	5	2	0	3	0.8	1	0.6
CG fault	3	1	2	4	0.7	0.6	0.8
ABCG fault	4	2	1	3	0.7	0.8	0.6
Normal signal	4	3	1	2	0.6	0.8	0.4

Fig. 6 Performance analysis of proposed DWT-RBFNN method in various faults

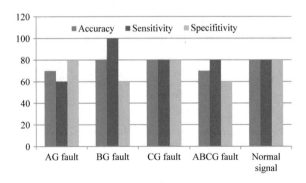

Fig. 7 Performance analysis of proposed DWT-FFNN method in various faults

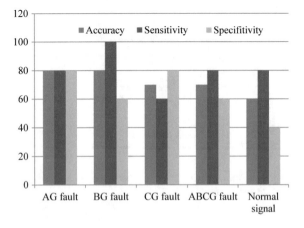

Fig. 8 Performance analysis of evaluation measures using different methods

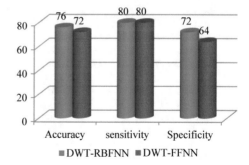

specificity of DWT-RBFNN are 76, 80, and 72%. Also, the accuracy, sensitivity, and specificity of DWT-FFNN are 72, 80, and 64%. The proposed methods are tested on the set 1, where, the fault location is 3 km.

Similarly, the types of faults are analyzed at different locations such as 5, 6, 8, and 10 km. Likewise, the DWT-FFNN and DWT-RBFNN performances are determined. The performance of accuracy, sensitivity, and specificity are determined and computed. From the comparative analysis, the accuracy, sensitivity, and specificity are analyzed at the different input.

5 Conclusion

By applying wavelet and FFNN/RBFNN, detection of fault location and types are examined in the distributed power system. Here, four types of faulty signals are analyzed in the five different segments, such as 3, 5, 6, 8, and 10 km, respectively. DWT is presented for extracting the features of the input signals, such as low-and high-frequency components. FFNN or RBFNN is used for classifying the signals whether it is faulty or not and recognize the fault position. The recommended technique for identifying the types of faults and locations was experimented with MATLAB/Simulink platform. The effectiveness of the proposed method was analyzed and evaluated. Using statistical parameters such as accuracy, sensitivity, and specificity, the performance of the suggested work was authenticated and compared with one another.

References

1. R.H. Salim, M. Resener, A.D. Filomena, K.R.C. De Oliveira, A.S. Bretas, Extended fault-location formulation for power distribution systems. IEEE Trans. Power Deliv. **24**(2), 508–516 (2009)
2. M. Mirzaei, M.Z.A. Ab Kadir, E. Moazami, H. Hizam, Review of fault location methods for distribution power system. Aust. J. Basic Appl. Sci. **3**(3), 2670–2676 (2009)

3. R.J. Patton, F.J. Uppal, C.J. Lopez-Toribio, Soft computing approaches to fault diagnosis for dynamic systems: a survey. Control Intell. Syst. **7**(3), 198–211 (2000)
4. L. Wei, W. Guo, F. Wen, G. Ledwich, Z. Liao, J. Xin, Waveform matching approach for fault diagnosis of a high-voltage transmission line employing harmony search algorithm. IET Gener. Transm. Distrib. **4**(7), 801–809 (2010)
5. A.T. Azar, S.A. El-Said, Superior neuro-fuzzy classification systems. Neural Computing and Applications **23**(Suppl 1), 55–72 (2012)
6. M. Jia, C. Zhao, F. Wang, D. Niu, A new method for decision on the structure of RBF neural network, in *Proceedings of the 2006 International Conference on Computational Intelligence and Security*, Nov 2006, pp. 147–150
7. A.L. Christensen, R. O'Grady, M. Birattari, M. Dorigo, Automatic synthesis of fault detection modules for mobile robots, in *IEEE Computer Society NASA/ESA Conference on Adaptive Hardware and Systems*, pp. 693–700 (2007)
8. R. Huang, L. Law, Y. Cheung, An experimental study: on reducing RBF input dimension by ICA and PCA, in *Proceedings of the 2002 International Conference on Machine Learning and Cybernetics*, vol. 4, Nov 2002, pp. 1941–1945
9. M. Gaouda, E.F. El-Saadany, M.M.A. Salama, V.K. Sood, A.Y. Chikhani, Monitoring HVDC systems using wavelet multi-resolution analysis. IEEE Trans. Power Syst. **27**(2), 1–9 (2001)
10. L. Shang, G. Herold, J. Jaeger, R. Krebs, A. Kumar, High-speed fault identification and protection for HVDC line using wavelet technique. Power Tech. Proc. IEEE Porto **3**(10), 1–5 (2002)
11. A.L. Christensen, R. O'Grady, M. Brittari, M. Dorigo, Exogenous fault detection in a collective robotic task. Inf. Sci. Technol., 1–10 (2004)
12. N. Rezaei, M.-R. Haghifam, Protection scheme for a distribution system with distributed generation using neural networks. Int. J. Electr. Power Energy Syst. **30**(4), 235–241 (2008)
13. K.R.C. De Oliveira, R.H. Salim, A. Shuck Jr., A.S. Bretas, Faulted branch identification on power distribution systems under noisy environment. Int. Power Syst. Trans. IPST **16**(5), 1–5 (2009)
14. A.M. El-Zonkoly, Fault diagnosis in distribution networks with distributed generation. Smart Grid Renew. Energy **2**(22), 1–11 (2011)
15. A.S. Bretas, K.C.O. Salim, R.H. Salim, Hybrid fault diagnosis formulation for unbalanced underground distribution feeders. Int. J. Power Energy Syst. **32**(1), 12–20 (2012)

A Mixed Strategy Approach for Fault Detection During Power Swing in Transmission Lines

Ch. D. Prasad and Paresh Kumar Nayak

Abstract In this paper, a mixed strategy approach is proposed for detecting faults during power swing. The instantaneous three-phase power signal is selected as actuating signal for the proposed fault detection unit based on the nature of variation with respect to time during power swing. Later, the signal is processed with Wavelet Transform to extract the approximated and detail coefficients at various frequency levels. Further, the absolute sum of these coefficients is calculated to detect the fault occurring during power swing. The performance of the proposed scheme is evaluated by generating data on a two-source equivalent test system through MATLAB/SIMULINK software. The observations in the results clearly show that using the proposed method a reliable and fast fault detection can be accomplished during power swing.

Keywords Fault detection unit · Power swing · Wavelet transform
Symmetrical fault · Relay

1 Introduction

The oscillations in rotor angles among the generators of a power system following the occurrences of disturbances (e.g., faults, switching on/off of large loads, disconnection of generators, switching of lines, etc.) may cause severe swings in power flow [1]. During power swing, the voltage and current vary as a function of rotor angle at the relay location. Such variations in voltage and current may cause the apparent impedance to enter into relay characteristics. This may be misinterpreted as a fault and can cause the relay to trip lines unnecessarily. False tripping

Ch. D. Prasad (✉) · P. K. Nayak
Department of Electrical Engineering, Indian Institute
of Technology (ISM), Dhanbad 826004, India
e-mail: chdpindia@gmail.com

P. K. Nayak
e-mail: nayak.pk.ee@ismdhanbad.ac.in

© Springer Nature Singapore Pte Ltd. 2018   597
S. S. Dash et al. (eds.), *Artificial Intelligence and Evolutionary Computations
in Engineering Systems*, Advances in Intelligent Systems and Computing 668,
https://doi.org/10.1007/978-981-10-7868-2_57

during power swing is avoided by incorporating power swing blocking (PSB) function in modern distance relays. However, if a fault occurs during PSB activation must be detected and the relay should be unblocked to trip the line. The variation of electrical quantities such as voltage, current, apparent impedance, active, reactive power, etc. are slow during power swing as these quantities vary as a function of machine rotor angle. However, these quantities vary suddenly at the inception of a fault. Conventional PSB schemes use this as the criteria for blocking or unblocking the distance relay during power swing [2].

In the recent years, significant research has been carried out for discriminating faults from power swing [3–18]. In [3], the swing center voltage (SVC) is used to detect power swing, where the frequency deviations of the system during power swing are not considered. The similar methodology of SVC is employed in [4] which take more time to react when compared to other methods. In [5], the rate of change of resistance is used to distinguish power swings from faults. In [6], low and high impedance faults are detected during power swing by using voltage phase angle at the relay location. Conventionally, voltage/current/frequency signals are selected as actuating signals for implementation of the fault detection units [3–6]. In [7], a new approach is introduced using sequential components to detect symmetrical faults during power swing. The method takes more time when compared to time–frequency transformation techniques. In [8], PSB function is implemented using WT with four levels of decomposition for all voltage and current signals. In [9], the instantaneous superimposed power signal is taken for assessment. But the usage of fast Fourier transform (FFT) for extraction of frequency components is slower when compared to WT. In this paper, the instantaneous three-phase active power signal is used as actuating signal for detecting faults during power swing, where features are extracted using WT technique for fast execution. The remaining part of the paper is organized as follows. The proposed method is described in detail in Sect. 2. In Sect. 3, the results for the simulation studies are presented. Finally, the conclusion of the paper is provided in Sect. 4.

2 Proposed Method

The proposed method can be explained in four steps as below.

- Selection of actuating signal for the relay logic
- Processing with WT to extract the approximation, detail coefficients
- Mathematical operation on coefficients
- Implementation of trip signal based on the threshold condition.

Here, three-phase instantaneous power is taken as actuating signal for implementation of relay logic because of its unique nature. During normal operation of the power system, the instantaneous power transferred from generation to load is constant at a given load angle "δ." In other words, the instantaneous power transferred between two areas is constant when the two areas operating at constant frequency mentioned in (1). During power swing, the instantaneous power oscillates and follows the sinusoidal curve as given (2). Further by approximating the slip frequency $(\omega_1 - \omega_2)$ as small the second and third parts of (2), it is rewritten as in (3) [9].

$$P_{3\varphi} = v_A(t) \cdot i_A(t) + v_B(t) \cdot i_B(t) + v_C(t) \cdot i_C(t) \tag{1}$$

$$P_{3\varphi} = P_1 + P_2 \cos[(\omega_1 - \omega_2)t] + P_3 \sin[(\omega_1 - \omega_2)t] \tag{2}$$

$$P_{3\varphi} = P + P_3(\omega_1 - \omega_2)t - \frac{P_2}{2}(\omega_1 - \omega_2)^2 t^2 \tag{3}$$

When a fault occurs at any section of transmission line during power swing, the three-phase instantaneous power can be expressed as in (4). Because of its variation of curve shape when compared to regular signals such as voltage and current signal, it is selected as actuating signal for fault detection unit during power swing. The pattern of variation of typical signals at different cases of power system is shown in Fig. 1.

$$P_{3\varphi}(t) = P_f + V_m \cdot e^{-(R/L)t} \cdot F(t) \tag{4}$$

Here $F(t)$ is function in "t" which can be expressed as

$$F(t) = [k_{1A} \cdot \cos(\omega t + \theta_A)] + k_{1B} \cdot \cos(\omega t + \theta_B) + k_{1C} \cdot \cos(\omega t + \theta_C)] \tag{5}$$

Under power swing condition, the superimposed instantaneous power is a regular curve and it deviates from curve nature with fault inception. Further, the three-phase instantaneous superimposed power signal is processed through wavelet transformation technique for obtaining additional information. In the proposed

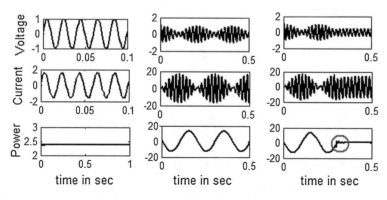

Fig. 1 Variation of voltage, current and power signals, **a** during normal condition, **b** during power swing and **c** fault during power swing

method, DWT is used for the extraction of additional features [10–12]. The extraction of approximated and detailed coefficients is taking place by processing the actuating signal through two filters known as low-pass filter and high-pass filter. Here as actuating signal changes its shape from curve to line with noise, i.e., the detecting part may undergo abrupt changes when a fault is incepted in line when the system is in power swing conditions. For detection of such break point analysis, Haar wavelet is used. As a conclusion, the wavelet transform analyses the signal at different frequency bands by separating the signal into approximated and detail coefficients at different resolutions. In earlier study, FFT was used but compared to FFT, the proposed approach is fast in operation and can also handle other typical cases. The mathematical expression of DWT is as

$$\varphi_{a,b}(t) = \frac{1}{\sqrt{a_o^p}} \left(\frac{t - q b_o a_o^p}{a_o^p} \right) \tag{6}$$

Here, "p" and "q" are the integers used in the frequency transformation process to control dilation and translation of the wavelet. In (6), a_0 and b_0 are fixed parameters which influence the process. In general, p and q are set of positive and negative numbers, respectively. The parameter a_0 is a fixed step parameter with a value always greater than one and the value of b_0 is greater than zero. By using wavelets, given function can be analyzed at various levels of resolution. The one-dimensional wavelet transform is given as

$$w_f(a, b) = \int_{-\infty}^{\infty} x(t) \cdot \varphi_{a,b}(t) \cdot dt \tag{7}$$

After extraction of approximated, detailed coefficients of Haar wavelet family, the difference of successive samples is taken and a threshold is fixed under swing condition using (8). When a fault occurs during power swing, the super imposed power signal deviates from its regular wave shape and the corresponding difference of successive samples of detailed coefficients of signal after processing to the DWT, absolute sum of coefficients is taken by using (9).

$$\text{Threshold}(\vartheta) = \text{Cdi}(K + 1) - \text{Cdi}(K) \tag{8}$$

$$\text{Abs}(K) = \max\{(\text{Cdi}(K) - \text{Cdi}(K - 1)) - \vartheta, 0\} + \text{Abs}(K - 1) \tag{9}$$

After mathematical operation on detailed coefficients, final step is implementation of trip signal and ideally it can be achieved by using (10).

$$\text{Abs}(K) > 0, \quad \text{Trip signal is activates and reset PSB Block} \tag{10}$$

As actuating signal undergo abrupt changes when a fault is initiated in the line during power swing. For detection of such break point analysis, Haar wavelet is used, after extraction of coefficients, the additional mathematical approach increases the overall reliability.

3 System Studied and Simulation Results

A 400 kV, 50 Hz, 2-bus interconnected power system as shown in Fig. 2 is considered for performance evaluation of the proposed method. The length of the transmission line is 150 km. The per km positive and zero-sequence impedance of the transmission line are $0.03 + j0.34\,\Omega$ and $0.28 + j1.04\,\Omega$, respectively. For Area-1, operating load angle is $10°$ whereas Area-2 is $0°$ such that the active power is transferred from Area-1 to Area-2 under normal operating conditions of the system. For power swing, one area is deviated from its operated frequency 50 Hz (ω_1) to a new frequency (ω_2) such that power angle between two areas oscillates and is characterized by "slip" frequency ($\omega_1 \sim \omega_2$). If the slip frequency is low (1–3 Hz) then slow swing and high (4–7 Hz) then it is fast swing [7]. Here, the performance of the proposed method is tested for five different fault scenarios during power swing. Results of the simulation studies are presented below.

Case-1: Analysis of proposed method during power swing

Sudden frequency variation in Area-2 from its nominal operating frequency leads large oscillations in power transferred between two areas and hence causes power swing. The variation of instantaneous active power during power swing is shown in Fig. 3a. The detailed coefficients of Haar wavelets transform are shown in Fig. 3b. As actuating signal follows a regular curve shape, the corresponding detailed coefficients are ideally zero and hence no trip signal is generated from relay block which obviously improves the reliability of the protection scheme during power swing. There are of course different PSB functions are available to support distance relay during power swing. The main objective of the present paper is to detect faults during power swing which are discussed below.

Case-2: Analysis of proposed method under unsymmetrical faults during power swing with slow slip frequency (1–4 Hz)

In this case, the performance of proposed technique is analyzed for occurrence unsymmetrical fault during power swing with slow slip frequency, i.e., of 2 Hz. Figure 4 shows the assessment results of the presented method for detection of unsymmetrical faults during slow power swing. A line to ground fault (AG-type) is

Fig. 2 Single line representation of the test system simulated in MATLAB/SIMULINK

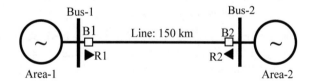

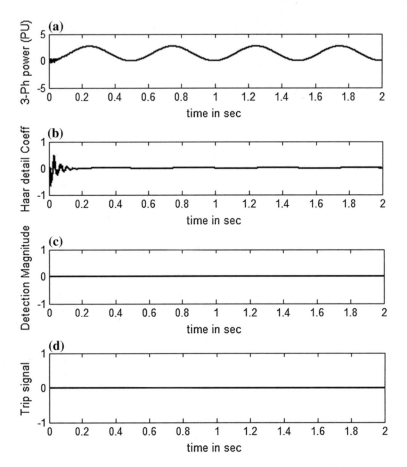

Fig. 3 Performance during power swing, **a** actuating signal, **b** detailed coefficients, **c** detection magnitude, and **d** trip signal during power swing

created at the middle of the line, i.e., at 75 km from the relay point at an inception angle corresponding to 0.8 s. Figure 4a shows the corresponding three-phase instantaneous active power, whereas Fig. 4b shows the version of detailed coefficients. In case of unsymmetrical faults, as disturbance is large for corresponding faulty phases, the magnitude of detection index is also very large (Fig. 4c) and hence, a quick action can be taken place immediately after the occurrence of the fault (Fig. 4c).

Case-3: Analysis of proposed method under symmetrical faults during power swing with slow slip frequency (1–4 Hz)

The performance of the proposed technique is tested for detecting symmetrical fault during power swing with slow slip frequency (=2 Hz). A symmetrical fault is created at the middle of the line, i.e., at 75 km from the relay point at an inception

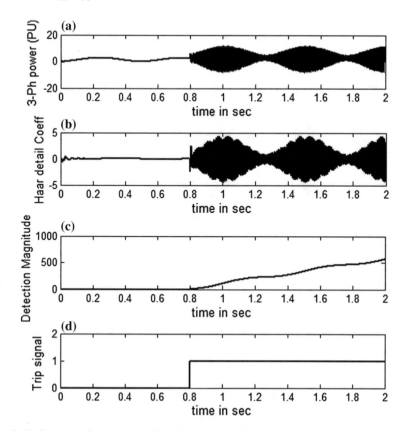

Fig. 4 Performance for unsymmetrical fault during slow power swing, **a** actuating signal, **b** detailed coefficients, **c** detection magnitude, and **d** trip signal

angle corresponding to 0.8 s. Figure 5 shows the assessment results of the presented method for detection of symmetrical faults during power swing. Figure 5a shows the corresponding three-phase instantaneous active power, whereas Fig. 5b shows the version of detailed coefficients. In case of symmetrical faults, as disturbance is small for corresponding faulty phases, the magnitude of detection index is also very small. Despite of small changes in the instantaneous power, the proposed fault detector index is able to detect the fault within half-cycle of its inception which is clearly evident from Fig. 5c.

Case-4: Analysis of proposed method under unsymmetrical faults during power swing with fast slip frequency (5–7 Hz)

Here, the performance of proposed technique is analyzed for occurrence unsymmetrical fault during power swing with high slip frequency. A line to ground fault (AG-type) occurs at the middle of the line, i.e., at 75 km from the relay point at an inception angle corresponding to 1.1 s. The slip frequency is taken 5 Hz under this

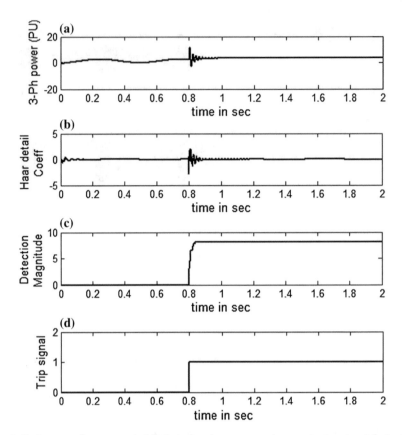

Fig. 5 Performance for symmetrical fault during slow power swing, **a** actuating signal, **b** detailed coefficients, **c** detection magnitude, and **d** trip signal

case. Figure 6 shows the assessment results of the presented method for detection of unsymmetrical faults during power swing. Here also, irrespective of the fault inception angle, location, and type of unsymmetrical fault nature, the detection magnitude index is comparatively high and a quick action is taking place because of generation of trip signal as evident from Fig. 6c, d.

Case-5: Analysis of Proposed method under symmetrical faults during power swing with fast slip frequency (5–7 Hz)

This case illustrates the performance of a proposed technique for detecting symmetrical fault during power swing with fast slip frequency (=5 Hz). A three-phase fault is created at the middle of the line, i.e., at 75 km from the relay point at a inception angle corresponding to 1.1 s. Figure 7 shows the assessment results of the presented method. Figure 7a shows the corresponding three-phase instantaneous active power, whereas Fig. 7b shows the version of detailed coefficients. In case of symmetrical faults at fast slip frequency, as disturbance is small for corresponding

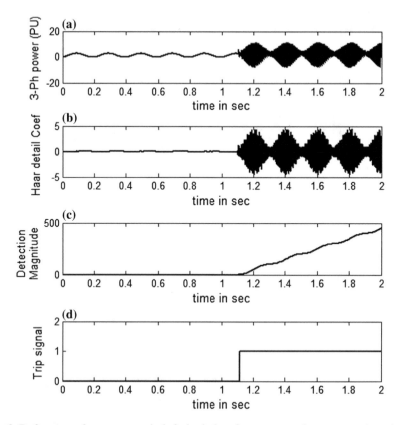

Fig. 6 Performance for unsymmetrical fault during fast power swing, **a** actuating signal, **b** detailed coefficients, **c** detection magnitude, and **d** trip signal

faulty phases, the magnitude of detection index is also very small. However, the proposed method effectively detects the symmetrical faults which are clearly observed from Fig. 7d.

Case-6: Generalized fault cases

The location of fault, inception angle at which fault occurs, fault resistances, type of fault are some of the miscellaneous cases studied in this paper. During power Swing, in Case-2 and Case-4 two different types of unsymmetrical faults are created with different fault inception angle at typical midpoint location and it is found that the proposed method can be able to produce trip signal in both cases. In Case-3 and Case-5, symmetrical faults are created with different fault inception angle at typical midpoint location. For these two cases also the proposed method is able to produce trip signal. Hence, the technique presented in this paper has the ability to detect faults during power swing irrespective of the fault type, fault location, and fault inception angle.

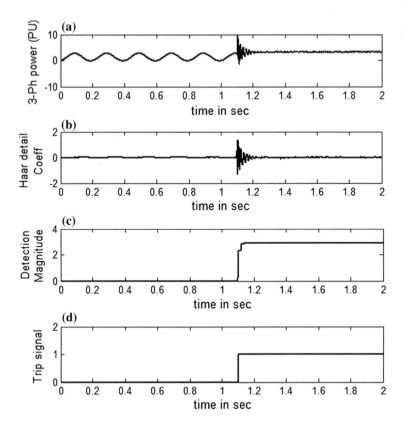

Fig. 7 Performance for symmetrical fault during fast power swing, **a** actuating signal, **b** detailed coefficients, **c** detection magnitude, and **d** trip signal

4   Conclusions

In this paper, a novel method is proposed to detect faults during power swing. Three-phase instantaneous power signal is employed as an excitation signal to accomplish the fault detection task during power swing. Sample difference calculation of instantaneous power with a suitable threshold is used as a mathematical approach for the generation of trip signal. After assessment of different case studies, it is concluded that the technique presented in this paper can effectively detect both symmetrical and unsymmetrical faults during power swing with wide variations in slip frequency.

References

1. P.M. Anderson, *Power System Protection* (IEEE Press & McGraw-Hill, 1999)
2. Power System Relay Committee: Power swing and out-of-step considerations on transmission lines, in *IEEE PSRC Working Group D6* (New York, 2005)
3. G. Benmouyal, D. Hou, D. Tziouvaras, Zero-setting power-swing blocking protection [Online]
4. B. Su, X.Z. Dong, Y.Z. Sun, B.R.J. Caunce, D. Tholomier, A. Apostolov, Fast detector of symmetrical fault during power swing for distance relay, in *Proceedings of the IEEE Power Engineering Society General Meeting*, pp. 604–609 (2005)
5. A. Mechraoui, D.W.P. Thomas, A new blocking principle with phase and earth fault detection during fast power swings for distance protection. IEEE Trans. Power Del. **10**(3), 1242–1248 (1995)
6. Q.X. Yang, Z.Y. Xu, L.L. Lai, Z.H. Zhang, N. Rajkumar, Fault identification during power swings with symmetrical component, in *Proceedings of the EMPD International Conference on Energy Management and Power Delivery*, pp. 108–111 (1998)
7. S.M. Brahma, Distance relay with out-of-step blocking function using wavelet transform. IEEE Trans. Power Del. **22**(3), 1360–1366 (2007)
8. S. Lotfifard, J. Faiz, M. Kezunovic, Detection of symmetrical faults by distance relays during power swings. IEEE Trans. Power Del. **25**(1), 81–87 (2010)
9. B. Mahamedi, J. Zhu, A novel approach to detect symmetrical faults occurring during power swing using frequency components of instantaneous three phase active power. IEEE Trans. Power Deliv. **27**(3) (2012)
10. A.H. Osman, O.P. Malik, Transmission line distance protection based on wavelet transforms. IEEE Trans. Power Del. **19**(2), 515–523 (2004)
11. *Wavelet Toolbox User's Guide* (The Math Works Inc., Natick, MA, 2005)
12. J.A. Jiang, P.L. Fan, C.S. Yu, J.Y. Sheu, A fault detection and faulted phase selection approach for transmission lines with Haar wavelet transform, in: *IEEE Conference on transmission and distribution*, vol. 1, Apr 2003, pp. 285–289
13. P.K. Nayak, A.K. Pradhan, P. Bajpai, A fault detection technique for the series compensated line during power swing. IEEE Trans. Power Delivery **28**(2), 714–722 (2013)
14. R. Dubey, S.R. Samantaray, Wavelet singular entropy based symmetrical fault-detection and out-of-step protection during power swing. IET Gener. Transm. Distrib. **7**(10), 1123–1134 (2013)
15. S.R. Samantaray, R.K. Dubey, B. Chitti Babu, A novel time–frequency transform based spectral energy function for fault detection during power swing. Electr. Power Compon. Syst. **40**, 881–897 (2012)
16. C. Pang, M. Kezunovic, Fast distance relay scheme for detecting symmetrical fault during power swing. IEEE Trans. Power Del. **25**(4), 2205–2212 (2010)
17. S. Lofifard, J. Faiz, M. Kezunovic, Detection of symmetrical faults by distance relays during power swings. IEEE Trans. Power Del. **25**(1), 81–87 (2010)
18. P.K. Nayak, J.G. Rao, P. Kundu, A.K. Pradhan, P. Bajpai, A comparative assessment of power swing detection techniques, in *2010 Joint International Conference on Power Electronics, Drives and Energy Systems* (2010)

Optimal Placement and Sizing of DG in a Distributed Generation Environment with Comparison of Different Techniques

T. C. Subramanyam, S. S. Tulasi Ram and J. B. V. Subrahmanyam

Abstract This area proposes about the optimal location for fixing fuel cells in a distribution system by an innovative technique. The innovation of this method is the combined performance of the Genetic Algorithm and Artificial Intelligence (RNN) technique, thus integrating GA in two stages and Artificial Intelligence (RNN) technique. The optimum placement of fuel cell is attained by the GA first stage (Mohammadi and Nasab in Res J Appl Sci Eng Technol 3:838–842, 2011 [1]). The RNN is suitably trained by the target fuel cell size and the corresponding inputs such as load variation and bus number. The main objective helps to enhance the bus voltage profile and reduce power loss (Zayandehroodi et al. in Int J Phys Sci 6:3999–4007, 2011 [2]). Thus, this approach is implemented in MATLAB/simulink. Its performance is evaluated by comparing different methods like GA, PSO and other hybrid PSO techniques (Mohammadi and Nasab in Res J Appl Sci Eng Technol 2:832–837, 2011 [3]). The comparison result is explicitly demonstrated the supremacy of this method and confirm its sterling potentiality to solve the problem (Chowdhury et al. in IEEE Trans Ind Appl 39:1493–1498, 2003 [4]).

Keywords Fuel cell · GA · RNN · PSO · Voltage · Real power

T. C. Subramanyam (✉)
JNTUCE, Hyderabad, Telangana, India
e-mail: tcsubramanyam@gmail.com

S. S. Tulasi Ram
Department of EEE, JNTUCE, Hyderabad, India

J. B. V. Subrahmanyam
TKREC, Meerpet, Saroornagar, Hyderabad, India

© Springer Nature Singapore Pte Ltd. 2018   609
S. S. Dash et al. (eds.), *Artificial Intelligence and Evolutionary Computations in Engineering Systems*, Advances in Intelligent Systems and Computing 668, https://doi.org/10.1007/978-981-10-7868-2_58

1   Introduction

The positioning of distributed generators (DG) within distribution system gives more useful with many benefits could possibly to enhance the electric power resources. The particular distributed generation (DG) can be comprised surrounded by a new emphasis for the electrical power generation [2].

Distributed Generation defined as any kind of generating power that's bundled in the distribution process [5]. Distributed generation has extensively found their place in power process which satisfies the organization needs on reliability cost and quality fulfillment [6]. Consequently, installation of DG introduction in distribution system has important significance on power flow and voltages.

This paper proposes an optimal location for fixing fuel cells in a distribution system by an innovative technique [7]. The innovation of recommended approach is the joined performance of the Genetic Algorithm (GA) and Artificial Intelligence (Recurrent Neural Network RNN) technique GA first stage utilizes the load flow data at different loading conditions for determining the optimum location. The RNN is aptly trained by the target fuel cell size and the corresponding inputs such as load variation and bus number. During the testing time, the RNN provides the fuel cell capacity according to the load variation and bus number. By using the attained fuel cell capacities, the GA second stage optimizes the fuel cell capacity to decrease the power loss and voltage deviation [8]. The objective function mainly helps to enhance the power loss reduction and bus voltage profile.

2   Literature Survey: A Recent Related Work

Mohammadi et al. have suggested a method PSO approach for placing DG in the distribution generation environment to reduce the power loss to improve process reliability. Hybrid goal purpose is employed with the maximum DG position.

Heydari et al. [9] have suggested a proposal and remedy to locate and dimensions of DGs perfectly. The placement along with size regarding DGs are generally optimized having a Genetic Algorithm (GA). To analyze the recommended algorithm, the particular IEEE 34 buses distribution feeder is employed.

The distributed generator is now frequently located in power system to improve the overall performance of the distribution system. Main advantages of using DG in the distribution system are: power loss reduction, enhance the voltage profile, and many more.

3　Problem Formulation

The placement of fuel cell at best location ultimately leads to various factors, such as loss reduction, enhanced voltages. The optimum location and sizing of the fuel cell are the important problems with nonlinear impartial function having respective constraints like power balance, voltage, and fuel cell constraints [10]. The most aim of this projected approach is reducing the precise power loss, load equalization, and voltage deviation of the system at the maximum load condition [8]. The multi-objective function is developed as per the subsequent Eq. (1).

$$J = \text{Min}\{f_1, f_2\} \tag{1}$$

Here, f_1 and f_2 are the power loss and voltage deviation.

3.1　Power Loss (f_1)

The necessary précised loss of the power distribution system is evaluated by the subsequent equation.

$$f_1 = P_\text{L} = \sum_{i=1}^{n} I_i^2 R, \tag{2}$$

where P_L is the power loss of the system, R the resistance between buses i and j, and I_i is the line current.

3.2　Voltage Deviation (f_2)

When the fuel cell is placed in a distribution generation environment, the voltage profile is modified. The specified voltage discrepancy equation is represented in the coming equation.

$$f_2 = \sum_{i=1}^{N} (V_i - V_\text{rated})^2, \tag{3}$$

where, V_rated is the specified voltage; V_i the bus voltage and N being number of buses.

4 Optimal Location and Optimal Capacity of Fuel Cell Using Projected Approach

The optimal placement and size of the fuel cell are developed by multi-objective constrained optimization problems. This area explains combined GA and RNN for determining the optimal location and capacity of the fuel cell. Here, the proposed hybrid method utilizes GA in two stages, with the combination of RNN technique.

4.1 GA First Phase-Based Optimal Location Determination

Steps of the GA first phase

Step 1: Run the load flow equation for normal condition and different types of loading conditions.

Step 2: Initialize the required parameters of GA such as radial distribution network bus dataset as N number of buses, bus voltage, real, and reactive power losses.

Step 3: Generate the random population of load value and apply it to the buses.

$$X = [X_i^1, X_i^2 \ldots X_i^d],$$

where, $i = 1, \ldots n$, and d is specified as the dimensions of the population space.

Step 4: Set the count $k = k + 1$.

Step 5: Find the fitness as follows

$$\text{fitness} = \text{Min}(P_L).$$

Step 6: Select the chromosome X_i^{best}, which has the minimum fitness.

Step 7: Apply the load changes and go to step 4, until the required termination criteria is achieved.

At the end of the process, the GA first phase develops the optimum location to place the fuel cell.

4.2 RNN-Based Fuel Cell Ratings Prediction

Recurrent neural network (RNN) is, in fact, analogous to the feedforward neural network (FNN), though RNN comprises feedback loop around the neuron. The feedback loop, on the other hand, also comprises unit delay operator (z^{-1}) [3]. The RNN structure is explained in Fig. 1. The supervised learning process is utilized for training the RNN.

From the concept of Supervised Learning and Training Process, the error function is minimized with following Eq. (4).

$$E = \frac{1}{2}(W_a - W_t)^2 = \frac{1}{2}e_s^2, \tag{4}$$

where, W_a is actual output of the given network, W_t be the target output of the network and E the error function. The weight update and error calculation are explained as follows:

Layer 1:

This is used to update the weight of the w_{bc}^3. Here updated weight is given by Eq. (5).

$$w_{bc}^3(N+1) = w_{bc}^3(N) + \eta_{bc}\Delta w_{bc}^3(N), \tag{5}$$

where, $\Delta w_{bc}^3 = -\frac{\partial E}{\partial R_c^3} = \left[-\frac{\partial E}{\partial R_c^3}\frac{\partial R_c^3}{\partial \text{Net}_c^3}\right] = \delta_c R_b$ with $\delta_c = \frac{\partial E}{\partial R_c^3} = \left[-\frac{\partial E}{\partial e_s}\frac{\partial e_s}{\partial R_c^3}\right]$ is propagates the error term, η_{bc} is the learning rate for adjusting the parameter w_{bc}.

Layer 2:

This layer discharges the multiplication operation and the updated rule for w_b^2 and w_{ab}^2 is given by the following Eqs. (6) and (7).

$$w_b^2(N+1) = w_b^2(N) + \eta_b\Delta w_b^2(N), \tag{6}$$

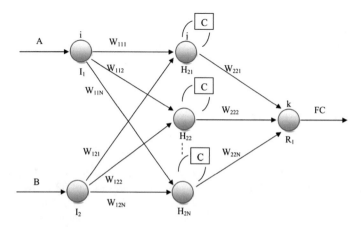

Fig. 1 Structure of the RNN

$$w_{ab}^2(N+1) = w_{ab}^2(c) + \eta_{ab}\Delta w_{ab}^2(N), \tag{7}$$

where, $\Delta w_b^2 = -\frac{\partial E}{\partial w_b^2} = \left[-\frac{\partial E}{\partial R_c^3} \frac{\partial R_c^3}{\partial R_b^2} \frac{\partial R_c^3}{\partial R_b^2} \right] = \delta_c w_{bc}^2 P_b^2$ and $\Delta w_{ab}^2 = -\frac{\partial E}{\partial w_{ab}^2} = \left[-\frac{\partial E}{\partial R_c^3} \frac{\partial R_c^3}{\partial R_b^2} \frac{\partial R_b^2}{\partial w_{ab}^2} \right] = \delta_c w_{bc}^2 Q_{ab}^2.$

When this process is completed, the RNN is ready to give the fuel cell capacity. But the optimum capacity of the fuel cell to enhance the voltages and limit the power loss of the system is determined by the GA second phase.

4.3 Fuel Cell Capacity Optimization Using GA Second Phase

Steps for finding the size of fuel cell

Step 1: Parameters like required radial power distribution system bus data, such as bus voltage, power loss, fuel cell capacity limit, etc. are initialized.

Step 2: Set the time counter $t = 0$ $Y = [Y_i^1, Y_i^2, \dots Y_i^d]$.

Step 3: By using the objective function, determine the every chromosome in the starting population (1) and search for the most effective value of the objective function J_{best}. This setup adjusts the chromosome proportionality to the J_{best} as the best.

Step 4: Update the time counter $t = t + 1$.

Step 5: Generate a fresh population with repeating the Selection, Crossover, Mutation up to the completion of new population.

Step 6: Run the algorithm with the new set of population.

Step 7: If stopping criteria is achieved once, stop the operation, else go back to the step 3.

Once the above-mentioned steps are finished, it is ready to give the optimum size of the fuel cell with improved voltage profile and minimum power loss. This projected approach is enforced in the MATLAB through standard IEEE benchmark system and performance is evaluated by using the comparison studies [10].

5 Numerical Results and Discussion

The projected methodology is enforced in MATLAB/simulink. The IEEE 33 bus distribution system with 3.7 MW and 2.32 MVAr is utilized for testing of the proposed method. The mentioned testing system comprises of 32 branches and 33 nodes. The efficiency of the projected approach is analyzed with the help of comparative studies with the GA, PSO, and hybrid PSO techniques. The results are displayed as follows [1].

Figure 2 represents the IEEE standard 33 bus power distribution system voltage profile. The testing system line loss in normal condition is described in Fig. 3. The maximum line losses are present in the bus system is 58 kW. Figure 4 shows the bus voltage of the power distribution system at the faulty condition. Load raising at the range of 50%, the bus voltage is deviated from the normal condition. The power loss of the IEEE 33 bus radial power system at faulty condition is given in Fig. 5. It is observed that the power loss is raised due to variations in the load of the distribution system [2]. In faulty condition, power loss at bus is likely to go up to 62 kW. So it is essential to find the optimum location to place a fuel cell at right capacity. The proposed method is utilized for determining the optimal location to achieve minimum power loss by placing the optimum capacity of fuel cell. The bus voltage profile after locating the fuel cell is explained in Fig. 6, at the same time the line losses of the system are described in Fig. 7. From Figs. 6 and 7, it is crystal

Fig. 2 Normal bus voltage

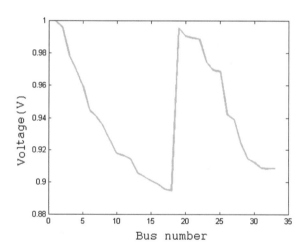

Fig. 3 Normal power loss

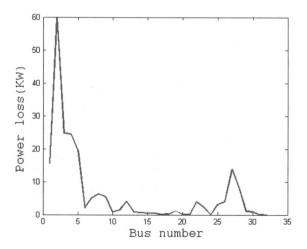

Fig. 4 Bus voltage at 50%
load variation

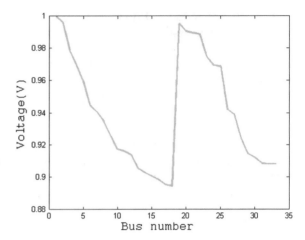

Fig. 5 Power loss at 50%
load variation

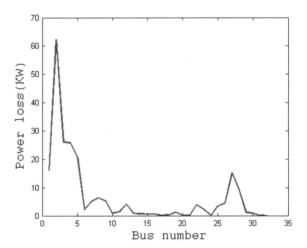

clear that this proposed method voltage profile is effectively maintained near the normal condition and the power loss is minimized at 58 kW.

The IEEE 33 bus system is allowed to meet the 50% load increment, which affects the power flow quantities of the system. During the load increment period, the system normal voltage profile gets collapsed, which is shown in Fig. 4. Similar load condition the system power loss is described in Fig. 5. In the situation, to restore the normal condition of the bus system with the help of optimum rating of fuel cell at the optimum location, the proposed method finds the placement and size of the fuel cell at the 50% load increment. After the fuel cell placement, the bus voltage profile is analyzed, which is shown in Fig. 6. Due to the voltage stability, the constraints are maintained in the stable limit, which reflects minimum power loss. The attained minimized power loss is explained in Fig. 7. Then the proposed

Fig. 6 Bus voltage using proposed method at 50% load variation

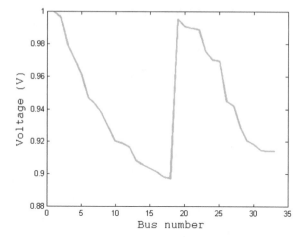

Fig. 7 Power loss using proposed method at 50% load variation

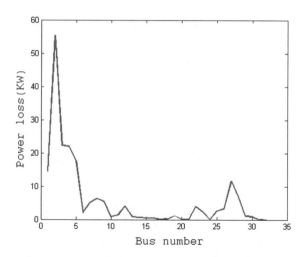

method results are distinguished to the other methods, such as Genetic Algorithm, Particle Swam Optimization, and hybrid PSO. Here, we find the placement and size of the fuel cell using the above-mentioned techniques for 50% load variation. With this comparison, the projected approach voltage profile mostly improved near the normal voltage profile comparing with the other methods. The power loss of the different methods is compared. Here, this method effectively reduces the power loss by selecting the optimum location and capacity of the fuel cell, it is essential to placement and size of the fuel cell minimizes the voltage deviation and power loss.

The above tables explain the performance analysis of the different techniques like GA, PSO, hybrid PSO, and the proposed method. Tables 1 and 2 show the performance analysis of both the GA technique and PSO technique, respectively. The hybrid PSO performance is analyzed in Table 3 and this proposed method

Table 1 Performance analysis of the GA technique

Load in %	Bus no.	Fuel cell capacity	Power loss in kW			Voltage	
			Normal power loss	After fault	After fuel cell placement	Min	Max
50	15	173	210.859	214.439	194.836	0.9047	1

Table 2 Performance analysis of the PSO technique

Load in %	Bus no.	Fuel cell capacity	Power loss in kW			Voltage	
			Normal power loss	After fault	After fuel cell placement	Min	Max
50	15	173	210.859	214.798	194.836	0.9047	1

Table 3 Performance analysis of the hybrid PSO technique

Load in %	Bus no.	Fuel cell capacity	Power loss in kW			Voltage	
			Normal power loss	After fault	After fuel cell placement	Min	Max
50	28	168	210.859	216.212	194.959	0.8974	1

Table 4 Performance analysis of the proposed method

Load in %	Bus no.	Fuel cell capacity	Power loss in kW			Voltage	
			Normal power loss	After fault	After fuel cell placement	Min	Max
50	29	160	210.859	220.950	191.509	0.8773	1

effectiveness is analyzed in Table 4. The above-mentioned techniques are tested against 50% load. The power loss, voltage deviation, and time for attaining the optimum results are tabulated. From the tables, we can conclude that this proposed approach effectively attains the minimum power loss and improved voltage profile, thereby, optimally selecting the location for fixing the fuel cell effectively.

6 Conclusion

In the projected approach, the optimal location for fixing the fuel cell is elucidated by the GA first phase and the fuel cell capacity is predicted by the Artificial Intelligence technique (RNN). The fuel cell size is attained by using the GA second phase [1]. The advantage of the proposed method is its enhanced capacity to achieve improved bus voltage profile, reduced power loss. This approach is proven in the IEEE standard power distribution benchmark systems, performance is analyzed by different algorithms. Here, the comparison studies are made between the radial distribution system power loss and the voltage at different conditions like natural and during the faulty condition. The projected approach is the best effective technique to identify the optimum placement and size of the fuel for the power distribution system, which is superior to the other methods.

References

1. M. Mohammadi, M.A. Nasab, DG placement with considering reliability improvement and power loss reduction with GA method. Res. J. Appl. Sci. Eng. Technol. **3**(8), 838–842 (2011)
2. H. Zayandehroodi, A. Mohamed, H. Shareef, M. Mohammadjafari, Impact of distributed generations on power system protection performance. Int. J. Phys. Sci. **6**(16), 3999–4007 (2011)
3. M. Mohammadi, M.A. Nasab, PSO based multiobjective approach for optimal sizing and placement of distributed generation. Res. J. Appl. Sci. Eng. Technol. **2**(8), 832–837 (2011)
4. A.A. Chowdhury, S.K. Agarwal, D.O. Koval, Reliability modeling of distributed generation in conventional distribution systems planning and analysis. IEEE Trans. Ind. Appl. **39**(5), 1493–1498 (2003)
5. A.T. Davda, M.D. Desai, B.R. Parekh, Impact of embedding renewable distributed generation on voltage profile of distribution system: a case study. ARPN J. Eng. Appl. Sci. **6**(6), 70–74 (2011)
6. M. Moeini-Aghtaie, P. Dehghanian, S.H. Hosseini, Optimal distributed generation placement in a restructured environment via a multi-objective optimization approach, in *16th Conference on Electrical Power Distribution Networks (EPDC)*, Iran, pp. 1–6 (2011)
7. F. Gharedaghi, M. Deysi, H. Jamali, A. khalili, Investigation of power quality in presence of fuel cell based distributed generation. Aust. J. Basic Appl. Sci. **5**(10), 1106–1111 (2011)
8. S.M. Rios, V.P. Vidal, D.L. Kiguel, Bus-based reliability indices and associated costs in the bulk power system. IEEE Trans. Power Syst. **13**(3), 719–724 (1998)
9. M. Heydari, A. Hajizadeh, M. Banejad, Optimal placement of distributed generation resources. Int. J. Power Syst. Oper. Energ. Manag. **1**(2), 2231–4407 (2011)
10. B. Singh, K.S. Verma, D. Singh, S.N. Singh, A novel approach for optimal placement of distributed generation & facts controllers in power systems: an overview and key issues. Int. J. Rev. Comput. **7**, 29–54 (2011)

An Improved CMOS Voltage Bandgap Reference Circuit

Bellamkonda Saidulu, Arun Manoharan, Bellamkonda Bhavani
and Jameer Basha Sk

Abstract Bandgap Reference (BGR) circuit plays a vital role in analog and digital circuit design. BGRs provide temperature-insensitive reference voltages subject to silicon bandgap (1.2 eV). Implementation of BGR circuit zero temperature coefficient (TC) by using P–N diodes and making temperature independent by combining proportional to absolute temperature (PTAT) and complimentary to absolute temperature (CTAT) voltages. PTAT generation and start-up circuits are the two basic elements of BGR to anticipate in nano-watt applications. Start-up circuit requires for PTAT to avoid undesirable zero bias condition. The start-up circuit consists a potential divider with resistors between supply rails. The DC current flows through a resistance path which is larger than leakage current to make the start-up circuit in stable operation. This work proposes bandgap reference circuit with PTAT generation by avoiding the need of start-up circuit and assumption made for PTAT as strong forward bias. This simulation is carried with cadence environment in UMC 180 nm technology. This proposed work results show low temperature coefficient, high accuracy in output voltage, and temperature is varied from −20 to 200 ppm/°C with 1.8 V supply voltage.

Keywords Bandgap reference · CTAT · PTAT · CMOS · Opamp
Low power

B. Saidulu (✉) · A. Manoharan
Vellore Institute of Technology, Vellore, India
e-mail: bellamkonda.saidulu@gmail.com

B. Bhavani
GATE Engineering College, Kodad, India

J. B. Sk
GITAM University, Hyderabad, India

© Springer Nature Singapore Pte Ltd. 2018
S. S. Dash et al. (eds.), *Artificial Intelligence and Evolutionary Computations
in Engineering Systems*, Advances in Intelligent Systems and Computing 668,
https://doi.org/10.1007/978-981-10-7868-2_59

1   Introduction

A reference is mandatory for integrated circuit (IC)-based analog and digital signal processing applications, and it may be either a voltage, current, or time. The reference creates a node which is stable that can be used by other subcircuits to produce expected and fair results. Fundamentally, the reference point should be constant as per operating conditions such as variations in supply voltage, temperature, and loading transients. These reference circuits are basically required in design of subsystems like data converters (ADCs and DACs), voltage converters (DC–DC, AC–DC), operational amplifiers, and linear voltage regulators. The reference conduction is measured by its variation and is narrated by its functional conditions. The reference specifications are line regulation, temperature drift, flow of bias current, input supply voltage range, and load conditions. The steady state changes in supply voltage and temperature which gives variations in reference voltage can be correlated with line regulation and temperature drift. The temperature coefficient (TC) is a metric for variations over temperature [1]. As a rule, TC is formulated in parts per million per degree Celsius (ppm/°C) [2].

$$TC_{ref} = \frac{1}{Reference} \cdot \frac{\partial Reference}{\partial Temperature} \tag{1}$$

Conventionally, voltage across the diode is taken as V_{CTAT} and the difference of two $V_{CTAT}s$ provide the V_{PTAT}. The BJT transistor with V_{BE} or P–N junction diode with forward bias condition exhibits a negative temperature coefficient (NTC).

$$\frac{dV_{BE}}{dT} = -1.5\,MV/K \tag{2}$$

If two bipolar transistors as shown in Fig. 1 operate at unequal current densities, then the difference between base-emitter voltages is directly proportional to absolute temperature.

$$\Delta V_{BE} = V_{BE1} - V_{BE2} \tag{3}$$

$$= V_T \ln \frac{nI_o}{I_{s1}} - V_T \ln \frac{I_o}{I_{s2}} \tag{4}$$

$$= V_T \ln (n) \tag{5}$$

$$\frac{\partial \Delta V_{BE}}{\partial T} = \frac{k}{q} \ln (n) \tag{6}$$

Fig. 1 Basic generation of
temperature-independent
voltage

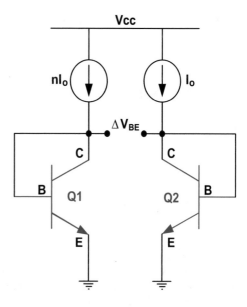

Fig. 2 Bandgap reference

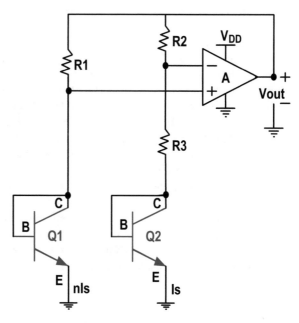

Basic bandgap reference is shown in Fig. 2; output voltage is given by

$$V_{\text{out}} = V_{\text{BE2}} + \frac{V_T \ln n}{R_3}(R_3 + R_2)$$ (7)

for $n = 31$ and taking ratio of R_2, $R_3 = 4$ assume resistors are temperature independent.

$$V_{\text{out}} = V_{\text{BE2}} + 17.2V_T$$ (8)

2 Related Works

Wan et al. [3] proposed a low-voltage second-order high-precision CMOS bandgap reference circuit. The achieved results are 2.5 ppm/°C, supply rejection of −53 dB, line regulation of 0.23 mV/V with 1 V supply voltage. Ji et al. [4] proposed a ultra-low-power BGR suitable for the future IoT applications. This ULP BGR is simulated in 0.18 μm technology and shows results of voltage and current reference generation values of 1.238 V and 6.64 nA. The temperature coefficients are 26, 283 ppm/°C, respectively. Quan et al. [5] implemented a low voltage and high PSRR BGR with Brokaw topology which is adopted for low output voltage. Wang et al. [6] proposed a PTAT with a current compensation method. In this work, emitter area ratios of differential pair npn transistors are used to produce a current which is proportional to PTAT. Hu et al. [7] presented a high-performance CMOS BGR which adopted a current mode architecture for low-voltage applications. Brokaw architecture is employed in portion of circuit, in which a three-stage opamp is selected to avail high PSRR. First-order temperature compensation method is employed to have low TC. Saidulu and Manoharan [8] presented a CMOS amplifier with high gain using a low-power methodology. This high gain amplifier helps to improve precision in BGR circuits in the place of error amplifier. Yi and Laleh [9] presented a new temperature coefficient compensation technique for Si–Ge reference circuits. This work employed two-step temperature compensation technique for the cancellation of major nonlinear temperature-dependent terms. This compensation technique provides an output voltage proportional to difference of bandgap voltages Si and Si–Ge.

3 Proposed Bandgap Reference

Analysis of bandgap references noted the currents in the bipolar junction transistors are proportional to the absolute temperature. PTAT is directly proportional to the diode voltage. The current in the diode is varied exponentially from fA to nA range as temperature changes. This PTAT voltage generated by small amount of current

Fig. 3 V_{PTAT} generation

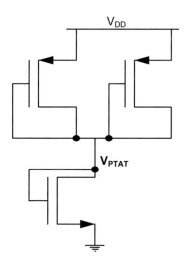

in a circuit at low temperature can be dominated by unwanted leakage current from gate to be connected to the PTAT node. The dependence of leakage current on temperature is very less. MOS-based V_{PTAT} generation is shown in Fig. 3.

$$LI_s = I_s\left(e^{\frac{V_{\text{PTAT}}}{V_T}} - 1\right) \tag{9}$$

$$V_{\text{PTAT}} = V_T \ln(L+1) \tag{10}$$

$$\frac{dV_{\text{PTAT}}}{dT} = \frac{K}{q}\ln(L+1) \tag{11}$$

Implementation of P–N diode, considered a npn bipolar junction transistor with twin-well technology is placed on the resistors which provide V_{CTAT} voltage is shown in Fig. 4.

$$\frac{dV_{\text{BE}}}{dT} = \frac{dV_{\text{CTAT}}}{dT} = -1.5\,\text{mV/K} \tag{12}$$

The schematic of proposed bandgap reference (BGR) is shown in Fig. 5. This BGR has three stages which are V_{PTAT}, error amplifier stage, and V_{CTAT} stage. The opamp sense the voltage at nodes of V_{PTAT} and node at voltage division between R_1, R_2 and fed to the PMOS transistor in the V_{CTAT} generation stage. This PMOS device draws the current from V_{DD} which is proportional to the error voltage fed by the opamp. This current passes through the resistors, provides the V_{PTAT} voltage. This error amplifier has NMOS differential pair input stage, and second stage is source follower-based level sifter to receive low V_{PTAT}. The current consumption in the opamp is less than 1 nA at room temperature. The temperature

Fig. 4 V_{CTAT} generation

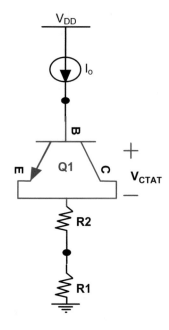

dependency can be eliminated by adjusting the ratio of resistors (R_1, R_2), and the reference voltage is derived as

$$V_{\text{REF}} = \left(1 + \frac{R_2}{R_1}\right) V_{\text{PTAT}} + V_{\text{CTAT}} \tag{13}$$

$$V_{\text{REF}} = (R_1 + R_2) \cdot \frac{V_{\text{PTAT}}}{R_1} + V_{\text{CTAT}} \tag{14}$$

$$V_{\text{REF}} = \left(1 + \frac{R_2}{R_1}\right) \cdot V_P + V_C \tag{15}$$

If temperature is increased by ΔT, then

$$V_{\text{REF}} = \left(1 + \frac{R_2}{R_1}\right)(V_P + \Delta V_P) + V_C - \Delta V_C \tag{16}$$

$$V_{\text{REF}} = \left(1 + \frac{R_2}{R_1}\right) \cdot V_P + V_C + \left(1 + \frac{R_2}{R_1}\right) \cdot \Delta V_P - \Delta V_C \tag{17}$$

Select proper R values to remove the effect of ΔT

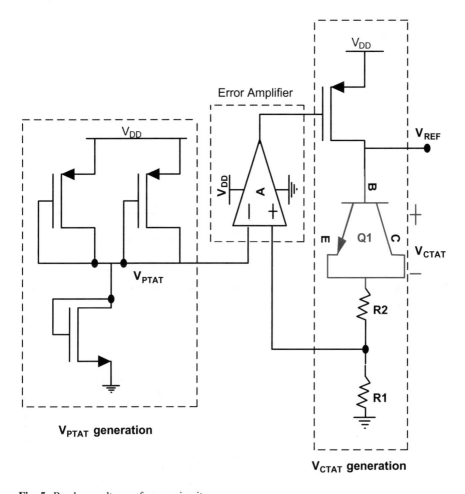

Fig. 5 Bandgap voltage reference circuit

4 Results

	V_{OUT}	Temp range (°C)	Power consumption
[10]	1.176	−10 to 110	28.7 nW
This work	1.695	−20 to 120	151.8 μW

Fig. 6 Reference voltage versus temperature

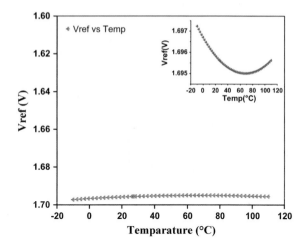

Fig. 7 Reference current versus temperature

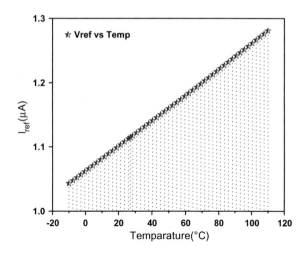

5   Conclusion

The proposed bandgap reference circuit is implemented in UMC 180 nm technology. The temperature that is swept from −20 to 200 °C is shown in Figs. 6 and 7 for variation in reference voltage, current, respectively. This BGR circuit achieves high accuracy in output voltage with the cost of 151.8 μW power consumption, with V_{DD} of 1.8 V at room temperature while compared to similar state of the art of BGR circuit.

Acknowledgements The authors are grateful for the support of UGC (Government of India) and Research Groups, VLSI Laboratory, VIT University, India.

References

1. G.A. Rincon-Mora, Voltage references (Wiley-Inter science, London, 2002)
2. B. Razavi, *Design of Analog CMOS Integrated Circuits* (McGraw-Hill, New York, 2001)
3. M. Wan, Z. Zhang, K. Dai, X. Zou, A 1-V 2.5-ppm/Â°C second-order compensated bandgap reference, in *ASIC (ASICON), 2015 IEEE 11th International Conference on IEEE*, 2015, pp. 1–4
4. Y. Ji, C. Jeon, H. Son, B. Kim, H.-J. Park, J.-Y. Sim, 5.8 A 9.3 nW all-in-one bandgap voltage and current reference circuit. in *Solid-State Circuits Conference (ISSCC), 2017 IEEE International,* IEEE, 2017, pp. 100–101
5. L. Quan, Y. Yin, X. Yang, H. Deng, Design of a high precision band-gap reference with piecewise-linear compensation. in *Anti-Counterfeiting, Security and Identification (ASID), 2012 International Conference on IEEE*, 2012, pp. 1–5
6. S. Wang, S. Wang, A bandgap reference circuit with temperature compensation. in *Microwave and Millimeter Wave Technology (ICMMT), 2016 IEEE International Conference on IEEE,* 2016, vol. 1, pp. 111–113
7. J. Hu, Y. Yin, H. Deng, Design of a high-performance brokaw band-gap reference. in *Anti-Counterfeiting Security and Identification in Communication (ASID), 2010 International Conference on IEEE*, 2010, pp. 126–129
8. B. Saidulu, A. Manoharan, Design of low power amplifier for neural recording applications using inversion coefficient methodology. Int. J. Control Theory Appl. **10**(6), 719–727 (2017)
9. H. Yi, N. Laleh, A precision SiGe reference circuit utilizing si and SiGe bandgap voltage differences. IEEE Trans. Electr. Dev. **64**(2), 392–399 (2017)
10. J.M. Lee, Y. Ji, S. Choi, Y. Cho, S. Jang, J.S. Choi, B. Kim, H. Park, J. Sim, A 29 nW bandgap reference circuit, in *Proceedings IEEE International Solid-State Circuits Conference* 2015, pp. 100–102

Modeling of an ANFIS Controller for Series–Parallel Hybrid Vehicle

K. Rachananjali, K. Bala Krishna, S. Suman and V. Tejasree

Abstract The main objective of the paper is to acclimatize the throttle accommodated in the internal combustion engine (ICE) to attain maximum output torque while alternating fuel consumed by adaptive neuro-fuzzy inference system (ANFIS) controller. This paper describes various techniques to acclimatize the throttle in ICE. The throttle manages the flow of fluid and can increase or decrease the engine's power, and the air fuel ratio is checked and adjusted. Constantly maintain the air fuel ratio of the vehicle to attain the improved performance of the maximum possible torque demand from the engine. ANFIS control technique is an optimal method for controlling the hybrid vehicles. Hybrid vehicles are propelled by two sources. In this paper, hybrid vehicle uses two sources, that is, ICE and battery which is being charged with wind turbine.

Keywords Adaptive neuro-fuzzy inference system (ANFIS) · Internal combustion engine (ICE) · Hybrid electric vehicle · Permanent magnet synchronous motor (PMSM) · Vector control · Artificial neural network (ANN) · Power train system

1 Introduction

Nowadays, the levels of green house gases are already high, usage of conventional vehicles may increase that limit. Conventional vehicles produce smog. By the introduction of hybrid vehicle, the effect of smog can be minimized. By the usage of hybrid vehicles, fuel consumption can be reduced, emission can be decreased, running cost is decreased, and many more. The throttle manages the flow of fluid and can increase or decrease the engine's power, and the air fuel ratio is checked and adjusted. Constantly maintain the air fuel ratio of the vehicle to attain the improved performance of the maximum possible torque demand from the engine.

K. Rachananjali (✉) · K. B. Krishna · S. Suman · V. Tejasree
Department of EEE, Vignan's Foundation for Science,
Technology & Research, Vadlamudi, Guntur, India
e-mail: rachananjali@gmail.com

© Springer Nature Singapore Pte Ltd. 2018
S. S. Dash et al. (eds.), *Artificial Intelligence and Evolutionary Computations in Engineering Systems*, Advances in Intelligent Systems and Computing 668,
https://doi.org/10.1007/978-981-10-7868-2_60

ANFIS control technique is an optimal method for controlling the hybrid vehicles. The benefits associated with ANFIS controller are as follows: (i) number of rules are determined automatically, (ii) computational time is reduced, (iii) faster learning, and (iv) minimizes error. A hybrid vehicle requires torque for driving and operating the onboard accessories which is bred by amalgamation of ICE and battery [1]. In this paper, different controllers, that is, PI, fuzzy, and ANFIS controller, are introduced to maintain constant air fuel ratio [2, 3].

In PI, PID controller parameters are fixed. The disadvantage associated with a PID controller is its assumption that we have exact knowledge about the system, which may not be true for most of the cases. To eradicate the disadvantage associated with PID controller, advanced control methods have been introduced. These methods use the principle of fuzzy logic. The design of a traditional fuzzy controller depends on input–output membership functions' number and input–output membership functions' shape.

ANFIS controller means adaptive neuro-fuzzy inference system, which uses the structure of IF-THEN rules [2]. They can be easily understood as it uses linguistic terms. Membership functions are assigned to neurons because it is a very efficient controlling technique. The proposed adaptive neuro-fuzzy inference system (ANFIS) is a type of neural network which applies the principle of Takagi–Sugeno fuzzy inference system. As Takagi–Sugeno fuzzy inference system amalgamates neural networks and fuzzy logic [3].

The proposed hybrid vehicle has two kinds of motive power sources: battery, which is charged with wind turbine and an ICE, to improve efficiency [1, 4–6].

2 Types of Hybrid Vehicle

This section describes the different configurations of hybrid vehicle. The three different configurations are (i) series hybrid vehicle, (ii) parallel hybrid vehicle. and (iii) series–parallel hybrid vehicle [7].

2.1 Series Hybrid Vehicle

It is similar to electric vehicle, and the vehicle is driven by only EM but it contains ICE for generating electric power to keep the batteries charged and propel the vehicle by means of the electrical power through the generator, motor, and battery [1, 4]. Figure 1 indicates the direction of energy flow in series hybrid vehicle.

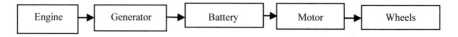

Fig. 1 Series hybrid vehicle

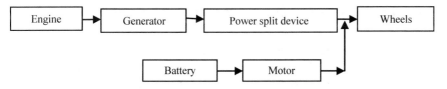

Fig. 2 Parallel hybrid vehicle

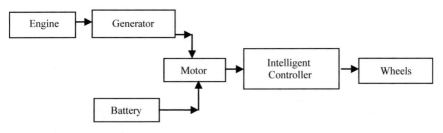

Fig. 3 Series–parallel hybrid vehicle

2.2 Parallel Hybrid Vehicle

This configuration consists of two separate energy sources, ICE and battery, that are connected in parallel [4]. Power split device comes into picture to check whether the engine and motor are working independently or simultaneously. The disadvantage associated with that is battery cannot be recharged at standstill. Figure 2 indicates the direction of energy flow in parallel hybrid vehicle [4].

2.3 Series–Parallel Hybrid Vehicle

As the name itself indicates, this configuration combines the advantages of both series and parallel hybrid vehicle. So the need of double connection arises between engine and battery. Figure 3 indicates the direction of energy flow in series–parallel hybrid vehicle.

3 System Model Description

This section describes the different parts which constitute a hybrid vehicle. Hybrid vehicle consists of Internal Combustion Engine (ICE), Wind turbine [1], PMSM generator.

3.1 Internal Combustion Engine

The different segments in a reciprocating internal combustion engine (ICE) are the cylinder and the piston. The piston seals the cylinder and disseminates gas pressure to the crankshaft via the connecting rod. The crankshaft passes the rotary motion through the gearbox to the wheels. So, an intake and exhaust manifold connected with the intake and exhaust valve, respectively, completes the spark ignition engine assembly [8–10].

3.2 Wind Turbine

To capture the energy generated by wind, wind turbines are used which converts the kinetic energy into mechanical energy [1]. Wind turbines are equipped with PMSM generators. The power of the wind is proportional to air density, area of the segment of wind being considered, and the natural wind speed. The main energy advantage realized through wind turbines is that power is proportional to the wind speed cubed [1].

$$P_{\mathrm{w}} = \frac{1}{2}MAu^3 \tag{1}$$

3.3 PMSM Generator

As the energy generated by wind is captured by using wind turbine. The output of the turbine is mechanical energy but in order to charge the battery, we need electrical energy [1]. So in order to convert mechanical energy into electrical energy there arises the need of generator. Depending upon the application, we can choose synchronous or asynchronous generator [4, 5].

4 Design of Optimal Controller

In order to control series–parallel hybrid vehicle, the ICE needs to be operated in peak efficiency region. When ICE is operated in peak efficiency region, the efficiency of the power train is improved. Operation of ICE depends on the speed, driver demand (road load), and the state of charge (SoC). These parameters give an idea what the controller needs. The three different inputs for the controller are as follows: the desired torque, driver demand, and SoC [11–14].

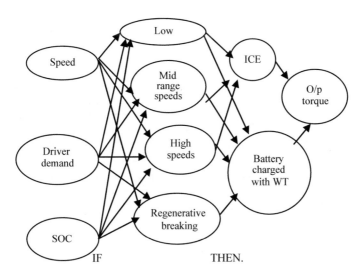

Fig. 4 ANFIS controller input and output and hidden layers

In Fig. 4, speed, driver demand, and SoC are input layers, and second and third layers indicate the hidden layers, fourth layer which is the output torque is output layer. In order to design a controller, we need five membership functions for inputs. The crisp output is generated from fuzzy inputs.

5 Operating Conditions

In the proposed hybrid vehicle, it is an amalgamation of ICE and battery charged with wind turbine. ICE and battery are controlled by using neuro-fuzzy controller. Neuro-fuzzy controller is used because it can work with nonlinear problems also.

Hybrid vehicle combines the benefits of gasoline and wind turbine to provide the improved fuel economy.

Operating regions: Hybrid vehicle having four operating regions

1. Starting condition (or) Low speed condition,
2. Mid-range speed condition,
3. High speed condition,
4. Regenerative braking condition.

Figure 5 indicates the block diagram of proposed system.

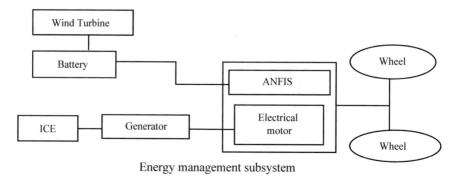

Fig. 5 Proposed hybrid vehicle

5.1 Starting Condition

At starting condition (0–4 min), an internal combustion engine cannot produce high torque in the low rpm range at that condition battery will provide energy to the vehicle. Figure 6 shows the direction of energy flow in proposed system at starting condition.

5.2 Mid-range Condition

In this condition in between 4 and 8 min, hybrid vehicles are propelled by the electrical motor, which is energized by battery which is charged with wind turbine. Figure 7 shows the direction of energy flow in proposed system at mid-range condition.

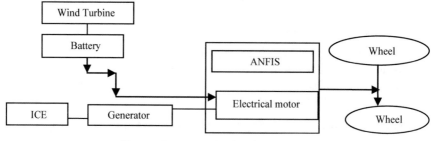

Fig. 6 Proposed system at starting condition

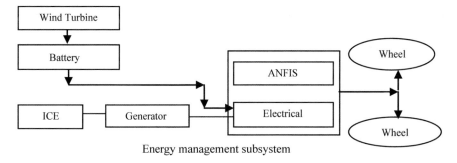

Fig. 7 Proposed systems at mid-range condition

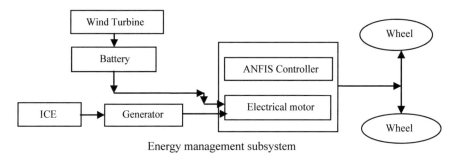

Fig. 8 Proposed systems at high speed condition

5.3 High Speed Condition

In this speed range battery (which is charge with wind turbine) and generator, motor will provide energy for propel the vehicle. i.e. both energy source are used in that condition. Figure 8 shows the direction of energy flow in proposed system at high speed condition.

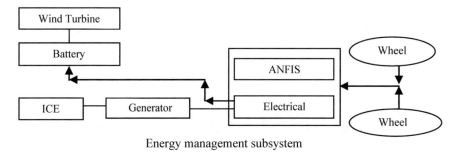

Fig. 9 Proposed systems during regenerative braking condition

5.4 Regenerative Braking Condition

In this range, motor act as generator, that provides power to the battery, that is, at breaking condition, KE of wheels is converted into electrical energy and it is stored into battery. And ICE and generator are off. Figure 9 shows the direction of energy flow in proposed system during regenerative braking condition (Fig. 10).

6 Simulation Circuits

6.1 Simulation Circuit of Hybrid Vehicle

Simulation circuit of proposed system consists of ICE, electrical subsystem, and energy management system. Figure 11 shows the simulation circuit of proposed hybrid vehicle.

6.2 Simulation Circuit of Energy Management System

See Fig. 12.

7 Comparison of Results

The throttle manages the engine speed, maximum possible torque demand from the engine. Throttle body controls the amount of air entering into the engine, so here controller controls the throttle signal to maintain the constant air fuel ratio. Here, ANFIS and PI controllers are compared for accurate results.

7.1 Simulation circuit of PI controller in energy management system

See Fig. 13.

7.2 Simulation result of PI controller throttle signal

See Fig. 14.

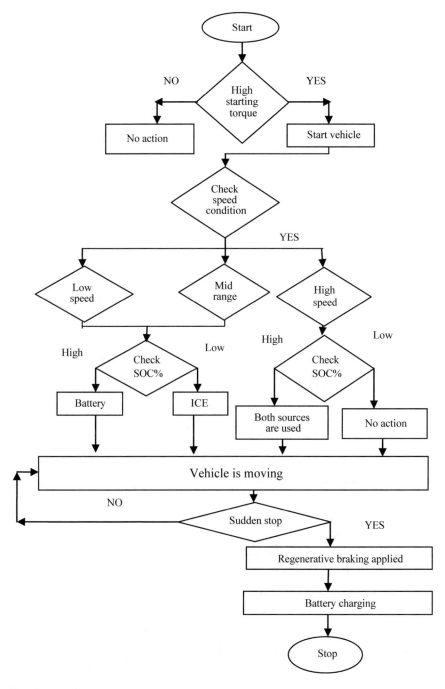

Fig. 10 Flowchart representing the proposed control

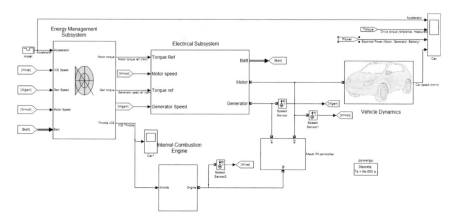

Fig. 11 Proposed hybrid vehicle

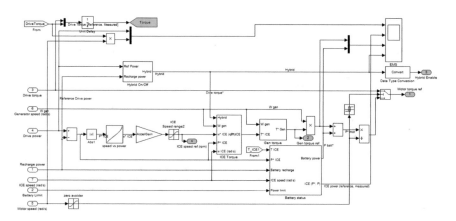

Fig. 12 Simulation circuit of energy management system

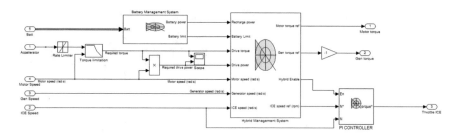

Fig. 13 PI controller in energy management subsystem

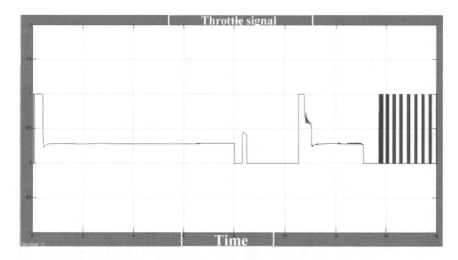

Fig. 14 Simulation result of PI controller throttle signal

7.3 Simulation circuit of ANFIS controller

See Fig. 15.

7.4 Simulation results of an ANFIS controller throttle signal

See Fig. 16.

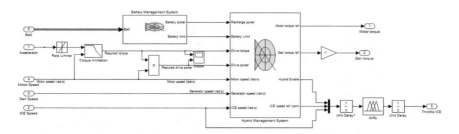

Fig. 15 Simulation circuit of ANFIS controller placed in energy management subsystem

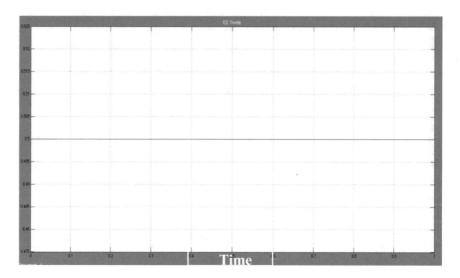

Fig. 16 Simulation result of ANFIS controller throttle signal

Table 1 Operating mechanisms of different controllers	Operating cycles (s)	Throttle (adjust the air fuel ratio maintain constant)	
		PI	ANFIS
	Starting (0–4)	(0–0.5 s)–(0–1) (0.5–4 s)–0.3	0.5
	Mid range (4–8)	0.3	0.5
	High speed (8–12)	Varying (0–1)	0.5
	Regenerative braking (12–16)	Varying (0–1)	0.5

8 Conclusion

To achieve the required operating performance, a number of electrical systems need to be interconnected as they are complex dynamic systems. To implement this strategy, the accelerator and brake pedal inputs of the driver are converted to a driver requiring torque at first. In this paper, different controllers like PI, fuzzy logic, and ANFIS are implemented. ANFIS control technique is an optimal method for controlling the hybrid vehicles and it is proved. Hybrid vehicle requires torque for driving and operating the onboard accessories and is generated by a combination of internal combustion engine and battery which is charged by wind turbine. On comparison of results, ANFIS controller performs the best compared to other controllers. Table 1 compares the result of controllers for different operating regions

References

1. S. Sowmini priya, S. Rajakumar, An energy storage system for wind turbine generators-battery and supercapacitor. Int. J. Eng. Res. Appl. (IJERA) (ISSN: 2248–9622) (2013)
2. M. Mohebbi, M. Charkhgard, M. Farrokhi, Optimal neuro-fuzzy control of parallel hybrid electric vehicles, in *Proceedings of IEEE Conference on Vehicle Power and Propulsion* (2005), pp. 26–30
3. D. pelusi, Genetic neuro-fuzzy controllers for second order control system, in *2011 UKsim 5th European Symposium on Computer Modeling and Simulation* (2011)
4. A. Boyali, M. Demirci, T. Acarman, L. Guvenc, O. Tur, H. Ucarol, B. Kirey, E. Ozatey, Modeling and control of a four wheel drive parallel hybrid electric vehicle, in *Proceedings of IEEE International Conference on Control Applications* (Munich, Germany, 4–6 Oct 2006)
5. K. Zhang, J. Li, M. Ouyang, J. Gu, Y. Ma, Electric braking performance analysis of PMSM for electric vehicle applications, in *2011 International Conference on Electronic & Mechanical Engineering and Information Technology* (2011)
6. J. Osorio, A. Molina, P. Ponce, D. Romero, A supervised adaptive neuro-fuzzy inference system controller for a hybrid electric vehicle's power train system, in *9th IEEE International Conference on Control and Automation (ICCA)* (Santiago, Chile, Dec 2011) pp. 19–21
7. A. Ntirnberger, D. Nauck, R. Kruse, L. Merz, A neuro-fuzzy development tool for fuzzy controllers under MATLAB/SIMULINK, in *Proceedings of the 5'th European Congress on Intelligent Techniques & Soft Computing (EUFIT'97)* (Aachen, Germany, 1997)
8. Y. Mihoub, B. Mazari, S. Fasla, *Neuro-Fuzzy Controller Used to Control, the Speed of an Induction Motor* (University of science and technology of Oran Mohamed Bodied Faculty of Electrical Engineering-Electro technical Department B.P1505,31000 Oran ElMnaouar Algeria) (2007)
9. C. Abbey, G. Joos, Supercapacitor energy storage for wind energy applications. IEEE Trans. Ind. Appl. **43**(3), 769–776 (2007)
10. W. Li, G. Xu, Z. Wang, Y. Xu, A hybrid controller design for parallel hybrid electric vehicle, in *Proceedings of IEEE International Conference on Integration Technology* (Shenzhen, China, 20–24 Mar 2007)
11. L.K Butler, M. Ehsani, P. Kamath, A mat lab-based modeling and simulation package for electric and hybrid electric vehicle design. IEEE Trans. Veh. Technol. **48**(6) (Nov 1999)
12. H. Lee, E. Koo, S. Sul, J. Kim, Torque control strategy for a parallel-hybrid vehicle using fuzzy logic, in *IEEE Industry Applications Magazine* (Nov 2000)
13. C. Xia,* C. Zhang, Power management strategy of hybrid electric vehicles based on quadratic performance index. Energies **8** pp. 12458–12473 (2015)
14. D. Sukumar, J. Jithendranath, S. Saranu, Three level Inverter fed Induction motor drive performance improvement with neuro-fuzzy space vector modulation. Electr. Power Compon. Syst (Taylor and Francis) (ISSN 1532-5008) pp. 1633–1646 (2014)

A Couple of Novel Stochastic Estimators Designed and Tested to Promote the Usage of Towed Arrays on the Regular Basis for Passive Tracking

D. V. A. N. Ravi Kumar, S. Koteswara Rao and K. Padma Raju

Abstract Hull mounted array (HA) and Towed array (TA) are two fundamental types of sensor arrays used nowadays to deal with the underwater bearings-only tracking issue. Despite of proving less estimation errors, the TA sensors are not widely used for tracking applications as compared to HA. The practical difficulties in handling TA are responsible for this. In this letter, a couple of novel stochastic estimation algorithms are designed for towed arrays to encourage or promote the usage of (TA) on the regular basis. These estimators can bring down the estimation errors to such an extent, that the feasibility issues of TA can be overlooked. The first proposed technique is named as the measurement conditioning technique (MC) while the second one is called the fusion of estimates technique (FE). The MC technique works on the principle of soothing the measurements prior to the estimation while the FE uses the concept of merging the estimates given by a set of unscented Kalman filters (UKF) which are driven by TA measurements. Monte Carlo simulations in MATLAB proved the superiority of TA over HA when used along with the proposed methods. The results also show that, the superiority is even better at long ranges and thick noise conditions.

Keywords Hull mounted array · Towed array · Bearings-only tracking Estimation errors · Unscented Kalman filter

D. V. A. N. R. Kumar (✉)
GVP College of Engineering for Women, Madhurawada, Visakhapatnam, India
e-mail: ravikumardwarapu@gvpcew.ac.in

S. Koteswara Rao
KL University, Vaddeswaram, Guntur, India

K. Padma Raju
JNT University, Kakinada, India

© Springer Nature Singapore Pte Ltd. 2018
S. S. Dash et al. (eds.), *Artificial Intelligence and Evolutionary Computations in Engineering Systems*, Advances in Intelligent Systems and Computing 668,
https://doi.org/10.1007/978-981-10-7868-2_61

1 Introduction

Tracking is a vital task performed by the signal processing module of the ownship in war environments. This is done with the assistance provided by the RADAR for tracking missiles or targets over the ground or in air and SONAR for tracking underwater torpedoes. SONAR again operates in two modes, namely active and passive. Active mode operation uses the range and bearing data of the target and suffers from the problem of providing self-information to the enemy before processing enemies signal while, passive mode uses only bearing data captured from the propellers of the enemy's vehicle. The passive mode, in turn, uses a single moving sensor or an array of sensors. The sensor arrays are again classified into Hull mounted (HA) array and Towed array (TA). As the name suggests, the HA comprises of the sensors located on hull or ships body while towed array comprises of the sensors which are towed by the ownship. TA is having the advantage of having measurements which are less corrupted by the noise compared to HA because of having sensors far from the noise generating propellers of ownship. Despite of this advantage, TA is not widely used because of the problems that arise while handling Kilometer long sensor array. In this paper, two new stochastic algorithms are proposed exclusively to process the TA measurements. These algorithms provide considerable improvements over HA in terms of estimation errors and error convergence times such that, the handling issues associated with TA can longer stop them from their wide usage.

The literature published so far on tracking supports a single moving sensor [1] or HA sensors. They include the following. Very good books on tracking using advanced stochastic algorithms like [2], papers on Kalman filter (KF) based algorithms like extended Kalman filter (EKF), modified gain extended Kalman filter (MGEKF), unscented Kalman filter (UKF), maximum likelihood estimator (MLE), pseudo linear estimator (PLE), cubature Kalman filter (CKF), gauss hermite Kalman filter (GHKF), interactive multiple model filter (IMM), particle filter (PF), probability hypothesis density (PHD) filter, and hybrid filters. The examples of such papers included the following. Reference [3] uses debiasing along with KF. Reference [4] uses EKF, Ref. [5] are based on MGEKF, Ref. [6–8] are based on UKF, [9] is based on MLE, [10, 11] are based on PLE, [12] are based on CKF, Ref. [13] is based on GHKF, Ref. [2] are based on PF, Ref. [14] is based on PHD filter, Ref. [8] is based on hybrid algorithms and Multisensor-based algorithms like [15]. As mentioned above, all the work published uses single maneuvering sensor or HA measurements but none of the papers published exclusively for TA measurements to defend their existence. This paper tries to do this by processing the TA measurements directly using a couple of novel algorithms to achieve much superior results than HA measurements. This paper can help in promoting the usage of TA on regular basis in future, despite of the handling difficulties.

Sections 2 and 3 introduce the two novel stochastic methods namely MC and FE to process TA measurements. Section 4 is comprised of a mathematical representation of a moving target, a mathematical representation of towed array sensor measurements and performance metrics are defined to quantify the superiority of the proposed algorithms. Section 5 deals with the simulation and analysis of results, finally, the paper is concluded with Sect. 6.

2 Measurement Conditioning (MC) Technique

In estimation problems, small errors in sensor measurements can cause a huge error in the dependent parameters (parameters which are obtained by transforming the sensor measurements). The best example of this can be seen in the Bearings-only tracking (BOT) problem. Here, a minute error in bearings (few rad) causes an enormous range error (in km). So if we reduce the variance of noise in the sensor measurements even by a small amount, that can account for a considerable minimization of the estimation error of dependent parameters. An effort is made in this paper by developing the measurement conditioning (MC) technique to bring down the variance of noise in the sensor measurements. The technique used is to average a set of received measurements, one of the measurements being the present instant measurement while the others are projected past readings. The projection is done with the help of the measurement database. The technique is expected to bring down the error variance to an appreciable amount, based on the fact that, averaging will cancel the measurement errors. In simple words, the MC technique will condition the measurements before applying to the conventional estimators.

2.1 Second-Order Measurement Conditioner

The present instant conditioning is obtained by averaging the present and one projected previous measurement. The block diagram of second-order measurement conditioner is shown in Fig. 1. The conditioned measurements at kth instant is denoted by Y_k^{MC} is obtained from the sensor measurements Y_k by the following equations. $Y_1^{MC} = Y_1$, $Y_2^{MC} = 0.5(Y_2 + Y_1^{1p})$, $Y_3^{MC} = 0.5(Y_3 + Y_2^{1p})$, $Y_4^{MC} = 0.5(Y_4 + Y_3^{1p})$. Y_1^{1p}, Y_2^{1p} and Y_3^{1p} are the one step projected values of the measurements at time instants 1, 2, and 3, respectively. These are computed with the help of data base of the measurements collected so far. $Y_1^{1p} = Y_1 + (Y_2 - Y_1)/1$, $Y_2^{1p} = Y_2 + [(Y_2 - Y_1) + (Y_3 - Y_2)]/2$, $Y_3^{1p} = Y_3 + [(Y_2 - Y_1) + (Y_3 - Y_2) + (Y_4 - Y_3)]/3$. Similarly the Lth order measurement conditioning is done as per Table 1.

Fig. 1 Block diagram of
second order measurement
conditioner (MC)

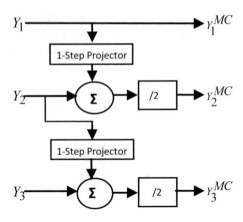

Table 1 Measurement
conditioning (MC) algorithm
(Lth order)

Inputs: received measurements Y_k
Outputs: conditioned measurements Y_k^{MC}
The m-step projector output $Y_k^{mp} = \left(\frac{1}{k-m} \sum_{s=1}^{k-m} \left[Y_{m+s} - Y_s \right] \right)$
$Y_k^{MC} = \frac{1}{k} \left(Y_k + \sum_{i=1}^{k-1} Y_{k-i}^{ip} \right) \quad k < L$
$Y_k^{MC} = \frac{1}{L} \left(Y_k + \sum_{i=1}^{L-1} Y_{k-i}^{ip} \right) \quad k \geq L$

3   Estimate Fusion (FE) Technique

In this technique, the response of different filters estimating the same parameter are
merged together using some stochastic algorithm to give the best possible estimate
from the available measurements. As the estimates are fused together the algorithm is
called fusion of estimates or estimate fusion (FE) algorithm (Table 2). In this paper the
basic estimators used are Unscented Kalman filters(UKF) which are driven by different sensors of towed array and outputs are fused using a fuser called weighted least
squares estimator(WLSE). The method is expected to give considerable satisfactory
performance based on the fact that joint decision of a state is much better than any
individual decision. Added to the above point, the feedback of better signal to individual filters can drag the estimation error at much quicker pace compared to the
traditional estimators. The block diagram of the FE algorithm is given in Fig. 2. If the
measurements received by "p" different sensors at kth instant are denoted by Y_{j_k} with
$j = 1, 2, \ldots, p$ are applied as inputs to p different estimators to get p number of
estimates ($\hat{X}_{j_{k+}}$ with $j = 1, 2, \ldots, p$.) and covariances $P_{j_{k+}}$ of the same parameter. The
fuser such as the weighted least squares estimator accepts and fuses the p estimates to
give a better estimate $\hat{X}_{k+}$ and P_{k+} as the associated covariance matrix.
$\hat{X}_{j_{k+}} = \begin{bmatrix} \hat{X}_{j_{k+}^1} & \hat{X}_{j_{k+}^2} & \cdots & \hat{X}_{j_{k+}^n} \end{bmatrix}^T$. This equation shows that the estimate given
by each estimator is a combination of n elements. The covariance matrix corre-
sponding to the estimate is in the form of $P_{j_{k+}} = \text{diag} \begin{bmatrix} P_j^1 & P_j^2 & \cdots & P_j^n \end{bmatrix}$. Where

Pj^i is the variance of error in the estimation of the ith element provided by jth estimator.

$$\begin{bmatrix} \hat{X}1^i_{k+} & \hat{X}2^i_{k+} & \cdots & \hat{X}p^i_{k+} \end{bmatrix}^T = \begin{bmatrix} 1 & 1 & \cdots & 1 \end{bmatrix}^T X^i_{k+} + \begin{bmatrix} l1^i_k & l2^i_k & \cdots & lp^i_k \end{bmatrix}^T \tag{1}$$

Equation (1) shows how the ith element of the individual estimators and the true value are related. lj^m_k Notation in the above equation mean, the error in the estimation of mth element of x given by the jth estimator at kth instant. Equation (1) in simplified form is written as

$$Y^i_{k+} = HX^i_{k+} + L^i_{k+} \tag{2}$$

where $Y^i_{k+} = \begin{bmatrix} \hat{X}1^i_{k+} & \hat{X}2^i_{k+} & \cdots & \hat{X}p^i_{k+} \end{bmatrix}^T$, $H = \begin{bmatrix} 1 & 1 & \cdots & 1 \end{bmatrix}^T$ $L^i_{k+} = \begin{bmatrix} l1^i_k & l2^i_k & \cdots & lp^i_k \end{bmatrix}$. The matrix L^i_{k+} is the error matrix with the covariance $R^{(i)} = \text{diag} \begin{bmatrix} p1^i & p2^i & \ldots & pp^i \end{bmatrix}^T$ and pj^i notation in the matrix above mean the variance in the estimate of ith element given by jth estimator. The optimal value of estimate of X^i_{k+} can now be obtained from the Y^i_{k+} with the help of WLSE. It is as follows $\hat{X}^i_{k+} = \left(\sum_{j=1}^{p} \frac{1}{(pj^i)^2} \right)^{-1} \left(\sum_{j=1}^{p} \frac{\hat{X}j^i_{k+}}{(pj^i)^2} \right)$. The Estimate of the state is a composite of individual element estimates, i.e., $\hat{X}_{k+} = \begin{bmatrix} \hat{X}^1_{k+} & \hat{X}^2_{k+} & \ldots & \hat{X}^n_{k+} \end{bmatrix}^T$. The covariance matrix associated with the above estimate is $P_{k+} = \frac{1}{p} \sum_{j=1}^{p} Pj_{k+}$.

4 Mathematical Modeling

4.1 Mathematical Representation of a Moving Target

The state vector (X) for the moving target tracking problem is commonly comprised of the position px, py, and velocity components vx, vy as $X = \begin{bmatrix} px & py & vx & vy \end{bmatrix}^T$

Table 2 Fused estimate (FE) algorithm	Input: outputs of individual estimators $\hat{X}j_{k+}, Pj_{k+}$ with $j=1, 2, 3… p$
	Output: fused estimate and covariance $\hat{X}_{k+}, P_{k+}$
	$\hat{X}^i_{k+} = \left(\sum_{j=1}^{p} \frac{1}{(pj^i)^2} \right)^{-1} \left(\sum_{j=1}^{p} \frac{\hat{X}j^i_{k+}}{(pj^i)^2} \right)$. $\hat{X}^i_k$ is the ith element of $\hat{X}j_{k+}$
	pj^i the (i, i) element of Pj_{k+}
	$\hat{X}_{k+} = \begin{bmatrix} \hat{X}^1_{k+} & \hat{X}^2_{k+} & \ldots & \hat{X}^n_{k+} \end{bmatrix}^T, P_{k+} = \left(\frac{1}{p} \sum_{j=1}^{p} Pj_{k+} \right)$

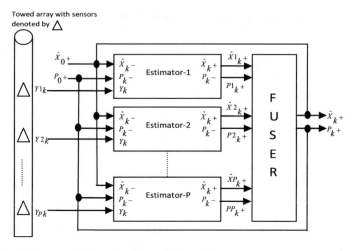

Fig. 2 Block diagram of fusion of estimates (FE) technique

and the state equation for a non-maneuvering case is in the form as follows $X_{k+1} =$

$$FX_k + w_k \quad k = \text{time instant} \quad F = \begin{bmatrix} 1 & 0 & \Delta T & 0 \\ 0 & 1 & 0 & \Delta T \\ 0 & 0 & 1 & 0 \\ 0 & 0 & 0 & 1 \end{bmatrix}$$

and w_k represents the process noise with covariance matrix Q.

$$Q = E\left[w_k \, w_k^T\right] = \begin{bmatrix} \Delta T^3/3 & 0 & \Delta T^2/2 & 0 \\ 0 & \Delta T^3/3 & 0 & \Delta T^2/2 \\ \Delta T^2/2 & 0 & \Delta T & 0 \\ 0 & \Delta T^2/2 & 0 & \Delta T \end{bmatrix} q$$

q is the standard deviation of the acceleration errors.

4.2 Mathematical Model of the TA Sensor Measurements

The azimuth measurements captured by two sensors located at (six, siy) and (sjx, sjy) denoted by $(\text{Bmi}_k, \text{Bmj}_k)$ constitute the elements of a measurement vector Y_k. The measurement equation can now be written as follows $Y_k = h(X_k) + m_k$ with, $Y_k = [\text{Bmi}_k \quad \text{Bmj}_k]^T$, $h(X_k) = \left[\arctan\left(\frac{py_k - \text{siy}}{px_k - \text{six}}\right) \quad \arctan\left(\frac{py_k - \text{sjy}}{px_k - \text{sjx}}\right) \right]^T m_k = [\text{Bni}_k \quad \text{Bnj}_k]^T$ is the measurement error matrix with a mean of 0 and covariance matrix R. $R = E[m_k \, m_k^T] = \text{diag}[\sigma_i^2, \sigma_j^2] \cdot \sigma_i^2$ and σ_j^2 are variances of noise in measurements at the two sensors.

4.3 Performance Metrics

1. RMS Position error (m):

$$\text{RMS Position error } (m) = \sqrt{\frac{1}{N}\sum_{i=1}^{N}\left(px_k^{(i)} - \hat{px}_k^{(i)}\right)^2 + \left(py_k^{(i)} - \hat{py}_k^{(i)}\right)^2}$$

N is the number of Montecarlo (MC) runs and i is the indexing variable, ^ over a variable indicates the estimated value it.

2. Convergence time (s):

It is defined as the time spent by the filter (Estimator) to pull down the error in the estimation of position to 500 m or below.

5 Simulation Results and Analysis

To quantify the superiority of the TA measurements (after applying the proposed techniques) over HA measurements, the five algorithms namely H-UKF (HA measurements processed by UKF), T-UKF (TA measurements processed by UKF), MC-UKF (TA measurements processed by MC and UKF), FE-UKF (TA measurements processed by FE and UKF), MCFE-UKF (TA measurements processed by MC, FE, and UKF) are applied for tracking a target which is moving as per scenario given in Table 3.

Table 3 Simulation scenario

Target parameters	Observer parameters
Initial range (R) = 15 km,	Towed array sensor count = 3
Initial bearing (B) = 300°,	Positions = [(0,0),(0, 500 m), (0, 1000 m)],
Velocity (vt) = 10 m/s,	Measured parameter = bearing,
Course (Tcr) = 90°	Mean of measurement noise = 0;
Sigma (σ) of acceleration errors = 0.01 m/s^2	Sigma (σ) of measurement noise = 0.005 −0.05 (rad)

Filter parameters
Basic filter = UKF,
Initial estimate = [7.5 km, −13 km, 0,10],
Covariance initialization = Diag [10000, 10000, 10, 10],
Scaling parameters of UKF α = 0.001, β = 2, Ka = 0,
Fuser = WLSE,
Weights assigned to fusing elements = equal,
MC order = 2, Montecarlo run count = 50

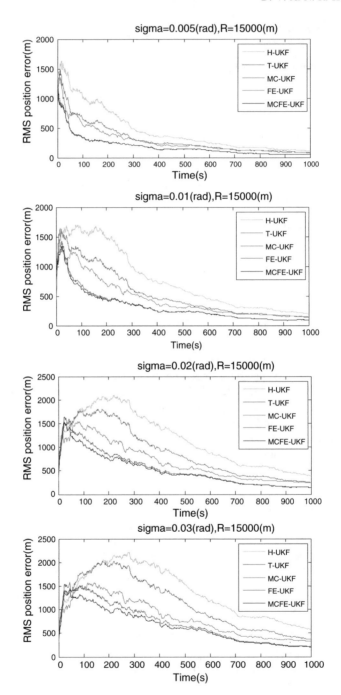

Fig. 3 RMS position error versus time of H-UKF, T-UKF, MC-UKF, FE-UKF and MCFE-UKF estimators at $\sigma = 0.005$, 0.01, 0.02 and 0.03

(a) **Performance comparison of the algorithms with the varying input noise intensities.**

RMS position error versus Time graph of the five estimators in different noise environments [σ = 0.005, 0.01, 0.02, and 0.03 (rad)] while tracking a target which is moving according to the scenario given in Table 3 is shown in Fig. 3. The average position error versus noise intensity and time of convergence versus noise intensity graphs of five algorithms is shown in Fig. 4.

The same is represented in quantitative form in Tables 4 and 5.

(b) **Performance comparison of the algorithms with the varying ranges.**

The comparison data is presented in Fig. 5.

From Figs. 3, 4, 5, and Tables 4, 5 the observed points are

1. The estimation error in position for the five algorithms is

$$\text{MCFE-UKF} < \text{FE-UKF} < \text{MC-UKF} < \text{T-UKF} < \text{H-UKF}.$$

2. TA-based algorithms contribute less estimation error compared to HA-based algorithm and among the TA algorithms the MCFE-UKF is the superior.
3. The time of convergence of the five estimators is

$$\text{MCFE-UKF} < \text{FE-UKF} < \text{MC-UKF} < \text{T-UKF} < \text{H-UKF}.$$

4. The time of convergence associated with the TA-based algorithms is less than that of HA algorithm and it is least for MCFE-UKF among the TA algorithms.

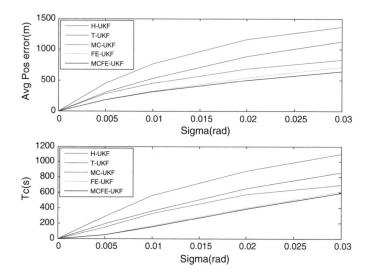

Fig. 4 Avg. position error and Tc versus sigma for different techniques

Table 4 Avg. position error of different estimators for different σ

Estimator	$\sigma = 0.005$ (rad)	$\sigma = 0.01$ (rad)	$\sigma = 0.02$ (rad)	$\sigma = 0.03$ (rad)
H-UKF	452	766	1160	1368
T-UKF	308	535	888	1128
MC-UKF	281	453	684	828
TE-UKF	185	319	541	718
MCTE-UKF	184	309	501	645

Table 5 Time of convergence of different estimators for different σ

Estimator	$\sigma = 0.005$ (rad)	$\sigma = 0.01$ (rad)	$\sigma = 0.02$ (rad)	$\sigma = 0.03$ (rad)
H-UKF	291	563	878	1100
T-UKF	194	361	652	859
MC-UKF	149	329	575	699
TE-UKF	47	163	401	612
MCTE-UKF	46	152	384	596

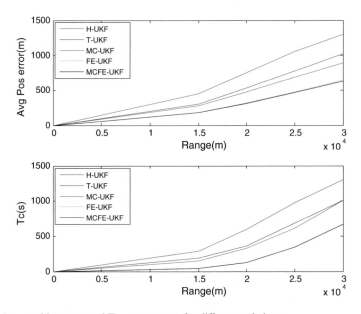

Fig. 5 Avg. position error and Tc versus range for different techniques

5. The superiority of TA algorithms over HA-based algorithm increases with increasing bearing noise intensities.
6. The superiority of TA algorithms over HA-based algorithm increases with increasing range.

6 Conclusion

Underwater bearings-only passive target tracking can be performed in a better way by using TA measurements instead of using HA measurements. The superiority is enhanced by the proposed novel stochastic estimators namely MC and FE techniques to an extent that, the feasibility difficulties associated with the TA can be overlooked. The superiority is even higher in complicated situations of the long-range tracking and thick noise situations. The significant improvements achieved by using the MC and FE suggests that the towed arrays can no longer be kept aside and should be used on regular basis for BOT problem, despite of the associated handling difficulties. The work is being continued to improve the performance of TA further, by using higher order MC for BOT problem.

References

1. S.C. Nardone, A.G. Lindgren, K.F. Gong, Fundamental properties and performance of conventional Bearings-only target motion analysis. IEEE Trans. Autom. Control **29**(9), 775–787 (1984)
2. B. Ristick, S. Arulampalem, N. Gordon, *Beyond the Kalman Filter–Particle Filters for Tracking Applications* (Artechhouse DSTO, London, 2004)
3. G. Zhou, M. Pelletier, T. Kirubarajan, T. Quan, Statically fused converted position and doppler measurement Kalman filters. Aerosp. Electr. Syst. IEEE Trans. **50**(1), 300–318 (2014)
4. T.L. Song, J.L. Speyer, A stochastic analysis of a modified gain extended Kalman filter with applications to estimation with bearings-only measurements. IEEE Trans. Autom. Control **30** (10), 940–949 (1985)
5. P.J. Galkowski, M.A. Islam, An alternative derivation of the modified gain function of song and speyer. IEEE Trans. Autom. Control **36**(11), 1323–1326 (1991)
6. D.V.A.N.R. Kumar, S.K. Rao, K.P. Raju, A novel stochastic estimator using pre-processing technique for long range target tracking in heavy noise environment. Optik **127**(10), 4520–4530 (2016)
7. D.V.A.N.R. Kumar, S.K. Rao, K.P. Raju, Integrated unscented Kalman filter for underwater passive target tracking with towed array measurements. Optic **127**(5), 2840–2847 (2016)
8. D.V.A.N.R. Kumar, S.K. Rao, K.P. Padma Raju, Underwater bearings-only passive target tracking using estimate fusion technique. Adv. Mil. Technol. **10**(2), 31–44 (2015)
9. S.K. Rao, Maximum likelihood estimator for bearings-only passive target tracking in electronic surveillance measure and electronic warfare systems. Def. Sci. J. **60**(2), 197–203 (2010)
10. V.J. Aidala, S.C. Nardone, Biased estimation properties of the pseudolinear tracking filter. IEEE Trans. Aerosp. Electr. Syst. **18**(4), 432–441 (1982)
11. S.K. Rao, Pseudo-linear estimator for bearings-only passive target tracking. IEE Proc. Radar Sonar Navig. **148**(1), 16–22 (2001)
12. I. Arasaratnam, S. Haykin, Cubature Kalman filters. IEEE Trans. Autom. Control **54**(6), 1254–1269 (2009)

13. I. Arasaratnam, S. Haykin, R.J. Elliott, Discrete-time non linear filtering algorithms using Gauss-Hermite quadrature. Proc. IEEE **95**(5), 953–977 (2007)
14. M. Beard, V. Ba-Tuong, V. Ba-Ngu, Multitarget filtering with unknown clutter density using a bootstrap GMCPHD filter. Signal Process. Lett. IEEE **20**(4), 323–326 (2013)
15. J.L. Deok, Nonlinear estimation and multiple sensor fusion using unscented information filtering. Signal Process. Lett. IEEE **15**, 861–864 (2008)

Fuzzy Rotor Side Converter Control of Doubly Fed Induction Generator

Karthik Tamvada and S. Umashankar

Abstract This paper presents control scheme for the rotor side converter of a grid connected doubly fed induction generator by the way of fuzzy logic control for combined vector and direct power control (CVDPC) scheme. The resulting scheme, FLC-DPC is analyzed with respect to steady state and variable wind conditions. Simulation was carried on a grid connected 9-MW DFIG-based wind farm and consequent analysis confirm the improved performance of FLC-DPC over CVDPC.

Keywords Control · DFIG · Electric grid · Fuzzy logic · Rotor side converter (RSC)

1 Introduction

Decline in availability of conventional sources of energy, caused researchers to consider the renewable resources and their various aspects [1, 2]. Year 2020 may witness wind power accounting for nearly 10% of world's electricity supply, which might double by the year 2040 [3]. In this context, the DFIG is of particular interest due to its converters' cost [4, 5], low power loss, and control aspects [6]. Stator real power (P_s), stator reactive power (Q_s), active power (P), and reactive power (Q) are the outputs of the DFIG that are of control interest. The DC-link capacitor acts as the DC-link voltage source. The power convertor of the wind turbine generator contains the rotor converter to control the generator speed and grid convertor to inject reactive power in the grid. The instantaneous powers can be defined as follows:

K. Tamvada (✉) · S. Umashankar
School of Electrical Engineering, VIT University, Vellore 632014, Tamil Nadu, India
e-mail: tamvadaka@gmail.com

S. Umashankar
e-mail: umashankar.s@vit.ac.in

© Springer Nature Singapore Pte Ltd. 2018
S. S. Dash et al. (eds.), *Artificial Intelligence and Evolutionary Computations
in Engineering Systems*, Advances in Intelligent Systems and Computing 668,
https://doi.org/10.1007/978-981-10-7868-2_62

$$P_s = \frac{3}{2}(V_{ds}I_{ds} + V_{qs}I_{qs}) \tag{1}$$

$$Q_s = \frac{3}{2}(V_{qs}I_{ds} - V_{ds}I_{qs}) \tag{2}$$

$$P_g = \frac{3}{2}(V_{ds}I_{dg} + V_{qs}I_{qg}) \tag{3}$$

$$Q_g = \frac{3}{2}(V_{qs}I_{dg} - V_{ds}I_{qg}) \tag{4}$$

Mohammadi et al. [7] puts forth combined vector and direct power control (CVDPC) scheme for controlling RSC. It has advantages of VC and DPC and overcomes the respective drawbacks. Pulse generation is by sector selection utilizing switching table and hysteresis current regulators. The aim of this paper is to eliminate switching table and hysteresis current regulators utilised in CVDPC scheme. This is achieved by introducing space vector modulator and consequent synthesis of the fuzzy logic controller for DPC. This work presents simulation and performance comparison of CVDPC and FLC-DPC schemes under steady state and variable wind conditions. The schematic for FLC-DPC is represented in Fig. 1. The advantages of the control scheme presented in this work are it eliminates feedforward terms, sector-based switching table and hysteresis current regulators. The paper organization is as follows. Section 2 gives a synthesis of FLC-DPC for rotor side converter (RSC) control of the grid connected DFIG. Section 3 provides the simulation results and associated analysis of the control scheme, with Sect. 4 serving as the conclusion.

2 Rotor-Side Converter Control

Wind power penetration exceeding 10% has a bearing on the economic operation of a power system [8, 9]. It impacts on voltage and energy profile too [10]. Selection of appropriate control strategy is necessary for regulation of active power, reactive power, and DC-link voltage. This following section outlines the synthesis of FLC-based DPC for RSC control of the DFIG.

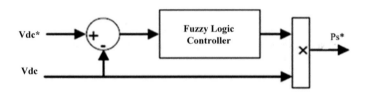

Fig. 1 Fuzzy controller structure for RSC control

Table 1 Fuzzy rule base for FLC-DPC

E	de/dt				
	NB	NM	Z	PM	PB
NB	Z	PS	PB	PB	PB
NM	NS	Z	PM	PB	PB
Z	NB	NM	Z	PM	PB
PM	NB	NB	NM	Z	PS
PB	NB	NB	NB	NS	Z

2.1 Fuzzy Logic Controller (FLC)-Based DPC

The Fuzzy Logic Controller generates new parameters from parameters of the PI controller to fit all operating conditions. This makes the PI controller adaptable to nonlinear systems. The inputs of fuzzy controller are: error (*e*) and its derivative (d*e*/d*t*). The error signals are fuzzified using five triangular MFs. The fuzzy sets have been defined as: NB-Negative Big, NM-Negative Medium, Z-Zero, PB-Positive Big, PM-Positive Medium. 25 rules are chosen for all the MFs as shown in Table 1 and equal weights are chosen for all the rules.

3 Simulation Results and Discussion

This section provides the simulation of grid connected 9-MW DFIG-based wind farm for evaluating the performance of FLC-DPC for steady state and variable wind conditions. Simulation is implemented on the system shown in Fig. 2, where the wind farm connects to a 120-kV grid utilizing 30-km, 25-kV transmission line.

Figure 3a–c depicts the simulation results under assumed constant wind speed of 15 m/s and zero references reactive power at the stator. In Fig. 3a, it can be seen that the FLC-DPC provides a rotor speed nearer to the reference value of 1.2 p.u. in a shorter duration compared to the CVDPC. The active power response is illustrated in Fig. 3b for both CVDPC and FLC-DPC that inherently follows the rotor speed of the DFIG, providing similar responses with FLC-DPC reaching a constant value in

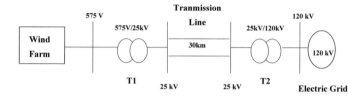

Fig. 2 Simulated system

Fig. 3 Steady-state response of **a** rotor speed **b** active power **c** reactive power

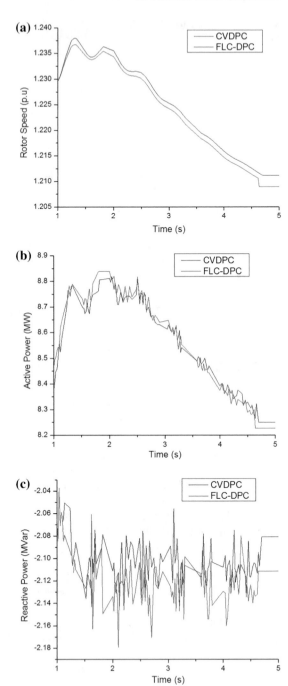

Fig. 4 Transient response of **a** rotor speed **b** active power **c** reactive power

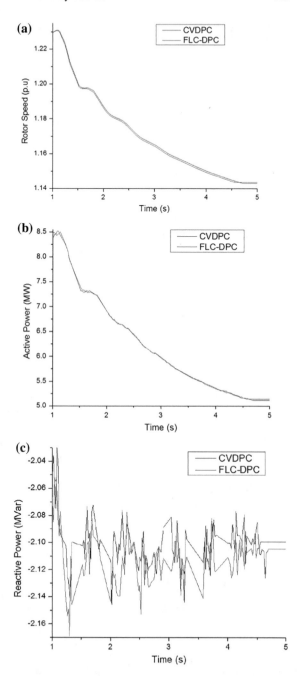

shorter time duration. Figure 3c shows that reactive power being supplied to the electric grid is higher for FLC-DPC scheme compared to CVDPC scheme.

Wind speed fluctuations are complex and stochastic in nature. The variable wind speed conditions are emulated by a step change in wind speed from 15 to 10 m/s, initiated at 1 s. Similar to the steady-state response there is a correlation between the rotor speed of DFIG and the corresponding amount of active power generated as is clear from Fig. 4a, b. The decreased speed gives a reduced power generation of 5.1 and 5.2 MW, respectively, for CVDPC and its fuzzy counterpart. From Fig. 4c, one can see that there is a decrease in the amount of reactive power being supplied to the electric grid with CVDPC scheme providing lower reactive power when compared to its fuzzy counterpart.

The results show the performance of the FLC-DPC with respect to wind speed variations to be nearly as fast as CVDPC, while providing an overall improved performance under both constant and variable speed wind conditions.

4 Conclusion

This work aims to improve RSC control scheme for grid connected DFIG. This is achieved by utilizing fuzzy logic controller for SVM based direct power control (DPC). The work is validated through simulation studies on a grid connected 9 MW DFIG wind farm. The performance of the designed controller is evaluated under constant and variable wind speed conditions. Under constant conditions, FLC-DPC shows better performance with respect to the operating variables of DFIG-based WT than CVDPC. In the variable conditions, the FLC-DPC gives an overall improved performance in response to wind speed variations. Simulation results confirm the improved performance of the FLC-DPC over CVDPC. These results demonstrate the possibility of utilizing proposed structure based on the FLC-DPC as a solution for DFIG-based wind energy systems.

References

1. R. Haas et al., A historical review of promotion strategies for electricity from renewable energy sources in EU countries. Renew. Sustain. Energy Rev. **15**(2), 1003–1034 (2011)
2. R. Saidur et al., Environmental impact of wind energy. Renew. Sustain. Energy Rev. **15**(5), 2423–2430 (2011)
3. World Wind Energy Report (2010), http://www.WWindEA.org
4. A. Gonzalo et al., *Doubly Fed Induction Machine: Modeling and Control for Wind Energy Generation* vol. 85. Wiley (2011)
5. R. Datta, V.T. Ranganathan, Variable-speed wind power generation using doubly fed wound rotor induction machine-a comparison with alternative schemes. IEEE Trans. Energy Convers. **17**(3), 414–421 (2002)
6. J.F. Manwell, J.G. McGowan, A.L. Rogers, *Wind Energy Explained: Theory, Design and Application*. Wiley (2010)

7. J. Mohammadi et al., A combined vector and direct power control for DFIG-based wind turbines. IEEE Trans. Sustain. Energy **5**(3) 767–775 (July 2014)
8. Y. Zou, M.E. Elbuluk, Y. Sozer, Simulation comparisons and implementation of induction generator wind power systems. IEEE Trans. Ind. Appl. **49**(3), 1119–1128 (2013)
9. G.M.J. Herbert et al., A review of wind energy technologies. Renew. Sustain. Energy Rev. **11**(6), 1117–1145 (2007)
10. T. Ackermann (ed.) *Wind Power in Power Systems*. Wiley (2005)

A New Multilevel Inverter Using Switched Capacitor Unit with Reduced Components

P. Venugopal and V. Sumathi

Abstract In the present paper, single phase multilevel inverter using switched capacitor units is proposed. The switched capacitor unit (SCU) is used in the multilevel inverter will increase the dc supply voltage at the input without using transformer by operating the capacitors in parallel and in series. The SCU comprises one dc source, two semiconductor switches, power diode, and capacitor. The proposed topology does not require the charge balancing operations, which results reducing the cost of the overall circuit. In addition to that, it does not required H-bride module to alter the polarity of the output voltage, which will lead to minimum required semiconductor power switches. The proposed topology is compared with the conventional similar architectures in terms of power semiconductor switches, power diodes, dc power supplies. Finally, to confirm the proposed architecture performance, simulation results are presented.

Keywords Multilevel inverter · dc voltage sources · Power switches
Output voltage levels · Charging and discharging

1 Introduction

Nonconventional energy sources such as wind turbine and photovoltaic systems have been accepted as a valuable replacement for fossil fuels for the reason that their consistent response and commercial profits in recent years. So as to manage boundless utilizing these renewable energy sources, the overall system output performance need to be improved through increasing the quality of power output, reducing the total losses, eliminating the filter at output, and reducing the size of the

P. Venugopal (✉)
School of Electronics Engineering, VIT University, Vellore, Tamil Nadu, India
e-mail: venugopal.p@vit.ac.in

V. Sumathi
School of Electrical Engineering, VIT University, Chennai, Tamil Nadu, India
e-mail: vsumathi@vit.ac.in

© Springer Nature Singapore Pte Ltd. 2018
S. S. Dash et al. (eds.), *Artificial Intelligence and Evolutionary Computations
in Engineering Systems*, Advances in Intelligent Systems and Computing 668,
https://doi.org/10.1007/978-981-10-7868-2_63

transformer [1]. To achieve this, multilevel inverters have gain more attention because of their high quality of output power, reduced harmonic distortion, higher magnitude of the fundamental component, high efficiency, lower switching losses, and less dv/dt. These advantages mentioned above are the motivation for the changeover from the conventional two-level converter to multilevel converters [2, 3]. In actuality, the multilevel inverter (MLI) intends to produce the stepwise voltage waveform at the output by integrating dc source values connected with its terminals. The output voltage levels increases by increasing the number of dc links at the input [4]. Considering the different topologies of MLI, the dc power supplies can be isolated or interconnected. Three primary structures of the multilevel inverters have been exhibited: diode clamped, flying capacitor, and cascaded multilevel inverter [5]. In spite of the majority of the specified benefits, MLIs have a few disadvantages over the classical two-level inverter topology. In multilevel inverter architectures, increasing the output voltage levels results increasing the circuit complexity, which decreases reliability and efficiency [6, 7]. The fundamental condition for producing high number of voltage levels is to utilize various dc voltage sources,for example, transformers or capacitors with combination of several switching components [8].

In the literature, Researchers have attempted to overcome these previously mentioned limitations by introducing the recently proposed MLI structures [9]. In spite of that, producing more number of voltage levels at the output, with least number of separated dc sources and other components, for example, gate driver circuit and power semiconductor switches in like manner, is considered a key features for investigators [10, 11]. A best approach to lessening the number of essential dc voltage sources is to utilize the capacitors. However, the previous mentioned approach requires a separate voltage balancing circuit for preclude the problem of discharging [12, 13]. Discharging problems can be reduced by switching states repetitions. In this approach, large number of semiconductor switches is required to achieve more number of voltage levels [14, 15]

To reduce the count of essential switches and gate drivers, Dargahi [16] presented a novel technique in FCMCs to balance the charge. These FCMCs can generate output voltage of 19-level, power semiconductor switches of 18, 5 capacitors, and two dc power supplies. However, this architecture requires no less than two number of voltage sensors to follow the voltage performance of every capacitor under working conditions. Moreover, this design generates a abnormal voltage ripples with increased levels of voltage at the output and is not possible to increase the output levels further as desired. To resolve the problem of discharging the alternative methodology is to use the modulation methods accompanied by lessening of duty cycle rate for every capacitor. Nevertheless such designs are only apt for a definite amount of voltage levels and the ability of extending the topology is not possible [17, 18].

The other alternative approach is, using the switched capacitor in multilevel inverter, in which charge balancing processes for eliminating the additional dc sources are not required, consequently overall cost will reducing [19–21]. This approach, not only transfers the large power from input to output, but also produces more number of output voltage levels. Because of this advantage, a few enhanced architectures were introduced by various authors [22–26]. Notwithstanding, H-bridges are used to change the polarity of the voltage waveform in these topologies and the cascaded procedure keeping in mind the end goal to generalized the topologies and achieve the more number of the levels at the output voltage which results high conduction losses in power switches [27, 28].

Taking into account the study situation previously mentioned, a novel MLI using switched capacitor converter is presented. This topology generates increased number of output voltage levels compared to the topologies mentioned in the literature. Moreover, this architecture can be extended to further levels by adding the basic units. H-bridges are not used in this topology, which results reduce the cost, complexity of the circuit. A far reaching, comparison has been done with the recently proposed structures, which can demonstrate the upsides of proposed MLI using switched capacitor structure in various perspective, such as the switch count, number of dc sources, and voltage levels at output. To examine the proposed structure, simulation has been done using MATLAB/SIMULINK and the results are verified to confirm performance of the proposed structure.

2 Proposed MLI Using SCU

In order to generate more number of levels compared to [15–17], a novel structure has been introduced, which utilizes the switched capacitor unit. The basic structure of switched capacitor converter is shown in Fig. 1a. The switched capacitor converter contains a capacitor, power diode, dc power supply, and two power semiconductor switches. The capacitor will be charged to the voltage Vdc when turned ON the switch Sb. The capacitor will be discharged when turn ON the switch Sa. Figure 1b shows the operations of charging and Fig. 1c shows the operation of discharging of capacitor.

The basic architecture of proposed multilevel Inverter is shown in Fig. 2, which contains two Switched capacitor units, six power semiconductor switches. The number of switches and power diodes required in the proposed topology can be stated as following equations;

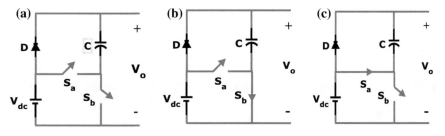

Fig. 1 **a** Basic switched capacitor unit **b** capacitor charging circuit **c** capacitor discharging circuit

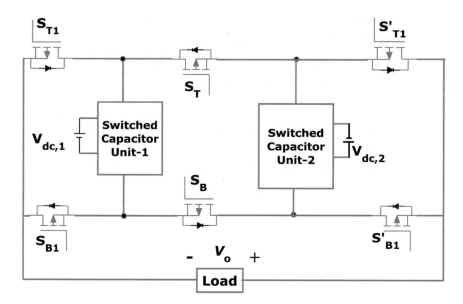

Fig. 2 The basic proposed switched capacitor multilevel inverter

$$N_{\text{Switch}} = 8k + 2 \tag{1}$$

$$N_{\text{diode}} = 2k \tag{2}$$

In the basic proposed topology, 2 switched capacitor converters are used, therefore 2 dc sources and $2k$ (k is the half of the isolated dc sources) capacitors required. To generate maximum number of output levels 2 dc sources ae in asymmetric in nature. The mathematical expression for the dc source of second switched capacitor unit is given by,

$$V_{\text{dc},2} = (1 + 2^k)V_{\text{dc},1} \tag{3}$$

The asymmetrical proposed topology will produce maximum voltage levels at output can be expressed as following,

$$N_{\text{level}} = 1 + 2^{k+2} + 2^{2k+1} \tag{4}$$

The seventeen levels proposed inverter is shown in Fig. 3, which has 10 power switches, 2 capacitors, 2 diodes, and 2 isolated power supplies. The value of DC sources of first SC unit and second SC unit are Vdc and 3 Vdc. The switching pattern of the s17 level topology is given in Table 1. For example, if the value of Vdc is 10 V, and then the maximum voltage obtained at load is 80 V. Each step size is 10 V. To get 80 V at the output side S_{B1}, S_1, S_T, S'_1, S'_{B1} should be turned

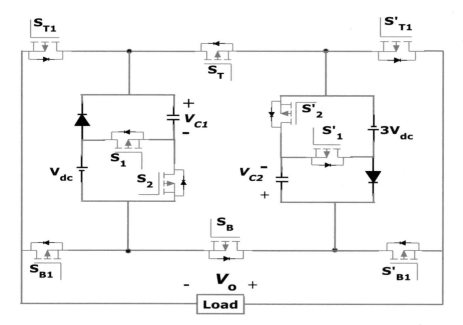

Fig. 3 Proposed 17-level multilevel inverter

ON. During this period both the capacitors are in discharging mode. When switches S_{B1}, S_2, S_T, S'_1, S'_{B1} are turn ON simultaneously, the voltage at output will be 70 V with capacitor C1 charging and C2 discharging.

The proposed topology can be extended to further levels by adding SCUs and switches. For instant, by adding the one SC unit and two switches to the basic topology, will produce 49 level output. Hence, 49 level inverter require 14 switches, 3 DC sources, 3 capacitors, and 3 diodes. 49-level inverter is shown in Fig. 4. For example, the switches S_{B1}, S_{T2}, S_T, S'_{T1} alongside the internal power switches of SCU-2 and SCU-3 and internal power switch of SCU-1 must be turn ON. So as to generate 9th level of output, switches of S_{B1}, S_{B2}, S_T, S'_{T1} along with internal power switches of SCU-3 and SCU-2 and turn ON the SCU-1 switches which are in parallel. The switching sequence is given in Table 2. To create higher number of voltage levels at the output side with regard to further levels, the value of additional dc power sources of individually SC unit can be implemented by the given expressions:

$$V_{dc,j} = 2(V_{dc,j-1} + V_{dc,r(j-1)}) + 1. \tag{5}$$

$$V_{dc,rj} = 2(V_{dc,j} + V_{dc,r(j-1)}) + 1. \tag{6}$$

Table 1 Switching sequence and capacitor charging and discharging states of seventeen level multilevel inverter

Switching states	ON switches	Vo	C1	C2
1	S_{B1}, S_1, S_T, S'_1, S'_{B1}	4Vdc + vc1 + vc2	D	D
2	S_{B1}, S_2, S_T, S'_1, S'_{B1}	4Vdc + vc2	C	D
3	S_{T1}, S_2, S_T, S'_1, S'_{B1}	3Vdc + vc2	C	D
4	S_{B1}, S_1, S_T, S'_2, S'_{B1}	4Vdc + vc2	D	C
5	S_{B1}, S_2, S_T, S'_2, S'_{B1}	4Vdc	C	C
6	S_{T1}, S_2, S_T, S'_2, S'_{B1}	3Vdc	C	C
7	S_{B1}, S_1, S_T, S'_2, S'_{T1}	Vdc + vc1	D	C
8	S_{B1}, S_2, S_T, S'_2, S'_{T1}	Vdc	C	C
9	S_{B1}, S_2, S_B, S'_2, S'_{B1}	0	C	C
10	S_{T1}, S_2, S_B, S'_2, S'_{B1}	−Vdc	C	C
11	S_{T1}, S_1, S_B, S'_2, S'_{B1}	−Vdc−vc1	D	C
12	S_{B1}, S_2, S_B, S'_2, S'_{B1}	−3Vdc	C	C
13	S_{T1}, S_2, S_B, S'_2, S'_{B1}	−4Vdc	C	C
14	S_{T1}, S_1, S_B, S'_2, S'_{B1}	−4Vdc−vc2	D	C
15	S_{B1}, S_2, S_B, S'_1, S'_{B1}	−3Vdc−vc2	C	D
16	S_{T1}, S_2, S_B, S'_1, S'_{B1}	−4Vdc−vc2	C	D
17	S_{T1}, S_1, S_B, S'_1, S'_{B1}	−4Vdc−vc1−vc2	D	D

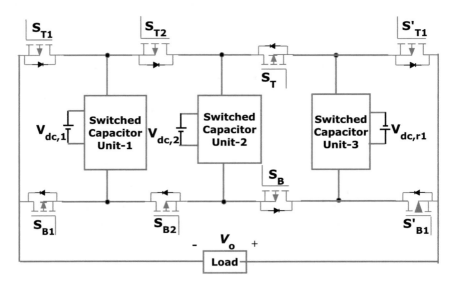

Fig. 4 49-level multilevel inverter

Table 2 Different states of switching of 49 level inverter

Switching states	ON switches	V_o
1	$S_{B1}, S_{T2}, S_T, S'_{T1}$	1, 2
2	$S_{T1}, S_{T2}, S_T, S'_{B1}$	3, 6
3	$S_{B1}, S_{T2}, S_T, S'_{B1}$	4, 5, 7, 8
4	$S_{B1}, S_{B2}, S_T, S'_{T1}$	9, 18
5	$S_{T1}, S_{B2}, S_T, S'_{B1}$	10, 11, 13, 14, 19, 20, 22, 23
6	$S_{B1}, S_{B2}, S_T, S'_{B1}$	12, 15, 21, 24
7	$S_{T1}, S_{B2}, S_T, S'_{T1}$	16, 17

3 Comparative Study

The proposed multilevel inverter using switched capacitor structure is compared with some of flying capacitor multilevel inverter topologies which have been introduced recently. Comparison is done based on the voltage levels at output, number of power switches and dc sources and required number of capacitors. The proposed-structure is compared with the present topologies [15–17] from various observations is provided in Table 3. Clearly, as for the measures of extent of count of levels of the voltage at output over the number of required devices, the recommended structure requires minimum number of devices in contrast with their relevant architectures. For example, 18 and 20 semiconductor switching components to produce 17, 19, 7 levels of voltage at output, respectively, while proposed architecture requires just 10 control switches for its 17-level topology. In addition to that, recommended structure in [15–17] need 16, 18, and 20 power diodes to produce 17, 19, 7 levels of voltage at output, respectively, while recommended architecture needs just 12 control switches for its 17-level topology.

Table 3 Comparison of 17-level inverter with topologies [15–17]

Key parameters	[15]	[16]	[17]	Proposed SCMLI
Nlevel	17	19	7	17
NIGBT	16	18	20	10
Ndiode	16	18	20	12
Ncap	4	4	4	2
Nsource	1	2	2	2
Charge balance requirement	Yes	Yes	Yes	No

4 Simulation Results

To analyze the performance of recommended switching capacitor based multilevel inverter, the simulation and experimental results of 17-level inverter are presented. MATLAB/Simulink has been used for simulation. In addition to that, the hardware prototype of 17-level inverter with 80 V has been developed. The FPGA spartan-3e has been used to produce the gating pulses. For the present structure fundamental switching frequency method is used. The MOSFETs used are IRF-840, 500 V, 8A, and Ron is 0.85 Ω and Power diodes used are FR107, 700 V, and 50 Ω. Capacitors have been used with 4700 μF and 50 V. The prototype has been tested on R-L load with the magnitude of 0.5 mH and 500 Ω for all the studies. In this topology, the values of dc sources are unequal and are 10 and 30 V. According to this present topology produces the maximum voltage at the output will be 80 V. R-L load have been used for both simulation and experimental tests. The simulation results of

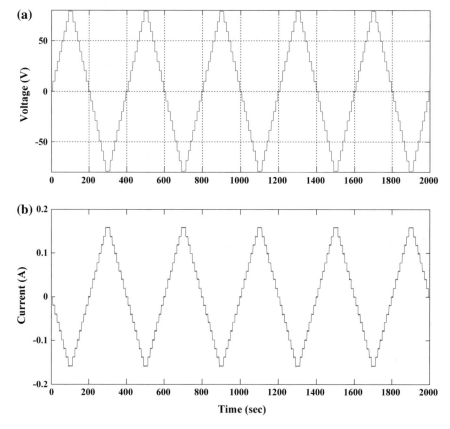

Fig. 5 Simulation waveforms of 17-level multilevel inverter **a** output voltage wave form **b** current waveform

proposed multilevel inverter which produces 17-level have shown in Fig. 5a, b. It is observed that the output voltage is 80 V and current is 950 mA.

5 Concluding Remarks

In this paper, a novel topology using switched capacitor converter has been proposed for multilevel inverters to produce seventeen voltage levels at the output. The basic topology can be extended to any number of levels at the output for example by adding one switched capacitor converter unit to the seventeen level multilevel inverter, the voltage levels at the output are 49. Similarly, to produce 137 levels at the output, two switched capacitor converter units should be added to the basic proposed multilevel inverter. The proposed topology has been compared with several existing topologies in the literature from the various points of observation. Based on these comparisons, the proposed structure requires less number of power switches, DC sources, and diodes. Consequently, the size and cost of the proposed structure will be reduced in comparison with the conventional similar topologies. At last, the viability and performance of recommended 17-level switched capacitor multilevel inverter have been confirmed through simulation results.

References

1. J. Chavarria, D. Biel, F. Guinjoan, C. Meza, J.J. Negroni, Energy balance control of PV cascaded multilevel grid-connected inverters under level-shifted and phase-shifted PWMs. IEEE Trans. Ind. Electron. **60**, 98–111 (2013)
2. A.L. Batschauer, S.A. Mussa, M.L. Heldwein, Three-phase hybrid multilevel inverter based on half-bridge modules. IEEE Trans. Ind. Electron. **59**, 668–678 (2012)
3. S. Kouro, M. Malinowski, K. Gopakumar, J. Pou, L. Franquelo, B. Wu, J. Rodriguez, M. Perez, J. Leon, Recent advances and industrial applications of multilevel converters. IEEE Trans. Ind. Electron. **57**, 2553–2580 (2010)
4. E. Babaei, S. Alilu, S. Laali, A new general topology for cascaded multilevel inverters with reduced number of components based on developed H-bridge. IEEE Trans. Ind. Electron. **61**, 3932–3939 (2014)
5. M.R. Banaei, E. Salary, New multilevel inverter with reduction of switches and gate driver. Energy Convers. Manage. **52**, 1129–1136 (2011)
6. A. Ajami, M.R.J. Oskuee, A. Mokhberdoran, H. Shokri, Selective harmonic elimination method for wide range of modulation indexes in multilevel inverters using ICA. J. Central South Univ. **21**, 1329–1338 (2014)
7. R. Stala, A natural DC-link voltage balancing of diode-clamped inverters in parallel systems. IEEE Trans. Ind. Electron. **60**, 5008–5018 (2013)
8. K.K. Gupta, A. Ranjan, P. Bhatnagar, L.K. Sahu, S. Jain, Multilevel inverter topologies with reduced device count: a review. IEEE Trans. Power Electron. **31**, 135–151 (2015)
9. A. Mokhberdoran, A. Ajami, Symmetric and asymmetric design and implementation of new cascaded multilevel inverter topology. IEEE Trans. Power Electron. **29**, 6712–6724 (2014)
10. K.K. Gupta, S. Jain, Comprehensive review of a recently proposed multilevel inverter. IET Power Electron. **7**, 467–479 (2014)

11. K.M. Tsang, W.L. Chan, Single DC source three-phase multilevel inverter using reduced number of switches. IET Power Electron. **7**, 775–783 (2014)
12. M. Khazraei, H. Sepahvand, K.A. Corzine, M. Ferdowsi, Active capacitor voltage balancing in single-phase flying-capacitor multilevel power converters. IEEE Trans. Ind. Electron. **59**, 769–778 (2012)
13. K. Sano, H. Fujita, Voltage-balancing circuit based on a resonant switched-capacitor converter for multilevel inverters. IEEE Trans. Ind. Appl. **44**, 1768–1776 (2008)
14. B.P. McGrath, D.G. Holmes, Analytical modeling of voltage balance dynamics for a flying capacitor multilevel converter. IEEE Trans. Power Electron. **23**, 543–550 (2008)
15. P. Roshankumar, R.S. Kaarthic, K. Gupakumar, J.I. Leon, L.G. Franquelo, A seventeen-level inverter formed by cascading flying capacitor and floating capacitor H-bridge. IEEE Trans. Power Electron. **30**, 3471–3478 (2015)
16. V. Dargahi, A.K. Sadigh, M. Abarzadeh, S. Eskandari, K. Corzine, A new family of modular multilevel converter based on modified flying capacitor multicell converters. IEEE Trans. Power Electron. **30**, 138–147 (2015)
17. V. Dargahi, A.K. Sadigh, M. Abarzadeh, M.R.A. Pahlavani, A. Shoulaie, Flying capacitor reduction in an improved double flying capacitor multicell converter controlled by a modified modulation method. IEEE Trans. Power Electron. **27**, 3875–3887 (2012)
18. H. Sepahvand, J. Liao, M. Ferdowsi, K. Corzine, Capacitor voltage regulation in single dc source cascade H-bridge multilevel converters using phase shift modulation. IEEE Trans. Ind. Electron. **60**, 3619–3626 (2013)
19. A.M.Y.M. Ghias, J. Pou, V.A. Agilidis, Voltage balancing method for stacked multicell converters using phase disposition PWM. IEEETrans. Ind. Electron. **62**, 4001–4010 (2015)
20. B. Axelrod, Y. Berkovich, A. Ioinovici, A cascade boost switched capacitor-converter two-level inverter with an optimized multilevel output waveform. IEEE Trans. Circuits Syst. I. **52**, 2763–2770 (2005)
21. M.S.W. Chan, K.T. Chau, A newswitched-capacitor boostmultilevel inverter using partial charging. IEEE Trans. Circuits Syst. II. Exp. Briefs **54**, 1145–1149 (2007)
22. Y. Hinago, H. Koizumi, A switched-capacitor inverter using series/parallel conversion with inductive load. IEEE Trans. Ind. Electron. **59**, 878–887 (2012)
23. J. Liu, K.W.E Cheng, Y. Ye, A cascaded multilevel inverter based on switched-capacitor for high-frequency ac power distribution system. IEEE Trans. Power Electron. **22**, 4219–4230 (2014)
24. E. Babaei, F. Sedaghati, Series-parallel switched-capacitor based multilevel inverter. in *Proceedings of International Conference on Electrical Machines and Systems*, 2011, pp 1–5
25. Y. Ye, K. Cheng, J. Liu, K. Ding, A step-up switched-capacitor multilevel inverter with self voltage balancing. IEEE Trans. Ind. Electron. **61**, 6672–6680 (2014)
26. E. Babaei, S.S. Gowgani, Hybrid multilevel inverter using switchedcapacitor units. IEEE Trans. Ind. Electron. **61**, 4614–4621 (2014)
27. R.S. Alishah, D. Nazarpour, S.H. Hosseini, M. Sabahi, Reduction of power electronic elements in multilevel converters using a new cascade structure. IEEE Trans. Ind. Electron. **62**, 256–569 (2015)
28. E. Babaei, M.F. Kangarlu, M. Sabahi, Extended multilevel converters: an attempt to reduce the number of independent dc voltage sources in cascade multilevel converters. IET Power Electron. **7**, 157–166 (2014)

Improved Performance of a Dynamic Voltage Restorer Using Hybridized Cascaded Multilevel Inverter for Solar PV Grid Connected System

C. Dhanamjayulu and S. Meikandasivam

Abstract This paper presents the enhanced execution of a "Dynamic voltage restorer" (DVR) by utilizing Hybridized Cascaded 17 level symmetric Multilevel. In this work, dynamic voltage restorer (DVR), for improving voltage quality at three-phase fault conditions is presented. The DVR acts as voltage controller for protecting the load voltage during short circuit fault condition in photovoltaic fed grid connected system, in addition to this, the H-Bridge yields a 5-level output from the 3-level input which it achiever by a bidirectional switch which acts as the middle motivation between twofold DC source. This paper presents the design of 17-level MLI is designed to execute the Sag or swell are identified in a transmission line and the voltage is infused from the inverter through a transformer, DVR is validated on MATLAB-SIMULINK. It is clear from the results that a multilevel inverter can produce a higher number of output voltage levels with much lower THDs. This, in turn, leads to energy saving and better power quality.

Keywords Dynamic voltage restorer · Hybridized cascaded symmetric multilevel inverter (HCSMLI) · Power quality improvement · Voltage sag
Voltage swell

1 Introduction

With the rapid exhaustion of fossil fuels, the reliance of renewable sources of energy is on a rise. These renewable sources are being considered as the potential sources of power for the future. To address these issue photovoltaic cells (PV cells) have been used in this proposed power system distribution. Photovoltaic cells are meant to utilize the solar energy that is obtained from the sun and convert it into electrical

C. Dhanamjayulu (✉) · S. Meikandasivam
School of Electrical Engineering, VIT University, Vellore, Tamil Nadu, India
e-mail: dhanamjayulu.c@vit.ac.in

S. Meikandasivam
e-mail: meikandasivam.s@vit.ac.in

© Springer Nature Singapore Pte Ltd. 2018
S. S. Dash et al. (eds.), *Artificial Intelligence and Evolutionary Computations in Engineering Systems*, Advances in Intelligent Systems and Computing 668,
https://doi.org/10.1007/978-981-10-7868-2_64

energy. This results in the generation of electricity without affecting the environment in a negative way. The two important advantages of PV cells are: it is relatively very static, the cost of maintain an array of PV Cells is very low. However, most of the loads used on a daily basis require a smooth flow of voltage supplied to them. However, the nonlinearity of most loads present poses a hurdle to this cause. This very hurdle is the cause of all power quality-related issues. One of the most repeating problems in using a PV Cell array is the problem of voltage sag. Voltage sag is basically the phenomenon due to which the output voltage of a source can fall from 90 to 10% of the actual rated output value for a very small period of time. This time period can be from a few seconds up to about half-a-cycle [1]. The issues regarding the quality of the power are beaten by the huge use of gadgets like appropriation generators, shunt capacitance, and element Dynamic voltage restorers. Dynamic voltage restorer is the best custom power gadget its working depends on voltage source converter innovation. Amid the supply of voltages, the DVR mechanical assembly is utilized to maintain or reestablish an operational electric load when droops or spikes will show up in modern regions, the sudden voltage drop will happen because of this reasons harms and misfortune will show up [2]. Control circuit topologies, logical plan, and capacity of DVR are utilized for the diminishment of voltage drop [2]. DVR reproduce the voltage waveform and guarantee settled load voltage. In a prior stage, the DVR works in light of the arrangement infusion of stage voltage levels yet from that point the outline has been changed as indicated by the requirements. The voltage source inverter topology is utilized as a part of the outline of DVR [3]. The multilevel inverter (MLI) topologies are characterized into diode clipped, flying capacitor and fell H-connect multilevel inverters. The mixture topologies are developed in light of their current multilevel topologies [3]. When making the inverter topologies the real thought of cost, parts and unwavering quality are imperative. By the best possible choice of exchanging gadgets, the cost esteem will be decreased. The work with lessened number of exchanging gadgets is gotten utilizing the H-connect topology [4]. The proposed work is the plan of 17-level symmetric HCMLI inverter for DVR with lessened number switches.

2 Proposed Methodology for Power Quality Change Utilizing 17-Level SCMLI-Based DVR

A conventional DVR shows in Fig. 1 and proposed DVR with PV grid connected systems show in Fig. 2. The proposed display plans a 17-level SCMLI for DVR to expand control quality. The outline of inverter utilizing control Technics. The power quality limitations considered in this paper is clarified in the under area.

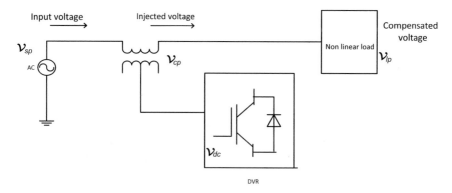

Fig. 1 DVR in power system

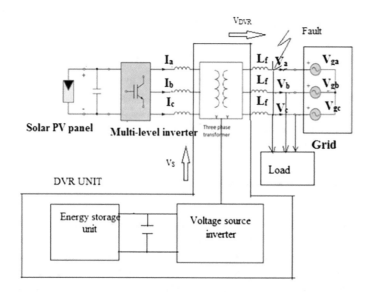

Fig. 2 Proposed model of DVR with HCMLI inverter for solar grid connected system

2.1 Constraints

The proposed design solved the power quality problems such as.

2.1.1 Voltage Sag

Voltage hang is considered as the most difficult issue of force quality. Voltage lists are brief length decrease in RMS voltage brought on by short circuits, overburdens and beginning of expansive engines. Voltage droop is portrayed by greatness and span. This voltage list lessens framework unwavering quality and decreases execution of the loads.

2.1.2 Voltage Swell

The start for voltage swell is exchanging of expansive capacitors, line-to-ground blame condition. The zero arrangement impedance esteem increments amid single line to ground blame. Voltage swell can be lessened by utilizing consistent voltage transformer or other sort of quick acting voltage controller. The proposed show managed alleviation of voltage list and swell utilizing Control techniques-based symmetric Cascaded Multilevel Inverter encouraged element voltage restorer [5].

3 Design of Symmetric Hybridized Cascaded Multilevel Inverter

The inverter is one of the most vital components of the DVR. Inverters with various topologies might be used. In a DVR structure, we generally use MLIs since they are suitable for high voltages. The output of the MLI is a multilevel one, and that increases the quality of output in shallow as well as deep voltage sags.

3.1 Introduction to Hybridized Cascaded Multilevel Inverter

A hybridized H-Bridge circuit topology can be used in the implementation of either symmetrical or asymmetrical configuration in cascaded form as can be seen in Fig. 3. Generally, H-Bridges, give five levels of output when connected with 2 DC sources, however, if we wanted to obtain more than five levels of output, the symmetrical or asymmetrical topology has to be incorporated with the basic H-bridge topology. A basic topology generally works in two phases: In the first phase a bidirectional switch connected to the second leg on the H-bridge gives the basic levels viz. $+V_1$, 0, $-V_1$. In the second phase, normal H-bridge operation using only two legs generates the peak levels of $+2V_1$, 0, $-2V_1$. The behavior of the same can be seen in the switching state table as given in Table 1. From, here we can hence conclude that asymmetric topology requires various different DC voltages whereas the symmetric topology requires equal DC voltages.

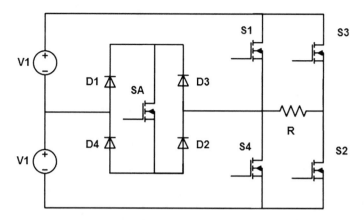

Fig. 3 Basic hybridized H-bridge topology

Table 1 Basic hybridized MLI—State table

Voltage levels	States of switches				
Switches	S1	S2	S3	S4	SA
+Vdc	1	1	0	0	0
+Vdc/2	0	1	0	0	1
0 Vdc	0	1	0	1	0
0 Vdc	1	0	1	0	0
−Vdc/2	0	0	1	0	1
−Vdc	0	0	1	1	0

3.2 Proposed Symmetric 17-Level Hybridized Cascaded Multilevel Inverter

The proposed 17-level symmetric hybridized MLI is represented in Fig. 5. In this configuration, the operation of each h-scaffold is same yet just contrast is there no man's land in exchanging designs. In essential voltage level era, for positive pinnacle, the elements in conduction along with the second DC source are the diodes D1 and D2 and the MOSFETs SA, S1, and S2. Although in pinnacle level era the MOSFETs S1, S2 will direct the two DC source. In this inverter system, which is symmetric, heartbeat is created utilizing unipolar phase disposition sinusoidal PWM [6] as used for MLI. The results of a simulation that includes the model, stage output voltages, output voltage, and current and output voltage are shown in Figs. 4, 5, 6, 7, and THD is shown in Table 2 respectively.

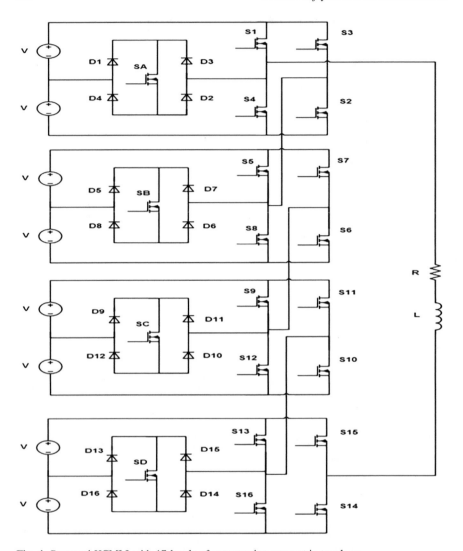

Fig. 4 Proposed HCMLI with 17-levels of output using symmetric topology

4   Proposed Model of DVR with Symmetric Hybridized Cascaded Multilevel Inverter

The grid connected photovoltaic system with multilevel inverter is shown in the Fig. 6. In this system, when fault is introduced, DVR detects and injects required compensated voltage with the use of series injection transformer. A structure associated control framework is an exceptionally complex structure. It is more critical to evacuate any framework issues or variances so that whatever remains of

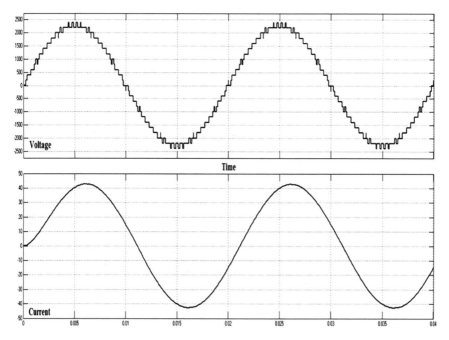

Fig. 5 Output voltage and current graphs for the HCMLI with 17-levels of output using symmetric topology

the power circulation system is not delayed or harmed [7]. At the point, when a sudden blame or an anomaly happens some place in a power circulation organizes, the voltage is endured all through the power framework. Among different power quality issues, the lion's share of variations from the norm is brought about by voltage list or a voltage swell. In many different transmission and distribution systems, dynamic voltage restorer is used a static-VAR tool. This dynamic voltage restorer acts as a series compensation device that protects the sensitive load from power quality problems [8, 9]. DVRs have been connected to ensure basic loads in utilities, semiconductor, and sustenance preparing. The different units of DVR are clarified as takes after. The Fig. 6 is the representation of the dynamic voltage restorer with proposed symmetric multilevel inverter and PI controller. The dynamic voltage restorer is connected in series with the power distribution system. It is a compensation device which is connected at the Load. The voltage restorer comprises of inverter, PWM device, and a controlling technique. Here, PI controller is used as the controller technique. Through this control technique, the error is minimized. For the error calculation, the voltage parameter is considered. The voltage value is measured for the error calculation the voltage parameter is considered. The voltage value is measured form the Load. An error signal is generated which is given as the input of the PI controller [3] after comparing the voltage value with the reference voltage.

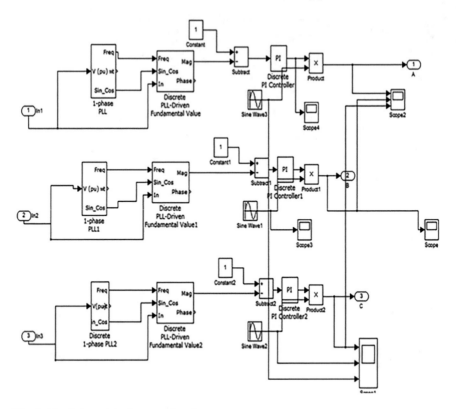

Fig. 6 Simulink model of proposed system

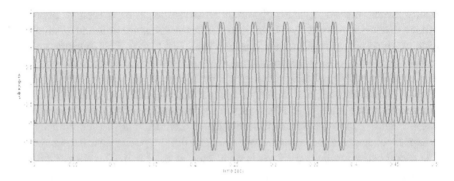

Fig. 7 Source side output voltage

Table 2 Input and output parameters

Inverter configuration	Input parameters	Output parameters
17-level symmetric cascaded multilevel inverter	Input DC supply = 200(4 stages) Load−R = 50 Ω, L = 60 mH Modulation indices−Ma = 0.95 & Mf = 50 Modulation scheme−Unipolar PD sine PWM	Peak voltage −Output = 1511 V RMS voltage −Output = 1069 V RMS current −Output = 28.43 A Voltage THD (%) = 7.54

5 Simulation and Results

The proposed DVR model is repeated and striven for an L–G fault in the transmission line and input and output parameters are presented in Table 3. Figure 7 shows source side voltage and Fig. 8 shows input signals to PWM. The reenactment yield for produced framework voltage with blame and repaid voltage. The voltage (v) is signified in Y-pivot and day and age (s) is demonstrated in X-hub. The voltage list is happening between day and age of 0.5–0.7 s. The droop voltage and the voltage swell happens between day and age of 0.2–0.4 s. amid the time of hang and swell, the infusion voltage is sustained to the line and the framework voltage is adjusted. The repaid voltage makes the framework to accomplish the ostensible voltage.

The inverter yield for the A stage is given in the Fig. 8. The last repaid stack side voltage is given in the Fig. 9. The other two stages additionally have the comparative yields as indicated by their reference signals given. As appeared in the Fig the controller of DVR is sending the information beats for PWM of the HCMLI.

The inverter yield for the A stage is given in the Fig. 8. The last repaid stack side voltage is given in the Fig. 9. The other two stages additionally have the comparative yields as indicated by their reference signals given. As appeared in the Fig the controller of DVR is sending the information beats for PWM of the HCMLI.

Table 3 Specifications of proposed DVR

Sl. no	Specification	Value
1	Supply voltage	11 kV
2	DC source voltage	5.5 kV
3	L,C filter values	10 mH, 20 μF
4	Load R,L	10 Ω, 1 mH
5	PI controller gain, Kp	0.9
6	Carrier frequency	2500 Hz
7	Fault	L–G

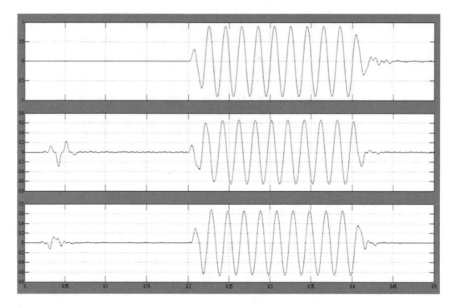

Fig. 8 Input signals to the PWM

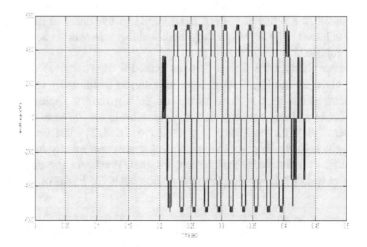

Fig. 9 HCMLI output voltage at phase A

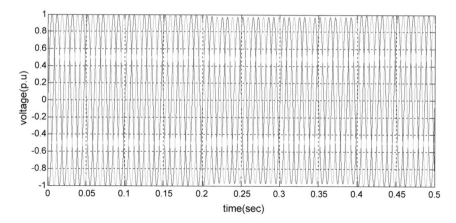

Fig. 10 Compensated signal

6 Conclusion

The Fig. 10 shows the represents the compensated output waveform which is free from sag and swell, The DVR is independently capable of compensating voltage sags to compensate for faults on the grid, The upside of using the SCML in the DVR is that it has a reduced number of changes when appeared differently in relation to the standard multilevel inverters and meanwhile, it diminishes the essential of the degree of capacitor as we can get the nearby sinusoidal. Solar energy has been utilized as a renewable energy source to satiate the power requirements. Thus, to diminish the effect of sag and swell in the generated voltage the stored energy from solar field is utilized. The DVR compares the difference between the obtained voltage and the expected voltage.

References

1. G.A. de Almeida Carlos, E.C. dos Santos, C.B. Jacobina, J.P.R.A. Mello, Dynamic voltage restorer based on three-phase inverters cascaded through an open-end winding transformer. IEEE Trans. Power Electron. **31**(188) (2016)
2. Z. Shuai, P. Yao, Z.J. Shen, C. Tu, F. Jiang, Y. Cheng, Design considerations of a fault current limiting dynamic voltage restorer (FCL-DVR). IEEE Trans. Smart Grid **6**(1) (2015)
3. K.K. Gupta, A. Ranjan, P. Bhatnagar, L.K. Sahu, S. Jain, Multilevel inverter topologies with reduced device count: a review. IEEE Trans. Power Electron. **31**(1) (2016)
4. J. Dixon, L. Morán, High-level multistep inverter optimization using a minimum number of power transistors. IEEE Trans. Power Electron. **21**(2) (2006)
5. S.J. de Mesquita, F.L.M. Antunes, S. Daher, A high resolution output voltage multilevel inverter topology with few cascade-connected cells, in *IEEE Applied Power Electronics Conference and Exposition*, IEEE, vol. 1, no. 1 (2014)

6. M.A. Jaoda, P.S. Kumar, Design of multilevel inverter with less number of power electronic components fed to induction motor, vol. 3 (TJPRC Publication, Chennai, 2013), p. 189
7. A. Chatterjee, Identification of photovoltaic source models. IEEE Trans. Energy Convers. **26** (3) (2011)
8. S.V. Jamuna, B. Subalakshmi, A. Rajeshwari, H.M. Sahul Hameed, R. Muralidharan, Enhancing power quality on solar grid system using multilevel inverter and dynamic voltage restorer. Int. J. Future Innovative Sci. Eng. Res. (2016)
9. I. Colak, R. Bayindir, E. Kabalci, Design and analysis of a 7-level cascaded multilevel inverter with dual SDCSs, in *IEEE International Symposium on Power Electronics, Automation and Motion Conference*, p. 180 (2010)

Grammar Rule-Based Sentiment Categorization Model for Tamil Tweets

Nadana Ravishankar, R. Shriram, K. B. Vengatesan, S. B. Mahajan, P. Sanjeevikumar and S. Umashankar

Abstract The widespread of social media is growing every day where users are sharing their opinions, reviews, and comments on an item or product. The aim is to develop a model to mine user tweets collected from Twitter. In this paper, our contribution on user tweets to find the sentiments expressed by users about Tamil movies based on the grammar rule. Tamil movies domain is selected to confine our scope of the work. After preprocessing, N-gram approach is applied to classify tweets into different genres. This work intends to find the polarity of Tamil tweets in addition to genre classification. In this work, it is also shown how to collect user tweets which comes as data stream using modified N-gram approach to predict the sentiments of the users in the dataset. Results suggest that N-gram model not only remove the complexity of natural language process but also help to improve the decision-making process.

N. Ravishankar (✉) · R. Shriram
Department of CSE, B.S Abdur Rahman University, Chennai, India
e-mail: nadanaravishankar@gmail.com

R. Shriram
e-mail: shriram@bsauniv.ac.in

K. B. Vengatesan
Department of Computer Science Engineering (CSE),
Marathwada Institute of Technology (MIT), Aurangabad, India
e-mail: vengicse2005@gmail.com

S. B. Mahajan · P. Sanjeevikumar
Department of Electrical and Electronics Engineering,
University of Johannesburg, Auckland Park, South Africa
e-mail: sagar25.mahajan@gmail.com

P. Sanjeevikumar
e-mail: sanjeevi_12@yahoo.co.in

S. Umashankar
School of Electrical Engineering, VIT University, Vellore, Tamil Nadu, India
e-mail: umashankar.s@vit.ac.in

© Springer Nature Singapore Pte Ltd. 2018
S. S. Dash et al. (eds.), *Artificial Intelligence and Evolutionary Computations in Engineering Systems*, Advances in Intelligent Systems and Computing 668,
https://doi.org/10.1007/978-981-10-7868-2_65

687

Keywords Data mining · Sentiment analysis · Social media
Natural language processing · Big data

1 Introduction

Social media is an interactive site where users sharing their views and comments on different items every day. The item can be a product, a topic, a movie, or an object. Twitter streams help government, business organization or any other groups to identify what people are thinking about. The popularity of twitter is growing since the data is publically available to analyze the user opinions.

Over the past few years, sentiment analysis from social media has gained much more progress to extract people emotion, in particular, Twitter feeds [1]. Most of the researcher's working on opinion mining use machine learning tools to detect opinion events, since there are a lot of challenges to be solved. The major problem in machine learning techniques is ambiguity since tweets do not reveal typical expression style, and appear in separation from other tweets. However, social platforms like Facebook or LinkedIn sharing information only to people that are friends. Tweets also make frequent use of hash tags to indicate subject, emoticons to express the feelings, and abbreviations which are technically challenging for machine learning techniques to detect the individual opinion and requires the development of new, automated techniques [2]. Machine learning tools are working well for event detection task, since it is trained for only one application. To the best of our knowledge there is no work on grammatical structure identification in user tweets.

Users freely sharing information over microblogging sites and microblogs such as tweets only hold short information and most of the tweets related to a particular subject only. Most of the user tweets tend to be varied in their grammatical style than longer conversation. Grammar rule-based approach is developed to sentiment analysis task with limited grammar rules. It is also required to analyze the user emotions in sentence level rather than document level for calculating accurate sentiments of user tweets. All the tweets data are collected from Twitter timeline for our experiment to analyze the opinions of different users about Tamil movies domain. Our objective is to combine grammar rules with N-gram model to analyze and mine the user tweets available in the corpus.

The paper is organized as follows. In the next section related work is described. Section 3 describes the structure of our proposed model. In Sect. 4, detail analysis of experimental results. Finally, Sect. 5 gives the concluding remarks and extension of the work.

2 Related Work

Opinion mining, also known as sentiment analysis [3, 4], is a branch of natural language processing that is first experimented in [5], they use SVM classifier that accept semantic role labeling, syntactic information, and lexical information as input to detect sentential accompaniments of the people. Choi et al. [6] experiment with conditional random field (CRF)-based methods and semantic role labeling to extract sources and opinions at the same time. The first method to deal with opinion classification [7] focus on adjectives and phrases where adjectives are connected with conjunction words such as "but", "or", "not". They developed a linear regression model to check whether two adjectives have the same orientation. Significant research work on opinion mining is carried in [8] to summarize positive or negative statements about product reviews written by customers on websites.

Social platform, such as Facebook, Twitter quite often used as an instructional tool and changed the way of association with researchers and students [4, 9]. However, Facebook and Twitter alone are not the social media. There are other alternatives, such as LinkedIn, blogs, and wikis. Scientists should effectively use the social media to describe their work to a wider audience to the popularization of science [10].

Sentiment analysis plays a major role in decision-making process of business, politics, and is itself a wide research area to detect user emotions [11]. Supervised machine learning technique such as suffix-tree structure is used for data [12] to characterize the syntactic relations between polarity scores and keywords in a user sentence to calculate the overall opinion. To solve this problem of modeling syntactic structures requires the development of new sentiment techniques that understand the conceptual relationships in a domain-specific application.

Twitter-like short-form communications can be leveraged in, as a source of indexing and retrieval information for movie/product recommendation. In [13], a political dataset which is collected from tweets is used to predict the politician character Volde-mort from the Harry Potter series of books. However, the unique feature of tweets is that they have a huge amount of metadata associated with each tweets. This tweets and its metadata can be used for opinion categorization and future decision-making process including training purposes. It is found that there is no work in Tamil to classify tweets (movie) into genres like action, comedy, commercial, love, and sentiment. However, it is very difficult to classify the tweets the reason may be unavailability of data and sentiment lexicons for natural language than English.

3   Proposed Methodology

Figure 1 shows the architecture of sentiment analysis model for Tamil tweets classification. In this work, tweets have been collected for 100 Tamil movies using API and create an unlabeled dataset to find sentiments of the user. Though the idea seems to be simple to implement, it is challenging task since Tamil tweets suffers from various linguistic and grammatical errors [14]. The major contribution of this research is to apply grammar-based N-gram approach to tweets stream to enhance the prediction results. It is quite difficult to acquire the information from tweets automatically using simple classifier or unigram models.

3.1   Preprocessing

The proposed model is built using own corpus and all the data are collected from twitter in the first week of August 2016. The following simple preprocessing methods are used; (i) remove reply/retweets, (ii) remove any external URL links present and any other user-specified symbols. The reason for this is to capture all the words in dataset in order to enhance the final outcomes.

3.2   TF-IDF Ranking

When developing a language model for Tamil, N-grams are used to generate not only unigram model (syntactic) but also to develop bigram and trigram models,

Fig. 1 Architecture of the categorizer model for classification of tweets

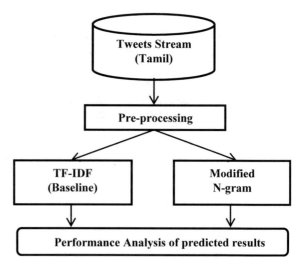

etc. [15]. The idea is to use tokens such as trigrams in the feature space instead of just unigrams (TF-IDF). A syntactic model like TF-IDF uses the number of occurrences of particular keyword and does not understand the context of the word. TF-IDF model will check for the presence or absence of keywords and its related terms to classify them. For example, consider a sentence "# கபாலி அதிரடி சண்டை படம் இல்லை" (Kabali is not an action movie). TF-IDF classifies this sentence into the action category (the words 'அதிரடி' and 'சண்டை' are related to genre of action). But it is failed to consider the negation term present in the sentence that will invert the polarity as well the category.

3.3 Modified N-Gram Approach

N-grams are essentially a set of co-occurring words within a user tweet and when computing the N-grams typically move one word forward. In this work, most common rules for negation words and adjective rules are developed with Tamil tweets on the mind. To identify a set of useful N-grams,

- Set of N-grams can be made out of consecutive words.
- Negation terms such as "இல்லை" (No) are attached to a word which follows or precedes it.
- For example: "காமெடி இல்லை கடி" has two bigrams:

  "காமெடி + இல்லை (comedy + illai)", "இல்லை + கடி (illai + kadi)"

  (illai + kadi)".
- So the performance of the sentiment prediction model improves by using language models, because extraction of negation terms is important in the field of sentiment analysis.

If W = Number of words in a given sentence S, the number of N-grams for sentence S would be:

$$\text{N-grams} = W - (n - 1)$$
$$\text{Where } n = 3 \text{ and } n < W \tag{1}$$

Example: அனைத்து பெண்களும் பார்க்க வேண்டிய படம். If n = 3, the N-grams would be:

அனைத்து பெண்களும் பார்க்க (Anaithu pengalum parkka)
பெண்களும் பார்க்க வேண்டிய (pengalum parkka vendiya)
பார்க்க வேண்டிய படம் (parkka vendiya padam).

So there are three N-grams in this case. Similarly, trigrams are generated for complex sentences in order to reduce the morphological richness of the natural

Table 1 Tweets length statistics for movie Kabali

Movie name	Word length	Number of tweets
கபாலி (Kabali) (254 tweets)	1	12
	2	29
	3	34
	4	25
	5	18
	6	26
	7	23
	8	18
	>8	69

language. However, small sentences less than four words in length are directly classified using grammar rules. To produce trigrams, the following steps are used.

1. Read the user tweets stream from the corpus.
2. If tweet length is ≤ 8, go to next step; otherwise ignore it.
3. Generate the trigrams for $W \leq 8$ and different values of n = 3.
4. Find the polarity and category of different N-grams using negation rules and adjective rules.

For this work, tweets stream is manually analyzed and find that most of the user tweets are short in length (less than eight words) as shown in Table 1. Most of the sentiment related words occur within first few words of a tweet. From the results, it is proved that it is good enough to extract the tweets with less word to find the sentiment of a particular movie. For example, consider only 185 tweets (254 total tweets—69 omitting) for the movie Kabali and classify them.

4   Results and Discussion

For all experimental run, our own dataset is built which contains 7418 Tamil tweets, manually annotated by various professors (domain experts) and researchers. The use of unlabeled dataset for our experimental design and do not have predefined sentiment lexicons for Tamil with focus on tweets. To overcome this problem, this approach leverages the use of API provided by Tamil Agarathy (http://agarathi. com/api/dictionary#) in order to derive the sentiments of user tweets. The API provides more than 1 lakh Tamil words and its related words are grouped based on their conceptual relations.

After formatting the data, the sentiment categorizer algorithms are applied into the Tamil tweets present in the corpus and classify them into the different categories. Polarity scores of the user tweets are also identified. To validate our model with manually annotated data using accuracy as a metric. The accuracy of the user sentiments is calculated as the sum of correctly predicted tweets and correctly predicted not to be in a class divided by the total number of tweets.

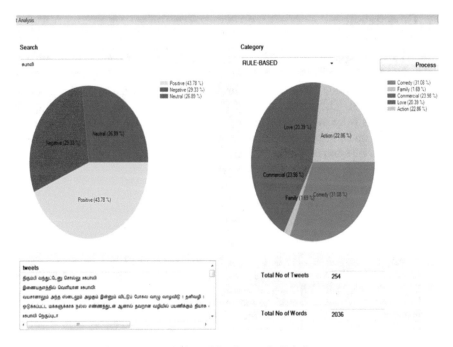

Fig. 2 Screenshot of sentiment analysis tool for the movie Kabali

Figure 2 shows the screenshot of the sentiment model designed for the classification of Tamil tweets on focus. For each selected movie and algorithm, the tool provides the percentage scores of each genres and polarity of the tweets. Finally, the predicted class results are analyzed with manually annotated results. Four professors and their scholars are voluntarily involved in this experiment. The accuracy values of the proposed model are shown in Table 2.

Table 3 provides the average accuracy of the sentiment analysis tool for all 100 movies in the dataset. It is observed that the proposed model produce better results than syntactic model. N-gram model removes the complexity of the sentences and achieves good accuracy with much smaller linguistic rules. However, the accuracy is not further improving for different values of n (4, 5 etc.). The reason is when the language complexity increases, the accuracy value will decrease.

Table 2 Sentiment analysis results for the movie Kabali

Model	Accuracy
TF-IDF	26.41
N-grams	60.18

Table 3 Overall performance analysis of our system

Model	Average accuracy
TF-IDF	29.87
N-grams	61.29

5 Conclusion

In this paper, the proposed modified N-gram model for sentiment classification of Tamil tweets. The contribution was put on the development of dataset for Tamil tweets and the various sentiment algorithms are developed based on the modified N-gram approach. The performance of N-gram model proves that extracting the user sentiments from the tweets dataset promising results. On this belief, the future work focuses on developing the algorithm for real-time tweets as well as for different domain of interest. Further, there is also a plan to develop an algorithm for complex user sentences to enhance the performance.

References

1. A. Pak, P. Paroubek, Twitter as a corpus for sentiment analysis and opinion mining, in *Proceedings of the 7th International Conference on Language Resources and Evaluation LREC* (Valletta, Malta, 2010), pp. 1320–1326
2. R. Colbaugh, K. Glass, Estimating sentiment orientation in social media for intelligence monitoring and analysis, in *Proceedings of IEEE International Conference on Intelligence and Security Informatics* (Vancouver, Canada, 2010), pp. 135–137
3. B. Pang, L. Lee, Opinion mining and sentiment analysis. Found. Trends Inf. Retr. 2(1–2), 1–135 (2008)
4. B. Liu, Sentiment analysis and subjectivity, in *Handbook of Natural Language Processing*, 2nd edn. (CRC Press, Taylor and Francis Group, Boca Raton, FL, 2010), pp. 627–666
5. S. Bethard, H. Yu, A. Thornton, V. Hatzivassiloglou, D. Jurafsky, Automatic extraction of opinion propositions and their holders, in *Proceedings of the AAAI Spring Symposium on Exploring Attitude and Affect in Text: Theories and Applications* (2004)
6. Y. Choi, E. Breck, C. Cardie, Joint extraction of entities and relations for opinion recognition, in *Proceedings of the International Conference on Empirical Methods in Natural Language Processing (EMNLP)* (Sydney, AU, 2006), pp. 431–439
7. V. Hatzivassiloglou, K. Mckeown, Predicting the semantic orientation of adjectives, in *Proceedings of the 8th International Conference on European Chapter of the Association for Computational Linguistics*, (Stroudsburg, PA, USA, 1997), pp. 174–181
8. M. Hu, B. Liu, Mining and summarizing customer reviews, in *Proceedings of the 10th ACM SIGKDD International Conference on Knowledge Discovery and Data Mining* (Chicago, IL, 2004), pp. 168–177
9. J. Bollen, H. Mao, X. Zeng, Twitter mood predicts the stock market. J. Comput. Sci. 2(1), 1–8 (2011)
10. M.J. Kumar, Expanding the boundaries of your research using social media: stand-up and be counted. IETE Tech. Rev. 31(4), 255–257 (2014)
11. A.B. Sayeed, J. Boyd-Graber, B. Rusk, A. Weinberg, Grammatical structures for word-level sentiment detection, in *Proceedings of the 2012 Conference of the North American Association of Computational Linguistics* (Montreal, CA, 2012), pp. 667–676
12. S.G. Esparza, M.P. O'Mahony, B. Smyth, Mining the real-time web: a novel approach to product recommendation. J. Knowl. Based Sys. 29(1), 3–11 (2012)
13. D. Maynard, A. Funk, *Automatic Detection of Political Opinions in Tweets*. ESWC Workshop (LNCS 7117, May 2011), pp. 88–99

14. S. Rajendran, S. Arulmozi, B. Kumara Shanmugam, S. Baskaran, S. Thiagarajan. Tamil WordNet, in *Proceedings of the First International Global WordNet Conference* (CIIL, Mysore, 2002), pp. 271–274
15. W. Schmit, S. Wubben, Predicting ratings for new movie releases from twitter content, in *Proceedings of the 6th Workshop on Computational Approaches to Subjectivity, Sentiment and Social Media Analysis (WASSA 2015)* (Lisboa, Portugal, 2015), pp. 122–126

Performance Enhancement for Detection of Myocardial Infarction from Multilead ECG

Smita L. Kasar, Madhuri S. Joshi, Abhilasha Mishra, S. B. Mahajan
and P. Sanjeevikumar

Abstract Computer-aided diagnosis have emerged as additional help to the medical domain. Over the years ECG signal being simple, cheap, and noninvasive, is explored for the diagnosis of heart diseases. Multilead simultaneously acquired ECG improves the accuracy in diagnosis of heart diseases. The paper focuses on diagnosing Myocardial Infarction from multilead ECG using Multilayer Perceptron Model. In the present work, the proposed feature vector used for the classification includes QRS point score as one of the feature along with the other morphological features. The study is an attempt to discuss the utility of point score as a feature in the feature vector for classification of Myocardial Infarction disease from ECG signal to enhance the performance of classification. The results show significant improvement when the point score is used in the feature vector. The model is evaluated with 34 ECG signals of normal subjects and 33 ECG signals of MI patients from PTB database, collected from physionet. The classification accuracy is above 95% including point score feature and the same is less than 85% excluding

S. L. Kasar (✉) · M. S. Joshi
Department of Computer Science and Engineering,
Marathwada Institute of Technology, Aurangabad, India
e-mail: smitakasar@gmail.com

M. S. Joshi
e-mail: madhuris.joshi@gmail.com

A. Mishra
Department of Electronics and Telecommunication Engineering,
Maharashtra Institute of Technology, Aurangabad, India
e-mail: abbhilasha@gmail.com

S. B. Mahajan
Department of Electrical and Electronics Engineering,
University of Johannesburg, Auckland Park, South Africa
e-mail: sagar25.mahajan@gmail.com

P. Sanjeevikumar
Department of Electrical and Electronics Engineering,
University of Johannesburg, Auckland Park, South Africa
e-mail: sanjeevi_12@yahoo.co.in

© Springer Nature Singapore Pte Ltd. 2018
S. S. Dash et al. (eds.), *Artificial Intelligence and Evolutionary Computations
in Engineering Systems*, Advances in Intelligent Systems and Computing 668,
https://doi.org/10.1007/978-981-10-7868-2_66

the point score in all the leads. The inclusion of point score as a feature for diagnosing Myocardial infarction results in better accuracy.

Keywords ECG · Myocardial infarction · Multilayer Perceptron QRS · Score

1 Introduction

The ECG is a noninvasive technique essential for the diagnosis and management of abnormal cardiac rhythms [1]. The ECG signal is represented by different parts and named with arbitrarily chosen letters P, Q, R, S, and T. The shape of the normal ECG is shown in Fig. 1. The electrical change accompanying the contraction of atria is small due to small muscle mass. Contraction of atria is associated with ECG wave called "P". With the large ventricular mass, there is large deflection during depolarization represented by "QRS" complex. The T wave in the ECG is associated with the resting electrical state. The ST segment indicates the time between S wave and the commencement of the T wave.

The ST-segment lies between the QRS complex and T wave. This segment should be isoelectric, at the same level as the part between the T wave and the next P wave. During abnormality, the ST segment may be elevated or depressed. Elevation of the ST segment is an indication of acute myocardial injury. The leads in which the elevation occurs indicate the part of the heart that is damaged [2]. Horizontal depression of the ST segment associated with an upright T wave is usually a sign of ischemia as opposed to infarction [3].

The categorization of the ECG signals can be done into three classes according to the probable medical significance. One class includes ECG with the minor discrepancy, other includes ECG of indeterminate clinical importance, and the third class includes ECG with probable clinical importance. Sinus tachycardia, Sinus Bradycardia, etc. belong to the class I whereas Ventricular Hypertrophy, Left bundle branch block are in class-II. Class III includes Atrial Fibrillation, Myocardial Ischemia, Acute Myocardial infarction, etc. [4]. The study demonstrating the failure to detect ECG abnormalities emphasized the significance of system change to improve the accuracy of ECG interpretation [5]. The elevated ST segment indicates the infarction and depressed ST segment is usually a sign of ischemia. The changes in ST segment are complicated to predict earlier in the course of the disease and is challenging even to experienced physicians. In [6] the author has proposed QRS point score for detecting myocardial infarction and has

Fig. 1 Electrocardiogram signal

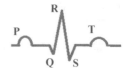

been validated in 21 patients where the correlation between infarct size and the point score has been found to be 0.80. Similarly, [7] evaluated a similar QRS scoring system and achieved 98% specificity. The present study is conducted to find out the diagnostic accuracy of QRS point score as an additional feature in the feature vector for classifying patients with myocardial infarction using Multilayer Perceptron Model.

2 Literature

ECG filtering is a difficult task since the actual signal value is very less, approximately 0.5 mV in a situation of 300 mV surrounding. ECG is corrupted by a variety of noise like Baseline Wander, Power line Interference, Muscle contractions, Electrode contact noise, etc. which can impact the accuracy of the signal. The exclusion actual cardiac signal from a noisy ECG is challenge since the further analysis depends upon the accuracy of the recorded cardiac signal. The various noises, overlap with the cardiac components, especially in the range of 0.01– 100 Hz numerous methods are projected to eliminate the unwanted noises from the ECG signal. The ECG signal preprocessing is followed by the detection of highest amplitude R peaks and the QRS wave, often called as morphological features in ECG. In [8], a real-time QRS detection algorithm was proposed using digital bandpass filter with adaptive threshold mechanism. This algorithm is called as Pan Tompkins algorithm. The literature available on QRS detection algorithms is compared with respect to the noise sensitivity in [9]. A distinct feature of Power-Line Interference (PLI) detection and suppression algorithm is its ability to detect the presence of PLI in the ECG signal before applying the PLI suppression algorithm. A PLI detector that employs an optimal Linear Discriminant Analysis (LDA) algorithm is used to make a decision for the PLI presence [10]. The paper [11] presents an algorithm for QRS detection using the first differential of the ECG signal and its Hilbert transformed data to locate the R peaks in the ECG. The differentiation of R waves from large, peaked T, and P waves is achieved with a high degree of accuracy. Muscle noises, Baseline wander and motion artifacts are minimized. Due to the large size, the ECG signal can be condensed into some other form of representation. The patterns available in the signal are converted into features containing only significant information. The methods for extracting features could be based on the syntactic descriptions or statistical characteristics. The paper [12] classifies the ECG signals into normal and arrhythmic using automatic extraction of time interval and morphological features. Artificial neural networks (ANN) are used for classification. In [13], redundant ECG data is reduced using ICA. An independent subspace analysis model for ECG analysis is developed allowing applicability to any random vectors available in an ECG dataset. The paper [14] extracts features using RR intervals with Independent Component Analysis (ICA) and Power spectrum. Independent component analysis (ICA) is used for separating independent components from ECG complex signals, whereas principal component analysis (PCA) is used to reduce dimensionality and for feature extraction of the ECG data prior to or at times after performing ICA in special circumstances [15]. The different methods proposed in recent years for ECG classification include Artificial Neural Network, Bayesian, digital signal analysis, Fuzzy Logic methods, Self-Organizing Map, Hidden Markov Model, Support Vector Machines, Genetic Algorithm, and many more with each method demonstrating benefits and drawbacks. Artificial neural networks (ANNs) have been applied to a number of real-world complex problems. Over the last decade, the ANNs are used for the classification of heart diseases using ECG signal. Once feature extraction is done, ANNs can be trained with the feature vector to classify the patterns. A methodology for the automated creation of fuzzy expert systems, applied in ischemic and arrhythmic beat classification is presented in [16]. In [17] the fuzzy-hybrid neural network is used for electrocardiogram beat classification. Autoregressive model coefficients, higher order cumulant and wavelet transform variances are used as features. The use of combined neural network model for model selection in order to classify ECG beats is presented in [18]. Recent development in the classification of Myocardial infarction includes considering the parameters from QRS complex for enhancement in classification compared to ST segment changes only. In [19] Vector cardiogram and ECG parameters are proposed to identify ischemic patients from healthy patients. The data consisted of 80 ischemic patients and 52 healthy subjects. For the given data, VCG parameters like Volume, Planar Area, Ratio between Area and Perimeter, Perimeter, and Distance between Centroid and Loop computed on QRS-loop were analyzed. Three conventional ECG ST-T parameters ST Vector Magnitude, ST segment level and, T-wave amplitude are calculated. It is shown that healthy and ischemic subjects have significant differences in the VCG and ECG parameter. In [20] the intraindividual and interindividual variations of TSV index, in order to determine reliable limits of significant repolarization variability due to an ischemic cardiac process is analyzed. In the paper [21], cross wavelet transform is used and the parameters proposed are used to classify normal and cardiac patients. In [22] the author worked with artificial neural networks, both RBF and MLP with genetic algorithm, where MLP with Genetic algorithm outperformed for the diagnosis of acute myocardial infarction.

3 Proposed Methodology

Detail process for classification model is as follows.

Step 1: Data Collection and Signal Processing

Physionet, funded by NIH, is considered as unique web-based resource for complex physiologic signals. In the present study, ECGs are referred from PTB (Physikalisch-Technische Bundesanstalt), the National Metrology Institute of Germany. The PTB database contains ECG records for both myocardial infracted and healthy subjects. Each ECG record includes 15 simultaneously measured signals including the conventional 12 leads and the 3 Frank lead ECGs. Each signal is digitized at 1000 samples per second [23]. It is required to eliminate the unwanted distortions prior to the processing. This preprocessing should be carried out leaving the original signal unchanged. In the present work, a second-order IIR notch filter is applied for reducing the distortions.

Step 2: Feature Extraction

In the morphology of the ECG wave, the R peak is higher in amplitude in few leads and often used as a reference to detect the beats. Another important feature is the QRS complex due to its specific shape. To reduce the dimensionality of a large dataset, statistical technique, principal component analysis is used. The set of correlated variables is converted into a few variables termed as principal components [24]. Principal Component Analysis is applied to the 12 lead ECG signal. The different features extracted from the signal are P amplitude, Q amplitude, R amplitude, S amplitude, T amplitude, P duration, Q duration, R duration, S duration, T duration, ratio for Q amplitude & R amplitude, ratio for R amplitude and S amplitude, ST segment. The t-test is used to select the set of features to be used as feature vector.

With the features extracted, an additional feature QRS point score is calculated. The scoring system given in [25] is followed in the current study. The Point score is calculated by adding the individual scores generated from the respective leads meeting the specified criteria. If the generated sum of scores is greater than 4 then the infarction cannot be ruled out. For a total score greater than 6 there is the possibility of infarction and if the score is greater than 8 then the signal has a definite infarction. The Proposed feature vector consists of Amplitude ratio QR & RS, Q & R duration,

P,S,T amplitude, Point score, ST segment (Tables 1 and 2).

Table 1 Point score calculation for a normal signal from PTB database

Leads		Amplitude ratio (Q/R)	Q duration	T amplitude	Score	Remark
Anterior	v2	0.17	20	0.35	**0**	*Score = 0* Normal signal
	v3	0.18	32	0.49		
	v4	0.13	36	0.47		
Lateral	I	0.13	29	0.07	**0**	
	v5	0.09	33	0.25		
	v6	0.12	29	0.09		
Inferior	II	0.17	34	0.15	**0**	
	III	0.15	31	0.08		
	avF	0.16	30	0.12		

Table 2 Point score calculation for MI signal from PTB database

Leads		Amplitude ratio (Q/R)	Q duration	T amplitude	Score	Remark
Anterior	v2	8.90	40 (3)	0.20	**9**	*Score = 19*
	v3	6.11	40 (3)	0.20		≥ 8 Definite infarction
	v4	4.21	40 (3)	0.12		
Lateral	I	0.54	38 (3)	0.03	**6**	
	v5	2.08	40 (3)	0.11		
	v6	0.27	30	0.03		
Inferior	II	19.93	31 (2)	0.12	**4**	
	III	0.12	23	0.07		
	avF	20.76	29 (2)	0.09		

Step 3: Signal Classification

Artificial neural network (ANN) has outstanding characteristics of machine learning, fault, tolerant, parallel reasoning, and processing nonlinear problem abilities. In the present study, total 66 ECG signals (33 ECG signals with myocardial infarction and 34 normal ECG signals) were evaluated to test the accuracy of the proposed feature vector for the efficacy of the point score feature (Fig. 2).

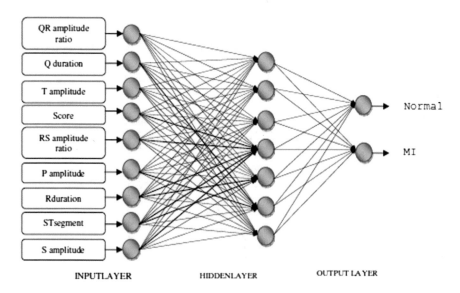

Fig. 2 Multilayer perceptron classifier

4 Results and Discussion

In order to maintain patient independent classification, tenfold cross-validation technique is used to generate the training and test set data. Figure 3 shows the classifier accuracy including and excluding the point score feature. It is observed that when multilayer perceptron is trained and tested using point score as one of the feature in the proposed feature set, the diagnostic accuracy is high. The model is also tested with the feature vector without the point score, but lower accuracy was resulted.

The sensitivity is the parameter used to identify the diseased class. It is also called as True positive rate. The specificity is used to identify the normal class. It is also called as True negative rate. Receiver operating Characteristic (ROC) is a plot of the true positive rate against false positive rate for the different possible cut points of a diagnostic test as shown in Fig. 4.

Fig. 3 Accuracy with MLP classifier

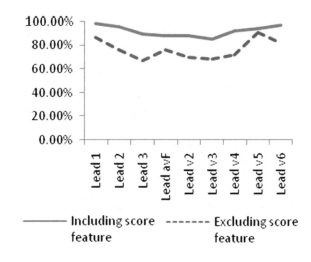

Fig. 4 Area under ROC using MLP with the score feature

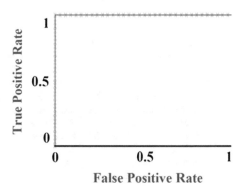

Table 3 Parameters including and excluding the point score feature using MLP

Lead I

Parameters	Including point score feature	Excluding point score feature
Accuracy (%)	98.5	92.53
Sensitivity (%)	96.96	81.81
Precision	0.98	0.86
Area under ROC	1	0.88

The following table shows the performance of the MLP with point score as one of the feature and without point score (Table 3).

5 Conclusion

There is a significant increase in cardiovascular diseases these days leading to the requirement of proficient automated disease classification. The most important benefit of Electrocardiogram signals is that it is easily available, reasonably priced and noninvasive. The morphology of the ECG wave changes depending on the lead position and have different significances. Multilayer Perceptron with improved feature vector is used for classification. The point score is calculated using QRS complex and T amplitude from nine leads. The feature vector with point score as one of the feature enhances the performance to classify the ECG signals with better accuracy for Myocardial infarction signals. It is found that the combination of both point score as a feature and Multilayer Perceptron predicts with better accuracy. To conclude it is found that the model gives good predictive utility in diagnosing myocardial infarction from multilead ECG signal with the proposed feature vector. However, still more large scaled analysis is needed to assess the practical utility of the classifier in diagnosing myocardial infarction.

References

1. J. Malmivuo, R. Plonsey, *Bioelectromagnetism: Principles and Applications of Bioelectric and Biomagnetic Fields* (Oxford University Press, New York, 1995)
2. J.R. Hampton, *The ECG Made Easy*, (Elsevier Edition, UK, 2013)
3. G.D. Clifford, F. Azuaje, P.E. McSharry, *Advanced Methods and Tools for ECG Signal Analysis* (Artech House, Boston, London, 2006)
4. E.R. Snoey, B. Housset, P. Guyon, S. Elhaddad, J. Valty, P. Hericod, Analysis of emergency department interpretation of electrocardiogram. J. Accid. Emerg. Med. **11**, 149–153 (1994)
5. F.A. Masoudi, D.J. Magid, D.R. Winson, A.J. Tricomi, E.E. Lyons, L. Crounse, P.M. Ho, P.N. Peterson, J.S. Rumsfeld, Implications of the failure to identify high risk electrocardiogram findings for the quality of care of patients with acute myocardial infarction: results of the emergency department quality in myocardial infarction (EDQMI) study. Circ. J. Am. Heart Assoc. **114**, 1565–1571 (2006)

6. R.E. Ideker, G.S. Wagner, W.K. Ruth, D.R. Alonso, S.P. Bishop, C.M. Bloor et al., Evaluation of a QRS scoring system for estimating myocardial infarct size. II. Correlation with quantitative anatomic findings for anterior infarcts. Am. J. Cardiol. **49**, 1604–1614 (1982)
7. G.S. Wagner, C.J. Freye, S.T. Palmeri, S.F. Roark, N.C. Stack, R.E. Ideker et al., Evaluation of a QRS scoring system for estimating myocardial infarct size. Circ. J. Am. Heart Assoc. **65**, 342–347 (1982)
8. J. Pan, W.J. Tompkins, A real time QRS detection algorithm. IEEE Trans. Biomed. Eng. BME. **32**(3), 230–236 (1985)
9. G.M. Friesen, T.C. Jannett, M.A. Jadallah, S.L. Yates, S.R. Quint, H.T. Nagle, A comparison of the noise sensitivity of nine QRS detection algorithms. IEEE Trans. Biomed. Eng. **37**, 85–98 (1990)
10. Y.D. Lin, Y.H. Hu, Power-line interference detection and suppression in ECG signal processing. IEEE Trans. Biomed. Eng. **55**, 354–357 (2008)
11. D. Beniteza, P.A. Gaydeckia, A. Zaidib, A.P. Fitzpatrickb, The use of the Hilbert transform in ECG signal analysis. Comput. Biol. Med. **31**, 399–406 (2001)
12. C. Alexakis, H.O. Nyongesa, R. Saatchi, N.D. Harris, C. Davies, C. Emery, R.H. Ireland, S.R. Heller, Feature extraction and classification of electrocardiogram (ECG) signals related to hypoglycaemia, in *IEEE Computers in Cardiology* pp. 537–540 (2003)
13. M.P.S. Chawla, Detection of indeterminacies in corrected ECG signals using parameterized multidimensional independent component analysis. Taylor and Francis, Comput. Math. Methods Med. **10**, 85–115 (2009)
14. T.M. Nazmy, H. El-Messiry, B. Al-Bokhity, *Adaptive Neuro-Fuzzy Inference System for Classification of ECG Signals*, 7th International Conference on Informatics and Systems (INFOS), pp. 1–6 (2010)
15. M.P.S. Chawla, PCA and ICA processing methods for removal of artifacts and noise in electrocardiograms: a survey and comparison. Appl. Soft Comput. **11**, 2216–2226 (2011)
16. T.P. Exarchos, M.G. Tsipouras, C.P. Exarchos, C. Papaloukas, D.I. Fotiadis, L.K. Michalis, A methodology for the automated creation of fuzzy expert systems for ischaemic and arrhythmic beat classification based on a set of rules obtained by a decision tree, in *Elsevier, Artificial Intelligence in Medicine* (2007)
17. M. Engin, ECG beat classification using neuro-fuzzy network. ScienceDirect, Pattern Recognit. Lett. **25**, 1715–1722 (2004)
18. I. Gulera, E.D. Ubeyl, ECGbeat classifier designed by combined neural network model. ScienceDirect, Pattern Recognit. **38**, 199–208 (2005)
19. Y. Wang, C.J Deepu, Y. Lian, *A computationally efficient QRS detection algorithm for wearable ECG sensors*, 33rd Annual International Conference of the IEEE Engineering in Medicine and Biology Society (EMBC), pp. 5641–5644 (2011)
20. R. Correa, P.D. Arini, M. Valentinuzzi, E. Laciar, Study of QRS-loop parameters and conventional ST-T indexes for identification of ischemic and healthy subjects, in *IEEE Computing in Cardiology, 2012*, pp. 649–652
21. S. Banerjee, M. Mitra, Application of cross wavelet transform for ECG pattern analysis and classification. IEEE Trans. Instrum. Meas. **63**(2), 326–333 (2014)
22. J. Kojuri, R. Boostani, P. Dehghani, F. Nowroozipour, N. Saki, Prediction of acute myocardial infarction with artificial neural networks in patients with nondiagnostic electrocardiogram. J. Cardiovasc. Dis. Res. **6**(2), 51–59 (2015)
23. A.L. Goldberger L.A. Amaral L. Glass, J.M. Hausdorff, P.C. Ivanov, R.G. Mark, J.E. Mietus, G.B. Moody, C.K. Peng, H.E. Stanley, PhysioBank, PhysioToolkit, and PhysioNet: components of a new research resource for complex physiologic signals. Circulation **101** (23), e215–e220 (2000). [Circulation Electronic Pages; http://circ.ahajournals.org/cgi/content/full/101/23/e215]
24. I.T. Jollife, *Principal Component Analysis*, 2nd Edition (Springer, Berlin, 2002)
25. M. Okajima, N. Okamoto, M. Yokoi, T. Iwatsuka, N. Ohsawa, Methodology of ECG interpretation in the Nagoya program. Methods Inf. Med. **29**, 341–345 (1990)

Fuzzy Controller-Based Intelligent Operation of Grid-Connected DFIG During Recurring Symmetrical Faults

G. V. Nagesh Kumar, D. V. N. Ananth, D. Deepak Chowdary and K. Appala Naidu

Abstract The grid-connected doubly fed induction generator (DFIG) can get adapted to modern grid rules to maintain synchronism and stability during disturbances if controlled by good control strategy. Every country framed certain grid rules such that in general the DFIG has to be in synchronism for approved time period during low voltages is called low voltage ride through (LVRT). Hence, for better power transfer capability and guarantee transient and dynamic stability margin improvement, enhanced FOC (EFOC) is proposed in the RSC of DFIG converter. The inner fast control loops are controlled using fuzzy controller with an aim to better voltage compensation, electromagnetic oscillations damping, and surge currents limit. This further leads to continued operation of DFIG under voltage sags. The proposed EFOC method helps in superior reactive power control with enhanced stability of current and voltage waveforms from rotor and stator to grid disturbance. The system behavior with recurring symmetrical low-voltage fault with decrease in voltage by 30 and 60% of the rated voltage happening at the point of common coupling (PCC) between 1–1.5 s and 2.5–3 s is analyzed using simulation studies.

Keywords DFIG · LVRT · Fuzzy controller · Field-oriented control (FOC)

G. V. Nagesh Kumar (✉) · K. Appala Naidu
Vignan's Institute of Information Technology, Visakhapatnam,
Andhra Pradesh, India
e-mail: drgvnk14@gmail.com

D. V. N. Ananth
K L Educational Society, Vaddeswaram, Vijayawada 522502,
Andhra Pradesh, India

D. D. Chowdary
Dr. L. Bullayya Engineering College for Women, Visakhapatnam, India

© Springer Nature Singapore Pte Ltd. 2018
S. S. Dash et al. (eds.), *Artificial Intelligence and Evolutionary Computations
in Engineering Systems*, Advances in Intelligent Systems and Computing 668,
https://doi.org/10.1007/978-981-10-7868-2_67

1 Introduction

Although single symmetrical faults are more severe, as per latest grid rules, wind generator system needs to withstand more symmetrical or asymmetrical faults. There are few situations in which faults may not be cleared in single reclosing and is found to occur multiple times. This type of situation is called recurring faults. In general, for this, automatic re-closures will open the circuit permanently and will protect the system from damage. With new modern grid code demand for wind energy system, the authors in [1–6] studied the behavior of DFIG WT system for single or multiple more severe faults at grid.

LVRT capability improvement with robust controllers like fuzzy controller is used to sustain rotor and stator windings over-currents and the dc side capacitor over-voltages during and after the fault [7]. To limit the rotor surge currents during faults, a combination of demagnetization control and virtual resistance is used in [8]. Its drawbacks are huge fault inrush surge rotor winding current and electromagnetic torque and power oscillations, etc., which are unavoidable. The parameters of PI controller for RSC are tuned effectively using genetic algorithm in [9] to have a better operation during different types of the grid faults. In this paper, an EFOC [10, 11] with fuzzy controller (FC) is used to improve the situation to much better value for symmetrical faults.

The system with proposed strategy is verified for two case studies under 30 and 60% decreases in the grid voltage. Two symmetrical recurring faults occur at PCC between 1–1.5 s and 2.5–3 s respectively. In Sect. 2, EFOC converters design is explained. In Sect. 3, mathematical analysis of the DFIG is dealt during symmetrical faults. In Sect. 4, rules for fuzzy controller are described. Section 5 describes the results with a voltage dip of 30 and 60% with FC in MATLAB environment. Section 6 summarizes the findings in the conclusion section.

2 Modeling of EFOC for RSC Control

The demand in the reactive power supply to the grid during normal and abnormal situations is important and can be improved using rotor side controller (RSC). The RSC's general role is to extract maximum possible power from the DFIG and will be available when rotor runs at the most favorable optimal speed. The reactive and active power control of stator and rotor is possible when i and i_{dr} components are controlled effectively. The stationary reference frame rotor voltage [5] is given by

$$V_r^s = V_{0r}^s + R_r i_r^s + \sigma L_r \frac{di_r^s}{dt} - j\omega\, i_r^s \tag{1a}$$

where ω is the rotor angular speed, $\sigma = 1 - \frac{L_{sm}^2}{L_s L_r}$,

$$V_{0r}^s = \frac{L_m}{L_s}\left(\frac{d}{dt} - j\omega_s\right)\Phi_s^s \tag{1b}$$

The stator and rotor flux components are expressed as

$$\Phi_s^s = L_s i_s^s + L_m i_s^s \tag{2}$$

$$\Phi_r^s = L_r i_r^s + L_m i_r^s \tag{3}$$

The direct and quadrature axis rotor voltage equations are given by

$$V_{dr} = \frac{d\Phi_{dr}}{dt} - (\omega_s - \omega)\Phi_{qr} + R_r i_{dr} \tag{4}$$

$$V_{qr} = \frac{d\Phi_{qr}}{dt} + (\omega_s - \omega)\Phi_{dr} + R_r i_{qr} \tag{5}$$

Now, reorganizing the terms, then

$$V_{dr} = \left(R_r + \frac{dL_r'}{dt}\right)i_{dr} - s\omega_s L_r' i_{qr} + \frac{L_m}{L_s}V_{ds} \tag{6}$$

$$V_{qr} = \left(R_r + \frac{dL_r'}{dt}\right)i_{qr} + s\omega_s L_r' i_{dr} + \frac{L_m}{L_s}\left(V_{qs} - \omega\,\Phi_{ds}\right) \tag{7}$$

where speed of stator flux is $\omega_{\Phi s}$ with synchronous speed ω_s. These Eqs. 6 and 7 are modified and become decoupled parameters.

$$\sigma V_{dr} = \sigma L_r \frac{dI_{dr}}{dt} - \omega_s\Phi_{qr} + \frac{L_m}{L_s}\left(V_{ds} - R_s I_{ds} + \omega_1\Phi_{qs}\right) \tag{8}$$

$$\sigma V_{qr} = \sigma L_r \frac{dI_{qr}}{dt} + \omega_s\Phi_{dr} - \frac{L_m}{L_s}\left(R_s I_{qs} + \omega_1\Phi_{ds}\right) \tag{9}$$

The RSC controller is modeled based on the developed Eqs. 8 and 9. The ω_s will change to a new synchronous speed ω_1 during any abnormal decrease in the rotor voltage. Generally, during healthy conditions, the reference stator d-axis flux Φ_d^* is zero and will have maximum value for q-axis Φ_q^*. The rotor transient current in dq axis is represented by Eqs. 10a and b as

$$\frac{di_{dr}}{dt} = \frac{-R_r}{\sigma L_r}i_{dr} + s\omega_s i_{qr} + \frac{1}{\sigma L_r}V_{dr} \tag{10a}$$

$$\frac{di_{qr}}{dt} = \frac{-1}{\sigma}\left(\frac{R_r}{L_r} + \frac{R_s L_m^2}{L_s^2 L_r}\right)i_{qr} + s\omega_s i_{dr} + \frac{1}{\sigma L_r}V_{qr} \tag{10b}$$

The rotor reference *dq* axis voltages are rewritten from Eq. 8 as in 11a and b. These are the derived voltage equations during ideal and transient situations in rotor windings.

$$V_{qr}^* = \left(i_{dr}^* + \frac{1}{\sigma} \left(\frac{R_r}{L_r} + \frac{R_s L_m^2}{L_s^2 L_r} \right) i_{qr} + s\omega_s i_{dr} \right) \sigma L_r \tag{11a}$$

$$V_{qr}^* = \left(i_{dr}^* + \frac{1}{\sigma} \left(\frac{R_r}{L_r} + \frac{R_s L_m^2}{L_s^2 L_r} \right) i_{qr} + s\omega_s i_{dr} \right) \sigma L_r \tag{11b}$$

The RSC and GSC block diagram is shown in Fig. 1a, b. Equations 1a–11b help in appreciating the DFIG performance during steady and transient states. The accuracy of RSC depends on *d*- and *q*-axis voltage control.

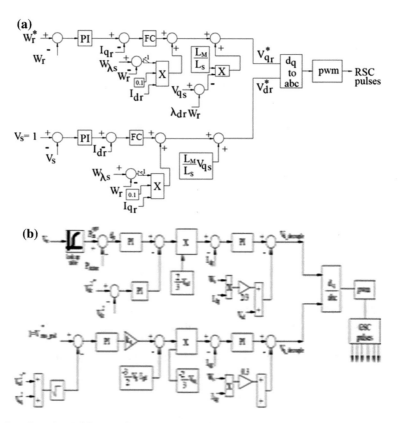

Fig. 1 a Complete RSC controller design **b** GSC for DFIG

3   Analytical Analysis of RSC and GSC Under Symmetrical Faults

The stator voltage of grid-connected DFIG will reach a magnitude of zero value when symmetrical fault occurs and its flux decreases to zero magnitude slowly. The decay in flux is not rapid like voltage, and its delay is because of inertia time lag $\tau_s = \frac{L_s}{R_s}$ forcing the rotor-induced emf V_{0r}. The stator flux at the time of symmetrical fault is

$$\Phi_{sf}^s = \Phi_s^s e^{-t/\tau_s} \tag{12}$$

and the first derivative $\frac{d\Phi_{sf}^s}{dt}$ is negative, it means that the stator flux is decaying with time. Substituting (12) in (1b)

$$V_{0r}^s = -\frac{L_m}{L_s}\left(\frac{1}{\tau_s} + j\omega\right)\Phi_s^s e^{-t/\tau_s} \tag{13}$$

Simplifying Eq. 12 in rotor reference frame as

$$V_{0r}^s = -\frac{L_m}{L_s}(j\omega)\Phi_s^s e^{-j\omega t} \tag{14}$$

At the first instant during the fault, Φ_s will not change instantly (14). At super synchronous speed with slip (s) of DFIG at -0.2 pu, at fault, the rotor speed advances on the term $(1 - s)$ described by (14). For dynamic stability improvement, the proposed EFOC technique helps to decrease stator and rotor flux magnitude and damps perturbations at the fault.

4   Rules for FLC for DFIG-Based System for Improved LVRT Operation

The torque and stator flux will reduce to a smaller value when a fault occurs to a grid-connected DFIG system. These parameters' behavior can be improved if inner control loop of RSC is made faster and accurate with duty cycle. For this, a general fuzzy controller is used with memberships as shown Fig. 2. It can act quickly to limit the inrush current flow during transients so that ripples and surges are minimized.

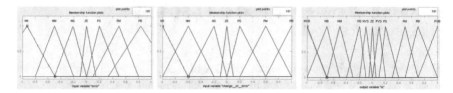

Fig. 2 Input error, input rate of change of error, and output variable membership for FLC

5 Result Analysis

5.1 Case 1: 30% Decrease in Grid Voltage with Symmetrical Faults at 1 and 2.5 s

The DFIG grid-connected system is shown in Fig. 3. The performance of EFOC-based system with fuzzy controller (FC) for recurring faults is shown in Figs. 4 and 5. A first fault between 1 and 1.5 s and a second fault between 2.5 and 3 s occur at PCC with grid voltage decreasing from 440 to 300 V, a 30% decrease compared to normal as in Fig. 4a. The results of EFOC with PI are available in [10] and with IMC and PIR [5]. The proposed EFOC-based system results can be compared with [1, 3–6], and our system has a better performance.

The current waveforms of stator and rotor are shown in Fig. 4b. The stator current is nearly constant with small deviation at the instants of fault occurring and clearing. The stator current increased from 24 A during steady state to 30 A during fault and regains to 24 A once fault is cleared at both fault instants. There is an increase in rotor current from 20 A at steady state to 28 A during fault with small increase in frequency. Based on Eqs. 10a–14, the rotor voltage increases exponentially to a certain value due to control in stator and rotor flux change during the fault by changing the rotor speed to a smaller value. It is observed that rotor current

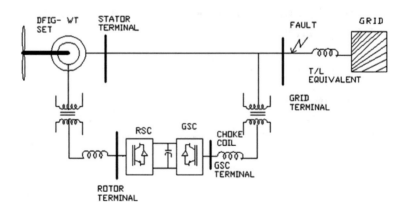

Fig. 3 Grid-connected DFIG showing the location of under-voltage fault

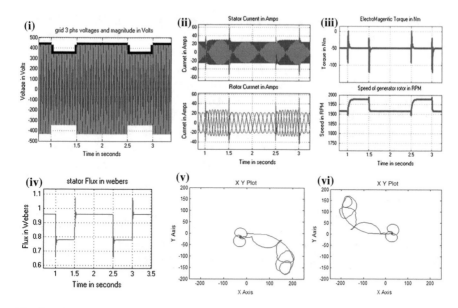

Fig. 4 **a** Grid voltage, **b** Stator and rotor current, **c** EMT and rotor speed, **d** Stator flux, **e** Ids versus Iqs stator current, **f** Idr versus Iqr rotor current

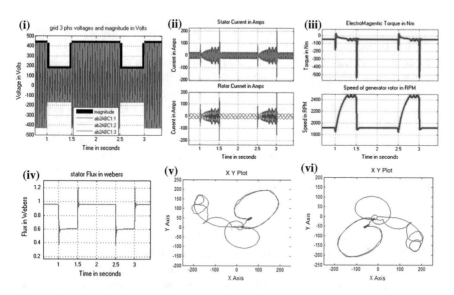

Fig. 5 **a** Grid voltage, **b** Stator and rotor current, **c** EMT and rotor speed, **d** Stator flux, **e** Ids versus Iqs stator current, **f** Idr versus Iqr rotor current

and the stator current rise as grid voltage drops. As shown in Fig. 4c, the surge in electromagnetic torque before the fault is −50 Nm. At fault instants 1 and 2 s, it reached −90 Nm within 1.08 s, and during fault, EMT is again −50 Nm. At fault relieving instants 1.5 and 3 s, EMT is having −120 Nm surge and regained to normal very quickly without oscillations. The rotor speed increased from 1920 to 2020 rpm at 1–1.5 s and 2.5–3 s and reaches again 1920 rpm after fault clearing.

The resultant stator flux magnitude shown in Fig. 4d has a dip from 0.96 to 0.82 Wb and remained constant during fault. The decay in flux is controlled and is maintained without any flux oscillations with EFOC compared to conventional techniques proposed in the literature. The stator and rotor d- and q-axis current waveform is shown in Fig. 4e, f. The d–q current waveforms are constant under state conditions. But during transient state, there occurs some deviation from normal and reaches to some external points and will get the shape of a circle at the instant of fault occurrence and clearance. Slowly during fault, the waveforms reach steady state and will become a straight line and reach a constant point. Finally, after clearance of fault, the d–q axis locus reaches its pre-fault state immediately.

5.2 Case 2: 60% Decrease in Grid Voltage with Symmetrical Faults at 1 and 2.5 s

In this case, more than two times a severe fault took place at PCC, hence the grid voltage dropped to 180 V from 440 at 1–1.5 s and 2.5–3 s as shown in Fig. 5a. This decrease in nearly 60% is compared to rated voltage without fault. But still, stator and rotor currents are continuous and maintained steady state of 24 A even during fault as shown in Fig. 5b. But at fault instant, stator and rotor current surges are produced at 1 and 2.5 s with amplitude of 90 A and immediately decayed to small value within 0.1 s of fault occurrence at both times of fault. The fault current slowly increased to 50A at 1.3 s and maintained uniformly till the fault is cleared at 1.5 s for the first time fault. The same holds good for the second time occurring fault also. After the fault is cleared naturally, a big surge produced with a peak value of 200 A can be observed.

The EMT at fault instants 1 and 2.5 s has surge value of 180 Nm and reaches steady state during fault to −50 Nm at 1.1 and 2.6 s, and later, when the fault is cleared, a severe surge of about 550 Nm can be observed in Fig. 5c. Immediately after the fault is cleared at 1.5 and 3 s, EMT is restored. For a conventional technique or using FFTC proposed by earlier authors, with such severe fault, the EMT decays and reaches to nearly zero value with many oscillations and also reaches steady state after fault clearing with oscillations. With our proposed FC-based EFOC technique, the torque oscillations are eliminated and stability was improved. The rotor speed increased from 1900 rpm to 2400 rpm during the two fault instants (1 and 2 s) respectively. The speed again reached its pre-fault value of 1900 rpm once the fault is cleared. The summary of DFIG parameters under normal and fault conditions are shown in Table 1.

Table 1 Summary of the parameters during normal and for the two cases

Parameter consideration	Normal system	Grid voltage 30% drop at the two fault instants	Grid voltage 60% drop at the two fault instants
Grid voltage (V)	440	300	200
EMT (Nm)	−50	Surge −120 at start and −160 at clearing and During fault −50	Surge −180 at start and −560 at clearing During fault −50
Speed (rpm)	1910	During fault 2020	During fault 2450
(Ids, Iqs) Amp	(−22, 0)	Surge at fault clearing instant (−55, −18) During fault (−32, −20)	Surge at fault clearing instant (−100, −200) During fault (−50, −50)
(Idr, Iqr) Amp	(22, 0)	Surge at fault clearing instant (50, −40) During fault (28, 5)	Surge at fault clearing instant (80, −60) During fault (50, 20)
(Ist, Irot) 3 phase Amps	(24, 20)	Surge at fault clearing instant (70, 60) During fault (38, 10)	Surge at fault clearing instant (200, 180) During fault (50, 50)

The resultant stator flux magnitude shown in Fig. 5d has a dip from 0.96 to 0.6 Wb and remained constant during fault. The stator and rotor d- and q-axis current waveform is shown in Fig. 5e, f. The d–q current waveforms are constant under state conditions. Compared to previous case with 30% dip in grid voltage, the deviation from normal value is higher as severity of fault increased.

Table 1 summarizes the variation in different parameters during normal and at the instant of the fault. As fault level is low in Case 1 with 30% decrease in the grid voltage, the stator voltage dropped by 30% giving 300 V while for 60% drop is 200 V from 440 V. The EMT during normal operation is −50 Nm. Due to the fault, the EMT has a surge upto −120 Nm at fault instant and reached its normal value of −50 Nm even the fault still exists. During fault, they became 0.5 and −0.79 Wb. At the instant of voltage dip by 60%, stator d- and q-axis flux are −0.25 and −0.15 Wb and reaches 0.2 and 0.2 Wb during the fault.

6 Conclusion

In this paper, recurring symmetrical faults (2 times) are studied with the proposed EFOC technique. For the inner control loop of RSC to act quickly, the conventional PI controller is replaced with a general Mamdani-based fuzzy controller (FC). A three-phase fault occurs at PCC creating the grid voltage to dip to 30 and 60% in the two cases. The grid voltage and current were decreased to a small value during symmetric fault and were recovered from falling with the proposed EFOC technique than with conventional methods. The reduction in torque value or gradual increase in rotor speed, maintaining winding currents, and thereby overall stability are promising with the EFOC method. With FC, the torque ripples and surge current and torque are limited effectively. The rotor speed is also within limits and maintained at a particular speed during fault. However, in conventional technique, EMT will have high frequency oscillations with the speed going-on increasing during fault. The decay in the flux is controlled effectively using the proposed control strategy. In this, the flux decay is limited by changing respectively its reference synchronous speed value during the faults. Under normal conditions, this synchronous speed is described by the grid frequency. But, when fault occurs, with dip in the voltage, reference synchronous speed value decreases accordingly and vice-versa. The above all are the major contributions with the proposed strategy.

References

1. W. Chen, F. Blaabjerg, N. Zhu, M. Chen, in Comparison of control strategies for doubly fed induction generator under recurring grid faults, *Proceedings of ICECE* (2014), pp. 398–404
2. W. Chen, F. Blaabjerg, N. Zhu, M. Chen, in Doubly fed induction generator based wind turbine systems subject to recurring grid faults, Proceedings on ICECE (2014), pp. 3097–3104
3. D.V.N. Ananth,, G.V. Nagesh Kumar, Performance of grid connected DFIG during recurring symmetrical faults using internal model controller based enhanced field oriented control. J. Electr. Syst. **12**(2), 406–418 (2016)
4. Q. Huang, X. Z, Donghai Zhu, Yong Kang, Scaled current tracking control for doubly fed induction generator to ride-through serious grid faults. IEEE Trans. Power Electron. **31**(3), 2150–2165 (2016)
5. D.V.N. Ananth, G.N. Kumar, Performance evaluation of DFIG during asymmetrical grid disturbances using internal model controller and resonant controller. Int. J. Electr. Eng. Inf. **8** (3), 494–517 (2016)
6. L. Zhou, J. Liu, S. Zhou, Improved demagnetization control of a doubly-fed induction generator under balanced grid fault. IEEE Trans. Power Electron. **30**(12), 6695–6705 (2015)
7. J.P. da Costa, H. Pinheiro, G. Arnold, Robust controller for DFIGs of grid-connected wind turbines. IEEE Trans. Ind. Electron. **58**(9), 4023–4038 (2011)
8. R.G. Almeida, J.A.P. Lopes, J.A.L. Barreiros, Improving power system dynamic behavior through doubly fed induction machines controlled by static converter using fuzzy control. IEEE Trans. Power Syst. **19**(4), 1942–1950 (2004)
9. J.P.A. Vieira, M.V.A. Nunes, U.H. Bezerra, A.C. Nascimento, Designing optimal controllers for doubly fed induction generators using a genetic algorithm. IET Gen. Transm. Distrib. **3**(5), 472–484 (2009)
10. V.N. Duggirala, V. Gundavarapu, Dynamic stability improvement of grid connected DFIG using enhanced field oriented control technique for high voltage ride through. J, Renew. Energy **2015**, 1–14 (2015)
11. V.N. Ananth Duggirala, V. Nagesh Kumar Gundavarapu, Improved LVRT for grid connected DFIG using enhanced field oriented control technique with super capacitor as external energy storage system. Int. J. Eng. Sci. Technol. **19**(4), 1742–1752 (2016)

Evaluation of a Simultaneous Localization and Mapping Algorithm in a Dynamic Environment Using a Red Green Blue—Depth Camera

Ardhisha Pancham, Daniel Withey and Glen Bright

Abstract Simultaneous localization and mapping (SLAM) assumes a static environment. In a dynamic environment, the localization accuracy and map quality of SLAM may be degraded by moving objects. By removing these moving objects SLAM performance may improve. Oriented FAST (Features from Accelerated Segment Test) and Rotated BRIEF (Binary Robust Independent Elementary Features) (ORB)-SLAM (Mur-Artal et al. in IEEE Trans Rob 31:1147–1163, 2015 [1]) is a state-of-the-art SLAM algorithm that has shown good performance on several Red Green Blue—Depth (RGB-D) datasets with a moving camera in static and dynamic environments. ORB-SLAM is robust to moderate dynamic changes (Mur-Artal et al. in IEEE Trans Rob 31:1147–1163, 2015 [1]). However, ORB-SLAM has not been evaluated with a moving RGB-D camera and an object moving at a range of specific linear speeds. This paper evaluates the performance of ORB-SLAM with a moving RGB-D camera in a dynamic environment that includes an object moving at a range of specific linear speeds. Results from experiments indicate that a moving object at lower speeds, in the range tested, degrades the performance of ORB-SLAM and by removing the moving object the performance of ORB-SLAM improves.

A. Pancham (✉) · D. Withey
MIAS, MDS, Council for Scientific and Industrial Research, Pretoria, South Africa
e-mail: apancham@csir.co.za

D. Withey
e-mail: dwithey@csir.co.za

A. Pancham · G. Bright
Department of Mechanical Engineering, University of KwaZulu-Natal,
Durban, South Africa
e-mail: brightg@ukzn.ac.za

© Springer Nature Singapore Pte Ltd. 2018
S. S. Dash et al. (eds.), *Artificial Intelligence and Evolutionary Computations
in Engineering Systems*, Advances in Intelligent Systems and Computing 668,
https://doi.org/10.1007/978-981-10-7868-2_68

Keywords ORB-SLAM · SLAMIDE · SLAM · Dynamic environment
RGB-D camera

1 Introduction

Simultaneous localization and mapping (SLAM) enables a mobile robot to construct a map of an unknown, static environment, and localize itself simultaneously
[2]. Real-world environments, however, are dynamic and contain moving objects
that may lead to localization errors and reduce map quality. The performance of
SLAM In Dynamic Environments (SLAMIDE) may be improved by the removal of
moving objects or detecting and tracking these moving objects.

Expensive sensors, such as laser scanners [3] have been used to solve
SLAMIDE. Recently, research has involved the use of low-cost sensors, such as
cameras [4, 5], and RGB-D cameras [1], which generate both color and depth data.
However, questions remain regarding sensor type, methods for differentiating stationary and moving objects [6], and how best to track moving objects and predict
their positions over time.

Oriented FAST (Features from Accelerated Segment Test) and Rotated BRIEF
(Binary Robust Independent Elementary Features) (ORB)-SLAM [1] is a
state-of-the-art SLAM algorithm that has shown good performance on several Red
Green Blue—Depth (RGB-D) datasets with a moving camera in static and dynamic
environments. ORB-SLAM is robust to moderate dynamic changes [1]. However,
ORB-SLAM has not been evaluated with a moving RGB-D camera and an object
moving at a range of specific linear speeds.

This paper evaluates the performance of ORB-SLAM with a moving RGB-D
camera in a dynamic environment that includes an object moving at a range of
specific linear speeds. A Vicon motion capture system is used to determine ground
truth positions for the moving camera. Experiments show that a moving object at
lower speeds, in the range tested, degrades the performance of ORB-SLAM, and by
removing the moving object the performance of ORB-SLAM improves.

The remainder of this work is organized as follows. Section 2 explains
ORB-SLAM, Sect. 3 describes the experimental methods, Sect. 4 contains the
experiments and results, and Sect. 5 concludes the paper.

2 ORB-SLAM

A detailed explanation of ORB-SLAM is given in [1]. It builds on Parallel Tracking
and Mapping (PTAM) [7] and other algorithms. ORB features are used as they are
computationally efficient and rotation invariant [8]. ORB-SLAM comprises of three
parallel threads: tracking, local mapping, and loop closing, as shown in Fig. 1. The
tracking thread performs camera localization and new keyframe decision. The local

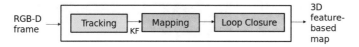

Fig. 1 ORB-SLAM threads, where KF is the keyframe

mapping thread carries out new keyframe processing, local bundle adjustment, and redundant keyframe removal. The loop closing thread performs loop detection and closure [1].

Extensive evaluations of ORB-SLAM have demonstrated its excellent accuracy and robustness [9]. It is robust to moderate dynamic changes [1], not affected by brightness variations and offers computational efficiency. However, it is unsuitable for environments without features, similar features may cause incorrect loop closures and drift arises without loop closures [10].

3 Experimental Methodology

3.1 Experimental Setup

In the experiments, a moving object, a checkerboard (0.21 m × 1.27 m) mounted on a Pioneer 3DX robot [11], was moved at a range of speeds (0.01, 0.1, 0.25, 0.5 m/s), in a straight line, in front of a feature-rich, scene. An RGB-D camera, an Asus Xtion Pro Live [12], was moved by hand in a straight line opposite to the direction of motion of the moving object, such that it captured the scene and the moving object, which moved to the center of the scene. Figure 2 shows the test environment and the moving object.

3.2 RGB-D Data

The Robot Operating System (ROS) [13] was used to capture RGB-D camera images of the experiments in ROS bag files [14]. The bag files together formed the dataset for the experiments and served as the inputs to ORB-SLAM.

A Vicon motion capture system was used to record the ground truth of the camera. The ground truth camera pose was computed at the average time between the color and depth image pairs from the ASUS.

Fig. 2 Test environment and
moving object

3.3 *Evaluation Measure*

Errors in the camera pose were produced by motion of the moving object through
the camera field of view. RMS Error (RMSE) was used to evaluate the translation
error of the camera pose. RMSE is defined in (1), where n is the number of
estimates, $\hat{x}_i$ is the ith estimate, and x_i is the corresponding ground truth.

$$\text{RMSE} = \sqrt{\frac{1}{n}\sum_{i=1}^{n}\left(\hat{x}_i - x_i\right)^2} \tag{1}$$

4 Results

The experiments were conducted with the ROS on a Linux computer, with an Intel
Core i7—3720 CPU, with 2.6 GHz clock speed and 8 GB of RAM.

The open-source implementation of ORB-SLAM [15] was evaluated with a
moving RGB-D camera in a dynamic environment that included an object moving
at a range of specific linear speeds.

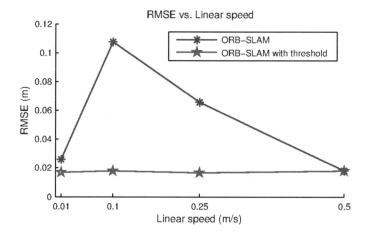

Fig. 3 RMSE versus linear speed

The performance of ORB-SLAM was compared to a modified version where a distance threshold was used to remove the moving object from its known position in the environment. The modified version is referred to as ORB-SLAM with threshold.

Default parameters of ORB-SLAM were used. ORB-SLAM and ORB-SLAM with threshold were executed three times on each bag file in the experimental dataset, and the RMSE was calculated for the camera poses produced.

Figure 3 shows the RMSE of ORB-SLAM and ORB-SLAM with threshold for the linear speeds tested. The error of ORB-SLAM increases as the speed decreases of the moving object from 0.5 to 0.1 m/s. The error is highest at 0.1 m/s (0.11 m) and lowest at 0.5 m/s (0.02 m).

ORB-SLAM with threshold has lower error than ORB-SLAM because the moving object is removed from the SLAM process. By removing the moving object, ORB-SLAM's performance improves by around 50% or more for linear speeds 0.01, 0.1, and 0.25 m/s.

Figure 4 shows the camera trajectories for one execution of ORB-SLAM, ORB-SLAM with threshold and the corresponding ground truths for 0.1 m/s.

Spurious measurements from the moving object were included in the maps at all speeds for ORB-SLAM but were not visible in the map for ORB-SLAM with threshold as shown in Figs. 5 and 6.

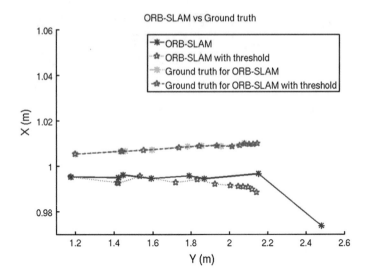

Fig. 4 Camera trajectories for one execution of ORB-SLAM, ORB-SLAM with threshold and ground truth for 0.1 m/s. The markers (stars and asterisks) for ORB-SLAM and ORB-SLAM with threshold indicate selected keyframes. Corresponding points are also marked on the ground truth trajectory

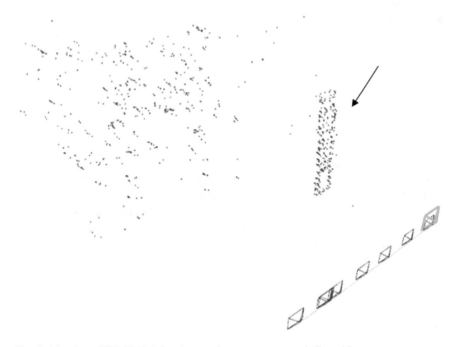

Fig. 5 Map from ORB-SLAM showing spurious measurements indicated by arrow

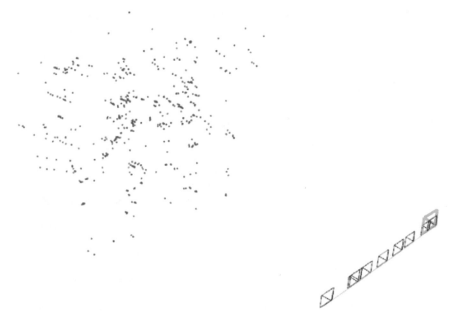

Fig. 6 Map from ORB-SLAM with threshold showing no spurious measurements

5   Conclusion

This paper evaluated the performance of ORB-SLAM with a moving RGB-D camera in a dynamic environment that included an object moving at a range of specific linear speeds. Results showed that a moving object at lower speeds degraded localization performance, and spurious measurements from the moving object were included in the map. By removing the moving object, the performance of ORB-SLAM improved by around 50% or more for linear speeds 0.01, 0.1 and 0.25 m/s.

In the tests performed, the environment was known and, in this case, the moving object could be removed by using a distance threshold. In an unknown, dynamic environment, methods for moving object removal and/or methods for detecting and tracking moving objects should be applied to improve SLAM performance.

References

1. R. Mur-Artal, J. Montiel, J. Tardos, ORB-SLAM: a versatile and accurate monocular SLAM system. IEEE Trans. Rob. **31**, 1147–1163 (2015)
2. H. Durrant-Whyte, T. Bailey, Simultaneous localisation and mapping (SLAM): part I the essential algorithms. Robot. Autom. Mag. **13**, 99–110 (2006)

3. C.-C. Wang, C. Thorpe, S. Thrun, M. Hebert, H. Durrant-Whyte, Simultaneous localization, mapping and moving object tracking. Int. J. Robot. Res. **26**(9), 889–936 (2007)
4. D. Migliore, R. Rigamonti, D. Marzorati, M. Matteucci, D. Sorrenti, Use a single camera for simultaneous localization and mapping with mobile object tracking in dynamic environments, in *Proceedings of International Workshop on Safe Navigation in Open and Dynamic Environments Application to Autonomous Vehicles* (2009)
5. S. Perera, A. Pasqual, Towards realtime handheld monoslam in dynamic environments, in *International Symposium on Visual Computing* (2011), pp. 313–324
6. D. Wolf, G. Sukhatme, *Mobile Robot Simultaneous Localization and Mapping in Dynamic Environments* (2005), pp. 53–65
7. G. Klein, D. Murray, Parallel tracking and mapping for small AR workspaces, in *Mixed and Augmented Reality, 2007. ISMAR 2007. 6th IEEE and ACM International Symposium on IEEE, 2007* (2007), pp. 225–234
8. E. Rublee, V. Rabaud, K. Konolige, G. Bradski, ORB: an efficient alternative to SIFT or SURF, in *Computer Vision (ICCV), 2011 IEEE International Conference on IEEE 2011* (2011), pp. 2564–2571
9. R. Mur-Artal, J. Tardos, Probabilistic semi-dense mapping from highly accurate feature-based monocular SLAM, in *Robotics: Science and Systems*, p. 2015
10. C. Bove, A. Wald, W. Michalson, M. Donahue, J. LaPenta, *Collaborative Robotics Heads-Up Display Major Qualifying Project* (2016)
11. Omron Adept MobileRobots, *PioneerP3DX*. Available at. http://www.mobilerobots.com/researchrobots/PioneerP3DX.aspx
12. ASUSTeK Computer Inc: Xtion pro live. Available at https://www.asus.com/3D-Sensor/Xtion/
13. ROS: ROS.org. Available at. http://wiki.ros.org/
14. ROS: ROS.org, Bags/Format. Available at. http://wiki.ros.org/Bags/Format
15. R. Mur-Artal, *Raulmur/ORB_SLAM2*. Available at. https://github.com/raulmur/ORB_SLAM2

Assessment of Texture Feature Extraction to Classify the Benign and Malignant Lesions from Breast Ultrasound Images

Telagarapu Prabhakar and S. Poonguzhali

Abstract Detection of Breast cancer which is growing fast is crucial. Mammography procedures are uncomfortable involving ionizing radiation. Ultrasound, a broadly popular medical imaging modality which is noninvasive, real time, convenient, and of low cost is preferable. But speckle noise corrupts, diminishing the excellence of ultrasound image. curvelet, shearlet, and tetrolet methods are a prerogative to reduce the speckle noise and enhance signal-to-noise ratio (SNR) and contrast resolution of the image before segmentation. After segmentation, to classify the breast lesion, a set of 15 texture features are extracted from the original image and also from curvelet, shearlet, and tetrolet filtered images. Optimal features are selected to increase the classification performance by using a Releiff algorithm and four best features are taking into account for feature ranking. These features used to classify the lesions from breast ultrasound images by using SVM with polynomial kernel and FKNN algorithms. The proposed algorithm tested on 178 samples, of which 85 samples are benign masses, and 93 samples are malignant masses. The results show that SVM classifier outperforms Fuzzy KNN with 87.39% accuracy, 88.89% sensitivity, and specificity of 88.46% for the texture features from the tetrolet transform.

Keywords Breast ultrasound image · Curvelet · Shearlet · Tetrolet
Feature extraction and classification

T. Prabhakar (✉)
Department of Electronics and Communication, GMR Institute of Technology,
Rajam, Srikakulam District, India
e-mail: prabhakar.t@gmrit.org

S. Poonguzhali
Department of Electronics and Communication, College of Engineering,
Guindy, Anna University, Chennai, India

© Springer Nature Singapore Pte Ltd. 2018 725
S. S. Dash et al. (eds.), *Artificial Intelligence and Evolutionary Computations in Engineering Systems*, Advances in Intelligent Systems and Computing 668,
https://doi.org/10.1007/978-981-10-7868-2_69

1 Introduction

Most of the women are suffering from breast cancer and many young women died due to this significant problem [1]. Early identification of breast cancer is vital to reduce the mortality rate. A common screening and detection technique for breast cancer is mammography. Breast cancer diagnosis and detection at an early stage is possible with mammography, which uses X-rays and are harmful. It is also uncomfortable, painful, and can be embarrassing for the women. These disadvantages of mammography can be overcome by using ultrasound imaging techniques. Ultrasound is normally considered safe and can be used for imaging the breast and other soft tissues. But, the ultrasound imaging is affected by speckle noise results in poor image quality. This noise affects the segmentation and feature extraction, hence results in poor classification. Some studies have conducted on denoising the speckle noise and many filtering architectures are proposed [2]. Recently, many segmentation methodologies have been recommended [3, 4]. The breast tumor severity and type can be recognized by the variations of the texture in ultrasound images [5]. The texture features calculated using a spectral, statistical, structural, block variations of local correlation coefficients [2], fractal dimensions [6]. These techniques are widely used to extract texture features in spatial domain that is difficult to segregate from ultrasound images. The majority of spatial domain methods does not have the multiscale properties. Numerous multiscale geometric analysis algorithms, including ridgelet [7], brushlet [8], curvelet [9], contourlet [10], and shearlet [11], has the capability for extracting the texture, morphological, and fractal features from a different type of images. These algorithms accomplish the promising performance [12–14]. To overcome the current problems in the widespread segmentation and feature extraction approach, this research work is focused on developing the algorithm for filtering and feature extraction to classify the tumor in the ultrasound image.

2 Materials and Method

In this research work, breast ultrasound images using linear transducer arrays are taken. The proposed algorithm is tested on 178 samples, of which 85 samples are benign masses, and 93 samples are malignant masses. This study focused on developing filtering techniques using curvelet, shearlet, and tetrolet for speckle noise reduction. An automatic active contour procedure [15] was adopted to detect the breast lesion. After segmentation, the Region of Interest (ROI) of without filter image and from Curvelet, Shearlet and Tetrolet filtered images the texture features are extracted and these are given as input to feature selection algorithm for selecting the optimal features. These features are used to classify the breast ultrasound tumor images by support machine vector (SVM) with polynomial kernel and Fuzzy k-Nearest Neighbor (FKNN) classifiers. Figure 1 represents the block description of filtering and feature extraction process.

2.1 Speckle Reduction and Automatic Segmentation

2.1.1 Speckle Reduction

The ultrasound images are low-contrast and affected by speckle noise. Hence, speckle noise should be reduced without destroying any features. Here, the filtering techniques are implemented using the curvelet, shearlet, and tetrolet transforms. The Curvelet [16] filter represents singularities of the curve and it has better directional characteristics than ridgelet. Coefficients of curvelet are obtained by unevenly sampling with Fourier coefficients of an breast ultrasound image. Observed that curvelet transform covers the entire frequency spectrum so that there is no loss of information. The shearlet transform [7, 8] combine direction and multiscale analysis. Initially, a Laplacian pyramid is utilized to divide the noisy breast ultrasound tumor image into low-and high-frequency components, and next, direction filtering is used to obtain various subbands and direction coefficients of shearlet. Here direction filtering is achieved by using a shear matrix. Tetrolet transform is in the form of adaptive Haar Wavelet Transform [9]. In this algorithm, in each block, the input image can be divided into 4×4 blocks to obtain the tetrolet representation with sparsest. Now change the position of the high-pass and low-pass filter coefficients in every block into a 2×2 block for storing the high-pass part coefficients and, lastly, applying the above process to the low-pass image as well.

2.1.2 Automatic Segmentation

The automatic segmentation algorithm is based on the Active contour model method [17]. This method was adopted to detect the breast lesion [15]. In this method, seed points are identified automatically from the image using statistical

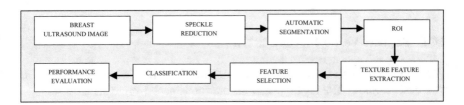

Fig. 1 Block diagram of a filtering and feature extraction process

features. Active contour algorithm-based segmentation uses the seed point, which acts as an initial contour for segmenting the lesion region accurately. The initial contour contracts when it is outside the object or grows when it is inside the region of interest.

2.2 Feature Extraction

2.2.1 Texture Features

GLCM features are used for measure textures in an image, GLCM matrices are two-dimensional histograms. The element P (m, n, d, θ) of the GLCM matrices. This is the joint probability of the "m" and "n" gray levels separated by the distance "d", the Manhattan distance along the direction of "θ" as 0, 45, 90, and 135°. Here, we calculated 15 features [18] from GLCM.

2.3 Feature Selection

For classification, all the features derived from each subband is considered, it leads to computational complexity. It reduces the accuracy of classification and is also time-consuming. Feature selection methods are proposed to avoid such problems with improved accuracy. Feature selection is the process of reducing redundant, irrelevant, or noise feature to get the better accuracy of classification. In this paper, feature selection method is Relieff algorithm [19] is used to reduce the dimensionality. Relieff-based feature selection is effective even the data are noisy and missing.

2.4 Classification

The reduced features are given to the classifier to classify the tumor as benign or malignant. In this paper, SVM and Fuzzy K-Nearest Neighbor (FKNN) classifiers are used to classify the breast lesion.

2.4.1 Support Vector Machine (SVM)

Vapnik [20] proposed support vector machine. It is based on the principle of statistical learning. The SVM works on the fundamental principle of inserting a hyperplane between the classes, and it will keep at the maximum distance from the nearest data points. Data points appear at the closest to the hyperplane is called as Support Vectors. Through a kernel function, SVM can be used to solve nonlinear classification problems. The kernel function maps two classes of data points in a lower dimensional feature space onto a higher dimensional space. Two sets of data points can be separated using a hyperplane. The kernel function converts the nonlinear classification to the linear classification. The popular kernels are the Linear kernel, the Polynomial kernel of degree "d", the Gaussian radial basis function (RBF), and Neural Nets (sigmoid). In this work, the Polynomial kernel is used.

2.4.2 Fuzzy K-Nearest Neighbor Classifier (FKNN)

The FKNN assigns class membership to a sample vector preferable to assigning the vector to a particular class. The benefit is that this classifier makes no arbitrary assignments. The origin of this classifier is to allow membership as a function of the vector's distance from its K-Nearest Neighbors and of those neighbors' memberships in the possible classes. The Fuzzy algorithm is related to the crisp version in the sense that it also has to search the labeled sample set for the K-Nearest Neighbors. The procedures differ considerably, apart from obtaining these K samples. The mean distance is calculated from K-points to testing data. The minimized mean distance based on the output class values are stored [19].

3   Results and Discussions

The proposed algorithm tested on 178 breast ultrasound lesion images of which 85 samples are found to be benign masses, and 93 samples are malignant masses are acquired from the hospital. The implementation of the method is done using MATLAB 13a Software and run on Intel Core i5-4200 M CPU @ 2.50 GHz processor. The speckle noise reduced in breast ultrasound images by using different filtering methods like curvelets, shearlets, and tetrolets Transforms. Breast ultrasound benign and malignant images and Speckle noise reduced by Tetrolet filter technique segmentation result as shown in Fig. 2. In this study, an automatic segmentation algorithm using active contour model was employed [15]. The seed points are identified automatically, which act as an initial contour and the active contour algorithm segmented the lesion region accurately as shown in Fig. 4. After segmentation, the 15 texture features are extracted from Input image, curvelet, shearlet, and tetrolet filter images obtain texture features values from each subimage of the ROI of size 10 × 10 pixel with 8-bit resolution. GLCM-based texture features extracted from the segmented regions. Observed that Autocorrelation, Cluster shade, Cluster Prominence, Variance, Sum Average, and Sum variance and has the highest mean and standard deviation values for malignant mass. This characteristic creates a problem for classification. There are totally 15 features extracted from without filtered region of image, curvelet, shearlet, and tetrolet filter images. But some feature characteristics creates a problem for classification. Optimal features are selected to increase the classification performance by using Releiff algorithm and four best features are taking into account for feature ranking. In this paper, the training set consists of 177 samples and the test sample is classified based on the trained data. Fuzzy KNN and SVM are exploited to assess the reliability of the feature set. In SVM, the polynomial kernel is helpful in detecting the optimal hyperplane. From Table 1, it is observed that the features extracted from tetrolet-based filtered image with SVM classification gives high performance in classification accuracy, sensitivity, and specificity. In SVM classifier, the training dataset defines the optimal hyperplane and classification only depends on the support vectors, not on all the features.

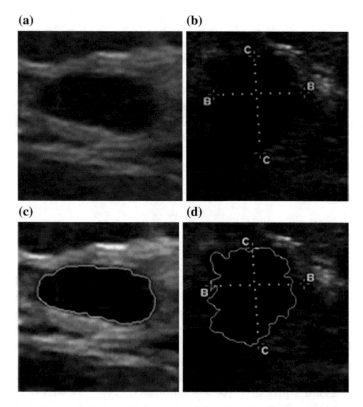

Fig. 2 a Benign image. **b** Malignant image. **c** and **d** Tetrolet filtered and automatic segmentation for benign & malignant

Table 1 Performance of different classifiers

	Without filter		Curvelet filter		Shearlet filter		Tetrolet filter	
	SVM	FKNN	SVM	FKNN	SVM	FKNN	SVM	FKNN
Accuracy	74.55	72.73	78.18	75.45	81.82	80	**87.39**	82.73
Sensitivity	74.07	72.22	77.78	74.55	81.48	79.63	**88.89**	84.91
Specificity	75	73.21	78.57	76.36	82.14	80.36	**88.46**	83.64

SVM Support Vector Machine; *FKNN* Fuzzy K-Nearest Neighbor

$$\text{Accuracy} = \frac{TP + TN}{TP + TN + FP + FN}$$
$$\text{Sensitivity} = \frac{TP}{TP + FN} \qquad (1)$$
$$\text{Specificity} = \frac{TN}{TN + FP}$$

TP—Malignant correctly classified as Malignant; TN—Benign correctly classified as Benign; FP—Benign incorrectly classified as Malignant; FN—Malignant incorrectly classified as Benign.

Table 1 shows the performance of different classifiers. From the table, it is observed that the features extracted from tetrolet-based filtered image with SVM classification gives high performance in classification accuracy, sensitivity, and specificity.

4 Conclusion

In this paper, filtering and feature extraction algorithm is developed for classifying the breast tumors from ultrasound images as malignant and benign lesions. At first, the speckle noise is reduced by using curvelet, shearlet, tetrolet filters. Then, the breast tumors are segmented utilizing a statistical feature-based active contour method. The segmented region is used to extract texture features and these are given as input to feature selection algorithm for selecting the optimal features. Optimal features are selected using Relieff algorithm and these features acts as input to Fuzzy KNN classifiers and SVM classifiers. This process leverages the performance of the classifiers for better diagnosis. It is clearly evident that from the results, classification performance of Tetrolet feature are found to be better when compared to other features. The proposed tetrolet feature with SVM classification shows a high performance than Fuzzy KNN with the accuracy of 87.39%, Sensitivity as 88.89% and Specificity as 88.46%. This proposed approach could be useful for the detection and classification of breast lesions accurately.

Acknowledgements We would like to thank the Government Stanley Hospital, Chennai, India for providing breast ultrasound images and clinical information.

References

1. R. Siegel, D. Naishadham, A. Jemal, Cancer statistics, 2013. CA Cancer J. Clin. **63**, 11–30 (2013)
2. H. Cheng, J. Shan, W. Ju, Y. Guo, L. Zhang, Automated breast cancer detection and classification using ultrasound images: a survey. Pattern Recogn. **43**, 299–317 (2010)
3. J. Noble, D. Boukerroui, Ultrasound image segmentation: a survey. IEEE Trans. Med. Imaging **25**, 987–1010 (2006)
4. B. Liu, H. Cheng, J. Huang, J. Tian, X. Tang, J. Liu, Fully automatic and segmentation-robust classification of breast tumors based on local texture analysis of ultrasound images. Pattern Recogn. **43**, 280–298 (2010)
5. D. Chen, R. Chang, W. Kuo, M. Chen, Y. Huang, Diagnosis of breast tumors with sonographic texture analysis using wavelet transform and neural networks. Ultrasound Med. Biol. **28**, 1301–1310 (2002)
6. S. Poonguzhali, B. Deepalakshmi et al., Optimal feature selection and automatic classification of abnormal masses in ultrasound liver images, in *International Conference on Signal Processing, Communications and Networking* (2007)
7. G. Easley, D. Labate, W. Lim, Sparse directional image representations using the discrete shearlet transform. Appl. Comput. Harmonic Anal. **25**, 25–46 (2008)
8. G.R. Easley, D. Labate, W.Q. Lim, Optimally sparse image representations using shearlets, in *Fortieth Asilomar Conference on Signals, Systems and Computers, 2006 ACSSC'06* (IEEE, 2006), pp. 974–978
9. J. Krommweh, Tetrolet transform: a new adaptive Haar wavelet algorithm for sparse image representation. J. Vis. Commun. Image Represent. **21**, 364–374 (2010)
10. P. Telagarapu, S. Poonguzhali, Analysis of contourlet texture feature extraction to classify the benign and malignant tumors from breast ultrasound images. Inter. J. Eng. Technol. **6**(1), 293–305 (2014)
11. S. Zhou, J. Shi, J. Zhu, Y. Cai, R. Wang, Shearlet-based texture feature extraction for classification of breast tumor in ultrasound image. Biomed. Signal Process. Control **8**, 688–696 (2013)
12. A. Alvarenga, W. Pereira, A. Infantosi, C. Azevedo, Complexity curve and grey level co-occurrence matrix in the texture evaluation of breast tumor on ultrasound images. Med. Phys. **34**, 379 (2007)
13. T. Prabhakar, S. Poonguzhali, Denoising and automatic detection of breast tumor in ultrasound images. Asian J Inform Tech **15**(18), 3506–3512 (2016)
14. T. Prabhakar, S. Poonguzhali, Filtering based feature extraction to classify the benign and malignant lesions from breast ultrasound images. J. Med. Imag. Health. Inform. **6**(6), 1469–1474 (2016)
15. T. Prabhakar, S. Poonguzhali, Feature based active contour method for automatic detection of breast lesions using ultrasound images. AMM. **573**, 471–476 (2014)
16. E. Candès, L. Demanet, D. Donoho, L. Ying, Fast discrete Curvelet transforms. Multiscale Modeling Simul. **5**, 861–899 (2006)
17. T. Chan, L. Vese, Active contours without edges, in *IEEE Transactions on Image Processing* (2001)
18. R. Haralick, K. Shanmugam, I. Dinstein, Textural features for image classification. IEEE Trans. Syst. Man, and Cybern. **3**, 610–621 (1973)
19. O. Castillo, P. Melin, E. Ramírez, J. Soria, Hybrid intelligent system for cardiac arrhythmia classification with Fuzzy k-Nearest Neighbors and neural networks combined with a Fuzzy system. Expert Syst. Appl. **39**, 2947–2955 (2012)
20. V. Vapnik, *The Nature of Statistical Learning Theory* (Springer, New York, 2000)

Author Index

Printed in the United States
By Bookmasters